SEPARATION AND PRECONCENTRATION METHODS IN INORGANIC TRACE ANALYSIS

ELLIS HORWOOD SERIES IN ANALYTICAL CHEMISTRY
EDITORS: *Dr. R. A. Chalmers & Dr. Mary Masson, University of Aberdeen*

"I recommend that this Series be used as reference material. Its Authors are among the most respected in Europe". *J. Chemical Ed., New York.*

Applications of Ion-selective Membrane Electrodes in Organic Analysis
  F. BAIULESCU & V. V. COȘOFREȚ, Polytechnic Institute, Bucharest
Handbook of Organic Microanalysis
  S. BANCE, May and Baker Research Laboratories, Dagenham
Ion-selective Electrodes in Life Sciences
  D. B. KELL, University College of Wales, Abserystwyth
Inorganic Reaction Chemistry
Volume 1: The Reactions of the Elements and their Compounds
Volume 2: Systematic Chemical Separation
  D. T. BURNS, Queen's University of Belfast, A. TOWNSHEND, University of Hull, A. G. CATCHPOLE, Kingston Polytechnic
Quantitative Chemical Analysis
  R. A. CHALMERS, M. CRESSER, University of Aberdeen
Handbook of Process Stream Analysis
  K. J. CLEVETT, Crest Engineering (U. K.) Inc.
Automatic Methods in Chemical Analysis
  J. K. FOREMAN & P. B. STOCKWELL, Laboratory of the Government Chemist, London
Theoretical Foundations of Chemical Electroanalysis
  Z. GALUS, Warsaw University
Laboratory Handbook of Thin Layer and Paper Chromatography
  J. GASPARIČ, Charles University, Hradeč Kralove
  J. CHURAČEK, University of Chemical Technology, Pardubice
Handbook of Analytical Control of Iron and Steel Production
  T. S. HARRISON, Group Chemical Laboratories, British Steel Corporation
Handbook of Organic Reagents in Organic Analysis
  Z. HOLZBECHER et al., Institute of Chemical Technology, Prague
Analytical Applications of Complex Equilibria
  J. INCZÉDY, University of Chemical Engineering, Veszprém
Particle Size Analysis
  Z. K. JELÍNEK, Organic Synthesis Research Institute, Pardubice
Operational Amplifiers in Chemical Instrumentation
  R. KALVODA, J. Heyrovský Institute of Physical Chemistry and Electrochemistry, Prague
Atlas of Metal-ligand Equilibria in Aqueous Solution
  J. KRAGTEN, University of Amsterdam
Gradient Liquid Chromatography
  C. LITEANU & S. GOCAN, University of Cluj
Titrimetric Analysis
  C. LITEANU & E. HOPÎRTEAN, University of Cluj
Statistical Theory and Methodology of Trace Analysis
  C. LITEANU & I. RÎCĂ, University of Cluj
Spectrophotometric Determination of Elements
  Z. MARCZENKO, Warsaw Technical University
Separation and Preconcentration Methods in Inorganic Trace Analysis
  J. MINCZEWSKI et al., Institute of Nuclear Research, Warsaw
Handbook of Analysis of Organic Solvents
  V. ŠEDIVEC & J. FLEK, Institute of Hygiene and Epidemiology, Prague
Methods of Catalytic Analysis
  G. SVEHLA, Queen's University of Belfast, H. THOMPSON, University of New York
Handbook of Analysis of Synthetic Polymers and Plastics
  J. URBAŃSKI et al., Warsaw Technical University
Analysis with Ion-selective Electrodes
  J. VESELÝ & D. WEISS, Geological Survey, Prague
  K. ŠTULÍK, Charles University, Prague
Electrochemical Stripping Analysis
  F. VYDRA, J. Heyrovský Institute of Physical Chemistry and Electrochemistry, Prague
  K. ŠTULÍK, Charles University, Prague
  B. JULÁKOVÁ, The State Institute for Control of Drugs, Prague
Iso-electric Focusing Methods
  K. W. WILLIAMS, L. SODERBERG, T. LAAS, Pharmacia Fine Chemicals, Uppsala

SEPARATION AND PRECONCENTRATION METHODS IN INORGANIC TRACE ANALYSIS

J. MINCZEWSKI
Professor of Analytical Chemistry
Warsaw Technical University

J. CHWASTOWSKA
Institute of General Chemistry and Inorganic Technology
Warsaw Technical University

R. DYBCZYŃSKI
Department of Analytical Chemistry
Institute of Nuclear Research, Warsaw

Translation Editor:
MARY R. MASSON
University of Aberdeen

ELLIS HORWOOD LIMITED
Publisher   Chichester

Halsted Press: a division of
JOHN WILEY & SONS
New York · Chichester · Brisbane · Toronto

First published in 1982 by
ELLIS HORWOOD LIMITED
Market Cross House, Cooper Street, Chichester, West Sussex,
PO19 1EB, England

The publisher's colophon is reproduced from James Gillison's drawing of the ancient Market Cross, Chichester

Distributors:

Australia, New Zealand, South-east Asia:
Jacaranda-Wiley Ltd., Jacaranda Press,
JOHN WILEY & SONS INC.,
G.P.O. Box 859, Brisbane, Queensland 40001, Australia.
Canada:
JOHN WILEY & SONS CANADA LIMITED
22 Worcester Road, Rexdale, Ontario, Canada.
Europe, Africa:
JOHN WILEY & SONS LIMITED
Baffins Lane, Chichester, West Sussex, England.
North and South America and the rest of the world:
HALSTED PRESS, a division of
JOHN WILEY & SONS
605 Third Avenue, New York, N. Y. 10016, U.S A.

Translated by Andrzej Szafrański (ch. 1–4 and 6)
and Andrzej Skup (ch. 5 and 7) from the Polish
Analiza śladowa — Metody rozdzielania i zagęszczania
Published by Wydawnictwa Naukowo-Techniczne, Warsaw

© 1982 Wydawnictwa Naukowo-Techniczne/Ellis Horwood Limited

British Library Cataloguing in Publication Data
Minczewski, J.
Separation and preconcentration methods in inorganic trace analysis. –
(Ellis Horwood series in analytical chemistry)
1. Chemistry, Analytic – Technique
2. Separation (Technology)
3. Chemistry, Inorganic – Technique
I. Title II. Chwastowska, J.
III. Dybczyński, R.
543   QD75.2
Library of Congress Card No. 81-6349   AACR2
ISBN 0-85312-165-6 (Ellis Horwood Ltd., Publishers)
ISBN 0-470-27169-8 (Halsted Press)

All rights reserved. No part of this book may be reproduced, stored in a retrieval system or transmitted, in any form or by any means, electronic, mechanical, photocopying, recording or otherwise, without prior permission of the Publishers.

PRINTED IN POLAND

PREFACE

The evolution of methods of trace analysis has demanded development of suitable methods for separating and preconcentrating trace constituents, and suitable techniques for handling them.

The advances achieved in complexation, extraction, chromatography, and coprecipitation have substantially broadened the possibilities of trace analysis so that nowadays practically every element can be determined in any sample at concentrations of the order of 10^{-5}–$10^{-6}\%$ and sometimes even less, without resort to the costly and not generally available instrumental methods that are particularly suitable for trace analysis, viz. mass spectrometry and neutron activation.

While dealing with these topics in the Departments of Analytical Chemistry in the Institute for Nuclear Research and Warsaw Technical University, we have come to the conclusion that it would be useful to write a book on the methods available for separating and preconcentrating trace constituents (with due reference to the principles underlying the working techniques employed in trace analysis), which could serve as a guide for both experienced and uninitiated analysts engaged in this field.

We did not find it possible to cover all the separation and enrichment methods, so the book deals only with the methods based on precipitation and coprecipitation, volatility of constituents, extraction, ion-exchange and reversed phase chromatography—the methods which have been used in the above-mentioned laboratories.

The book does not contain detailed procedures, as such an approach would expand the book to a prohibitive size. What the book actually contains is the general information that is—in the authors' opinion—indispensable in handling traces, and a compilation of numerous illustrative applications which will facilitate a search for the relevant literature affording either a direct answer to a problem of interest or a guide to analogous solutions.

We have also deemed it necessary to present some theoretical principles underlying the methods discussed, at a length proportional to the depth of knowledge required to allow correct application. It is with just this end in view that the theoretical parts of the chapters on extraction and ion-exchange have been particularly extended.

The bibliographies attached to each chapter represent lists of carefully selected references deemed most essential for the topics discussed.

The Authors

TABLE OF CONTENTS

Preface	v
Chapter 1 Fundamental Problems in Trace Analysis	1
1.1 Introduction	1
1.2 Range of Concentrations in Trace Analysis	2
1.3 Separation and Preconcentration of Traces—General Considerations	3
1.4 Masking	5
References	6
Chapter 2 Working Techniques in Trace Analysis	9
2.1 General Principles of Trace Analysis	9
2.2 Materials to be Analysed	11
2.2.1 Inorganic Materials	11
2.2.2 Organic Materials	11
2.2.3 Other Materials	12
2.3 Sampling	12
2.3.1 Metals and Alloys	12
2.3.2 Geological Materials	15
2.3.3 Powdered Materials	15
2.3.4 Liquid Materials	15
2.3.5 Organic Materials	15
2.3.6 Other Materials	16
2.4 Contamination from the Atmosphere	16
2.4.1 Avoidance of Contamination	18
2.4.2 Laboratory Equipment	20
2.5 Comminution of Samples	21
2.6 Laboratory Vessels and Containers	22
2.6.1 Glass	24
2.6.2 Quartz	25
2.6.3 Platinum	25
2.6.4 Plastics	26
2.6.5 Cleaning	27
2.7 Reagents Used in Trace Analysis	30
2.7.1 Water	30
2.7.2 Important Reagents	31
2.7.3 Solutions	33
2.8 Volatilization of Substances	34
References	34

Chapter 3 Precipitation and Coprecipitation — 37

3.1 General Considerations — 37
3.2 Theory — 37
    3.2.1 Crystalline Precipitates — 37
    3.2.2 Colloidal Precipitates — 38
    3.2.3 Coprecipitation — 39
3.3 Application of Precipitation and Coprecipitation to Trace Analysis — 42
    3.3.1 Separation of the Major Sample Constituent (Matrix) by Precipitation — 42
    3.3.2 Coprecipitation of Trace Elements with Inorganic Collectors — 45
    3.3.3 Coprecipitation of Traces with Organic Collectors — 52
    3.3.4 Electrodeposition in Trace Analysis — 56
References — 59

Chapter 4 Volatilization of Substances — 66

4.1 General Remarks — 66
4.2 Direct Distillation of the Sample Matrix — 69
4.3 Distillation of the Sample Matrix — 70
4.4 Isolation of Trace Constituents — 71
4.5 Ashing of Organic Matrices — 76
    4.5.1 Simple and Additive-Modified Dry Ashing — 83
    4.5.2 Wet Ashing (Wet Digestion) — 85
    4.5.3 Conclusion — 91
References — 92

Chapter 5 Liquid–Liquid Extraction — 97

5.1 Fundamental Physical Chemistry of Extraction — 97
    5.1.1 The Parameters Characteristic of Extraction — 97
    5.1.2 Inorganic Extraction Systems — 102
    5.1.3 Classification of Extraction Systems — 103
5.2 Chelate Systems — 104
    5.2.1 Complex Compounds — 104
    5.2.2 Extraction as a Chemical Reaction — 106
    5.2.3 Factors Influencing the Value of the Distribution Coefficient — 110
        5.2.3.1 The Effect of the Metal Ion — 110
        5.2.3.2 The Effect of pH — 112
        5.2.3.3 The Effect of Reactions Occurring in the Aqueous Phase — 115
        5.2.3.4 The Effect of the Chelating Reagent — 118
        5.2.3.5 The Effect of Solvent — 121
        5.2.3.6 The Effect of Temperature — 128
    5.2.4 Extraction Kinetics — 129
    5.2.5 Synergism — 135
    5.2.6 Co-extraction — 145
5.3 Ion-Association Extraction Systems — 147
    5.3.1 Fundamentals — 147

	5.3.2	Interactions Between Elements in the Extraction Process (Co-Extraction and Suppression of Extraction)	153
	5.3.3	The Extraction of Co-ordinatively Solvated Salts	154
	5.3.4	Extraction with High Molecular-Weight Amines	163
5.4	Applications of Extractions in Separation Processes		167
	5.4.1	Extraction with β-Diketones	183
		5.4.1.1 Acetylacetone (AA)	183
		5.4.1.2 Thenoyltrifluoroacetone (TTA)	184
		5.4.1.3 Dibenzoylmethane (DBM)	184
		5.4.1.4 1-Phenyl-3-methyl-4-benzoyl-5-pyrazolone (PMBP)	193
	5.4.2	Extraction with 8-Hydroxyquinoline and its Derivatives	193
		5.4.2.1 8-Hydroxyquinoline (HOx, oxine)	193
		5.4.2.2 Derivatives of 8-Hydroxyquinoline	199
	5.4.3	Extraction with Nitrosoarylhydroxylamines	202
		5.4.3.1 Cupferron	202
	5.4.4	Extraction with Hydroxamic Acids	202
		5.4.4.1 *N*-Benzoyl-*N*-Phenylhydroxylamine (BPHA)	203
	5.4.5	Extraction with Dithizone (H_2Dz)	203
	5.4.6	Extraction with Dithiocarbamates	207
		5.4.6.1 Sodium Diethyldithiocarbamate (Na-DDTC)	209
	5.4.7	Extraction with Organophosphorus Compounds	210
		5.4.7.1 Organophosphorus Acids	210
		5.4.7.2 Trialkyl Phosphates and Phosphine Oxides	219
	5.4.8	Extraction with High Molecular-Weight Amines	227
	5.4.9	Extraction with Halides and Pseudohalides	227
References			244

Chapter 6 Ion-Exchange Chromatography   283

6.1	Introduction		283
6.2	General Information on Ion-Exchangers		283
	6.2.1	Organic Ion-Exchangers	284
		6.2.1.1 Cation-Exchangers	284
		6.2.1.2 Anion-Exchangers	302
		6.2.1.3 Selective Ion-Exchange Resins	303
		6.2.1.4 Amphoteric Ion-Exchangers	304
		6.2.1.5 Oleophilic Ion-Exchangers	304
	6.2.2	Inorganic Ion-Exchangers	304
		6.2.2.1 Zeolites	304
		6.2.2.2 Heteropoly-Acid Salts	305
		6.2.2.3 Hydrous Oxides and Insoluble Salts	306
		6.2.2.4 Other Inorganic Ion-Exchangers	307
	6.2.3	Liquid Ion-Exchangers	307
	6.2.4	Ion-Exchange Paper	307
6.3	Fundamental Characteristics of and Methods for Examination of Ion-Exchangers		307
	6.3.1	General Information	307

Table of contents

6.3.2	Particle Size	308
6.3.3	Ion-Exchange Capacity and Polyfunctionality	308
6.3.4	Determination of pK	311
6.3.5	Mechanical and Chemical Stability	311
6.3.5.1	Thermal Stability	312
6.3.6	Swelling	314
6.3.7	Ion-Exchange Resin Density and Bed Density	315
6.3.8	Other Properties	315
6.4	Ion-Exchange Equilibrium	316
6.4.1	Fundamental Definitions	316
6.4.2	Selectivity	320
6.4.2.1	Ways of Expressing Selectivity	320
6.4.2.2	Sources of Selectivity	323
6.4.3	Ion-Exchange in Concentrated Electrolyte Solutions	345
6.4.3.1	Anomalous Sorption	348
6.4.4	Ion-Exchange in Very Dilute Solutions	348
6.4.5	Ion-Exchange in Non-aqueous and Mixed Solvents	349
6.4.6	Prediction of Selectivity from Independent Data	352
6.5	Ion-Exchange Kinetics	355
6.6	The Column Process	355
6.6.1	Classification of Chromatographic Development Techniques	357
6.6.2	Theory of Break-through Curves	360
6.6.2.1	Effect of Ion-Exchange Reaction Equilibrium	361
6.6.2.2	Effect of Dynamics of the Column Process	363
6.6.3	The Elution Development Technique	364
6.6.3.1	Plate Theory	367
6.6.3.2	Rate Theory	370
6.6.4	Analysis of Chromatographic Separations	371
6.6.4.1	Resolution	371
6.6.4.2	Other Ways to Express the Quality of Separation	397
6.7	The Technique of Ion-Exchange Chromatography	399
6.7.1	Pretreatment of the Ion-Exchanger	399
6.7.2	Equipment	401
6.7.3	Effluent Analysis	405
6.7.4	The Elution Development Technique	407
6.7.4.1	Simple Elution	407
6.7.4.2	Stepwise Elution	408
6.7.4.3	Gradient Elution	409
6.8	Applications and Examples	412
6.8.1	Preconcentration of Trace Amounts of Elements	412
6.8.1.1	Selective Sorption of Traces	423
6.8.1.2	Selective Retention of the Matrix	428
6.8.2	Separation of Traces	430
6.8.2.1	Separation into Groups	431
6.8.2.2	Isolation of the Individual Components of a Mixture	436
6.8.3	Methods Utilizing Reversal of Elution Order	467
6.8.4	Qualitative Analysis	470
6.8.5	Quantitative Analysis	471
References		473

Chapter 7 Extraction Chromatography — 503

7.1 General — 503
7.2 Theory — 505
    7.2.1 The Distribution-Reaction Equilibrium — 505
    7.2.2 The Column Separation Process — 515
7.3 Techniques in Extraction Chromatography — 517
7.4 Applications — 518
References — 529

ndex — 535

Chapter 1

FUNDAMENTAL PROBLEMS IN TRACE ANALYSIS

1.1 INTRODUCTION

Nowadays the term trace analysis no longer conveys a novel idea, although this branch of analysis was not distinguished until about 25 years ago. The origin of trace analysis in the modern meaning of the term is associated with determinations of what are known as trace elements in plants, in order to elucidate their role in plant physiology.

The determination of trace concentrations of certain substances has long been of interest in health protection and forensic analysis. The institutions responsible for public health have long ago established threshold limit values for toxic substances: these concentrations must not be exceeded in household articles and foodstuffs, nor in industrial or laboratory atmospheres. Only a few substances had threshold limits at the level nowadays considered to be the trace concentration level, and the methods for determination of these were specially developed for particular cases and did not become an individual branch of analytical chemistry.

An early example of trace analysis is the Gutzeit method [1] for determining traces of arsenic, based on the qualitative Marsh test [2] which was reported in 1836. This is a rare case of a method which is easy to apply, specific and generally applicable, and with a limit of detection as low as $10^{-5}\%$.

Although the work on trace elements in plant ash was the first field of systematic study in trace analysis, this new branch of analysis gained essential momentum only when modern technology needed to use high-purity materials on a large scale.

It can be observed that the development of the whole of analytical chemistry—which is a true applied science—is directly related to the needs of technology. Reciprocally, the achievements of analytical chemistry in providing more accurate information about the composition of substances and solutions permit the further development of science and technology.

The first field of technology having needs which made trace analysis more important than other branches of analysis, was that of nuclear materials used for construction of reactors. The collective volume on the

Analytical Chemistry of the Manhattan Project [3] was the first extensive review of practical trace analysis of nuclear materials. The volume summarized the work carried out in the United States in the 1940s.

The interest of biologists in trace elements has also expanded to include the role of trace elements in animal and human physiology. In 1955, the first Symposium on Trace Analysis was organized by the Academy of Medicine, New York, and the proceedings were published in 1957 as *Trace Analysis* [4].

At about this time, a new period in the application of high-purity materials began, and the range of concentrations of interest was shifted from 10^{-3}–$10^{-5}\%$ (nuclear materials) to 10^{-6}–$10^{-8}\%$ and less (semiconductors).

At the same time, reviews [5–19c] and monographs [20–25b] on trace analysis and on separation and preconcentration methods began appearing, together with numerous papers dealing with particular aspects. In a review published in 1968, West and West [15] estimated the number of works dealing with micro and trace inorganic analysis published in 1966 and 1967 at about 5000; this very large number indicates the interest in and needs of this field.

The problems of trace analysis and of separation and preconcentration methods have been discussed at numerous scentific conferences and symposia [6,9,13,26–30a].

The most important fields in which trace analysis is used now include physics and technology of high-purity materials (reactor materials, semiconductors, ferromagnetics, luminophors, etc.), geochemistry, plant and animal biology, ecology in conjunction with the study of air, water and soil pollution, and forensic analysis.

1.2 RANGE OF CONCENTRATIONS IN TRACE ANALYSIS

The range of concentrations considered to be 'trace concentrations' is quite arbitrary and has changed as trace techniques have developed. In a pre-World War II standard text, (Berl-Lunge's *Chemisch-Technische Untersuchungsmethoden* [31]) traces were defined as amounts of the order of magnitude 10^{-1}, 10^{-2}, and only rarely $10^{-3}\%$. In the first edition of Rodden [3], traces are considered to be concentrations in the range 10^{-3}–$10^{-5}\%$, whereas by 1965 Alimarin [21] was already concerned mainly with concentrations as low as 10^{-6}–$10^{-8}\%$. Some interesting comments on trace analysis, its nomenclature and characterization have been made by Kaiser [30].

It has become common for trace concentrations to be expressed in parts per million (ppm) or per billion (ppb), which correspond to concentrations of $10^{-4}\%$ and $10^{-7}\%$, respectively. Unfortunately, the term 'billion' means 10^9 in America and 10^{12} in Europe; the term 'milliard' (symbol M) is preferable.

The prefixes micro, ultramicro, and submicro have been suggested [32] for denoting the following concentration ranges:

traces	10^{-1}–$10^{-3}\%$
micro traces	10^{-4}–$10^{-6}\%$
ultramicro traces	10^{-7}–$10^{-9}\%$
submicro traces	10^{-10}–$10^{-12}\%$

This classification is consistent with the major applications of trace analysis. In common technical materials trace impurities are usually contained within the 'trace' range; in reactor materials, they constitute mainly micro traces; in semiconductors, they are ultramicro traces; and radioactive substances contaminating our environment, when calculated as weight per cent concentrations, fall into the submicro-trace range.

1.3 SEPARATION AND PRECONCENTRATION OF TRACES – GENERAL CONSIDERATIONS

As a rule, methods for separating substances utilize differences between the distribution coefficients of the individual constituents of a mixture between two phases. A schematic diagram of separation methods with indication of the phases into which the material is distributed is presented in Table 1.1.

Trace analysis deals with the determination of trace constituents in a sample material. Three modifications [33] of the separation methods are possible.

(1) The major constituent is isolated from the sample, whereas the traces remain in solution—a macro–micro separation.

(2) Trace constituents are isolated from a solid or dissolved sample, major constituents being retained in solution—a micro–macro separation.

(3) Trace constituents are separated from one another after isolation—a micro–micro separation.

A method of separation is characterized by:

(a) the separation factors of the constituents of the mixture to be analysed,

(b) the specificity or selectivity of the operation,

(c) the rate of the process,

Table 1.1 Distribution between two phases—the principle underlying separation methods

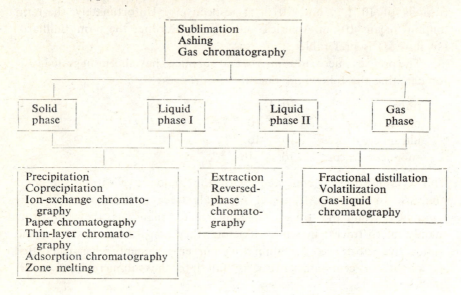

    (d) the ease of performance and availability of equipment,

    (e) the suitability for further treatment of the isolated fraction of the constituents to be determined,

    (f) the coefficient of enrichment.

These parameters vary considerably with the methods adopted and the groups of elements and compounds analysed; therefore the trace analyst will have to master as many separation techniques as possible in order to be able to develop the best separation scheme for a specific analytical task.

Regarding the process of separation from another point of view, we can distinguish two modifications of the separation process: one involving specific isolation of a single constituent from a mixture, and another providing for a group separation of all the trace elements concerned. The first modification will be, as a rule, associated with a macro–micro separation in which the macro constituent is specifically isolated with no trace elements carried along (for example, isolation of lead in trace analysis of lead materials). It may also happen that we want to separate specifically one trace element, especially when it is the only one to be determined. A good example is the Gutzeit method for arsenic traces, already mentioned [1], which allows trace amounts of this element in practically any sample to be separated and determined by a very simple procedure. Another example is the isolation of caesium from aqueous solutions on inorganic ion exchangers such as copper ferrocyanide or zirconium ferrocyanide [34].

Group separation of trace constituents (second modification) usually seeks to separate simultaneously all the trace elements to be determined, with the sample matrix (i.e. major sample constituent) retained unaffected. Examples include the isolation of trace elements from sea-water by precipitation with a mixture of 8-hydroxyquinoline, thionalide and tannic acid [35], water and sodium chloride being the matrix in this case, and separation of trace constituents from a platinum matrix on a cation-exchanger in a hydrochloric acid medium [36]. Group separation of trace elements has very often been applied when the traces thus separated are to be determined by spectrography or atomic-absorption spectrophotometry.

As already mentioned in the preface, this book deals only with methods of separation and preconcentration of trace elements. Methods for organic traces are not discussed. Also, as far as methods are concerned, we restrict ourselves to those most widely used, viz., precipitation and coprecipitation, methods making use of the volatility of substances, extraction, ion-exchange chromatography, and reversed-phase extraction chromatography.

1.4 MASKING

Before the individual methods for separating components are discussed it is worth mentioning a procedure frequently applied in the course of analysis (not only trace analysis) which can make a separation more of even completely specific. This is masking of constituents of the sample which may interfere with the separation.

A masking agent is a substance which reacts to form stable complexes with constituents which interfere with the main reaction of the component to be isolated or determined. For example, if uranium is to be freed from associated components, carbonate may be used as the masking agent. This converts uranium into a soluble carbonate complex while most other elements precipitate as hydroxides or carbonates.

In order to understand the masking operation, it is necessary to understand the chemical equilibria in the solution when the desired analytical reaction is being conducted. Cheng [37] and Hulanicki [38] have given methods for calculating the equilibrium in a medium to which a masking agent has been added, and data useful for theoretical prediction of the course of the masking reaction. For an extensive description of masking, the reader is referred to the monograph by Perrin [39].

Marczenko [40] has collated a very useful table of major masking agents most frequently used in everyday analytical practice (Table 1.2). Certain complexing agents which behave similarly are included in the table jointly. Thus, oxygen and hydroxyl, cyanide and ammine, citrate and tartrate complexes are referred to jointly. It should be borne in mind that citrates are usually more stable in acidic media and tartrates in alkaline media. It should be remembered however, that citrate and tartrate frequently give rise to kinetically stabilized mixed metal complexes.

Table 1.2 Some masking (complex-forming) agents

1	2	3	4	5	6	7	8	9	10	11	12	13	14	15	16
	Be eia											B ga			
	Mg de											Al adg	Si g	P a	
	Ca de	Sc de	Ti egh	V ahj	Cr aej	Mn aed	Fe dej	Co cdj	Ni cdj	Cu cd	Zn cad	Ga ade	Ge afj	As abf	Se a
	Sr de	Y de	Zr egh	Nb egh	Mo aeg		Ru acf	Rh cf	Pd cf	Ag c	Cd cdf	In dej	Sn abf	Sb abe	Te a
	Ba de	La de	Hf egh	Ta egh	W aeg	Re aej	Os acf	Ir cf	Pt cf	Au cf	Hg cf	Tl fej	Pb aed	Bi dej	
			Ce de	Th edi	U ieh										

Complexes:
a—oxo and hydroxo
b—sulphide
c—cyanide and ammine
d—EDTA
e—citrate and tartrate
f—halo (Cl$^-$, Br$^-$, I$^-$)
g—fluoro
h—peroxo
i—carbonate
j—oxalate

The data given in Table 1.2, along with available stability constants and theoretical considerations [38,39] make it possible to specify masking agents for most situations.

References

[1] Gutzeit, M., *Pharm. Ztg.*, **24**, 263 (1879).
[2] Marsh, J., *Edinburgh New Phil. J.*, **21**, 229 (1836).
[3] Rodden, C. J. (Ed.), *Analytical Chemistry of the Manhattan Project*, McGraw-Hill, New York, 1950.

References

[4]   Yoe, J. H. and Koch, H. J. (Eds.), *Trace Analysis*, Wiley, New York, 1957.
[5]   Specker, H. and Hartkamp, H., *Angew. Chem.*, **67**, 173 (1955).
[6]   Burriel-Martí, F., *Experientia Suppl.*, **5**, 71 (1956).
[7]   Lipis, L. V., *Usp. Fiz. Nauk*, **68**, 71 (1959).
[8]   Claassen, A., *Chem. Weekblad*, **38**, 33 (1962).
[9]   Alimarin, I. P., *Pure Appl. Chem.*, **7**, 455 (1963).
[10] Specker, H., *Z. Erzbergbau Metallhüttenwes.*, **17**, 132 (1964).
[11] Ehrlich, G. and Rexer, E., *Wissensch. Z. Techn. Hochschule Chem. Leuna-Merseburg*, **6**, 207 (1964).
[12] Minczewski, J., *Chim. Anal. (Paris)*, **47**, 401 (1965).
[13] Minczewski, J., *Pure Appl. Chem.*, **10**, 567 (1965).
[14] Konovalev, E. E. and Pelzulyaev, Sh. I., *Tr. Kom. Anal. Khim., Akad. Nauk SSSR*, **15**, 375 (1965).
[15] West, P. W. and West, F. K., *Anal. Chem*, **40**, 138R (1968).
[16] Zaduban, M., *Mikrochim. Acta*, **1970**, 433.
[17] Goryushina, V. G., *Zavodsk. Lab.*, **37**, 513 (1971).
[18] Rottschafer, J. M., Boczkowski, R. J. and Mark, H. B. Jr., *Talanta*, **19**, 163 (1972).
[19] Irving, H., *Z. Anal. Chem.*, **263**, 264 (1973).
[19a] Tölg, G., in *Wilson and Wilson's Comprehensive Analytical Chemistry*, Vol. III, G. Svehla (Ed.), Elsevier, Amsterdam, 1975.
[19b] Kuźmin, N. M., *Zavodsk. Lab.*, **43**, 1301 (1977).
[19c] Zolotov, Yu. A., *Pure Appl. Chem.*, **50**, 129 (1978).
[20] Iwantscheff, G., *Das Dithizon und seine Verwendung in der Mikro- und Spurenanalyse*, 2. Aufl., Verlag Chemie, Weinheim, 1972.
[21] Alimarin, I. P., (Ed.), *Metody analiza veshchestv vysokoi chistoty* (*Methods for Analysing High-Purity Materials*), Moscow, 1965.
[22] Alimarin, I. P., (Ed.), *Metody poluchenya i analiza veshchestv osoboi chistoty* (*Methods of Preparation and Analysis of High-Purity Materials*), Nauka, Moscow, 1970.
[23] Koch, O. G. and Koch-Dedic, G. A., *Handbuch der Spurenanalyse*, Springer Verlag, Berlin, 1973.
[24] Kokotov, Yu. A., *Radiometricheskie metody opredeleniya mikroelementov* (*Radiometric Methods for the Determination of Trace Elements*), Nauka, Moscow, 1965.
[25] Zolotov, Yu. A., *Ekstraktsionnye kontsentrirovanie. Khimiya* (*Extraction Concentration. The Chemistry*), Moscow, 1971.
[25a] Dilts R. V., *Analytical Chemistry, Methods of Separation*, Van Nostrand, New York, 1974.
[25b] Miller J. M., *Separation Methods in Chemical Analysis*, Wiley-Interscience, New York, 1975.
[26] Belcher, R., *Pure Appl. Chem.*, **25**, 681 (1971).
[27] Zolotov, Yu. A., *Reinststoffe Wiss. u. Techn., Symposiumsber., Dresden 1970* Akademieverlag, Berlin, 1972, p. 655.
[28] Pszonicki, L., *Colloq. Spectr. Int. 16, Heidelberg, 1971, Main Lectures and Proceedings*, London, 1972, p. 199.
[29] Chalmers, R. A., *Pure Appl. Chem.*, **31**, 569 (1972).
[30] Kaiser, H., *Pure Appl. Chem.*, **34**, 35 (1973).
[30a] Zolotov, Yu. A., *Analyst (London)*, **103**, 56 (1978).

[31] Berl, E. and Lunge, G., *Chemisch-Technische Untersuchungsmethoden*, V Ausg. Springer Verlag, Berlin 1938.
[32] Minczewski, J., *Acta Chim. Acad. Sci. Hung.*, **34**, 123 (1962).
[33] Minczewski, J., cf. *Trace Characterisation, Physical and Chemical*, N.B.S. Monograph No. 100, Meinke, W. W. and Scribner, B.F. (Eds.), Washington D.C., 1967, p. 385.
[34] Prout, W. E., Russell, E. R. and Groh, H. J., *J. Inorg. Nucl. Chem.*, **27**, 473 (1965).
[35] Silvey, W. D. and Brennan, R., *Anal. Chem.*, **34**, 784 (1962).
[36] Chwastowska, J., Dybczyński, R. and Kucharzewski, B., *Chem. Anal. (Warsaw)*, **13**, 721 (1968).
[37] Cheng, K. L., *Anal. Chem.*, **33**, 783 (1961).
[38] Hulanicki, A., *Talanta*, **9**, 549 (1962).
[39] Perrin, D. D., *Masking and Demasking of Chemical Reactions*, Wiley-Interscience, New York, 1970.
[40] Marczenko, Z., *Spectrophotometric Determination of Elements*, Horwood, Chichester, 1975.

Chapter 2

WORKING TECHNIQUES IN TRACE ANALYSIS

2.1 GENERAL PRINCIPLES OF TRACE ANALYSIS

In 1956 in his discussion of problems associated with trace analysis Burriel-Martí [1] suggested that any trace-analysis operation must be carried out under 'chemically aseptic' conditions. When elements as universally distributed as sodium, calcium, magnesium, silicon, boron or zinc are to be determined, their presence in the air and in the reagents used begins to affect determinations at the concentration level of 10^{-3}–$10^{-4}\%$. If concentrations of 10^{-6}–$10^{-7}\%$ are to be determined, the list of the elements occurring in the laboratory conditions in large enough amounts to distort the analytical response is extended to include iron, copper, nickel, lead, aluminium, titanium, and often other elements, too.

The influence of the environment on the results of trace analyses makes it necessary to take special precautions in all steps of the analytical procedure, including sampling. Such precautions have been very strongly recommended in all the surveys and books mentioned in Chapter 1, and they have been discussed in detail [2,2a]. Errors resulting from such effects are included in computer programs developed for error analysis of particular trace-analysis methods [3].

The successive operations of an ordinary trace analysis may be presented schematically as in Table 2.1 (central column). The scheme covers the most general case when a compact solid material is to be analysed. With powdered and liquid materials, some of the preliminary operations become redundant, and ashing is normally only required for organic materials.

The left-hand column lists the variables which have, or may have, a direct effect upon the course of analysis; these will be discussed later in more detail. The right-hand column lists the methods which may be applied at individual stages of the analytical process. Obviously, the analyst's dream is a direct non-destructive method which would allow the material to be examined in its original condition. Unfortunately, in trace analysis, such methods are rare.

In laboratory practice, it is only necessary to consider the preliminary treatment of a sample which is to be further examined by a direct method,

Table 2.1 General Outline of Trace Analysis

Factors affecting result of analysis	Analytical operations	Method of determination
Homogeneity, Storage	Material to be analysed →	Non-destructive
Tools, Atmosphere, Reagents, Changes in analyte composition	Sampling, Grinding, Mixing, Surface cleaning	
	↓ Laboratory sample →	Direct, for solids
	↓ Ashing	
	↓ Dissolved sample →	Direct, for solutes
Vessels, Reagents, Atmosphere, Volatility	Separation of matrix / Separation of traces ↓ Preconcentration	
	↓ Dissolved traces →	Simultaneous determination of many constituents
	↓ Separation ↓ Individual trace constituents →	Determination of individual constituents

such as emission spectroscopy or mass spectrometry. However, in most cases it is necessary to separate the trace elements from the sample matrix and analyse the resulting solution of the traces at a suitable concentration, e.g. by emission spectroscopy or atomic absorption. Very often even this procedure is impracticable and for some reason or other it is necessary to separate the preconcentrated traces into individual constituents, which are subsequently determined by simpler methods such as spectrophotometry, polarography, etc.

Although this book is primarily concerned with selected methods for separation of traces, the full analysis scheme given in Table 2.1 has been adopted as a basis for the treatment, and the description of trace-analysis techniques will include all the factors which can affect the analytical result.

2.2 MATERIALS TO BE ANALYSED

Materials to be analysed can be classified into two principal groups, inorganic and organic. The latter almost always have to be ashed (cf. Section 4.5) before the true trace analysis can be started.

For the purposes of sampling, materials can be classified, according to their form, as lumps, powders, liquids, pastes, and gases

On the basis of the two classifications presented above, we shall deal in succession with the problems associated with the trace analysis of the two main categories of materials—inorganic and organic. Gases are beyond the scope of this book.

2.2.1 Inorganic Materials

Materials in the form of lumps may be divided into two groups. One includes compact materials such as metals, alloys and single crystals, which, in terms of trace analysis, are usually high-purity materials. The second group encompasses geological materials and meteorites. These are of interest for geochemical studies, prospecting for rare and dispersed elements, and fundamental research on meteorites.

Powdered materials are usually obtained by mechanical size-reduction of lump materials, by chemical treatment of mineral raw materials, or as products of chemical synthesis. In terms of trace analysis they are of interest mainly as raw materials for preparing high-purity materials but they are found also, for example, in examination of soil for trace elements.

Liquid materials include solutions and liquid substances such as water and other solvents. In trace analysis, liquids are of interest primarily as the reagents employed in the course of analysis, as auxiliary materials used in the production of high-purity materials, but in certain studies on the biological and ecological significance of trace elements, i.e. in studying natural waters (surface, ground and sea-water), soil solutions, etc., they will have to be analysed.

2.2.2 Organic Materials

Organic materials may be classified into three groups: fossil materials, biological materials, and organic chemicals. The first group includes coals, bituminous shales, and crude oil: these are of interest because they contain trace elements. Such trace elements may actually be extracted for use (e.g. germanium from coals) or they may be important for other reasons (e.g. vanadium in crude oil).

Biological materials include plant materials of all kinds, body fluids and animal tissues which are analysed for trace elements because of the role of such elements in metabolic processes, or occasionally for forensic information.

Organic chemicals include solvents employed in analysis and in the technological processing of high-purity materials, and also other organic reagents used for similar purposes.

2.2.3 Other Materials

This category includes materials which cannot be classified in any other group. They may be mixtures of materials belonging to different categories, or they may be difficult to classify in a strict manner. An example is sea-water, which is, generally speaking, an aqueous solution of various salts, but it also contains various inorganic colloids and suspended solids, and a number of organic substances in solution or suspension. Such a material calls for special treatment.

2.3 SAMPLING

In sampling of materials and in storing the sample to be analysed for trace constituents, three aspects must be taken into account: homogeneity of material, size of analytical sample, and possible changes occurring in the sample in collection, handling and storing [4].

2.3.1 Metals and Alloys

The determination of the concentrations of trace constituents is extremely difficult because of the non-uniform distribution. Segregation occurs not only with alloying components but also with trace constituents: it can be revealed clearly by autoradiography. Non-uniform distribution of alloying components in a slowly cooled alloy is illustrated in Figs. 2.1 and 2.2 [5], which show the distributions of copper (dark field in Fig. 2.1) and nickel (dark field in Fig. 2.2) in aluminium–copper and aluminium–nickel alloys, respectively. Another well-known phenomenon is the non-uniform migration of impurities in the zone-refining [6] of metals and other substances. Differences in the distribution coefficients between the solid and liquid phases cause various constituents to behave differently on zone-melting, and this prevents widespread application of this technique, e.g. for concentrating trace constituents in trace analysis [7]. This means that particular difficulties have to be overcome in the determination of impurities in high-purity metals, and in a single-crystal metal block the concentrations of impurities have to be determined as local rather than average values.

 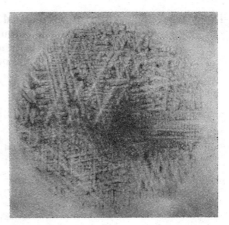

Fig. 2.1 Non-uniform distribution of copper in an aluminium–copper alloy

Fig. 2.2 Non-uniform distribution of nickel in an aluminium–nickel alloy

(from [5], by permission of the copyright holders, Berg-Akademie, Freiberg)

Metal and alloy specimens can be made homogeneous in two ways. Metallurgically, it can be achieved by repeated melting, mixing and rapid crystallization. These operations are usually employed in the preparation of standard samples for spectrography and, in general, for metallography [8,9]. Technical difficulties arise from the lack of suitable equipment for such metallurgical operations in a typical analytical laboratory, and an additional hazard is associated with their application to routine preparation of samples for trace analysis: it is very easy for secondary contamination to arise from the material of the vessels, furnaces, etc., in which the operations are carried out.

In a normal trace analysis, the procedure reduces to determining the average concentrations of trace constituents in a lump (or a suitable lot) of metal (or alloy). At various sites the test metal is drilled, chipped or otherwise treated, and the drillings or chips collected are mixed together, and, after any surface-borne secondary impurities originating from the tools used have been removed, analytical samples are collected. This procedure leads to satisfactory results provided that the necessary precautions are taken.

Table 2.2 lists [10] the normalized variances of spectrographic determinations of five impurities in a standard sample of aluminium of guaranteed homogeneity (V_S) and in a laboratory sample (V_L) of aluminium prepared by the method described above, i.e. by chipping. From a 200-g aluminium block about 20 g was chipped off, mixed, etched, washed and analysed. Fisher's test shows that the homogeneity thus obtained was quite sufficient.

Table 2.2 Comparison of normalized variances in the determinations of trace constituents in aluminium: standard sample (V_S) and sample homogenized in laboratory (V_L)
(from [10] and [10a], by permission of Polska Akad. Nauk and B. Strzyżewska)

Element	Sample variance		Fisher's test variance ratio, F	
	standard V_S	laboratory V_L	calculated $F = V_S/V_L$	tabulated
Fe	6·80	4·04	1·68	2·60
Si	6·92	11·71	0·59	2·60
Mg	2·76	5·12	0·54	2·60
Mn	2·51	1·76	1·43	2·60
Pb	8·99	4·64	1·94	2·60

In the preparation of a chipped metal sample, the final operation involves etching the chips with specially chosen reagents to remove the secondary impurities originating from the tools used to obtain the sample. The amount of these impurities increases with decreasing difference in hardness between the metal and the tool. However, even if the difference is quite considerable, contamination still occurs. Figure 2.3 shows autoradiographs of two aluminium blocks, one of which had been cut with a steel tool (Fig. 2.3a) and the other with a tungsten carbide tool (Fig. 2.3b). Dark spots indicate the residual material transferred from the tools.

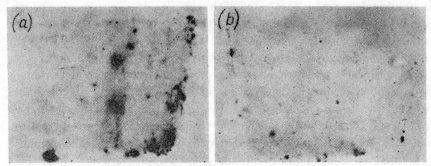

Fig. 2.3 Autoradiograms of aluminium blocks

The secondary impurities are removed by etching the chips with suitable chemical reagents, usually with acids such as hydrochloric, nitric, hydrofluoric or their mixtures (obviously the acids used must be of highest purity), followed by washing with water and possibly with volatile solvents. In performing the washing, it must be borne in mind that a badly chosen solvent or an excessively long treatment with solvent may result in selective dissolution of certain constituents of the sample and thus also affect the composition of the sample.

2.3.2 Geological Materials

Preparation of a sample which is truly representative of a geological stratum or, e.g., of a batch of iron-ore, is a very difficult problem, and the nature of the material and of the particular constituents to be determined must be considered. In practice, it is best to use a procedure which is known to provide a representative sample from similar materials, and to draw conclusions only from a sufficiently large number of results obtained from different samples.

Grinding and mixing usually offer only minor technical difficulties; since geological materials are usually examined for rare elements not present in the tools employed, contamination of the sample is unimportant.

There is a danger of stratification of the powdered material on mixing. Different densities of the individual minerals and different hardness values of the individual constituents make the preparation of a uniform sample difficult, particularly when the analysis samples are small.

2.3.3 Powdered Materials

Sampling of powdered materials prepared by reduction of the particle size of fossil materials is tricky for the reasons mentioned above. Samples of soils are particularly inhomogeneous and extremely difficult to homogenize. Powdered materials obtained in chemical processing or produced synthetically are usually much more homogeneous than those mentioned above and collection of a representative sample presents fewer difficulties.

2.3.4 Liquid Materials

It is not difficult to prepare a homogeneous representative sample of a liquid material, provided that the material is a real solution or a homogeneous liquid. However, storage of such liquids is problematical, on account of possible sorption of trace constituents from the liquid onto the vessel walls and also possible leaching of trace components from the walls of the vessel into the liquid. These problems will be dealt with at greater length later (Section 2.6).

2.3.5 Organic Materials

Fossil materials should be sampled similarly to minerals; they present no particular problems. Body fluids correspond, to a greater or lesser degree, to liquid samples.

Plant material and animal tissue are, by their very nature, considerably inhomogeneous; nevertheless, a correctly prepared ash readily yields representative samples. Conditions for ashing will be described later in more detail (Section 4.5).

Organic chemicals are easily sampled, but secondary contamination may occur on storage (Section 2.6) and on ashing (Section 4.5).

2.3.6 Other Materials

As already indicated, these are materials which cannot be unequivocally classified in any specific category. They include mixed inorganic–organic materials (e.g. bituminous shales) which after ashing correspond to inorganic materials. Mention should also be made of mixtures of liquids and sediments; usually the sediment separated by filtration and the filtrate are analysed separately.

Natural waters, whether surface, ground, or sea, present a special group of major importance from the viewpoint of trace analysis. Trace elements are an important factor in ecological investigations of aquatic media. The high non-homogeneity of such waters and the high sensitivity to the influences of storage vessels and atmosphere as extraneous factors render both collection of representative samples and analysis very difficult. The most difficult problem, so far unsolved in an unambiguous manner, is the separation of the insoluble or partially soluble fraction from the true aqueous solution. In this respect the composition is highly dependent upon the depth and location of the sampling site. Usually the sample is filtered through arbitrarily selected filters (these can be a source of secondary contamination), and then the filtrate and (if desired) the sediment are analysed separately [11].

2.4 CONTAMINATION FROM THE ATMOSPHERE

The air in any laboratory is a prolific source of contamination of analysis samples. It is contaminated with natural dust and also with metal particulates originating from laboratory equipment such as burners, taps, etc. Metal from these objects may be directly dispersed, e.g. on heating (burners), or it may be corroded under the action of the acids usually present in the laboratory atmosphere. The urban atmosphere itself is highly polluted and contains unexpectedly high air-borne concentrations of elements often determined as trace constituents in high-purity materials [12].

Tabor and Warren [13] have reported air-borne metal concentrations in the U.S. urban atmosphere (Table 2.3). The concentrations of elements as uncommon as vanadium, titanium, and chromium are surprisingly high.

Table 2.3 Air-borne metal concentrations in U.S. urban atmospheres
(from [13], by permission of the copyright holders, the American Medical Association)

Metal	Concentration, $\mu g/m^3$	Metal	Concentration, $\mu g/m^3$
Zinc	0.4–49.0	Vanadium	0.002–0.60
Iron	0.23–30.0	Titanium	0.01–0.24
Copper	0.05–30.0	Nickel	0.005–0.20
Lead	0.33–17.0	Chromium	0.002–0.12
Manganese	0.01–3.0	Cadmium	0.002–0.10
Barium	0.005–1.5	Bismuth	0.002–0.03
Tin	0.004–0.80		

Ehrlich et al. [14] found the dust content in a laboratory atmosphere (in a large city) to be of the order of 0.2 mg/l. The compositions of air-borne dusts of various origins are listed in Table 2.4. If the dust settling rate is taken as 10^{-9} g·cm^{-2}·sec^{-1} [17], the amount of dust that can fall into a beaker 6 cm in diameter is 100 µg per hour. This amount is quite commensurate with, and sometimes may even exceed, the contents of trace constituents determined in a high-purity material. Table 2.5 demonstrates the contamination of silicon and nitric acid exposed to the atmosphere in a normal laboratory environment. In the case of silicon analysis, the trace impurities were concentrated on graphite powder, and sample (A) was analysed spectrographically immediately after concentration; sample (B) was exposed for a week to the laboratory atmosphere before being analysed. In the analysis of nitric acid, duplicate determinations were made on the same sample. In case (A) all the operations were carried out in sealed boxes; in case (B) the operations were performed in the atmosphere of the laboratory with no protection. The differences between the concentrations of the trace components determined in the two cases are quite distinct.

Table 2.4 Chemical composition of air-borne particulates

Component	laboratory [15]	Content, %, in dust urban area [16] (Prague, Czechoslovakia)	laboratory [14]
SiO_2	38	45	10
Na	4.5	+Mg, 3	2
Ca	7.25	—	10
Pb	2.7	0.15	—
Zn	0.4	0.15	—
Cu	0.03	0.15	0.5
Fe	0.6	14	3
Mn	0.04	0.015	0.15

Table 2.5 Air-borne contamination after exposure to the atmosphere of an ordinary laboratory
(after data from [18])

Sample	Conditions	Contamination $\times 10^{-7}$%						
		Al	Fe	Ca	Cu	Mg	Mn	Ni
Silicon	A	10	10	10	0.4	10	<0.2	<0.5
	B	50	300	50	2	50	2	2
Nitric acid	A	1	8	3	<0.1	5	0.1	<0.1
	B	50	30	50	5	30	0.6	0.7

Air-borne pollutants can also affect the course of chemical reactions, e.g. Iwantscheff [19] has reported that physiologically imperceptible trace amounts of air-borne hydrogen sulphide can produce decomposition of numerous dithizonates.

2.4.1 Avoidance of Contamination

There are two major ways to avoid trace analyses [20] being affected by air-borne pollutants. One involves complete air-conditioning of the laboratory and the other involves carrying out the individual operations in glove boxes. The first requires complete isolation of the trace analysis laboratory from other laboratories and from the outside atmosphere and also efficient filtration of the air entering. The laboratory must be connected with other parts of the building by an air-lock where the personnel may change clothes. It should contain no metal fittings. Chemically aseptic rooms are employed in the electronics industry in the manufacture of semiconducting materials, but application to chemical work is not really justified. They are extremely costly and cause problems, particularly in research laboratories, where a large variety of samples have to be analysed or individual methods developed.

Trace analysis certainly needs special rooms, especially since it must be separated from possible major concentrations of the trace constituent to be determined. If, for example, the work is to be conducted simultaneously on the determination of traces of boron in silicon and on the trace analysis of high-purity boron, the two analyses cannot be performed in the same room. The trace analysis laboratory must be easy to keep clean. The surface of floors should be coated with a plastic layer and the surface of ceilings and walls painted with a washable paint. Wooden fume hoods are not so liable to corrosion as are iron hoods and do not give rise to a permanent risk of contamination of samples with iron or other metals (but perchloric acid cannot then be used).

Location of the laboratory in the terrain is also of major importance. In an area of sandy soil or in the vicinity of factories which emit dust into the atmosphere, all the air for laboratory ventilation must be filtered.

The laboratory equipment should be as simple as possible, and easy to keep clean.

It is much easier and equally efficient to avoid contamination by carrying out the individual operations of trace analysis in glove boxes. The use of these relatively simple enclosures has been borrowed by trace analysts from radiochemical laboratories where they are commonplace, e.g. in handling of plutonium. The box is made from clear plastic (Plexiglas) with or without use of some poly(vinyl chloride) (PVC). The size of the box is chosen according to the intended operations. At least two rubber gloves are mounted over holes in the box, and these allow the operator to manipulate equipment and perform operations in the box. Operations are performed in the glove box to allow them to be physically isolated from the laboratory atmosphere, or because a specially controlled atmosphere is required. The special atmosphere (a purified inert gas, e.g. nitrogen, or air) is admitted from a cylinder (or compressed-air network) through a train of suitable washers.

Results obtained with the aid of glove boxes are illustrated in Fig. 2.4, which shows the spread of spectrochemical analyses for iron, copper, magnesium, lead, and silicon in a high-purity aluminium ($5N$ Al) [10]. The observed blackening data, V_M, are presented as a function of the

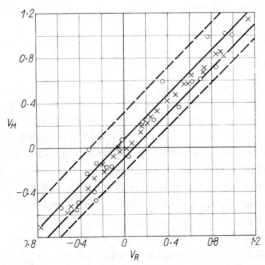

Fig. 2.4 Distribution of results of spectrochemical determination of trace contaminants in high-purity aluminium (from [10], by permission of Polska Akad. Nauk)

values V_R from the regression line. Circles denote the results of measurements exposed to an ordinary laboratory atmosphere, while crosses represent the results of measurements made in glove boxes. Evidently, isolation has halved the spread in the data.

Chow and McKinney [21] measured air-borne lead contamination introduced into the residue from 500 ml of hydrochloric acid evaporated in various ways. Their results (Table 2.6) show that evaporation under an inert atmosphere reduces the lead contamination by a factor of 10–40.

More detailed discussion of clean laboratories may be found in ref. [2a].

Table 2.6 Lead contamination of a sample solution after exposure to laboratory air
(from [21], by permission of the copyright holders, the American Chemical Society)

Beaker	Evaporation conditions	Laboratory	Time of evaporation, days	Lead, µg
Teflon	open	ordinary	8	4·07
				2·32
	nitrogen-flushed container	ordinary	8	1·13
	open	air-conditioned	8	0·44
	nitrogen-flushed container	air-conditioned	8	0·18
				0·13
	nitrogen-flushed container	air-conditioned	1	0·02
Borosilicate glass	nitrogen-flushed container	air-conditioned	1	0·03

2.4.2 Laboratory Equipment

In the trace analysis laboratory metal apparatus should be avoided. This is particularly important in operations carried out at high temperatures, i.e. use of any type of heating equipment. Porcelain water- or sand-baths are best, but if these are not available, the conventional electric hot-plate [22] should be shielded with a graphite cover. Suitable holes made in this cover to take a crucible or an evaporating dish ensure good evaporation. The graphite will protect the iron plate from acid vapours, and the sample against contamination with traces of iron which would enter if evaporation were carried out directly on the hot-plate.

Heating with quartz infrared heat lamps is advantageous, because surface heating prevents splashing of the sample.

Other equipment, including retort stand bases and funnel stands, should be made of plastic.

2.5 COMMINUTION OF SAMPLES

Grinding may give rise to considerable contamination of samples even when the difference in hardness between the sample and the material of the mortar is quite large [23,24]. The significance of the problem is well demonstrated in Tables 2.7 and 2.8, which list contamination introduced into oat grains on milling in various devices [25], and the increase in silica content on grinding several samples in various mortars [26].

Two guidelines should be followed when choosing suitable materials for grinding equipment. Such materials should either be very hard or they should contain elements other than those to be determined in the samples. Since numerous high-purity materials (e.g. those used in the electronics industry) are very hard, the second guideline is more often practicable.

Agate is the typical material from which mortars are made, but an agate mortar is inadmissible for grinding high-purity materials (unless relatively soft), because the sample is very likely to become contaminated with silicon, magnesium, aluminium, and sodium, i.e. elements often determined as trace constituents. A tungsten carbide mortar, which is harder and which can contaminate the sample only with tungsten, an element only rarely determined as a trace impurity, is much more useful.

A relatively new material used in the manufacture of mortars is single-crystal alundum (Al_2O_3). An alundum mortar is very hard and less abradable than a tungsten carbide one by a factor of about 1500 [27]. Even if some alundum is rubbed off, the sample will become contaminated only with aluminium.

Hard and friable samples are crushed in pure-metal devices; for example, Kalininkov and Shteinberg [28] have suggested a copper impact mortar for crushing tungsten carbide. Vasilevskaya et al. [29] have recommended a molybdenum mortar for grinding silicon. In this operation, the sample is reported to have been contaminated with about 1% of molybdenum, which does not interfere in the spectrochemical determination of boron in silicon, but does interfere in the determination of some other trace elements such as Al, Ni, Ca, Ti, Cu, In.

In certain cases, when friable materials are handled, good results are obtained by wrapping the sample in a polyethylene or Teflon foil to protect it from direct contact with the mortar or the pestle material.

The contaminants introduced into the sample on comminution can sometimes be removed by washing the sample first with extremely pure acids to dissolve surface impurities and then with pure water and occasionally ethanol. Such manipulations require sufficient testing because they may give rise to partial dissolution of the sample itself and thus to biased results.

Table 2.7 Trace-element contamination of oat-grains ground in various types
(from [25], by permission of the copyright holders, the American Chemical

Sample ground in	Fe content %	Fe contamination %	Zn content %	Zn contamination %	Cu content %	Cu contamination %
—	24·9	—	21·6	—	3·75	—
Porcelain mortar and pestle	24·9	0·0	18·8	0·0	4·0	5·0
Wiley mill	38·2	35·0	21·0	0·0	9·0	57·8
Hammer mill	43·3	42·5	17·3	0·0	4·4	13·6
Flint-ball mill	35·0	28·9	166·0	86·6	5·5	30·9
Porcelain-ball mill	159·5	84·4	196·0	89·0	9·3	59·1
Mullite-ball mill	365·0	93·2	140·0	85·4	13·3	71·4

Table 2.8 Abraded silica contamination of samples ground in various mortars
(from [26], by permission of Polska Akad. Nauk)

Sample	SiO_2 (%) abraded in mortar made of		
	sintered tungsten carbide	agate	porcelain
Fuse-cast alumina ('korvisit')	0·40	4·45	11·68
Electrocorundum A	0·51	4·60	8·70
Slip-cast alumina	1·07	2·64	5·32

2.6 LABORATORY VESSELS AND CONTAINERS

The most important materials of which laboratory vessels and containers are made include glass, quartz, platinum, and plastics, mainly polyethylene (high- and low-densities), polypropylene, and Teflon.

When brought into contact with the sample (especially in solution), these materials can affect the concentrations of the trace constituents to be determined, in three ways.

(1) Trace constituents may be sorbed on the vessel walls, thus diminishing their concentration in the solution.

(2) Constituents of the vessel material may be leached into the solution, thus increasing the concentrations of these constituents.

(3) Constituents adsorbed on the walls from a preceding solution may be desorbed into the solution.

At low concentrations the equilibrium between sorption, leaching and desorption usually sets in very slowly. Therefore, the three effects exercised by the vessel walls on the solution composition are much more

of equipment
Society)

Co		Na		Ca		S	
content %	contamination %	content %	contamination %	content %	contamination %	content %	contamination %
0·01	—	0·29	—	1·3	—	1·45	—
0·01	0·0	0·28	0·0				
0·01	0·0	0·32	9·4				
0·01	0·0	0·32	9·4				
0·03	66·7	0·42	30·9	1·2	0·0	1·40	0·0
1·12	99·1	1·07	72·9	1·5	13·3	1·49	2·7
0·30	96·7	2·23	87·0	2·4	45·8	3·35	56·8

pronounced in analytical operations of long duration such as evaporation and dissolution of samples at elevated temperatures, and when the vessels are used for storage of sample solutions, standard solutions, and reagent solutions. In short-duration operations such as pipetting, filtration, dilution in a standard flask, and measurement of absorbance, these effects may usually be ignored.

These problems have been discussed by numerous investigators, and ample data can be found in monographs and research works concerned with trace analysis, e.g. Zief and Mitchell [2a], Sandell [30], Koch and Koch-Dedic [31], West et al. [32], Robertson [33], Wichers et al. [34], and also in the literature pertaining to oceanographical research. The necessity for prolonged storage of sea-water samples collected for trace element analysis has made oceanographers aware of these problems and given rise to a number of investigations [35–40].

In the past dozen or so years, the study of sorption and desorption of radioactive elements present in solution at very low concentrations has become of particular interest, because of the relevance to decontamination of nuclear equipment. Kokotov [41] tried to establish a mathematical relationship between the concentration of isotope adsorbed onto the vessel wall and that remaining in the solution, as a function of the distribution coefficient, solution volume, and vessel surface area in contact with the solution. However, the phenomena are highly complicated and influenced by many factors, and quantitative predictions are still not possible. Direct application of the results of studies of the behaviour of radioactive isotopes to stable isotopes is inadvisable in view of the energy differences, which, at the concentrations occurring, may not be insignificant

Each of the materials of which vessels may be made will contain minor or major impurities, including unexpected elements. A useful study has been made by Robertson [33] of trace impurity concentrations in container materials and in some chemical reagents used in trace analysis. The materials may be arranged in the following order of decreasing usefulness for trace analysis: Teflon > polypropylene > high-density polyethylene > quartz > low-density polyethylene > platinum ≫ glass. This classification has been developed on the basis of the average properties of the materials and in individual cases the sequence may require considerable changes. Generally, it may be shown that for solutions at concentrations above $10^{-3}\%$, the material of the vessel is of no consequence. Concentration changes due to interaction with the vessel may affect the results considerably at concentrations of $10^{-5}\%$ and less.

2.6.1 Glass

Inspection of the series given above shows that glass is not very suitable for trace-analysis work. The constituents of glass pass into solution very easily, attaining concentrations of the order of $10^{-4}\%$. Knížek and Provazník [42] found that water boiled in a beaker covered with a watch-glass leaches up to 1 mg of SiO_2 per hour from the watch-glass; an ammonia solution leaches 1·4 mg of SiO_2 within 30 min. Even trace constituents of glass, such as lead, are leached into the solution. Chow and McKinney [21] have found that in the course of a normal analytical procedure as much as 1 mg of lead may be leached from glass. Table 2.9 lists data on leaching of aluminium and iron [43] from Jenaer Glas G20 (30 ml of solution was heated in a 35 ml beaker on a boiling water-bath and then allowed to cool for 1 hr). Heavy metals are readily adsorbed [44] onto a glass surface which may, especially after prolonged contact with water, act as an ion-exchanger.

If glass must be used, a silane coating can greatly improve its properties. A dried glass vessel is filled with a solution of chlorosilane or alkoxysilane and emptied, and the residual silane is washed out with benzene, carbon tetrachloride or ligroin. The vessel wall becomes coated with a film which is not wetted by water and aqueous solutions. The film results from a reaction of the silane with the layer of hygroscopic water always present on untreated vessel walls. Alkaline solutions slowly attack the film, and a wash with alcoholic potash or 10% aqueous hydrofluoric acid removes the film completely. Studies with radiophosphorus [45] have shown that a treated surface sorbs much less than does an untreated one, and that adsorbed phosphorus-32 can be readily washed off with

Table 2.9 Leaching of aluminium and iron from Jena glass
(from [43], by permission of the copyright holders, Springer-Verlag)

Solution	Heating time, min	Al, µg	Fe, µg
Water	30	0	0
Water	60	0	0
0·3M HCl	30	0·3	0
0·3M HCl	60	0·6	0
0·9M HCl	30	0·5	0
0·9M HCl	60	0·9	0
0·15M H_2SO_4	30	0·3	0
0·15M H_2SO_4	60	0·5	0
0·45M H_2SO_4	30	0·5	0
0·45M H_2SO_4	60	0·8	0
0·3M NaOH	30	38	3·2
0·3M NaOH	60	81	7·0
0·9M NaOH	30	53	4·2
0·9M NaOH	60	125	turbid (H_2SiO_3)
0·3M NH_3	30	7	0·3
0·3M NH_3	60	13	0·6
0·9M NH_3	30	11	0·7
0·9M NH_3	60	16	1·3

water, whereas that adsorbed on a normal glass surface is impossible to remove. Similar results have been obtained with glass treated with poly(vinyl trichlorosilane) or with a polyester resin [46].

2.6.2 Quartz

Quartz has much better properties than glass and is chemically more resistant. In trace analytical work a special grade of quartz produced for semiconductor purposes may be used. Table 2.10 lists the analytical data for four grades of quartz used for laboratory ware [47,48]. Quartz is particularly resistant to acid solutions (other than those containing hydrofluoric acid).

2.6.3 Platinum

Pure and iridium-alloyed platinum ware, commonly used in analytical laboratories, is not suitable for trace analysis work. In particular, operations with concentrated acid solutions will result in considerable metallic contamination of the solutions.

Table 2.10 Trace contaminants in various grades of quartz
(after data from [47, 48])

Species	Purest commercial quartz, %	Vitreosil, %	H.P.Q.* Vitreosil, %	Spectrosil †, %
Al	0.0053	0.005–0.007	0.005–0.006	$<2 \times 10^{-6} - <25 \times 10^{-6}$
As				$<0.02 \times 10^{-6}$
B		0.00005	0.00001	$<1 \times 10^{-6} - \ll 50 \times 10^{-6}$
Ca		0.00004	0.00004	$<10 \times 10^{-6}$
Cu		0.000001	0.000001	$<0.02 \times 10^{-6}$
Fe	0.0035	0.00007	0.000074	$<10 \times 10^{-6} - <10 \times 20^{-6}$
Ga				$<0.4 \times 10^{-6}$
K	0.022			$<0.4 \times 10^{-6}$
Mn		0.000003	0.0000026	$<0.1 \times 10^{-6} - 4 \times 10^{-6}$
Na	0.03	0.0004	0.0004	$<4 \times 10^{-6} - \ll 10 \times 10^{-6}$
P			0.000001	$<0.1 \times 10^{-6}$
Sb		0.00002	0.000023	$<0.01 \times 10^{-6}$

* High Purity Quartz † High-purity synthetic quartz

2.6.4 Plastics

Plastics, especially polyethylene, have become commonplace as reagent containers. Unlike the low-density grade, high-density polyethylene contains no inorganic catalysts and is a good material for handling traces at moderate temperatures (up to 60–70°C). However, it may exhibit sorptive properties. For example, alkali metals have been found to be liable to substantial sorption from solution at concentrations of $\sim 10^{-4}$ %.

Teflon is the most chemically and thermally resistant material; it is not attacked by concentrated sulphuric acid on evaporation. It is attacked only by fluorine and molten alkali metals. At present, Teflon is the best material for making equipment for trace analysis.

It is important to realize that plastic laboratory ware will not always have the same properties. Even when supplied by the same manufacturer it may differ in properties from batch to batch of polymer. Hence, it is necessary to check each lot bought, for suitability for its intended use.

Vasilevskaya [18] has reported spectrochemical analysis data on hydrofluoric acid, hydrochloric acid, and nitric acid after evaporation in Teflon, platinum, and quartz dishes (commercial platinum and quartz grades) (Table 2.11), which support the earlier generalizations. Evaporation in platinum raised the content of trace elements by 0.5–1.0 order of magnitude, and evaporation in quartz raised it even more.

Table 2.11 Contamination after the evaporation of mineral acids in Teflon, platinum, and quartz dishes
(from [18], by permission of the copyright holders, Nauka, Moscow)

Acid	Material	Elements, ng/ml										
		Al	Fe	Ca	Cu	Mg	Mn	Ni	Pb	Ti	Cr	Sn
HF	Teflon	3	3	1	<0.04	<3	0.1	<0.4	<0.1	0.1	<0.4	ND
	platinum	10	10	10	0.4	10	0.2	0.3	0.5	1	0.5	ND
HCl	Teflon	<4	3	5	0.2	3	0.1	ND	<0.4	ND	ND	ND
	platinum	2	2	10	1.0	6	0.2	0.6	<0.4	0.4	Tr	<0.4
	quartz	10	10	60	1.0	10	0.4	2	0.5	2	0.6	0.4
HNO$_3$	Teflon	2	8	4	<0.01	7	0.1	ND	ND	Tr	ND	ND
	platinum	20	20	30	0.4	20	0.6	Tr	1	0.8	ND	ND
	quartz	20	20	60	0.1	20	0.6	ND	1	0.3	ND	ND

ND = not detected   Tr = traces, not evaluated quantitatively

Analysis by spark-source mass spectrometry [48a] of high-purity acids and water after storage and/or evaporation in Teflon-FEP, polypropylene, Vycor, linear polyethylene and TPX (also called PMP, polymethylpentene) showed little evidence of inorganic contamination from polypropylene or TPX, but Teflon-FEP proved to be a source of iron, nickel, chromium, and manganese [48b].

Useful comments on various materials used in sampling devices and sample-storing containers are made in the International Atomic Energy Agency publication [11] already referred to. Although the information was originally intended to apply to the study of trace elements in sea-water, the findings are applicable to numerous other fields of trace analysis (Table 2.12).

2.6.5 Cleaning

Cleaning of laboratory ware is of particular significance in trace analysis. The classic mixture for cleaning glassware—sulphuric acid and an alkali metal dichromate—is inapplicable because of sorption of chromium onto glass. Butler and Johnston [53] found that Pyrex glass adsorbed 10 ng of Cr per cm^2 from the cleaning mixture, and Lang [54] long ago reported on the difficulties in washing off residual chromium from glassware thus cleaned. A mixture of concentrated sulphuric acid and nitric acid, as recommended by Thiers [55], gives much less risk of contamination. Häberli [56] has recommended the use of a hot ammoniacal EDTA solution, but he warned that, if traces of heavy metals are to be determined, cleanliness of the glassware washed with the EDTA solution must be checked by a rinse with dilute dithizone solution.

Table 2.12 Construction materials for sea-water samplers and storage containers
(from [11], by permission of the copyright holders, I.A.E.A. Vienna)

Material	Remarks and warnings	Recommended for use	Further studies advised	Reference
Metals				
brass	inapplicable to trace analysis for metals	never		[49]
bronze	superior to brass but still inadvisable	never		[50]
aluminium	corrosion resistance varies much with alloy; inadvisable for use in trace analysis unless coated with plastic or Teflon	plastic-coated only		[50]
stainless steel	liable to chloride corrosion; resistance to corrosion varies with composition; not useful for trace analysis unless plastic-coated	plastic-coated only		
other metals	platinum and gold are useful; Ag-plated and Ni-plated brasses and bronzes useful for preparing low-toxicity biological samples: lack of tests on application to trace element analysis		yes	[51]
Glass and ceramics				
soft glass	often recommended as sea-water container but unlikely to be useful in handling metal traces	unsafe		[32,36, 52]
Jena glass	neither adsorbs nor desorbs zinc during storage of sea-water	pretesting required	yes	
borosilicate glass	liable to chemical attack at the pH of sea-water; no losses by adsorption claimed	unsafe		
Vycor (USA)	liable to chemical attack at the pH of sea-water; variable trace metal content	unsafe		
quartz	variable composition; liable to alkali attack	pretesting required	yes	[32–34]

Laboratory Vessels and Containers [§2.6]

Material	Comments	Pretesting	Disposable	Refs.
natural and synthetic rubbers	each may be suspected of contamination of solutions with Zn, Sb, Co, Cr, and other metals; most can also adsorb from solution	never		[33,35]
silicones	reported to adsorb Zn, Ag, and Hg from solution	pretesting required		[32]
polyethylene (a) low-density	may contamine solutions with Sb, Co, Cr, and Fe; may adsorb Zn and other traces; varies much with batch	pretesting required	yes	[33,35,37,52]
(b) high-density	superior to low-density	pretesting required	yes	[32]
polypropylene	adsorbs Ba (and Sr?) from solution at pH > 3·5; also adsorbs lanthanides and silver	pretesting required	yes	[32]
polystyrenes	adsorb Ag from solution	pretesting required	yes	[32]
polyacrylates	often very low impurity level; adsorb Zn and Ag from solution	pretesting required	yes (samplers)	[32,33]
PVC	often high impurity level; may adsorb traces from solution	never	yes (tubes, etc.)	[33,35]
polycarbonates	no information	pretesting required	yes	[33]
Teflon (PTFE)	excellent versatile material; expensive and soft; pierceable coat; adsorbs Ag from solution	pretesting required	yes	
polytrifluoroethylene	similar to Teflon	?	yes	
polyolefins		unsafe	yes	
polyesters	may contaminate solutions with Zn	unsafe	yes	
Nylon	one sample examined very impure			[33]

2.7 REAGENTS USED IN TRACE ANALYSIS

2.7.1 Water

Pure water is the reagent most frequently used in the analytical laboratory. Usually a combination procedure is used for its preparation; treatment with a mixed-bed ion-exchanger is followed by double distillation in quartz or platinum. Such water is pure enough for most purposes. Some typical water analyses [31], are given in Table 2.13. Traces of silica, the most difficult to remove, can be removed [57] by a strongly alkaline anion-exchanger (Permutit ES). After the Permutit ES treatment, the SiO_2 concentration was $<1 \cdot 5$ µg/ml.

Table 2.13 Trace-element concentrations in distilled water
(from [31], by permission of the copyright holders, Springer-Verlag)

No.	Water	Elements, ng/ml						
		Cu	Zn	Mn	Fe	Mo*	Pb	Al
1	Distilled conventionally	200	20	—	—	—	55	
2	Distilled from tin-plated copper still	10	2	1	2	2		
3	No. 2 redistilled from Pyrex	1	0·12	0·2	0·1	0·002		
4	No. 2 doubly distilled from Pyrex	0·5	0·04	0·1	0·02	0·001		
5	No. 2 triply distilled from Pyrex	0·4	0·04	0·1	0·02	0·001		
6	Distilled from Jena glass still	0·1	3				3	
7	Redistilled once from Pyrex	1·6	0·00				2·5	
8	Doubly distilled	0·5	1·4		0·9			1

* Assayed microbiologically with *Aspergillus* species.

With ion-exchangers, it is always possible that a soluble substance may be leached out and interfere with the determination (e.g. by complexing the ions to be determined). Therefore, water which has been treated by ion-exchange should always be distilled.

Typically, water treated on mixed-bed ion-exchangers (Amberlite IR-120 and IR-410) has been found to contain Mg (2×10^{-4} µg/ml), Cr (2×10^{-5} µg/ml), Mo (2×10^{-5} µg/ml), Sr (6×10^{-5} µg/ml) and Ba (6×10^{-6} µg/ml), and traces of B, Si, Sn, Ag, and V [55].

An all-polyethylene still has been suggested for distillation of water. The polyethylene still is heated in a glycerol bath to 98°C, and the water is distilled under a stream of nitrogen and cooled in a polyethylene coil immersed in ice-water [58]. The output of the still is about 100 ml/hr. Polypropylene or Teflon should work better; the water could be heated to boiling, and higher outputs would be obtained.

2.7.2 Important Reagents

Acids and bases form an important group of reagents; those most frequently used are HCl, HF, HBr, H_2SO_4, HNO_3, CH_3COOH, NaOH, and NH_3 solution. These reagents are available in ultrapure grades, such as 'Aristar' (BDH Chemicals Ltd.) or 'Suprapur' (Merck). Table 2.14 lists some purity data on such chemicals, based either on manufacturers' specifications or reported analyses. The data illustrate that these extremely pure chemicals contain very low concentrations of many elements, but that if certain abundant elements such as iron, aluminium, copper, nickel and zinc are to be determined at trace levels of the order of 10^{-6}% or below, even an ultrapure reagent will result in significant or serious contamination of the sample, so the reagent would have to be further purified before use. In practice, special purification of such reagents as mineral acids is usually indispensable.

Irving and Cox [63] have reported a simple but effective method of preparing ultrapure hydrochloric acid and ammonia solutions, isopiestic distillation. About 500 ml of concentrated hydrochloric acid (s.g. 1·19) are placed in a large desiccator. The surface area of the acid should be ca. 300 cm^2. A polyethylene dish holding 250 ml of doubly distilled water is placed on a porcelain or glass-rod support. The desiccator is closed with an ungreased cover. In two days at room temperature, $2M$ and in four days, $4M$ hydrochloric acid is obtained. With only 50 ml of water placed in the dish, about $10M$ acid is obtained in 3 days. Kuehner *et al.* [64] give detailed procedures for purification and analysis of acids. Acids purified in this way should contain not more than 10^{-9} g/l of each of the 20 elements usually determined. Ammonia is analogously purified by placing 500 ml of a concentrated ammonia solution (s.g. 0·88) and 50 ml of water in the desiccator: $9·5M$ ammonia solution is obtained in two days. If 250 ml of water are used, $8·7M$ ammonia is obtained in four days. According to Häberli [56], a similar result is obtained by saturating pure water with gaseous ammonia (from a cylinder) prepurified in a train of three wash-bottles of ammoniacal EDTA solution.

Hydrofluoric acid is much more difficult to purify. Dabeka *et al.* describe a polypropylene still which can be used for sub-boiling distillation of hydrofluoric acid, as well as other acids and water [48b]. A sufficiently pure acid is obtained by slow distillation from a silver or palladium still at a rate of 20 ml/hr, with 50% of the distillate collected as a middle cut, and subsequent redistillation. Stegemann [58] developed a special procedure for removing traces of silica from hydrogen fluoride. Hydrogen fluoride gas from a cylinder is carried in a nitrogen stream through a train of polyethylene washers holding in succession: (1) a layer of polyethylene

Table 2.14 Trace-element concentrations in ultrapure reagents

Species	Found, ng/ml					Maximum permissible		
	HCl* ref. [59]	HNO$_3$* ref. [59]	HCl† ref. [60]	HNO$_3$† ref. [60]	HF† ref. [60]	HCl 'Aristar' [61]	HNO$_3$ 'Aristar' [61]	HF 'Aristar' [61]
Al	ND	ND	75	25	55	50	50	NL
As	ND	<2	ND	ND	ND	10	5	1000
Be	ND	ND	<0.2	<0.2	<0.2	NL	NL	NL
Bi	ND	ND	<2	<2	<2	NL	NL	NL
Cd	<0.5	<0.5	ND	ND	ND	5	5	10
Co	<0.5	<0.5	<2	<2	<2	5	5	10
Cr	<0.5	0.5	<2	12	3	NL	50	NL
Cu	1	5	ND	ND	ND	5	5	10
Fe	28	23	140	55	260	200	200	20
Ga	ND	ND	<2	<2	<2	NL	NL	NL
In	ND	ND	<2	<2	<2	NL	NL	NL
Mg	ND	ND	3	7	7	50	500	10
Mn	ND	ND	2.5	<2	3	5	5	10
Mo	ND	ND	<2	<2	<2	NL	NL	NL
Ni	1	2	5	2	<2	5	5	10
P	4.5	1.1	ND	ND	ND	30	30	150
Pb	<1	1	25	15	35	1	1	10
Ti	<1	1	6	<2	300	NL	NL	NL
Zn	27	31	40	25	23	20	40	10
Zr	ND	ND	<5	<5	<5	NL	NL	NL

* Mean of 3–5 spectrophotometric determinations. † Determined by spectrography.
ND = not determined. NL = no limit given in specification.
For clarity, all concentrations are expressed as ng/ml.

pellets immersed in pure water, (2) and (3) layers of polyethylene pellets immersed in an aqueous suspension of sodium fluoride, (4) air as a buffer volume, and (5) pure water to absorb the purified HF for making the desired reagent.

Pure sodium hydroxide is particularly difficult to prepare. Nazarenko and Flyantikova [65] have suggested the dissolution of sodium metal, purified by distillation, in high-purity water, and Fischer and Kunin [66] have described purification of a sodium hydroxide solution on ion-exchange resins. Mitchell [67] has described the preparation of anhydrous sodium carbonate, free from heavy metals, which he used for fusing plant ash for the determination of trace elements.

Ethylenediaminetetra-acetic acid (EDTA) can be purified [68] by dissolving the disodium salt in water, precipitating the free acid with hydrochloric acid and dissolving it in aqueous sodium hydroxide; this procedure is repeated several times. Finally, chloride ions are washed out with water, acetone and ether and the preparation is dried. It is kept

§2.7] Reagents 33

concentration, ng/ml					Typical, ng/ml	
HCl 'Suprapur' [62]	HNO$_3$ 'Suprapur' [62]	HF 'Suprapur' [62]	HCl Ultrex [62a]	HNO$_3$ Ultrex [62a]	HCl Ultrex [62a]	HNO$_3$ Ultrex [62a]
10	5	NL	50	50	9	4
5	1	NL	2	1	<1	1
NL	NL	NL	NL	NL	—	—
NL	NL	NL	NL	NL	—	—
NL	5	NL	NL	5	—	<5
5	1	NL	5	5	<5	<5
NL	NL	NL	10	50	<5	<5
5	5	10	5	5	3	0.8
20	10	20	50	50	7	5
NL	NL	NL	NL	NL	—	—
NL	NL	NL	NL	NL	—	—
NL	NL	10	50	50	5	1
NL	5	NL	10	10	5	<0.5
NL	NL	NL	NL	NL	—	—
5	5	NL	5	5	<5	<5
NL	3	NL	30	60	9	12
5	5	10	10	10	5	0.5
NL	NL	NL	NL	500	—	—
5	5	NL	5	5	5	5
NL	NL	NL	NL	NL	—	12

under nitrogen or argon. The acid thus purified contains metal impurities of the order of 10^{-5}–10^{-6} %.

A good procedure for purifying numerous reagents from traces of heavy metals is to treat them with dithizone. For particular instructions the reader is referred to monographs which describe the use of dithizone in spectrophotometric analysis [19,69,69a]. Instead of dithizone, other complexing agents may be used, and also ion-exchange resins [70]. Coprecipitation is also employed. Meites [71] and DeFord [72] have successfully used electrolysis at a mercury cathode for purification of reagents. Obviously, in this case only certain metals can be removed (cf. Section 3.3.4).

More detailed information is given in ref. [2a].

2.7.3 Solutions

Dilute standard solutions of metals and their storage over lengthy periods of time present a particular problem, because of sorption on vessel walls, as already mentioned. For example, the concentrations of 10^{-3}% solutions of molybdenum, manganese, vanadium, titanium and nickel, acidified with a strong acid (6%) and stored for 15 days in preleached Jenaer Glas bottles were found to drop to 20–40% of the original value [73].

Quartz vessels cause only minor changes in solution concentrations on storage. Vanadium (7 ppm) and manganese (1 ppm) solutions stored in polyethylene bottles showed no changes over a year [55]. On the other hand, the cations present in dilute sodium and potassium chloride solutions, also kept in polyethylene bottles, were almost entirely sorbed after a few months [73a]. Procedures for preservation of dilute mercury solutions have been given by Feldman [74].

The available data are confused, but they do show clearly that solutions must always be checked before use. It is best to store concentrated stock solutions and to dilute them directly before use.

The purity of the reagents used is crucial for attaining a low limit of detection, especially where this limit is controlled by the blank. This is of particular significance in chemical preconcentration involving large sample weights and large amounts of reagents; the elements of interest must be present in the reagents in amounts at least two orders of magnitude lower than in the analysis sample if the blank is to be sufficiently low.

2.8 VOLATILIZATION OF SUBSTANCES

It is important to remember the complications that may arise from losses of the traces to be determined, owing to their volatility. This may occur in dry ashing processes, and in concentration of solutions by prolonged evaporation.

Numerous metal salts, primarily halides, chelate compounds, free elements, and also some oxides, have boiling points low enough to allow their complete loss on evaporation or ashing at elevated temperatures. This must be borne in mind when evaporating solutions in hydrochloric acid and hydrofluoric acid (also hydrobromic and hydroiodic acids) and also on evaporating solutions of certain chelates in organic solvents; 8-hydroxyquinolinates and acetylacetonates are examples of volatile chelates.

Volatilization of substances will be discussed at a greater length in Chapter 4.

References

[1] Burriel-Marti, F., *Experientia Suppl.*, **5**, 71 (1956).
[2] Tölg, G., *Talanta* **19**, 1489 (1972).
[2a] Zief, M. and Mitchell, T. W., *Contamination Control in Trace Element Analysis*, Wiley, New York, 1976.
[3] Indelevich, I. G., Shelpakova, I. R. and Brusentsev F. A., *Zh. Analit. Khim*, **26**, 2075 (1971).
[4] Fair, J. R., Crocker, B. B. and Null, H. R., *Chem. Eng.*, **79**, 146 (1972).
[5] Radwan, M., *Freiberger Forschungshefte*, Heft B 111, Dez., 167 (1965).
[6] Pfann, W. G., *Zone Melting*, Wiley, New York, 1959.
[7] Downarowicz, J., *Chem. Anal. (Warsaw)*, **4**, 643 (1959).

References

[8] Gałązka, J. and Łobzowski, J., *Chem. Anal. (Warsaw)*, **8**, 695 (1963); **9**, 555, (1964).
[9] Doerffel, K., *Chem. Anal. (Warsaw)*, **8**, 333 (1963).
[10] Strzyżewska, B. and Radwan, Z., *Chem. Anal. (Warsaw)*, **11**, 979 (1960).
[10a] Strzyżewska, B., personal communication.
[11] *Reference Methods of Marine Radioactivity Studies*, IAEA, Technical Report Series No. 118, Vienna, 1970.
[12] Hwang, J. Y., *Anal. Chem.*, **44**, (14), 20 A (1972).
[13] Tabor, E. C. and Warren, W. N., *A.M.A. Areh. Ind. Health*, **17**, 145 (1958).
[14] Ehrlich, G., Gortbatsch, R., Jaetsch, K. and Scholze, H., *Reinstsoffe in Wissenschaft und Technik*, Akademie-Verlag, Berlin, 1963, p. 421.
[15] Dittrich, E. and Mohaupt, G., *Chem. Anal. (Warsaw)*, **17** (1972).
[16] Spurný, K., *Cesk. Hyg.*, **7**, 430 (1962).
[17] Cholak, J. and Schafer, L. J., in Zolotov, Yu. A., *Ekstraktsionnoe kontsentrirovanie (Extraction Concentration)*, Khimiya, Moscow, 1971.
[18] Vasilevskaya, L. S. and Alimarin, I. P. (Eds.), in *Metody analiza veshchestv vysokoi chistoty (Methods for Analysis of High-Purity Substances)*, Nauka, Moscow, 1965, p. 16.
[19] Iwantscheff, G., *Das Dithizon und seine Verwendung in der Mikro- und Spurenanalyse*, 2. Aufl., Verlag Chemie, Weinheim, 1972.
[20] Zaidel', A. N., Kaliteevskii, N. I., Lippe, L. V. and Chaika, M. P., *Emissionnyi spektral'nyi analiz atomnykh materialov (Emission Spectral Analysis of Atomic Materials)*, Goskhimizdat, Moscow, 1960.
[21] Chow, T. J. and McKinney, C. R., *Anal. Chem.*, **30**, 1499 (1958).
[22] Suzano, C. D., *Anal. Chem.*, **27**, 1038 (1955).
[23] Mields, M. and Schering, G., *Silikattechnik*, **6**, 241 (1955).
[24] Draignaud, P., *Ind. Céram.*, No. 492, 343 (1959).
[25] Hood, S. L., Parks, R. Q. and Hurwitz, C., *Ind. Eng. Chem., Anal. Ed.*, **16**, 202 (1944).
[26] Szmal, Z., *Chem. Anal. (Warsaw)*, **12**, 393 (1967).
[27] Fritsch, A. OHG, *Pulverisette Analysette*, Katalog, Idar-Obernstein.
[28] Kalininkov, V. T. and Shteinberg, A. N., *Zavodsk. Lab.*, **30**, 178 (1964).
[29] Vasilevskaya, L. S., Kondrashina, A. J. and Shifrina, T. T., *Zavodsk. Lab.*, **28**, 675 (1962).
[30] Sandell, E. B., *Colorimetric Determination of Traces of Metals*, 3rd Ed., Interscience, New York, 1959.
[31] Koch, O. G. and Koch-Dedic, G. A., *Handbuch der Spurenanalyse*, Springer-Verlag, Berlin, 1964.
[32] West, F. K., West, P. W. and Iddings, F. A., *Anal. Chem.*, **38**, 1566 (1966); *Anal. Chim. Acta*, **37**, 112 (1967).
[33] Robertson, D. E., *Anal. Chem.*, **40**, 1067 (1968).
[34] Wichers, E., Finn, A. N. and Clabaugh, W. S., *J. Res. Nat. Bur. Stand.*, **26**, 537 (1941).
[35] Robertson, D. E., Rancinelli, L. A. and Perkins, R. W., *Proc. Symp. Application of Neutron Activation Analysis in Oceanography*, Brussels, 1968, p. 143.
[36] Cooper, L. H. N., *J. Mar. Res.*, **17**, 128 (1958).
[37] Robertson, D. E., *Anal. Chim. Acta*, **42**, 533 (1968).
[38] Black, W. A. P. and Mitchell, R. L., *J. Mar. Biol. Assoc. U. K.*, **30**, 575 (1952).
[39] Schutz, D. F. and Turekian, K. K., *J. Geophys. Res.*, **70**, 5519 (1965).

[40] Riley, J. P., *Analytical Chemistry of Sea-Water, Chemical Oceanography*, Academic Press, London, 1965.
[41] Kokotov, Yu. A., *Radiometricheskie metody opredeleniya mikroelementov* (*Radiometric Methods for Determining Trace Elements*), Nauka, Moscow, 1965.
[42] Knížek, M. and Provazník, J., *Chem. Listy*, **55**, 389 (1961).
[43] Oelschlager, W., *Z. Anal. Chem.*, **154**, 329 (1957).
[44] King, W. G., Rodriguez, J. M. and Wai, C. M., *Anal. Chem.*, **46**, 771 (1974).
[45] Rubin, B. A., *Science*, **110**, 425 (1949).
[46] Boocock, G., Grimes, J. H. and Wilford, S. P., *Nature (London)*, **194**, 672 (1962).
[47] Kleinteich, R., *Glas-Instrumenten-Techn.*, **5**, 334 (1961).
[48] Jack, K. and Hetherington, G., *Glas-Instrumenten-Techn.*, **5**, 378 (1961).
[48a] Mykytiuk, A., Russell, D. S. and Boyko, V., *Anal. Chem.*, **48**, 1462 (1976).
[48b] Dabeka, R. W., Mykytiuk, A., Berman, S. S. and Russell, D. S., *Anal. Chem.*, **48**, 1203 (1976).
[49] Bogoyavlenski, A. N., *Tr. Inst. Okeanol. Akad. Nauk SSSR*, **79**, 49 (1965).
[50] Bodman, R. H., Slabaugh, L. V. and Bowen, V. T., *J. Mar. Res.*, **19**, 141 (1961).
[51] *Zooplankton Sampling*, UNESCO-ICO, Monographs on Oceanographic Methodology, 2 (1968).
[52] Murphy, J. and Riley, J. P., *Anal. Chim. Acta*, **14**, 318 (1956).
[53] Butler, E. B. and Johnston, W. H., *Science*, **120**, 543 (1954).
[54] Lang, E. P., *Ind. Eng. Chem., Anal. Ed.*, **6**, 11 (1934).
[55] Yoe, J. H., and Koch, H. J. (Eds.), *Trace Analysis*, Wiley, New York, 1957.
[56] Häberli, E., *Z. Anal. Chem.*, **160**, 15 (1958).
[57] Wickbold, R., *Z. Anal. Chem.*, **171**, 81 (1959).
[58] Stegemann, H., *Z. Anal. Chem.*, **157**, 267 (1957).
[59] Marczenko, Z., *Chem. Anal. (Warsaw)*, **8**, 849 (1963).
[60] Oldfield, J. H. and Bridge, E. P., *Analyst (London)*, **85**, 97 (1960).
[61] *Laboratory Chemicals Catalogue*, British Drug Houses, Poole, 1971, pp. 169, 170, 233.
[62] *General Catalogue*, E. Merck, Darmstadt, 1971, pp. 360, 491.
[62a] *Catalogue*, J. T. Baker Chemical Co., Phillipsburg, NJ 08865, USA.
[63] Irving, H. and Cox, J. J., *Analyst (London)*, **83**, 526 (1958).
[64] Kuehner, E. C., Alvarez, R. Paulsen, P. J. and Murphy, T. J., *Anal. Chem.*, **44**, 2050 (1972).
[65] Nazarenko, V. A. and Flyantikova, T. V., *Zavodsk. Lab.*, **24**, 663 (1958).
[66] Fischer, S. and Kunin, R., *Nature (London)*, **177**, 1125 (1956).
[67] Mitchell, R. L., *The Spectrographic Analysis of Soils, Plants and Related Materials*,. Commonwealth Bureau of Soil Science, Harpenden, England, 1948, p. 28.
[68] Barnard, A. J. Jr., Joy, E. F., Little, K. and Brooks, J. D., *Talanta*, **17**, 785 (1970).
[69] Marczenko, Z., *Spectrophotometric Determination of Elements*, Horwood, Chichester, 1975.
[69a] Irving, H. M. N. H., *Dithizone*, Chemical Society, London, 1977.
[70] Sharkovskaya, I. A. and Shevchenko, S. I., *Tr. Khim. Khim. Tekhnol. (Gorkii)*, 1969, 49.
[71] Meites, L., *Anal. Chem.*, **27**, 416 (1955).
[72] DeFord, D. D., *Anal. Chem.*, **28**, 660 (1956).
[73] Leutwein, B., *Zbl. Mineral. Geol. Paläontol.*, 1940a, 129.
[73a] Minczewski, J., unpublished work.
[74] Feldman, C., *Anal. Chem.*, **46**, 99 (1974).

Chapter 3

PRECIPITATION AND COPRECIPITATION

3.1 GENERAL CONSIDERATIONS

The oldest method of chemical separation is precipitation; this method underlies the classical scheme of qualitative analysis introduced by Fresenius [1] and popularized by Treadwell in his book [2], first published in 1899.

Precipitation methods are satisfactory for macro separations, but closer examination reveals that precipitates are usually contaminated with foreign ions present in the solution although these ions would not themselves have been precipitated under the given experimental conditions. This phenomenon is called coprecipitation, and it has been much studied by analysts anxious to precipitate very pure solids so as to achieve nearly complete isolation of the substance to be determined. Extensive radiochemical studies were initiated by the work on separation of radium from lead and barium, but the mechanism of coprecipitation has not yet been fully elucidated.

Developments in trace analysis and in the methods of separation and preconcentration of traces have led to study of the possibility of using precipitation techniques for separating traces. Direct separations are not possible, because the amounts involved are too tiny to handle, but separation of trace constituents by coprecipitation on a collector (carrier) intentionally added to the solution has been found to be practicable and extremely useful. Detailed information concerning formation and properties of precipitates is given in [2a].

3.2 THEORY

3.2.1 Crystalline Precipitates

Crystalline precipitates form in two or three distinct crystallization stages. The first stage involves nucleation, the second crystal growth, and the third, if it occurs, aging of the precipitate (on prolonged standing in contact with the mother liquor).

Nucleation [3] has an important influence on the nature of the precipitate. If precipitation is carried out so that many nuclei are formed, the resulting precipitate is finely crystalline, difficult to filter, and severely contaminated. Conversely, when relatively few nuclei are formed and they grow rather slowly, the resulting precipitate is coarsely crystalline, easy to filter and usually less contaminated.

The formation of a nucleus is associated with the attainment of supersaturation in the solution. Under conditions of slight supersaturation the number of nuclei formed is small and crystallization of the precipitate is slow. If a certain higher degree of supersaturation is attained, precipitation proceeds rapidly [4]. Impurities in the solution, especially those isomorphous with the precipitate, are believed to facilitate nucleation, which thus can start at a lower degree of supersaturation [5].

The mechanism of crystal growth is also highly complicated [5-7]. Growth proceeds until supersaturation no longer exists, but even then, the precipitate-solution equilibrium continues to be dynamic in nature, small crystals being more soluble than large ones [8]; therefore the primary precipitate allowed to stand in contact with the mother liquor undergoes changes. Small crystals keep dissolving and large ones keep growing at their expense. In analytical chemistry this stage of crystallization is discerned as the third stage of precipitation, known as aging of the precipitate.

These general principles of precipitation suggest that if traces are to be separated from a major constituent by preferential precipitation, precipitation should be conducted very slowly, to avoid excessive supersaturation. Such conditions can be procured by 'precipitation from homogeneous solution' (PFHS) developed by Gordon [9]. In this method the precipitant is slowly generated by a reaction proceeding in the solution, e.g. hydrolysis of a suitable reagent. The concentration of precipitant ion thus generated is very small and the degree of supersaturation is low (and may be controlled, e.g. by temperature), so the conditions are conducive to producing a very pure precipitate.

3.2.2 Colloidal Precipitates

The mechanism of formation of colloidal precipitates differs from that for crystalline precipitates. In colloidal precipitates, there is a continuous transition from particles characteristic of true molecular solutions, through those separable only by ultrafiltration, to macroscopic aggregates.

From the analytical aspect the difference between the two principal types of colloidal precipitates, hydrophilic and hydrophobic, is critical. A hydrophilic precipitate has a strong affinity for water, is reluctant to flocculate, difficult to wash, apt to incorporate numerous substances and hard to separate by filtration: examples are hydrated aluminium oxide and stannic and silicon oxides.

Hydrophobic colloids have a low affinity for water and produce aggregates which are easy to flocculate by adding a small amount of a suitable electrolyte. Hence, a hydrophobic precipitate is much purer and more readily filterable. Examples are silver chloride and arsenious sulphide.

The stability of colloidal solutions and colloidal precipitates depends upon the relation of the two opposing forces: van der Waals forces resulting in mutual attraction of molecules and the electric double layer forces around molecules, producing their mutual repulsion [10].

3.2.3 Coprecipitation

According to Kolthoff [11] coprecipitation is "the incorporation of impurities into a precipitate by substances which under experimental conditions are usually soluble in the liquid phase", but in trace analysis the term is used in a much broader sense. It is not confined to precipitation of traces soluble under experimental conditions, but also covers all instances when trace constituents are precipitated together with a collector.

The mechanism of coprecipitation is usually considered to include mixed-crystal formation, occlusion and adsorption. Mixed-crystal formation involves substitution of ions in the crystal lattice of the carrier by coprecipitating ions. When both ions are about the same size, mixed crystals can be formed at all ratios of the two ions, to form true mixed crystals. Mixed crystals can also be formed with ions considerably differing in size, but then the amount of the coprecipitated ion is limited and the crystals are described as anomalous.

Coprecipitation is favoured when the precipitation process attains equilibrium very slowly. The crystal forms by deposition of successive layers on the nucleus, and each individual layer of solid is in equilibrium with the surrounding solution at the instant when it is formed. Since the composition of the solution changes continuously throughout the precipitation, the composition of the crystal varies continuously from the centre to the external face.

In such heterogeneous solid-solution formation, the overall distribution of the coprecipitated ion between the precipitate and the solution obeys the logarithmic distribution law [12]

$$\log \frac{[Tr]_o}{[Tr]_p} = \lambda \log \frac{[C]_o}{[C]_p} \tag{3.1}$$

where $[Tr]_o$ and $[Tr]_p$ are the concentrations (in solution) of the coprecipitated ion before and after precipitation, λ is the logarithmic distribution coefficient, and $[C]_o$ and $[C]_p$ are the concentrations of the collector before and after precipitation.

Equation (3.1) is illustrated graphically in Fig. 3.1 for various values of λ. It is apparent that pure precipitates of the major sample constituent are obtained when the value of λ is low. It follows that if minor constituents are to be concentrated in the solution, the major constituent must be precipitated only to the extent, defined by λ, that there is no coprecipitation of the trace constituents.

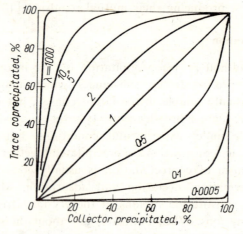

Fig. 3.1 Distribution of the coprecipitated ion between solution and precipitate at various λ-values

The reverse is true when all the trace elements are to be precipitated with a collector. It is then desirable that λ be large, because the higher the value of λ, the lower the amount of collector that has to be precipitated to achieve complete coprecipitation of traces. At $\lambda < 0.5$, complete precipitation of the collector cannot provide for complete coprecipitation of traces.

When precipitation is carried out so that complete equilibrium may be assumed to have been established between the precipitate and the solution (i.e. the precipitate recrystallizes so rapidly that it is always homogeneous), the coprecipitated ion is homogeneously distributed in the collector crystals and its distribution between the solution and the host crystals obeys Khlopin's law [13]

$$\left(\frac{[Tr]}{[C]}\right)_{solid} = D\left(\frac{[Tr]}{[C]}\right)_{solution} \tag{3.2}$$

where [Tr] is the concentration of the coprecipitated ion, [C] is the concentration of the collector, and D is the homogeneous distribution coefficient.

To establish whether coprecipitation is heterogeneous or homogeneous the distribution of the coprecipitated ion between the solution and the precipitate as a function of the percentage of the collector precipitated should be studied experimentally. In systems with homogeneous distribution, D is constant; in heterogeneous ones, λ is constant and D is variable.

Most of the data on coprecipitation comes from radiochemical studies; for more details, see Hahn's monograph [14].

A second mechanism of coprecipitation (occlusion) involves mechanical entrapment of foreign ions and solvent molecules at the surface of the precipitate during the rapid growth of its crystals. This mechanism is very frequently operative in the precipitation of colloidal precipitates because of the rapid formation of the precipitate. Slow precipitation, e.g. from homogeneous solutions, largely prevents occlusion and allows pure precipitates to be obtained.

The mechanism of adsorption depends on the surface of the resulting precipitate. Since a surface always has some electrical charge, because the ions at the crystal surface are in a more anisotropic electrical field than ions in the interior, it can interact with oppositely charged ions from the solution. It is still disputable whether this kind of coprecipitation is a typical adsorption process, or is more like an ion-exchange process. For example, the coprecipitation of traces on hydrated iron oxide has been treated by some investigators as a typical adsorption process [15–17], but by others as a typical ion-exchange process [18] or as chemisorption [19,20] or 'coordination coprecipitation' [21]. The mechanisms of coprecipitation on silver halides [22], aluminium phosphate [23], and of copper on various hydroxides [24,25] have also been studied.

The adsorption rule of Paneth, Fajans and Hahn was quoted by Kolthoff and Sandell [11] as "those ions whose compounds with the oppositely charged constituent of the lattice are slightly soluble in the solution in question are well adsorbed by the ionic lattice"; an alternative formulation is that, as a general principle, adsorption increases with decreasing solubility of the trace constituent to be adsorbed. Barium sulphate precipitated from a solution containing sulphate ions in excess will preferentially adsorb cations arranged in the following series of decreasing solubilities of the sulphates: $Na^+ < K^+ < Ag^+ < Ca^{2+} < Pb^{2+}$. When precipitated from a solution containing barium ions in excess, the barium sulphate will preferentially adsorb anions in order of diminishing solubilities of their barium salts, e.g. $Br^- < Cl^- < ClO_4^- < NO_3^-$.

Adsorption of trace constituents on a collector often follows the Freundlich adsorption isotherm

$$C = \alpha C_0^n$$

where C is the concentration of the trace constituent in the precipitate, C_0 is the original concentration of the trace constituent in the solution, α and n are constants characteristic of a given process.

Most frequently, $n < 1$. If $n = 1$, the adsorption isotherm is a straight line. Coprecipitation can be said to follow the Freundlich isotherm if plots of $\log C = f(\log C_0)$ are linear. Usually only systems with very low C_0 values, i.e. those corresponding to trace analysis, follow the Freundlich adsorption isotherm.

Salutsky [26] has produced a general list (Table 3.1) describing the effects of individual precipitation parameters on the quality of the resulting precipitate.

Table 3.1 Effect of precipitation conditions on the purity of precipitates
(from [26], by permission of the copyright holders, Interscience)

Condition		Form of impurity		
	mixed crystals	surface adsorption	occlusion and inclusion	post-precipitation
Dilute solution	0	+	+	0
Slow precipitation	+	+	+	−
Prolonged digestion	−	+	+	−
High temperature	−	+	+	0
Agitation	+	+	+	0
Washing the precipitate	0	+	0	0
Reprecipitation	+	+	+	+

+ increased purity, − decreased purity, 0 little or no change in purity

3.3 APPLICATION OF PRECIPITATION AND COPRECIPITATION TO TRACE ANALYSIS

3.3.1 Separation of the Major Sample Constituent (Matrix) by Precipitation

It follows from the foregoing discussion that in trace analysis precipitation (as opposed to coprecipitation) is useful only for macro–micro separations, i.e. for removing the major constituent of a sample, if this interferes with the subsequent determination of traces.

In agreement with expectation, the precipitation conditions can be chosen so that, contrary to a quite common opinion among analysts, a considerable amount of the major constituent can be precipitated without causing losses, by coprecipitation, of the trace constituents to be determined subsequently.

Karabash *et al.* [27] have studied the coprecipitation of trace constituents in the precipitation of lead sulphate from $6M$ nitric acid. With the aid of radiotracers they showed that many trace elements remained nearly quantitatively (85–100%) in solution, which is sufficient for trace-analysis purposes.

Degtyareva and Ostrovskaya [28] removed lead as the major constituent by precipitation with a 10% excess of hydrochloric acid. After the precipitate had been digested and filtered off, the filtrate was found to have retained quantitatively many trace elements, which could then be determined spectrographically. The overall error involved did not exceed $\pm 20\%$.

In analysing high-purity silver for trace constituents Marczenko and Kasiura [29] removed silver by precipitating silver chloride from nitric acid medium. Again, the traces were completely recovered.

Similarly, it has been found that macro amounts of bismuth can be removed as sparingly soluble bismuth iodide [30] or basic bismuth nitrate [31], mercury by reduction with formic acid in nitric acid solution [32], molybdenum as its benzoinoximate [33], zirconium as its mandelate [34], and aluminium as $AlCl_3 \cdot 6H_2O$ with hydrogen chloride gas [35].

Numerous noble metals and other elements such as selenium and tellurium can be isolated from solution in the elemental form by chemical reduction, but it has not yet been established whether coprecipitation occurs. Some work [36] suggests that, in certain cases, coprecipitation does take place (e.g. copper on platinum).

Precipitation from homogeneous solution can be used for separation of major constituents in very pure, compact form. For example, tin and aluminium were separated quantitatively from copper and lead by a PFHS procedure, and this allowed the copper and lead to be determined by differential pulse polarography [37]. However, such procedures have seldom been applied.

Table 3.2 lists examples of removal of the major constituent by precipitation before determination of trace constituents.

Table 3.2 Separation of major constituent by precipitation

Major constituent	Precipitated as	Conditions	Trace elements left in solution	Ref.
Aluminium	$AlCl_3 \cdot 6H_2O$	saturation with HCl		[35]
Bismuth	$Bi(OH)_n(NO_3)_m$	pH = 4	Ag, Al, Cd, Cu, Mg, Mn, Ni, Pb	[31]
Bismuth	BiI_3	$0.4M$ HNO_3	Al, Ba, Ca, Cd, Co, Cr, Fe, In, Mg, Mn, Mo, Ni, Sb, Te, Ti, V, Zn, (Ag, Au, Cu, Pb, Pt, Sn, Tl— partly coprecipitate)	[30]
Lead	$PbSO_4$	$6M$ HNO_3	Ag, Al, As, Bi, Ca, Cd, Co, Cr, Cu, Fe, In, Mg, Mn, Mo, Na, Ni, Sb, Sn, Te, Ti, V, Zn	[27]
Lead	$PbCl_2$	$1M$ HCl	Ag, Al, As, Au, B, Ba, Be, Bi, Ca, Cd, Co, Cr, Cu, Fe, Ga, In, K, La, Mg, Mn, Mo, Na, Ni, P, Pd, Pt, Rb, Sb, Si, Sn, Sr, Tl, U, Zn	[28]
Mercury	metal	5% HCOOH	Bi, Cd, Co, Cu, Fe, Mg, Mn, Ni, Pb, Tl, Zn	[32]
Molybdenum	α-benzoinoximate	$1.5–2M$ HNO_3 + HCl	Ag, Al, Ba, Bi, Ca, Cd, Co, Cr, Cu, Fe, Mg, Mn, Na, Ni, Pb, Sb, Sn, Ti, V, Zn	[33]
Palladium	dimethylglyoxime + flotation			[38]
Platinum	metal	HCOOH; pH = 2	Al, Cr, Fe, Mn, Ni, Pb, (Cu, Zn—highly coprecipitated)	[36]
Platinum	$(NH_4)_2PtCl_6$			[38]
Silver	AgCl	$1M$ HNO_3	Cd, Co, Cu, Ni, Zn	[29]
Silver	AgCl	$1M$ HNO_3	Co, Cr, Fe, Mn, Ni	[39]
Thallium	TlSCN		Cd	[40]
Thallium	TlI		Bi, Cd, Co, Cu, Fe, Ni, Pb, Th	[41]
Zirconium	tetramandelate	$2.5M$ HCl	Ag, Al, Ba, Bi, Ca, Cd, Co, Cr, Cu, Fe, In, Mg, Mn, Ni, Sb, Ti, V, Zn	[34]

3.3.2 Coprecipitation of Trace Elements with Inorganic Collectors

Coprecipitation of trace elements by inorganic collectors (carriers) may be used for two purposes: to separate a large group of trace elements, or to isolate one specific trace constituent. The former is commonly used, especially when the trace elements are to be determined subsequently by emission spectrography or atomic-absorption spectrophotometry or if the step is preliminary to an extensive separation scheme; the latter separation is used when only a single individual trace element is to be determined.

The success of the separation process depends on the choice of collector and precipitant. To be effective, the collector should meet the following requirements.

(i) The precipitate formed should be readily filtered off and washed.

(ii) It should not interfere in the further course of analysis; or it should be easy to remove. Thus, mercury, arsenic and tellurium are useful as collectors, as they can be readily removed by distillation, and organic carriers, which are easy to remove by ashing, have come to be used (see Section 4.5).

(iii) When coprecipitation is preparatory to emission spectrographic analysis, the collector should also be a spectrographic buffer, spectral carrier or internal standard.

(iv) The collector should be chosen so that the amount needed for complete coprecipitation of the trace elements is as small as possible.

(v) The amount of the collector should be chosen to make the resulting precipitate easy to collect by filtration or centrifugation and readily handled in further operations.

An experimental check is always advisable. For example, in precipitation of 43 µg of vanadium with various carriers, complete recovery was obtained with 20 mg of chromium(III) hydroxide, or 20 mg of hydrated manganese dioxide or 20 mg of hydrous iron(III) oxide, but with aluminium hydroxide as collector 50 mg was required [42].

Most often, group precipitants are reagents which react with collectors to give hydroxides or sulphides, or organic reagents which react with the groups of metals concerned. In the first case, group precipitants are mainly compounds containing oxygen and nitrogen as ligand atoms, and in the second they are compounds containing sulphur and/or nitrogen atoms.

A classic example of the use of sulphides as collectors is lead sulphide used as collector for copper or, conversely, copper sulphide employed as collector for lead [43]. Group II sulphides can be used as collectors for Group III cations [43a], although certain experiments [44] contraindicate this use. On the other hand, molybdenum sulphide has proved excellent as a collector for traces of arsenic, antimony, tin, cadmium, lead and other elements [45]. Marczenko and Kasiura [46] used lanthanum as a collector for sulphides. With a mixture of sodium sulphide and hydroxide as precipitant, they achieved complete precipitation of traces of indium, thallium, cadmium, zinc, nickel, cobalt and manganese. Lanthanum, which precipitates as the hydroxide, coprecipitates with sulphides and hydroxides and, if spectrophotometry is to follow, is convenient in that it gives few colour reactions and is only rarely determined as a trace element. Elemental sulphur has also been used as collector [47,48].

Examples of organic sulphide analogues are a complex of zinc with 1-pyrrolidinedithiocarbamic acid [49], which precipitates a number of Analytical Group II and III cations, thallium(I) as collector in conjunction with thioacetamide as precipitant [50] and ammonium pyrrolidinedithiocarbamate as collector with thionalide as precipitant [51]. Sulphide-type coprecipitations are summarized in Table 3.3.

Common hydroxide-type collectors include $Fe(OH)_3$ [61–67], $La(OH)_3$ [62,63,68,69], $Cd(OH)_2$ and $Zn(OH)_2$ [70], $Bi(OH)_3$ [71], $Al(OH)_3$ [72–75], etc. Zörner et al. [75] have described an interesting rapid separation with a mixed collector composed of equimolar amounts of Fe(II) and Fe(III) hydroxides which precipitates completely traces of aluminium, titanium, chromium, vanadium, zirconium and antimony. Since the mixture has ferromagnetic properties, an electromagnet placed under the flask and its contents assists the precipitate to settle rapidly. The supernatant liquid can be decanted quantitatively without difficulty. This procedure has been used in rapid routine determinations of the above-mentioned metals in steel.

Manganese dioxide, first suggested by Blumental [76], has been widely used as collector. Although its mechanism of action is still disputable, it has proved to be an ideal collector for precipitating and separating traces of antimony, bismuth and tin, in the presence of lead [77,78], and for isolating traces of thallium [79,80], molybdenum [81], and tungsten [82], or radiotraces of ruthenium, manganese, cobalt [83], etc.

Cupferron, an organic reagent which behaves similarly to hydroxides, was used by Strock and Drexler [84] to precipitate traces of titanium, vanadium, and zirconium with iron as collector, and 8-hydroxyquinoline was introduced by Scott and Mitchell [85] for group precipitation of trace elements with aluminium or iron as collector.

Table 3.4 summarizes coprecipitation of trace elements on hydroxide-type collectors.

Table 3.3 Coprecipitation of trace elements with sulphide-type collectors

Collector	Precipitant	Coprecipitated traces	Sample	Ref.
Ag	H_2S	Ga		[58]
Al	NH_4 thioglycollate	Be	silicates	[56]
Cd + C	H_2S + Na diethyldithiocarbamate	Ag, Al, As, Au, Bi, Co, Cr, Cu, Fe, Ga, Ge, Mn, Mo, Ni, Pb, Sb, Sn, Ti, Tl, V	carbonates and chlorides of alkali metals, alkaline earths and Mg; reagents	[57]
Cu	H_2S	Pb		[43, 52]
Cu	H_2S	Sb	silicates, soil	[53]
Cu	H_2S	Ru(IV)	solution, $0.01M\ H_2SO_4$	[60]
Fe	H_2S	Ni, Cu		[54]
La	Na_2S + NaOH	Cd, Co, In, Mn, Ni, Tl, Zn		[46]
Mo	H_2S	Ag, As, Bi, Cd, Cu, Ge, Sb, Sn, Tl	high-purity metals	[45]
Pb	H_2S	Cu		[43]
Pb	H_2S	Au	natural waters	[59]
S	$(NH_4)_2S + HNO_3$	Au, Hg, Pt, Rh	silver, copper lead	[47]
S	elemental S	Au	copper	[48]
Sb	H_2S	Cu, Pb	high-purity zinc	[55]
Tl	thioacetamide in alkaline medium	Ca, Cd, Co, Cr, Cu, Fe, Mg, Mn, Ni, Pb, Ti, Zn	high-purity aluminium	[50]
Tl	pyrrolidinedithiocarbamate + thionalide in acidic medium	Cd, Co, Cr*, Cu, Fe, Ga, Mn*, Mo, Ni, Pb*, Sb, Sn, Ti*, V, Zn	high-purity aluminium	[51]
Zn	Na pyrrolidinedithiocarbamate	Co, Cr, Cu, Fe, Mn, Mo, Ni, Pb, V	cellulose	[49]

* Incomplete coprecipitation.

Another group of insoluble substances which lend themselves as collectors for coprecipitation includes the sparingly soluble halides of silver, thallium(I) and mercury(I). These are often used for coprecipitation of traces of gold [98], silver [99], mercury [100] and copper [101]. Trace components are often coprecipitated as their charged complexes with organic reagents. Sparingly soluble sulphates are used as collectors for coprecipitation of traces of lead [102] and other elements. The phosphates of iron, lanthanum and calcium have been used for coprecipitation of aluminium [103], calcium and magnesium [104], and beryllium [105], respectively.

Table 3.4 Coprecipitation of trace elements with hydroxide-type collectors

Collector	Precipitant	Coprecipitated traces	Sample	Ref.
Al	$NH_3 \cdot aq$	rare earth elements	rocks, separation from Ca and Mg	[72]
Al	$NH_3 \cdot aq$	Co, Cr, Zn, Ru		[73]
Al_2O_3	ultrasonics	Sb(V)		[74]
Bi + C	KOH	Ag, Au, Cd, Co, Cr, Cu, Mn, Ni, Pb, Sn, Zn	gallium, arsenic, gallium arsenide	[71]
Cd	NaOH	As, Bi, Cr, Cu, Ga, Ge, In, Mo, Pb, Sb, Sn, Te, Ti, V, W, Zn	cadmium	[70]
Cu	8-hydroxyquino-line	Al, Fe, Mg, Mn	tungsten	[96]
Fe	$NH_3 \cdot aq$	As, P		[61, 63]
Fe	$NH_3 \cdot aq$	Bi, Pb	high-purity metals	[86]
Fe	$NH_3 \cdot aq$	Pb	water, milk, etc.	[62]
Fe	$NH_3 \cdot aq$	Ge	phosphorites	[64]
Fe	$(NH_4)_2CO_3$	rare-earth elements, Sc	separation from U, Zr, Th	[66]
Fe	$NH_3 \cdot aq$	Po		[67]
Fe + Mg	NaOH	Be, Bi, Cd, Co, Cu, Ni, Pb, Zn	natural waters	[65]
Fe	cupferron	Ti, V, Zr	mineral waters	[84]
Fe or Al	8-hydroxyquino-line	Ag*, Be*, Cd*, Co, Cr*, Cu, Ga, Ge*, Mo, Ni, Pb*, Sn*, Ti, Tl, V*	plant and biological materials, soil solutions	[85]
Fe	NaOH	Ga(III), Ru, Co		[85 a-c]
Fe	cupferron + tannic acid + Crystal Violet	Mo	soil solutions	[97]
Fe(II) + Fe(III)	$NH_3 \cdot aq$	Al, Co, Sb, Ti, V, W, Zr	steels	[75]
La	$NH_3 \cdot aq$	Pu	water, milk, etc.	[62]
La	$NH_3 \cdot aq$	Al, Au, Bi, Fe, Pb	silver	[68]
La	$NH_3 \cdot aq$	As, Bi, Fe, Sb, Se, Sn, Te	copper	[69]
Mg	NaOH	Sn(II), Sn(IV)	natural water	[93]
Mg	NaOH in presence of EDTA	Bi, Fe	lead	[94, 95]
$MnO_2 \cdot aq$		Bi, Sb, Sn, Pb		[77, 78]
$MnO_2 \cdot aq$		Sb, Sn		[87, 89]

Table 3.4 (continued)

Collector	Precipitant	Coprecipitated traces	Sample	Ref.
$MnO_2 \cdot$ aq		Te	metals, alloys	[79, 80]
$MnO_2 \cdot$ aq		Mo	sea-water	[81]
$MnO_2 \cdot$ aq		Ru-103, Ru-106, Mn-54, Co-60	sea-water	[83]
$MnO_2 \cdot$ aq		Bi	copper	[90]
$MnO_2 \cdot$ aq		Sn		[91]
$MnO_2 \cdot$ aq		W		[82]
$MnO_2 \cdot$ aq		Bi, Pb	nickel	[92]
Th	$NH_3 \cdot$ aq	Mo	sea-water	[88]
various	various	Ag, Bi, Pb, Sb, Se, Sn	semiconductors	[86]

* Non-quantitative coprecipitation

Examples of coprecipitations with highly specific collectors, are traces of silica with niobic acid [106], traces of chloride with barium sulphate [107], and traces of palladium with nickel dimethylglyoximate [108].

A special group of collectors includes the noble metals and selenium, tellurium and arsenic, which are readily reducible to the element, and can coprecipitate other metals of this group present in the solution. Coprecipitations of traces of silver [109] or gold [110] with tellurium, or traces of selenium and tellurium with arsenic [111] are examples. Separation of traces of noble metals by adding less noble metals to the solution, e.g. precipitation of ruthenium on antimony dust (separation from iridium) [112], or on copper powder [113], may be included in the same category. Examples are listed in Table 3.5.

Metallic silver, gold, mercury and amalgams can also be used as collectors. Although the mechanism is not coprecipitation, some examples are mentioned here. Silver and gold foil are used for collecting traces of mercury vapour [114–116]. On a drop of mercury, traces of silver [117, 118], selenium [119] and palladium [120] have been separated. Jackwerth [121] developed a method of enrichment of traces of silver and gold present in high-purity mercury. A large sample of the mercury was dissolved, but a small amount (ca. 100 mg) was left undissolved, and the trace components were concentrated in this portion. Trace elements in high-purity zinc, cadmium, aluminium, bismuth, indium, lead, gallium and

Table 3.5 Coprecipitation of trace elements with miscellaneous collectors

Collector	Precipitant	Coprecipitated traces	Sample	Ref.
Ag	I^-	Hg	high-purity silver	[100]
Ag	CN^-	Pd		[125]
Ag	Ar^-	Cu		[101]
Al	PO_4^{3-}	Cr, Fe, Mn, Ru, Zn	natural waters	[128]
As	reduced by H_3PO_2	Te, Se	high-purity lead	[111]
Ba	SO_4^{2-}	Pb		[127]
Ca	SO_4^{2-}	Pb		[102]
Ca	PO_4^{3-}	Be		[105]
Ca, Sr, Ba	F^-	rare-earth elements		[132]
Cu (powder)		Rh		[113]
Fe	CH_3COO^-	Al		[135]
Fe	PO_4^{3-}	Al		[103]
Fe, Mg	OH^-, PO_4^{3-}	As, Ba, Be, Cd, Co, Cu, Mo, Nb, Ni, Pb, V, W, Zn	natural waters	[129]
Hg_2^{2+}	Cl^-	Au	copper, iron	[98]
Hg_2^{2+} or Tl	Cl^-, Br^-	Au	cadmium, mercury, zinc tellurides and selenides	[85,99]
La	PO_4^{3-}	Ca, Mg		[104]
Mo	PO_4^{3-}	V	molybdenum	[130]
$Nb_2O_5 \cdot$ aq		SiO_2	water, reagents	[106]
Ni	dimethylglyoxime	Pd		[108]
Pb	PO_4^{3-}	Cl^-	high-purity water	[131]
Sb (powder)		Rh		[112]
SO_4^{2-}	$Ba^{2+} + Ag^+$	Cl^-		[107]
Sr	SO_4^{2-}	Pb	high-purity thallium	[85, 126]
Te	reduced by $SnCl_2$	Ag	silicates	[109, 136]
Te	reduced $H_2N \cdot NH_2 + SO_2$	Au		[110]
Y	$NH_3 \cdot$ aq	Sc		[134]
Zr	AsO_4^{3-}	Ti	separation from Fe, Cr, Mo, Ni, V	[133]

tin were enriched similarly [122,123]. The metal sample was amalgamated with high-purity mercury, and the amalgam formed was again incompletely dissolved; the traces were accumulated in the undissolved portion. Detailed descriptions of the procedures are given for zinc [124] and cadmium [125].

When collection of all precipitable traces is desired, a mixture of sulphide- and hydroxide-type organic reagents, along with an inorganic collector, usually Fe(III), Al(III) or In(III), is used. Such methods were developed for separation of trace elements for subsequent determination by emission spectrography. Mitchell and Scott [85] initially used 8-hydroxyquinoline with Al(III) and Fe(III) as collectors, but later they added thionalide and tannic acid [135–139]. Thionalide precipitates the sulphide group elements and tannic acid reacts with numerous elements to give sparingly soluble salts and also promotes the formation of more readily filterable precipitates.

Pickett and Hankins [140] studied the completeness of precipitation by precipitating trace amounts of some radioactive isotopes (Co, Cu and Mo). Their results indicate that all three carriers are required for complete recovery of the trace elements.

The method has been used widely when complete precipitation of trace elements and high enrichment factors (ca. 10^3) are required. Table 3.6

Table 3.6 Coprecipitation of trace elements with inorganic collectors and a mixture of 8-hydroxyquinoline, thionalide and tannic acid

Collector	Coprecipitated traces	Sample	Ref.
Al, Fe(III)	Ag, Be, Cd, Co, Cr, Cu, Ga, Ge, Mo, Ni, Pb, Sn, Ti, Tl, V, Zn	various materials	[137, 139, 149]
Al	Co, Cu, Mo, Zn	biological materials and soil extracts	[140]
Al	Cu, Fe, Mn, Ni, Pb, Sn, Zn	alkali metal nitrates and halides	[151]
Al	Ag, Co, Cr, Mn, Mo, Ni, Pb, Sn, V, Zn	rock salt	[154]
In	Ag, Al, Bi, Co, Cr, Cu, Fe, Mo, Ni, Pb, Sn, T, V, Zn	various materials	[150]
In	Al, Be, Bi, Cd, Co, Cr, Cu, Fe, Ga, Ge, Mn, Mo, Ni, Pb, Ti, V, Zn	natural waters	[152]
In	Ag, Au, Be, Bi, Cd, Co, Cr, Cu, Dy, Er, Eu, Fe, Ga, Gd, Ge, Ho, Lu, Mn, Mo, Ni, Os, Pb, Pd, Rh, Ru, Sb, Sc, Sm, Sn, Tb, Th, Ti, Tl, Tm, U, V, W, Yb, Zn, Zr	potassium chloride	[153]

gives examples of coprecipitation of trace elements with an 8-hydroxyquinoline–thionalide–tannic acid mixture and inorganic collectors.

A more recent method 'adsorption colloid flotation' is particularly useful for collecting trace elements by coprecipitation from larger volumes of solution. Usually the collector and precipitant are added to the solution, then the flotation reagent, and a fixed gas (air or nitrogen) is bubbled through the solution. The traces coprecipitated with the collector gather in the flotation foam and are readily separated. Zeitlin *et al.* [141–146] used this procedure to enrich traces of many elements in sea-water. The method has also been used for enrichment of traces in high-purity zinc [147] and for specific separation of trace amounts of uranium in sea-water [148].

3.3.3 Coprecipitation of Traces with Organic Collectors

Kuznetsov [155,156] introduced a method for concentrating traces of metals which is sometimes referred to as co-crystallization [157], involving coprecipitation with organic reagents. Two reagents are used together. One is able to form complexes with the metal ions to be separated; it may be a chelating agent or a simple anion (e.g. thiocyanate, chloride, bromide or iodide) which forms anionic complexes with the ions to be separated. The other must be an organic compound which is sparingly soluble in water. Several precipitation mechanisms are possible. In one, the second reagent is a bulky organic cation which forms ion-pairs with the anionic complexes formed in solution, and these precipitate together with the excess of water-insoluble reagent. An example is Kuznetsov's procedure [156] for concentrating traces of mercury by precipitating the iodide complex HgI_4^{2-}, together with a large excess of Crystal Violet iodide $(CV^+ \cdot I^-)$, as

$$[(HgI_4^{2-}) \cdot 2CV^+ + (CV^+ \cdot I^-)_{excess}]$$

In another mechanism, the reagent anion reacts with the ion to be separated to yield an insoluble salt which coprecipitates along with the excess of reagent in the acid form. An example is the precipitation of traces of potassium, rubidium and caesium with ammonium dipicrylaminate $(DPA^- \cdot NH_4^+)$ as the sparingly soluble salt $(DPA^- \cdot M^+)$ which coprecipitates along with water-insoluble dipicrylamine as

$$[(DPA^- \cdot M^+) + (DPA^- \cdot H^+)_{excess}]$$

Again, an acid reagent may react with the ion to be separated, to form a chelate complex which coprecipitates along with the precipitate formed by the excess of acid reagent and a bulky cationic organic reagent.

An example is the coprecipitation of zirconium and hafnium (M^{n+}) with 4-dimethylazobenzene-4'-arsonic acid (DMABAs$^-\cdot$H$^+$) and Crystal Violet (CV$^+$) [158]. The acid (DMABAs$^-\cdot$H$^+$) reacts with the two elements to form chelates [(DMABAs)$_n^-\cdot M^{n+}$] which coprecipitate along with the copious precipitate produced by reaction of (DMABAs$^-\cdot$H$^+$) with (CV$^+$), as

$$[(\text{DMABAs})_n^- \cdot M^{n+} + (\text{DMABAs}^- \cdot CV^+)_{\text{excess}}]$$

A fourth type of coprecipitation involves formation of an inner-complex compound by the ion to be separated, with an added organic reagent, and this complex coprecipitates with the same reagent added in excess. Alizarin [160] (Aliz) forms a plutonium complex which coprecipitates with excess of Alizarin as

$$\{[(\text{Aliz})_n\text{Pu}_m] + (\text{Aliz})_{\text{excess}}\}$$

Another type of coprecipitation with organic collectors involves first, formation of a chelate with an organic reagent added to the solution, and then sorptive coprecipitation with a water-insoluble neutral organic reagent.

A number of metal ions (M) which react with dithizone (Dz) can be coprecipitated in this manner, e.g. with phenolphthalein (PP) [159], as $\{[(\text{Dz})_n(M)] + (\text{PP})\}$.

It should be emphasized that it is a non-complexing organic reagent, such as phenolphthalein, β-naphthol, or 2,4-dinitroaniline that is employed as an additive.

The last type of coprecipitation, described by Myasoedova [160] as 'colloidal-chemical', involves sorptive, chemical, or any other sort of precipitation of metal ions with a mixture of organic reagents. For example, Kuznetsov [156] coprecipitated a number of metal ions with a mixture of Crystal Violet and tannic acid. Table 3.7 lists examples of coprecipitation of traces of various metals with miscellaneous organic reagents.

An advantage of this last method is that no foreign ions are added to the solution. The resulting precipitate is collected by centrifugation or filtration and then ashed. Marczenko [197] has raised objections, claiming that combustion of the precipitate may cause losses of the traces to be determined; such losses can, however, be avoided by adding, e.g. a drop of concentrated sulphuric acid during ashing (cf. Section 4.5.1).

The Kuznetsov method [198] often enables very high enrichment factors to be achieved. Precipitation has been achieved, for example, of americium [175] with a mixture of arsenazo I and Methyl Violet, potassium with triphenylboron [170], and uranium from sea-water with Methyl

Table 3.7 Coprecipitation of traces of metals with organic reagents

Collector	Elements coprecipitated
	as anionic complexes
Methyl Violet thiocyanate	Cu [156], Zn [156], [161], Mo [168], U [163]
Butyl rhodamine thiocyanate	V, Mo, W [164]
Diphenylguanidine thiocyanate	Nb, Bi, Re, Fe, Co [165]
Methyl Violet iodide	Cu [166], Cd [161], Hg [156], Pb [161], Sb [166], Bi [161]
Diphenylguanidine iodide	Tl(III) [165]
Methylene Blue iodide	In [167]
	as compounds formed with precipitants
Ammonium dipicrylaminate	K, Rb, Cs [168], [169]
Ammonium tetraphenylborate	K [170]
	as inner complexes formed with reagent–Methyl Violet or reagent–Crystal Violet precipitates
Arsenazo I	Sc [171], rare-earth elements [172], Am [173]
Stilbazo	W [156], Pu [174]
Eriochrome Black T	Cr [156], [164]
4-Dimethylazobenzene-4'-arsonic acid	Zr, Hf [158]
	as inner complexes formed with excess of complexant precipitate
p-Dimethylaminoazobenzenearsonic acid	Zr [175]
Alizarin	Pu [159]
Anthranilic acid	Zn [176]
2-Mercaptobenzimidazolone	Ag, Au, Hg, Sn, Ta [177]
1-Nitroso-2-naphthol	Zn, Ce, Zr, U, Fe, Co, Ru, Pu [178]
8-Hydroxyquinoline	Ce, Pr, Pu [179]
Thionalide	Au, Ag, Hg, Zn, Tl, In, Hf, Sn, Ta, W, Cr, Mn, Co, Os, Ru, Ir [180]
Potassium rhodizonate	Ba, Ce, La, Pr, Ra, Sr, Pu [181]
Cupferron	Ti, V, Zr [182]
p-Dimethylaminobenzylidenerhodanine	Ag [183]
Thiourea	Pt, Pd, Rh [184]

Table 3.7 (*continued*)

Collector	Elements coprecipitated	
	as inner complexes formed with neutral reagents	
	complexant	elements
2,4-Dinitroaniline	dithizone	Cu, Au, Ag, Zn, In, Sn, Pb, Co, Ni [164]
Phenolphthalein	dithizone	Ag, Cd, Co, Ni [159]
Phenolphthalein	8-hydroxyquinoline	Ag, Cd, Co, Ni [159]
2-Naphthol	8-hydroxyquinoline	Ag, Cd, Co, Ni [159]
Biphenyl	8-hydroxyquinoline	Ag, Cd [159]
Naphthalene	8-hydroxyquinoline	Ni [159]
p-Nitrotoluene	8-hydroxyquinoline	Ag, Cd, Ni [159]
Phenolphthalein (2-naphthol, thymolphthalein, diphenylamine)	8-hydroxyquinoline	Zr, Hf [185]
Phenolphthalein	thionalide	Ag, Cd [159]
Diphenylamine	thionalide	Rh [186]
Diphenylamine	thiobenzamide	Pt, Pd, Rh, Ir [187]
2,4-Dinitroaniline	α-furildioxime	Ni [188]
	as colloidal-chemical compounds	
Methyl Violet + tannic acid	Be [189], Ti, Sn, Zr, Hf, Nb, Ta [156], Mo, W [190]	
Tannic acid + gelatin	Ca, Ba, Ra, Pb, U [191], Ge [192]	
Activated carbon modified with 8-hydroxyquinoline, dithizone, or diethyldithiocarbamate	Cu, Pb [193]	
Activated carbon + chlorinated lignin	Ag, Au, Be, Bi, Co, Cr, Cu, Ga, Ge, In, La, Mo, Nb, Ni, Pb, Sb, Se, Sn, Ta, Te, V, W, Y, Yb, Zn [194]	
Powdered spectral-electrode carbon	Ag, Au, Bi, Ca, Co, In, Ni, Pb, Sb [195]	
Powdered spectral-electrode carbon	Ag, Al, Be, Bi, Ca, Cd, Co, Cr, Cu, Mg, Mn, Ni, Pb, Zn [196]	
Activated carbon, and zincon, dithizone or sulphide	Cu (Cd matrix), Ag, Bi, Cd, Co, Cu, In, Ni, Pb, Zn (Mg matrix), Cu, Cd, Co, Ni, Pb, (Al matrix) [196a]	
Activated carbon, and potassium ethyl xanthate	Ag, Bi, Cu, Co, Cd, In, Pb, Ni, Tl, Fe, Hg (Zn matrix) [196b]	
Activated carbon, and potassium ethyl xanthate	Bi, Cd, Co, Cu, Fe, In, Ni, Pb, Tl, Zn (Mn matrix) [196c]	
Activated carbon, and ammonium pyrrolidinedithiocarbamate	Ag, Bi (Co and Ni matrices) [196d]	

Violet thiocyanate [163], at dilutions of $1:10^{18}$, $1:10^8$, and $1:10^{10}$ respectively.

Another method of enrichment of traces, by using organic reagents only, is to add a solid chelating organic reagent to the solution and expose the mixture to an ultrasonic field. In this way traces of silver can be enriched on dithizone [199], and traces of iron, silver and cobalt on dithizone and α-nitroso-β-naphthol [200]. Braun and Farag [201] enriched traces of silver on plasticized open-cell polyurethane foam saturated with dithizone.

Activated charcoal and spectral-grade electrode carbon may also be included in the category of organic collectors, as they have often been used as carriers during the evaporation of a solution of isolated traces before emission spectrography. The activated carbon must first be freed from any trace metal impurities, e.g. by treatment with various organic reagents, such as 8-hydroxyquinoline, dithizone or diethyldithiocarbamate [193]. A mixture of activated carbon and chlorolignin [194] has been used for isolation of trace elements from natural waters. Examples are given in Table 3.7.

3.3.4 Electrodeposition in Trace Analysis

Electrolysis, whether normal or 'internal' may be used for precipitation of components of a solution.

From the practical viewpoint, if the constituent ions of a solution are to be separated electrolytically, the first method to be considered is electrolysis at a mercury cathode. The high overpotential of hydrogen on mercury enables numerous mercury-soluble metals to be deposited from an acid solution [202] in large or small quantities.

This technique has been used to remove the major components of steel so as to allow traces of boron to be determined [203], and to separate molybdenum from vanadium [204], etc. [205,206]. Vanadium cannot be deposited, and this fact was utilized by Schmidt and Bricker [207] in a method for determining trace impurities in vanadium by depositing them on the mercury cathode.

A related technique is electrochemical stripping analysis. This is a two-step technique in which, first, traces of the species to be determined are concentrated electrolytically from the solution onto the measuring electrode, then, secondly, the concentrated traces are 'stripped' back into solution by the reverse electrolytic process, which is monitored by any suitable electrochemical technique. The measuring electrode may be a hanging mercury drop [208], a thin-layer mercury electrode, or a solid electrode of platinum, silver, gold or graphite [209]. Stripping analysis

Table 3.8 Electrolytic separation of trace and major constituents

Elements deposited on electrode	Separation from	Medium or material	Electrode	Ref.
Ag, As, Au, Bi, Cd, Co, Cr, Cu, Fe, Ga, Ge, Hg, In, Ir, Mo, Ni, Pb, Pd, Pt, Re, Rh, Sb, Se, Sn, Te, Tl, Zn	Al, Ba, Be, Ca, Mg, P, Sr, Ti, U, V, Zr	dil. H_2SO_4	mercury	[202]
Fe and other steel constituents	B	steel	mercury	[203]
Mo	Fe, V			[204]
Co, Cu, Fe, Ni	Nb, Ti	magnesium alloys	mercury	[206]
Cd, Co, Cu, Fe, Ni, Pb (traces)	V	vanadium metal	mercury	[207]
Cd, Cu, Pb (traces)	Zn, ZnS	zinc, zinc sulphide	hanging Hg drop	[210]
Cl^-, Br^-, I^- (traces)		solutions	hanging Hg drop	[211]
Cu, Pb (traces)		HCl, HNO_3	hanging Hg drop	[212]
Cd, Cu, Pb, Zn (traces)		NaOH, CH_3COOH	hanging Hg drop	[213]
Cu, Pb (ultramicro traces)			pin-shaped	[214]
Ag, Au, Cu, Pb (ultramicro traces)			pin-shaped	[215]
Cd, Co, Cu, Fe, Ni, Pb (traces)	U	uranium and its compounds	mercury	[217]
Cu, Pb (traces)		urine	mercury	[219]
Pb, Tl	Cd		at the anode	[222]
Co (traces)		biological material	platinum	[223]

Table 3.8 (*continued*)

Elements deposited on electrode	Separation from	Medium or material	Electrode	Ref.
Mn (traces)	Zn		graphite anode	[226]
Ag, Cu (traces)	Pb, Zn	acid	carbon	[227]
Ag (traces)	Zn	HNO_3	carbon	[227]
Cu (traces)	Sb		graphite	[228]
Co, Cr, Cu, Hg, Ni, Zn		water	graphite	[229]
Pb		natural silicates	anode–cathode	[230]
Ag, Bi, Cd, Cu, Pb, Sb (traces)			internal electrolysis	[231, 232]
Co, Ni	Bi, Cd, Cu, Pb, Sb		internal electrolysis	[233]
Se, Te (traces)			internal electrolysis	[234]
Po		urine	internal electrolysis	[235]

is an extremely important technique, but since it is really beyond the scope of this book, the reader is referred to the monograph literature [208–213].

Mention must also be made of the work of Alber [214] and Brenneis [215], who succeeded in the 1930s in detecting and determining very small amounts of trace elements (10^{-9} g) by using needle-shaped mini-electrodes. By performing the deposition under special conditions, then examining the electrode and its deposit with an electron microscope, Wiesenberger [216] was able to detect 5×10^{-11} g of copper. Development of such techniques still continues [216a].

Electrodeposition has been employed for the determination of trace elements in uranium [217], traces of gold [218], copper and lead [219], uranium [220] in urine, lead [221], thallium [222] and cobalt-60 in biological [223] and other [224,225] materials.

Another procedure involves electrodeposition of trace elements on a graphite electrode which is then arced, and the traces are determined spectrographically. Examples are the determinations of traces of manganese [226] (deposited as MnO_2 at the anode), silver and copper [227,228].

Majumdar and Bhowal [231–234] have developed many methods that utilize internal electrolysis. By this method polonium-201 was deposited from urine [235] and other materials [236].

Examples of the electrolytic deposition of elements are listed in Table 3.8.

Welch and Ure used finely powdered aluminium to collect traces of many elements from solution by internal electrolysis. The aluminium was subsequently formed into an electrode for analysis by spark-source mass spectrometry [236a].

References

[1] Fresenius, C. R., *Anleitung zur qualitativen chemischen Analyse*, Vieweg, Braunschweig, 1841.
[2] Treadwell, F. P., *Kurzes Lehrbuch der analytischen Chemie*, Zürich, 1899.
[2a] Walton, A. G., *The Formation and Properties of Precipitates*, Interscience, New York, 1967.
[3] La Mer, V. K., *Ind. Eng. Chem.*, **44**, 1270 (1952).
[4] La Mer, V. K. and Dinegar, R. H., *J. Am. Chem. Soc.*, **73**, 380 (1951).
[5] Collins, F. C. and Leineweber, J. P., *J. Phys. Chem.*, **60**, 389 (1956).
[6] Buckley, H. E., *Crystal Growth*, Wiley, New York, 1951.
[7] Verma, A. R., *Crystal Growth and Dislocations*, Butterworths, London, 1953.
[8] Knapp, L. F., *Trans. Faraday Soc.*, **17**, 457 (1921).
[9] Gordon, L., Salutsky, M. L. and Willard, H. H., *Precipitation from Homogeneous Solution*, Wiley, New York, 1959.
[10] Overneck, J. Th. G., *Discuss. Faraday Soc.*, **18**, 9 (1955).
[11] Kolthoff, I. M. and Sandell, E. B., *Textbook of Quantitative Inorganic Analysis*, 3rd Ed., Macmillan, New York, 1952.
[12] Doerner, A. N. and Hoskins, W. M., *J. Am. Chem. Soc.*, **47**, 662 (1925).
[13] Chlopin, W. G., *Z. Anorg. Chem.*, **143**, 97 (1925).
[14] Hahn, O., *Applied Radiochemistry*, Cornell Univ. Press, Ithaca, New York, 1936.
[15] Kolthoff, I. M. and Overholzer, Z. G., *J. Phys. Chem.*, **41**, 629 (1937); **43**, 909 (1939).
[16] Pushkarev, V. V., *Zh. Neorgan. Khim.*, **1**, 170 (1956).
[17] Plotnikov, V. I., *Zh. Neorgan. Khim.*, **3**, 1761 (1958).
[18] Kolařík, Z. and Kouřím, V., *Collect. Czech. Chem. Commun.*, **25**, 1000 (1960).
[19] Chuiko, V. T. and D'yachenko, N. P., *Tr. Ternopol'sk. Med. Inst.*, **1**, 477 (1960); *Anal. Abstr.*, 1962, **9**, 4204.
[20] Korenman, Ya. I., *Tr. Khim. Khim. Tekhnol. Gorkii*, **1963** (1), 140, 146, 152; *Chem. Abstr.*, **60**, 7658 c,d,e (1964).
[21] Novikov, A. I., *Zh. Analit. Khim.*, **17**, 1076 (1962).
[22] Jackwerth, E. and Graffmann, G., *Z. Anal. Chem.*, **251**, 81 (1970).

[23] Chuiko, V. T., Kovaleva, N. V. and Kvartsova, A. A., *Zh. Analit. Khim.*, **27**, 703 (1972).
[24] Marov, I. N., Malofeeva, G. I. and Evshikova, G. A., *Zh. Analit. Khim.*, **28**, 246 (1973).
[25] Malofeeva, G. I., Ruzinov, L. P. and Andreeva, N. P., *Zh. Analit. Khim.*, **27**, 1087 (1972).
[26] Salutsky, M. L., Precipitates: Their Formation, Properties and Purity, in *Treatise on Analytical Chemistry*, Kolthoff, I. M. and Elving, P. J. (Eds.)., 1st Ed., Part I, Vol. 1, Ch. 18, Interscience, New York, 1959.
[27] Karabash, A. G., Bondarenko, L. S., Morozova, G. G. and Peizulaev, Sh. I., *Zh. Analit. Khim.*, **15**, 623 (1960).
[28] Degtyareva, D. F. and Ostrovskaya, M. F., *Zh. Analit. Khim.*, **20**, 814 (1956).
[29] Marczenko, Z. and Kasiura, K., *Chem. Anal. (Warsaw)*, **9**, 87 (1964).
[30] Krauz, L. S., Karabash, A. P., Peizulaev, Sh. I., Lipatova, V. M. and Moleva, V. S., *Tr. Kom. Anal. Khim., Akad. Nauk SSSR*, **12**, 175 (1960).
[31] Baronova, L. L. and Solobodnik, S. M., *Zh. Analit. Khim.*, **19**, 588 (1964).
[32] Mayer, J., *Z. Anal. Chem.*, **219**, 147 (1966).
[33] Karabash, A. G., Samsonova, Z. I., Smirnova-Averina, N. I. and Peizulaev, Sh. I., *Tr. Kom. Anal. Khim., Akad. Nauk SSSR*, **12**, 255 (1960).
[34] Sotnikova, N. P., Romanovich, L. S., Peizulaev, Sh. I. and Karabash, A. G., *Tr. Kom. Anal. Khim., Akad. Nauk SSSR*, **12**, 151 (1960).
[35] Karabash, A. G., Peizulaev, Sh. I., Slyusareva, R. A. and Mashkova, V. M., *Zh. Analit. Khim.*, **14**, 598 (1959).
[36] Chwastowska, J., Dybczyński, R. and Kucharzewski, B., *Chem. Anal. (Warsaw)*, **13**, 721 (1968).
[37] Hitchen, A., *Talanta*, **26**, 369 (1979).
[38] Babina, F.L., Karabash, A.G. and Smirenkina, I.I., *Zavodsk. Lab.*, **37**, 287 (1971).
[39] Viebrock, J. M., *Anal. Lett.*, **3**, 373 (1970).
[40] Nazarenko, V. A. and Flyantikova, G. V., *Zavodsk. Lab.*, **24**, 801 (1958).
[41] Jackwerth, E., Lohmar, J. and Schwark, G., *Coll. Spectr. Int. 16. Heidelberg 1971, Preprints*, Vol. 2, London, 1971, p. 49.
[42] Bock, R. and Gorbach, S., *Mikrochim. Acta*, **1958**, 593.
[43] Lucas, R. and Grassner, F., *Mikrochemie*, **1930**, 197.
[43a] Tikhomirova, A. A., Patin, S. A. and Morozov, N. P., *Zh. Analit. Khim.*, **31**, 282 (1976).
[44] Rudnev, N. A., *Zh. Analit. Khim.*, **8**, 3 (1953).
[45] Golden, G. S. and Atwell, H. G., *Appl. Spectrosc.*, **24**, 514 (1970).
[46] Marczenko, Z. and Kasiura, K., *Chem. Anal. (Warsaw)*, **10**, 449 (1965).
[47] Hirose, A. and Ishii, D., *Kogyo Kagaku Zasshi*, **1972**, 1996; *Anal. Abstr.* **28**, 1B 33, 1975.
[48] Podberezskaya, N. K. and Sushkova, V. A., *Zavodsk. Lab.*, **36**, 1048 (1970).
[49] Pohl, F. A. and Treiber, E., *Melliand Textilber.*, **34**, 747 (1952).
[50] Pohl, F. A., *Angew. Chem.*, **66**, 603 (1954).
[51] Pohl, F. A., *Z. Anal. Chem.*, **142**, 19 (1954).
[52] Willmer, T. K., *Arch. Eisenhüttenwes.*, **29**, 159 (1958).
[53] Onishi, H. and Sandell, E. B., *Anal. Chim. Acta*, **11**, 444 (1954).
[54] Pierrucini, R., *Mikrochem. Mikrochim. Acta*, **36/37**, 522 (1951).
[55] Rutkowski, W., *Chem. Anal. (Warsaw)*, **14**, 905, 1099 (1969).
[56] Sandell, E. B., *Anal. Chim. Acta*, **3**, 89, (1949).

References

[57] Pevtsov, G. A. and Manova, T. G., *Tr. Vses. Nauchno-Issled. Inst. Khim. Reactivov Osobo. Chist. Khim. Veshchestv*, **29**, 7 (1966); *Chem. Abstr.*, **67**, 87419y (1967).
[58] Rudnev, N. A., Tuzova, A. M. and Malofeeva, G. I., *Zh. Analit. Khim*, **26**, 886 (1971).
[59] Schiller, P. and Cook, G. B., *Anal. Chim. Acta*, **54**, 364 (1971).
[60] Sinitsyn, N. M., Borisov, V. V. and Dobronravov, S. A., *Radiokhimiya*, **15**, 741 (1973).
[61] Marczenko, Z. and Mojski, M., *Chem. Anal. (Warsaw)*, **14**, 495 (1969).
[62] *Quick Methods for Radiochemical Analysis*, IAEA, Technical Report Series No. 95, Vienna, 1969.
[63] Pierrucini, R., *Spectrochim. Acta*, **4**, 189 (1950).
[64] Lebedeva, N. V. and Vinarova, L. I., *Zavodsk. Lab.*, **39**, 798 (1973).
[65] Lebedinskaya, M. P. and Chuiko, V. T., *Zh. Analit. Khim.*, **26**, 863 (1973).
[66] Upor, E. and Nagy, Gy., *Acta Chim. Acad. Sci. Hung.*, **68**, 313 (1971).
[67] Ampelogova, N. I., *Radiokhimiya*, **15**, 487 (1973).
[68] Marczenko, Z. and Kasiura, K., *Chem. Anal. (Warsaw)*, **9**, 87 (1964).
[69] Reichel, W. and Bleakley, B. G., *Anal. Chem.*, **46**, 59 (1974).
[70] Tiptsova-Yakovleva, V. G. and Dvortsam, A. G., *Zavodsk. Lab.*, **37**, 676 (1971).
[71] Rudnev, N. A., Pavlenko, L. I., Malofeeva, G. I. and Simonova, L. V., *Zh. Analit. Khim.*, **24**, 1223 (1969).
[72] Pitwell, L. R., *Analyst (London)*, **86**, 137 (1961).
[73] Strohal, P., Molnar, K. and Bačić, I., *Mikrochim. Acta*, **1972**, 586.
[74] Fukuda, K. and Mizuike, A., *Anal. Chim. Acta*, **51**, 527 (1970).
[75] Zörner, A., Krath, E. and Feucht, H., *Techn. Mitt. Rheinhausen*, **1955**, 4, 237.
[76] Blumenthal, H., *Z. Anal. Chem.*, **74**, 33 (1928).
[77] Reynolds, G. F. and Tyler, F. S., *Analyst (London)*, **89**, 538, 579 (1964).
[78] Burke, K. E., *Anal. Chem.*, **42**, 1536 (1970).
[79] Geilmann, W. and Neeb, K., *Z. Anal. Chem.*, **165**, 251 (1959).
[80] Luke, C. L., *Anal. Chem.*, **31**, 1680 (1959).
[81] Pilipchuk, M. F. and Volkov, I. I., *Geokhimiya*, **1967**, 8.
[82] Vinogradov, A. V., Dronova, M. I. and Lopatina, N. N., *Zavodsk. Lab.*, **39**, 150 (1973).
[83] Yamagata, Y. and Iwashima, T., *Nature (London)*, **200**, 52 (1963).
[84] Strock, L. W. and Drexler, S., *J. Opt. Soc. Amer.*, **31**, 167 (1941).
[85] Scott, R. O. and Mitchell, R. L., *J. Soc. Chem. Ind.*, **62**, 4 (1943).
[85a] Musić, S. and Wolf, R. H. H., *Mikrochim. Acta*, **1979** I, 87.
[85b] Musić, S., Gessner, M. and Wolf, R. H. H., *Mikrochim. Acta*, **1979** I, 95.
[85c] Musić, S., Gessner, M. and Wolf, R. H. H., *Mikrochim. Acta*, **1979** I, 105.
[86] Angermann, W., in *Reinststoffprobleme. II Reinststoffanalytik*. Akademie-Verlag, Berlin, 1966, p. 541.
[87] Marczenko, Z., *Mikrochim. Acta*, **1965**, 281.
[88] Kim, S. and Zeitlin, H., *Anal. Chim. Acta*, **51**, 516 (1970).
[89] Downarowicz, J. and Zagórski, Z., *Chem. Anal. (Warsaw)*, **4**, 445 (1959).
[90] Yao, Yu-Lin, *Ind. Eng. Chem., Anal. Ed.*, **17**, 114 (1945).
[91] Luke, C. L., *Anal. Chem.*, **28**, 1276 (1956).
[92] Blakeley, St. J. H., Manson, A. and Zatka, V. J., *Anal. Chem.*, **45**, 1941 (1973).
[93] Portretnyi, V. P., Malynta, V. F. and Chuiko, V. T., *Zh. Analit. Khim.*, **28**, 1337 (1973).

[94] Okochi, H. and Sudo, E., *Bunseki Kagaku*, **22**, 431 (1973).
[95] Ogata, N., *Bull. Soc. Sea Water Sci. Japan*, **24**, 197 (1971).
[96] Proházková, V. and Jára, V., *Z. Anal. Chem.*, **161**, 251 (1958).
[97] Sapek, B., *Chem. Anal. (Warsaw)*, **15**, 651 (1970).
[98] Dziedzianowicz, W., *Chem. Anal. (Warsaw)*, **5**, 827 (1960).
[99] Tiptsova-Yakovleva, V. G. and Dvortsan, A. G., *Zh. Analit. Khim.*, **24**, 1141 (1969).
[100] Jackwerth, E., Döring, E. and Lohmar, J., *Z. Anal. Chem.*, **253**, 195 (1971).
[101] Jackwerth, E. and Döring, E., *Z. Anal. Chem.*, **255**, 194 (1971).
[102] Lynch, L. T., Slater, R. H. and Osler, T. G., *Analyst (London)*, **59**, 787 (1934).
[103] Cholak, J., Hubbard, D. M. and Story, R. V., *Ind. Eng. Chem., Anal. Ed.*, **15**, 57 (1943).
[104] Marczenko, Z. and Kasiura, K., *Chem. Anal. (Warsaw)*, **7**, 775 (1962).
[105] Teribara, T. Y. and Chen, P. S. (Jr.), *Anal. Chem.*, **24**, 539 (1952).
[106] Marczenko, Z. and Kasiura, K., *Chem. Anal. (Warsaw)*, **9**, 321 (1964).
[107] Chwastowska, J., Marczenko, Z. and Stolarczyk, U., *Chem. Anal. (Warsaw)*, **8**, 517 (1963).
[108] Marczenko, Z. and Krasiejko, M., *Chem. Anal. (Warsaw)*, **9**, 291 (1964).
[109] Novikov, V. N. and Bondarenko, V. K., *Zavodsk. Lab.*, **34**, 1080 (1968).
[110] Onishi, H., *Mikrochim. Acta*, **1959**, 9.
[111] Luke, C. L., *Anal. Chem.*, **31**, 572 (1959).
[112] Westland, A. D. and Beamish, F. E., *Mikrochim. Acta*, **1956**, 1474.
[113] Tertipis, G. G. and Beamish, F. E., *Anal. Chem.*, **32**, 486 (1960).
[114] Fishman, M. I., *Anal. Chem.*, **42**, 1462 (1970).
[115] Anderson, D. H., Evans, J. H., Murphy, J. J. and White, W. W., *Anal. Chem.*, **43**, 1511 (1971).
[116] O'Gorman, J. V., Suhr, N. H. and Walker, P. L. (Jr.), *Appl. Spectrosc.*, **26**, 44 (1972).
[117] Mizuike, A., Fukuda, K., Sakamoto, T., *Bull. Chem. Soc. Japan*, **46**, 3596 (1973).
[118] Mizuike, A., Sakamoto, T. and Sugishima, K., *Mikrochim. Acta*, **1973**, 291.
[119] Toporova, V. F., Polyakov, Yu. N., Medvedeva, L. N. and Mindina, L. N., *Zh. Analit. Khim.*, **27**, 2041 (1972).
[120] Mizuike, A., Sakamoto, T. and Onishi, N., *Mikrochim. Acta*, **1971**, 783.
[121] Jackwerth, E. and Kulok, A., *Z. Anal. Chem.*, **257**, 28 (1971).
[122] Jackwerth, E., *Z. Anal. Chem.*, **256**, 128 (1971).
[123] Jackwerth, E., Döring, E., Lohmar, J. and Schwark, G., *Z. Anal. Chem.*, **260**, 177 (1972).
[124] Jackwerth, E., Höhn, R. and Koos, K., *Z. Anal. Chem.*, **264**, 1 (1973).
[125] Jackwerth, E., *Z. Anal. Chem.*, **251**, 353 (1970).
[126] Nazarenko, V. A., Fuga, N. A., Flyantikova, G. V. and Esterlis, K. A., *Zavodsk. Lab.*, **26**, 131 (1960).
[127] Marczenko, Z. and Kasiura, K., *Chem. Anal. (Warsaw)*, **10**, 449 (1965)
[128] Bacic, I., Rodaković, N. and Strohal, P., *Anal. Chim. Acta*, **54**, 149 (1971).
[129] Lebedinskaya, M. P. and Chuiko, V. T., *Zh. Analit. Khim.*, **28**, 2413 (1973).
[130] Paganok, L. P., Chuiko, V. T., Reznik, B. E., Mazan, P. K. and Stets, V. T., *Zavodsk. Lab.*, **39**, 169 (1973).
[131] Rodabaugh, R. D. and Upperman, G. T., *Anal. Chim. Acta*, **60**, 434 (1972).
[132] Steinberg, R. H., *Appl. Spectrosc.*, **7**, 163 (1953).
[133] Anduze, R. A., *Anal. Chem.*, **29**, 90 (1957).

References

[134] Fischer, W., Steinhauser, O., Hohmann, E., Bock, E. and Borchers, P., *Z. Anal. Chem.*, **133**, 57 (1951).
[135] Oelschläger, W., *Z. Anal. Chem.*, **154**, 321 (1957).
[136] Sandell, E. B. and Neumayer, J. J., *Anal. Chem.*, **23**, 1863 (1951).
[137] Mitchell, R. L. and Scott, R. O., *J. Soc. Chem. Ind.*, **66**, 330 (1947).
[138] Mitchell, R. L. and Scott, R. O., *Mikrochem. Mikrochim. Acta*, 36/37, 1042 (1951).
[139] Mitchell, R. L. and Scott, R. O., *Spectrochim. Acta*, **3**, 367 (1968).
[140] Pickett, E. E. and Hankins, B. E., *Anal. Chem.*, **30**, 47 (1958).
[141] Kim, Y. S. and Zeitlin, H., *Chem. Commun.*, **1971**, 672.
[142] Kim, Y. S. and Zeitlin, H., *Anal. Chem.*, **43**, 1390 (1971).
[143] Kim, Y. S. and Zeitlin, H., *Sep. Sci.*, **7**, 1 (1972).
[144] Leung, G., Kim, Y. S. and Zeitlin, H., *Anal. Chim. Acta*, **60**, 229 (1972).
[145] Matsuzaki, C. and Zeitlin, H., *Sep. Sci.*, **8**, 185 (1973).
[146] Voyce, D. and Zeitlin, H., *Anal. Chim. Acta*, **69**, 27 (1974).
[147] Mizuike, A. and Hiraide, M., *Anal. Chim. Acta*, **69**, 231 (1974).
[148] Sekine, K. and Onishi, H., *Anal. Chim. Acta*, **62**, 468 (1972).
[149] Pohl, F. A., *Spectrochim. Acta*, **6**, 19 (1953).
[150] Heggen, G. E. and Strock, L. W., *Anal. Chem.*, **25**, 859 (1953).
[151] Dehne, R. L., Dunn, W. G. and Loder, E. R., *Anal. Chem. (Warsaw)*, **33**, 607 (1961).
[152] Silvey, W. D. and Brennan, R., *Anal. Chem.*, **34**, 784 (1962).
[153] Farquhar, M. C., Hill, J. A. and English, M. M., *Anal. Chem.*, **38**, 208 (1966).
[153a] Thomas, A. D. and Smythe, L. D., *Talanta*, **20**, 469 (1973).
[154] Zaremba, J., *Chem. Anal. (Warsaw)*, **12**, 305 (1967).
[155] Kuznetsov, V. I., *Zh. Analit. Khim.*, **9**, 199 (1954).
[156] Kuznetsov, V. I., *Sessiya Akad. Nauk SSSR po ispolzovaniyu atomnoi energii v mirnykh tselyakh (Session of the USSR Academy of Sciences on the Peaceful Uses of Atomic Energy)*, Otd. Khim. Nauk. Izd. Akad. Nauk SSSR, Moscow, 1955, p. 301.
[157] Koch, O. G. and Koch-Dedic, G. A., *Handbuch der Spurenanalyse*, Springer-Verlag, Berlin, 1964.
[158] Tuzova, A. M. and Nemodruk, A. A., *Zh. Analit. Khim.*, **13**, 674 (1958).
[159] Pasternak, A., *Nukleonika (Warsaw)*, **6**, 113 (1961).
[160] Myasoedova, G. V., *Zh. Analit. Khim.*, **21**, 598 (1966).
[161] Yakovlev, P. Ya., Razumova, G. P. and Malinina, R. D., *Zavodsk. Lab.*, **25**, 1039 (1959).
[161a] Buzzelli, G. and Mosen, A. W., *Talanta*, **24**, 383 (1977).
[162] Kuznetsov, V. I. and Myasoedova, G. V., *Tr. Kom. Anal. Khim., Akad. Nauk SSSR*, **9**, 89 (1958).
[163] Kuznetsov, V. I. and Akimova, T. G., *Radiokhimiya*, **2**, 426 (1960).
[164] Ivanov, D. I., Ivanova, N. N. and Orlova, L. P., *Tr. Kom. Anal. Khim., Akad. Nauk SSSR*, **15**, 306 (1965).
[165] Kuznetsov, V. I. and Myasoedova, G. V., *Primenenie fotometricheskikh metodov analiza pri kontrole materialov (The Use of Photometric Analysis Methods in Materials Control)*, Moscow, 1963, p. 16 (referred to in Myasoedova, G. V., *Zh. Analit. Khim.*, **21**, 598 (1966)).
[166] Amano, H., *Nippon Kinzoku Gakkaishi*, **23**, 621 (1959).

[167] Kral, J., Jambor, J. and Sommer, L., *Chem. Listy*, **63**, 1036 (1969).
[168] Weiss, H. V. and Lai, M. G., *J. Inorg. Nucl. Chem.*, **17**, 366 (1961).
[169] Korenman, I. M. and Shatalina, G. A., *Zh. Analit. Khim.*, **13**, 299 (1958).
[170] Nevin'sh, A. F. and Peinberg, M. Ya., *Izdelya Akad. Nauk Latviiskoi SSR*, **1959**, 85 (referred to in Myasoedova, G. V., *Zh. Analit. Khim.*, **21**, 598 (1966)).
[171] Kuznetsov, V. I., Ni Uzhe-Min, Myasoedova, G. B. and Okhanova, L. A., *Hua Hsueh Hsueh Pao*, **27**, 74 (1961).
[172] Kuznetsov, V. I., Myasoedova, G. V., *Tr. Kom. Anal. Khim., Akad. Nauk SSSR*, **9**, 76 (1958).
[173] Kuznetsov, V. I. and Akimova, T. G., *Radiokhimiya*, **3**, 737 (1961).
[174] Kuznetsov, V. I. and Akimova, T. G., *At. Energ.*, **8**, 148 (1960).
[175] Grimaldi, F. S. and White, C. E., *Anal. Chem.*, **25**, 1886 (1953).
[176] Korenman, I. M. and Baryshnikova, M. I., *Zh. Analit. Khim.*, **12**, 690 (1957).
[177] Weiss, H. V. and Lai, M. G., *Anal. Chim. Acta*, **28**, 242 (1963).
[178] Weiss, H. V., Lai, M. G. and Gillespie, A., *Anal. Chim. Acta*, **25**, 550 (1961).
[179] Weiss, H. V. and Shipman, W. H., *Anal. Chem.*, **33**, 37 (1961).
[180] Lai, M. G. and Weiss, H. V., *Anal. Chem.*, **34**, 1012 (1962).
[181] Weiss, H. V. and Lai, M. G., *Anal. Chem.*, **32**, 475 (1960).
[182] Strock, L. W. and Drexler, S., *J. Opt. Soc. Am.*, **31**, 167 (1941).
[183] Zagórski, Z. and Kempiński, O., *Chem. Anal. (Warsaw)*, **4**, 423 (1959).
[184] Khitrov, V. G. and Belousov, G. E., *Zh. Prikl. Spektrosk.*, **14**, 5 (1971).
[185] Moroshkina, T. M. and Savinova, G. G., *Zh. Analit. Khim.*, **24**, 1165 (1969).
[186] Kuznetsov, V. I. and Marugin, V. A., *Zavodsk. Lab.*, **36**, 1320 (1970).
[187] Shlenskaya, V. I., Khvostova, V. P. and Bugakova, V. I., *Zh. Analit. Khim.*, **29**, 314 (1974).
[188] Pakhomova, K. S., Volkova, L. P. and Gorshkov, V. V., *Zh. Analit. Khim.*, **19**, 1085 (1964).
[189] Sudhalatha, K., *Talanta*, **10**, 934 (1963).
[190] Kuznetsov, V. I., Loginova, L. G. and Myasoedova, G. V., *Zh. Analit. Khim.*, **13**, 453 (1958).
[191] Keshishyan, G. O., Andreev, P. F. and Danilov, L. T., *Zh. Prikl. Khim.*, **35**, 2051 (1962).
[192] Andryanov, A. M. and Korynkova, V. P., *Zh. Prikl. Khim.*, **46**, 425 (1973).
[193] Zharikov, V. F. and Senyavin, M. M., *Tr. Gos. Okeanogr. Inst.*, (101), 128 (1970).
[194] Brodskaya, N. I., Vychuzhanina, I. P. and Miller, A. D., *Zh. Prikl. Khim.*, **40**, 802 (1967).
[195] Shvarts, D. Kh. and Portnova, V. V., *Spektralnyi analiz v tsvetnoi metallurgii* (*Spectral Analysis in Non-ferrous Metallurgy*), Sbornik, Moscow, 1960, p. 125.
[196] Yudelevich, I. G., Buyanova, L. M., Protopopova, N. P. and Dzhamakueva, B. K., *Zavodsk. Lab.*, **35**, 426 (1969).
[196a] Jackwerth, E., Lohmar, J. and Wittler, G., *Z. Anal. Chem.*, **266**, 1 (1973).
[196b] Kimura, M., *Talanta*, **24**, 194 (1977).
[196c] Berndt, H., Jackwerth, E. and Kimura, M., *Anal. Chim. Acta*, **93**, 45 (1977).
[196d] Kimura, M. and Kawanami, K., *Talanta*, **26**, 901 (1979).
[197] Marczenko, Z., *Spectrophotometric Determination of Elements*, Horwood, Chichester, 1976.
[198] Kuznetsov, V. I., *Radioisotopes Sci. Res. Proc. Intern. Conf.*, Paris, Sept. 1957, **2**, 264 (1958).

References

[199] Fukuda, K. and Mizuike, A., *Anal. Chim. Acta*, **51**, 77 (1970).
[200] Fukuda, K. and Mizuike, A., *Anal. Chim. Acta*, **67**, 207 (1973).
[201] Braun, T. and Farag, A. B., *Anal. Chim. Acta*, **69**, 85 (1974).
[202] Hillebrand, W. F., Lundell, G. E. F., Bright, H. A. and Hoffman, J. I., *Applied Inorganic Analysis*, 2nd Ed., Wiley, New York, 1953.
[203] Piper, E., Hagedorn, H., *Arch. Eisenhüttenwes.*, **28**, 373 (1957).
[204] Golubeva, I. A., *Zh. Analit. Khim.*, **24**, 467 (1969).
[205] Bock, R. and Hackstein, K. G., *Z. Anal. Chem.*, **138**, 339 (1953).
[206] Bagdasarov, K. N. and Osmanov, Kh. A., *Zavodsk. Lab.*, **34**, 1044 (1968).
[207] Schmidt, W. E. and Bricker, C. E., *J. Electrochem. Soc.*, **102**, 623 (1955).
[208] Kemula, W., in *Advances in Polarography*, Vol. *1*, Longmuir, I. S. (Ed.), Pergamon, New York, 1960.
[209] Vydra, F., Štulík, K. and Juláková, E., *Electrochemical Stripping Analysis*, Horwood, Chichester, 1976.
[210] Brainina, Kh. Z., *Stripping Voltammetry in Chemical Analysis*, Wiley, New York, 1974.
[211] Barendrecht, E., in *Electroanalytical Chemistry*, Vol. *2*, Bard, A. J. (Ed.), Arnold, London, 1967.
[212] Charlot, G., Badoz-Lambling, J. and Trémillon, B., *Electrochemical Reactions*, Elsevier, Amsterdam, 1962.
[213] Kolthoff, I. M. and Elving, P. J., *Treatise on Analytical Chemistry, Part I*, Vol. *4*, Interscience, New York, 1963.
[214] Alber, H., *Mikrochemie*, **14**, 219 (1934).
[215] Brenneis, H. J., *Mikrochemie*, **9**, 385 (1931).
[216] Wiesenberger, E., *Mikrochim. Acta*, **1960**, 946.
[216a] Olm, D. D. and Stock, J. D., *Mikrochim. Acta*, **1977** II, 575.
[217] Rodden, C. J., *Analytical Chemistry of the Manhattan Project*, McGraw-Hill, New York, 1950.
[218] Handley, T. H. and Cooper, J. M., *Anal. Chem.*, **41**, 381 (1969).
[219] Lucas, R. and Grassner, F., *Mikrochemie (Emich Festschrift)*, **1930**, 197.
[220] DeFord, D. D., *Anal. Chem.*, **28**, 660 (1956).
[221] Umland, F. and Kirchner, K., *Z. Anal. Chem.*, **143**, 259 (1954).
[222] Haupt, G. and Olbrich, A., *Z. Anal. Chem.*, **132**, 161 (1951).
[223] Ballentine, R. and Burford, D. D., *Anal. Chem.*, **26**, 1031 (1954).
[224] Smelik, J., *Prakt. Chem. (Wien)*, **5**, 86, 95, 133, 253 (1954); **6**, 91, 120, 129 (1955).
[225] Bagdasarov, K. N. and Osmanov, Kh. A., *Zh. Analit. Khim.*, **24**, 296 (1969).
[226] Brainina, Kh. Z. and Kiva, N. K., *Zavodsk. Lab.*, **29**, 526 (1963).
[227] Mizuike, A., Mitsuya, N. and Yamagi, K., *Bull. Chem. Soc. Japan*, **42**, 253 (1969).
[228] Kowalczyk, M., *Chem. Anal. (Warsaw)*, **10**, 395 (1965).
[229] Vassos, B. H., Hirsh, R. F. and Letterman, H., *Anal. Chem.*, **45**, 792 (1973).
[230] Arden, J. W. and Gale, N. H., *Anal. Chem.*, **46**, 2 (1974).
[231] Majumdar, D. K. and Bhowal, S. G., *Anal. Chim. Acta*, **35**, 206 (1966).
[232] Majumdar, D. K. and Bhowal, S. G., *Anal. Chim. Acta*, **36**, 399 (1966).
[233] Majumdar, D. K. and Bhowal, S. G., *Mikrochim. Acta*, **1967**, 1086.
[234] Majumdar, D. K. and Bhowal, S. G., *Anal. Chim. Acta*, **38**, 468 (1968).
[235] Bilkiewicz, J., Szepke, R. and Malinowski, Z., *Report No. CLOR-31*, Warsaw, 1964.
[236] Czakow, J., *Chem. Anal. (Warsaw)*, **12**, 981 (1967).
[236a] Welch, K. H. and Ure, A. M., *Anal. Proc.*, **17**, 8 (1980).

5 Separation

Chapter 4

VOLATILIZATION OF SUBSTANCES

4.1 GENERAL REMARKS

Separation of the major or trace constituents from samples by evaporation has long been known. A classic example is the determination of trace impurities in distilled water after the water as the major constituent has been distilled off; another example is the determination of trace amounts of arsenic in a sample by converting it into volatile arsine which is then absorbed in a small volume of a reagent and determined.

The volatility of substances is related to molecular structure, i.e. to the nature of the chemical bonds that hold the constituent atoms together. All elements are volatile. The volatility of inorganic compounds generally increases with the degree of covalency in their bonding. Certain chelate complexes are also volatile.

In trace analysis of inorganic materials, the volatility of substances has been used only for separating either the matrix or the trace elements, depending on which is the more volatile: they are never separated by fractional distillation. For the separation to be quantitative, the volatility of the matrix must differ quite considerably from that of the trace elements.

These separations can be classified as follows.

(1) Direct distillation of one or more trace constituents from the sample. This technique is employed mainly in the trace analysis of metals of widely differing volatilities, and also of liquid samples (e.g. water, silicon tetrachloride).

(2) Conversion of the constituents of the sample into chemical species which can be separated by virtue of the difference in their volatilities. Two procedures may be distinguished: (*a*) treatment with a gaseous reagent at such a temperature that either the major constituent or the traces can be distilled off *in situ*; (*b*) dissolution of the sample in such a way that the constituents concerned are simultaneously converted into volatile compounds. The most common reagents for both procedures are fluorine, chlorine and bromine, or the hydrogen halides either as dry gases or in solution, because the halides of numerous elements are volatile.

Determination of inorganic traces in organic materials usually requires the sample to be ashed as a preliminary step. This removes the interfering

Table 4.1 Volatilization of elements as chlorides and/or bromides by distillation at 200–220°C from $HClO_4$ and H_2SO_4 solutions [1]

Element*	Approximate percentage volatilized by distillation with					
	$HCl–HClO_4$ (Procedure 1)	$HBr–HClO_4$ (Procedure 2)	$HCl–H_3PO_4–HClO_4$ (Procedure 3)	$HBr–H_3PO_4–HClO_4$ (Procedure 4)	$HCl–H_2SO_4$ (Procedure 5)	$HBr–H_2SO_4$ (Procedure 6)
As(III)	30	100	30	100	100	100
As(V)	5	100	5	100	5	100
Au	1	0·5	0·5	0·5	0·5	0·5
B	20	20	10	10	50	10
Bi	0·1	1	0	1	0	1
Cr(III)	99·7	40	99·8	40	0	0
Ge†	50	70	10	90	90	95
Hg(I),(II)	75	75	75	75	75	90
Mn	0·1	0·02	0·02	0·02	0·02	0·02
Mo	3	12	0	0	5	4
Os‡	100	100	100	100	0	0
P	1	1	1	1	1	1
Re	100	100	80	100	90	100
Ru	99·5	100	100	100	0	0
Sb(III)	2	99·8	2	99·8	33	99·8
Sb(V)	2	99·8	0	99·8	2	98
Se(IV)	4	2–5	2–5	2–5	30	100
Se(VI)	4	5	5	5	20	100
Sn(II)	99·8	100	0	99·8	1	100
Sn(IV)	100	100	0	100	30	100
Te(IV)	0·5	0·5	0·1	0·5	0·1	10
Te(VI)	0·1	0·5	0·1	1	0·1	10
Tl(III)§	1	1	1	1	0·1	1
V	0·5	2	0	0	0	0

* The following elements were not volatilized in any of the procedures employed: Ag, Al, Ba, Be, Ca, Cd, Co, Cs, Cu, Fe, Ga, Hf, In, Ir, K, Li, Mg, Na, Nb, Ni, Pb, Pd, Pt, Rb, Rh, the rare earths, Si, Ta, Th, Tc, U, W, Zn, Zr.

† During the heating, GeO_2 precipitated; if HCl or HBr was added to the solution before heating, GeO_2 did not precipitate and the germanium easily passed into the distillate.

‡ At 200–220°C no Os was volatilized from the H_2SO_4-containing solution, however, at 270–300°C, the osmium was completely volatilized.

§ No Tl(I) (from reduction) was found in the distillate.

Procedure 1: The test element (25–100 mg), in solution as its chloride or perchlorate, was added to the distillation flask of the Scherrer apparatus, then 15 ml of 60% $HClO_4$ were added, and the distillate was collected in a beaker containing 100 ml of distilled water. A moderate stream of dry CO_2 was admitted, the flask was heated

Table 4.1 (*continued*)

to 200°C, and conc. HCl was added slowly to maintain the temperature at 200–220°C. After 15 ml of HCl had been added within 20–30 min, the distillation was stopped.
 Procedure 2: Same as Procedure 1, except that 40% HBr was substituted for HCl.
 Procedure 3: Same as Procedure 1, except that 5 ml of syrupy H_3PO_4 were added before the distillation.
 Procedure 4: Same as Procedure 3, except that 40% HBr was substituted for HCl.
 Procedure 5: Same as Procedure 1, except that 15 ml of conc. H_2SO_4 were substituted for $HClO_4$.
 Procedure 6: Same as Procedure 2, except that conc. H_2SO_4 was substituted for $HClO_4$.

Table 4.2 Preconcentration of trace elements on distillation of 50 ml of concentrated HCl
(from [2], by permission of Polska Akad. Nauk)

Element	Concentration range investigated, %	Recovery, %
Sb(V)	1×10^{-6}–$1 \cdot 4 \times 10^{-4}$	97·9–100·7
Ga(III)	1×10^{-6}–$3 \cdot 6 \times 10^{-5}$	96·5–99·6
Cu(II)	$1 \cdot 4 \times 10^{-6}$–$7 \cdot 2 \times 10^{-5}$	98·8–100·5
P(V)	$1 \cdot 4 \times 10^{-6}$–$1 \cdot 4 \times 10^{-5}$	99·2–99·8
Cr(III)	$1 \cdot 4 \times 10^{-6}$–$2 \cdot 4 \times 10^{-5}$	99·6–100·3
Sc(III)	$7 \cdot 2 \times 10^{-9}$–$7 \cdot 2 \times 10^{-7}$	98·3–101·1
Au(III)	$7 \cdot 2 \times 10^{-8}$–$3 \cdot 6 \times 10^{-6}$	97·2–99·7
Te(IV)	$3 \cdot 6 \times 10^{-6}$–$3 \cdot 6 \times 10^{-5}$	99·3–100·6
Zn(II)	$1 \cdot 8 \times 10^{-6}$–$1 \cdot 8 \times 10^{-5}$	99·3–99·4
Nd(III)	$3 \cdot 6 \times 10^{-5}$–$7 \cdot 2 \times 10^{-5}$	98·5–99·0
Ce(III)	$3 \cdot 6 \times 10^{-5}$–$7 \cdot 2 \times 10^{-5}$	99·5–99·9
Co(II)	$1 \cdot 8 \times 10^{-8}$–$1 \cdot 8 \times 10^{-7}$	97·4–99·9
Pt(IV)	$1 \cdot 8 \times 10^{-5}$–$1 \cdot 8 \times 10^{-4}$	98·0–100·6
Fe(III)	$3 \cdot 6 \times 10^{-5}$	99·4
Sm(III)	$3 \cdot 6 \times 10^{-5}$	99·5
W(VI)	$1 \cdot 4 \times 10^{-5}$	97·0
Na(I)	$3 \cdot 6 \times 10^{-7}$	99·5
As(V)	$3 \cdot 5 \times 10^{-6}$–$3 \cdot 3 \times 10^{-5}$	99·0–100·2
As(III)*	$6 \cdot 5 \times 10^{-7}$–$3 \cdot 3 \times 10^{-5}$	8·9–11·8
As(V)*	$6 \cdot 5 \times 10^{-7}$–$1 \cdot 3 \times 10^{-5}$	99·8–100·2

* In this case the HCl was diluted 1:1 with water.

organic matrix and concentrates the inorganic trace elements. Ashing dry or wet, transforms the organic constituents of a sample into readily volatile compounds ($C \rightarrow CO + CO_2$; $H \rightarrow H_2O$, etc.) either by burning in air or oxygen (dry ashing) or by oxidizing in solution or suspension (wet ashing with concentrated nitric acid, sulphuric acid, perchloric acid, hydrogen peroxide, etc.). Formally, this procedure falls in category (2), but in view of its specific features, ashing will be discussed separately (Section 4.5).

Useful information enabling the outcome of a potential volatilization separation to be predicted, has been collected by Hoffman and Lundell [1] (Table 4.1). These authors studied the behaviour of the chlorides and bromides of most of the elements in distillation in the presence of conc. perchloric acid and conc. sulphuric acid. It is evident from Table 4.1 that arsenic, chromium, osmium, germanium, rhenium, ruthenium, antimony and tin can be completely distilled off, with only very few elements partially co-distilling and most elements remaining quantitatively in the residue. Hoffman and Lundell's studies relate to milligram quantities of the compounds, so their conclusions may not be valid for the recovery of trace quantities.

Using a special flask fitted with a reflux condenser, Wąsowicz and Rutkowski [2] distilled 50-ml portions of hydrochloric acid containing trace amounts of radioactive tracers of various elements. Recoveries of the trace elements in the 0·5-ml residue are shown in Table 4.2. For the quantitative recovery of traces it is essential that the residue be not less than about 0·5 ml in volume, because evaporation of the sample to near or complete dryness results in losses of the trace elements to be determined, on account of local superheating in the flask.

4.2 DIRECT DISTILLATION OF THE SAMPLE MATRIX

This procedure has been used primarily for analysing certain high-purity metals and high-purity volatile compounds. The first case may be exemplified by the separation of zinc [3], cadmium [4], selenium [5] and sodium [6]. These metals can be distilled at temperatures of 500, 400, 460 and 350°C, respectively; at these temperatures, most trace impurities contained in the metals are non-volatile. An example of the second case is the evaporation of mercury selenide [7]; in a quartz beaker, at 400°C, the selenide sublimes completely, and the trace impurities remain intact in the beaker.

If liquids containing inorganic trace impurities are to be analysed, the usual procedure is to add a trace collector before evaporation of the major constituent. Carbon is convenient as collector, because usually the traces collected are to be determined by emission spectrography. Examples are the determination of several impurities in concentrated hydrochloric, nitric and sulphuric acids [8], in trichlorosilane [9], and in germanium tetrachloride [10]. The primary role of the carbon here is that of collector rather than that of a carrier for the subsequent spectrography, because it physically collects the traces during the evaporation and thus enables them to be transferred for further operations, whether spectrographic or not.

Instances are known where evaporation of the major constituent results in partial evaporation of the trace constituent to be determined. One counter-measure is to add a substance to complex the trace constituent and prevent its escape. For example, to prevent losses in the determination of traces of boron in hydrogen fluoride (BF_3 is volatile), mannitol [11] is added. Similarly, triphenylchloromethane [12–16] has been used to retain boron in the determinations of traces of boron in silicon tetrachloride, trichlorosilane and germanium tetrachloride.

4.3 DISTILLATION OF THE SAMPLE MATRIX

The data in Table 4.1 show that several elements such as arsenic, chromium, germanium, osmium, rhenium, ruthenium, antimony and tin can be distilled from solution as the chlorides or bromides. Most impurities remain in the distillation residue. This procedure has frequently been applied in the analysis of arsenic [17], germanium [18], chromium [19–21], antimony [22], silicon [23] and selenium [24].

Gaseous reagents have only rarely been used to volatilize the matrix. Shvarts and Portnova [25] treated tin with chlorine gas to produce tin(IV) chloride which was subsequently distilled off, and the residue was analysed for trace constituents. Similarly, chlorine gas has been used for chlorination of titanium [26], and fluorine gas [27] has been used to volatilize titanium and titanium oxide. Volatilization of metal oxides with hydrogen fluoride gas by forming metal fluorides or oxyfluorides has been studied

Table 4.3 Hydrofluorination of oxides
(from [28], by permission of Nauchno-Tekhnicheskoe Izdatelstvo, Moscow)

Oxide	Temperature, °C	Sample weight, mg	Time for volatilization of the major constituent, min
B_2O_3	200	250	80
V_2O_5	250	300	60
WO_3	280	200	300
GeO_2	250	300	20
SiO_2	200	200	90
MoO_3	250	200	90
SeO_2	140	500	120
Sb_2O_3	350	200	30
Ta_2O_5	400	220	480 (residue, 20 mg)
TiO_2	350	300	180

systematically [28]. Table 4.3 lists the metal oxides that can be converted by hydrogen fluoride gas into, and distilled off as, fluorides. The behaviour of other elements, later to be determined as trace constituents, under these conditions is not known. The operation reported was carried out by passing hydrogen fluoride gas from a cylinder over the metal oxides heated in Teflon or graphite tubes in an electric oven.

Distillation has often been used for removal of boron. Boric acid yields a readily volatile methyl ester, and the apparatus described by Ehrlich and Keil [29] allows the methanol to circulate and thus reduces the amount used. Boron, as the matrix of a sample, has been volatilized with hydrogen fluoride and removed as boron trifluoride [30]. Thallium and sulphur have been volatilized by converting the former into a volatile oxide [31] and burning the latter to sulphur dioxide [32]. Several elements can be readily converted into volatile hydrides; thus, for example, arsenic [33] as a major constituent and phosphorus [34] in phosphides have been treated with hydrogen bromide and distilled off as arsine and phosphine, respectively.

Shvarts [35] developed a highly complicated method for removing nickel. The powdered nickel specimen is autoclaved with carbon monoxide at a pressure of 150 atm to convert the nickel into the readily volatile nickel carbonyl.

Certain metal chelate complexes are also volatile [35a–35d]. Gallium has been volatilized as the 8-hydroxyquinolinate [31]; similarly, beryllium has been removed as the acetylacetonate [36]. Examples of removal by distillation of sample matrices are listed in Table 4.4.

4.4 ISOLATION OF TRACE CONSTITUENTS

Geilmann and Neeb [50,51] have studied the isolation of traces of metals by direct distillation in a stream of hydrogen at 1000–1100°C and collection in a cooled receiver. They found that antimony, arsenic, bismuth, germanium, indium, cadmium, lead, mercury, thallium and tellurium were quantitatively distilled. Tin, gallium, and silver distilled only partially. They determined indium, cadmium, lead, and thallium at concentrations of 10^{-6}–$10^{-2}\%$ in minerals, oxides, dust, and metals. At 10^{-9}–$10^{-4}\%$ levels they determined thallium and lead in organic materials, salts, and solutions. They applied the method for special determinations such as traces of zinc in other metals, traces of scandium in copper [52], and traces of thallium in biological and other materials [53].

Trace elements can be isolated in elemental form by the spectrographic technique of 'carrier distillation', developed by Scribner and Mullin [54].

Table 4.4 Separation of major constituents by volatilization

Material	Major constituent distilled as	Temperature of distillation, °C	Traces determined in residue	Medium or procedure	Ref.
Acids, volatile, high-purity	HCl, HNO$_3$, H$_2$SO$_4$, CH$_3$COOH, etc.		Ag, Al, Bi, Co, Cu, Ga, Mg, Mn, Ni, Pb, Sb, Sn, Zn	evaporate to dryness with carbon as collector	[8]
Antimony metal	SbBr$_5$		Al	HBr	[22]
Antimony pentachloride	SbCl$_5$		See AsCl$_3$		[38]
Arsenic, semiconductor-grade	AsCl$_3$ or AsBr$_3$		Al, Bi, Cr, Cu, Fe, Mg, Mn, Ni, Pb, Sb	HCl or HBr	[17]
Arsenic trichloride	AsCl$_3$		Ag, Al, Au, Bi, Cd, Cu, Fe, Mg, Mn, Ni, Pb, Sb, Tl, Zn	distil with carbon powder in dry N$_2$ atmosphere	[38]
Beryllium	Be acetylacetonate		Al, Co, Cd, Cu, Fe, Ga, Mg, Mn, Ni, Sn, Tl, Zn		[36]
Bismuth	BiOBr		Mn, Mo, Ta, W		[43]
Boron, high-purity	BF$_3$			H$_2$F$_2$, after conversion into B$_2$O$_3$	[30]
Boron tribromide	BBr$_3$		See AsCl$_3$		[38]
Cadmium, high-purity metal	Cd	400	Ag, Bi, Ni, Pb, Sn, Tl, Zn		[4]
Chromium, high-purity metal	CrO$_2$Cl$_2$			HCl	[20,21]
Gallium	Ga 8-hydroxyquinolinate				[31]
Gallium arsenide	GaAs		Ag, Al, Ba, Ce, Cr, Cu, Fe, Ge, In, Mg, Mn, Ni, Pb, Sb, Si, Sn, Sr, Ti	distil from carbon electrode in argon atmosphere	[37]
Germanium, semiconductor-grade	GeCl$_4$		Ag, Al, Au, Bi, Ca, Cr, Cu, Fe, Ga, In, Mg, Mn, Ni, Pb, Sb, Sn, Ta, Ti, Tl, Zn	HCl	[18,39,40]
Germanium tetrachloride, high-purity	GeCl$_4$		Ag, Al, Ca, Cr, Cu, Fe, Mg, Mn, Ni, Pb, Sn	distil with carbon powder as collector	[10]

§ 4.4] Isolation of Trace Constituents 73

Hydrochloric acid, high-purity	HCl		As, Au, Ce, Co, Cr, Fe, Ga, Na, Nd, P, Pt, Sb, Sc, Sm, Te, W	methane evaporate to 0·5 ml	[14] [2]
Hydrofluoric acid, high-purity	H_2F_2		B	distil with mannitol	[11]
Indium	$InBr_3$	225	Ag, Al, As, Au, Bi, Cd, Co, Cr, Cu, Fe, Ga, Mn, Ni, Pb, Su, Te		[41]
Indium phosphide, semiconductor-grade	PH_3		Al, Cd, Ca, Mg, Mn, Ni, Pb, Zn	HBr	[34]
Iodine, high-purity	I_2		Ag, Al, Au, Ca, Cr, Cu, Fe, Mg, Mn, Ni, Pb, Sb, Sn, Ti, Zn	sublime, with active carbon as collector	[42]
Mercury selenide	HgSe	400	Ag, Al, Bi, Cd, Cu, Fe, Mg, Mn, Ni, Pb, Sb	sublimation	[7]
Nickel, high-purity metal	$Ni(CO)_4$		See $AsCl_3$	autoclave at 150 atm	[35]
Phosphorus trichloride	PCl_3		See $AsCl_3$		[38]
Phosphorus oxychloride	$POCl_3$		Ag, Al, Bi, Cd, Cr, Cu, Fe, Ga, In, Mo, Nb, Ni, Pb, Sb, Sn, Ti, V		[38]
Rubidium and caesium arsenates	AsH_3		Ag, As, Re	HCl-HBr	[33]
Selenium, high-purity	Se	460	Ag, Al, Bi, Cd, Co, Cu, Fe, Mg, Mn, Ni, Pb, Te	directly from crucible	[5]
Selenium, high-purity	SeO_2	250–260			[32]
Selenium	$SeBr_4$			$HBr-Br_2$	[24]
Sulphur	SO_2		B	burn with In_2O_3 as collector	[31]
Silicon, semiconductor-grade	$SiBr_4$		Ag, Al, Bi, Ca, Cu, Fe, In, Mg, Mn, Ni, Pb, Ti, Tl, Zn	HBr	[23]
Silicon, semiconductor-grade	SiF_4			H_2F_2	[45,46]
Silicon, semiconductor-grade	SiF_4		Cl, Br, I		[47]

Table 4.4 (continued)

Material	Major constituent distilled as	Temperature of distillation, °C	Traces determined in residue	Medium or procedure	Ref.
Silicon tetrachloride	SiF_4		B	distil with triphenylchloromethane	[12–16]
Sodium	Na		U	vacuum	[6]
Sulphur	SO_2	350	Ag, Al, Bi, Co, Cu, Fe, Ga, Mg, Mn, Pb	burn with In_2O_3 as collector	[32]
Tin metal	$SnCl_4$			HCl	[48]
Tin, high-purity metal	$SnCl_4$		Ag, Al, Au, Bi, Co, Ga, Ni, Pb	in stream of chlorine	[25]
Thallium	Tl_2O_3	500	Si	directly from crucible	[31]
Thallium	$TlNO_3$			vacuum	[44]
Titanium, high-purity metal	$TiCl_4$		Ag, Al, Au, As, Bi, Cd, Co, Cu, Fe, Ga, In, Ni, Pb, Sb, Sn, Te	in stream of chlorine	[26]
Titanium and titanium dioxide, high-purity	TiF_4		Al, Bi, Ca, Cd, Cr, Mg, Mn, Ni, Pb—complete retention; Fe—partial retention; Sn, V, Mo—volatilize with Ti	in hydrogen fluoride stream	[27]
Trichlorosilane, high-purity	$SiHCl_3$		Al, Ca, Cr, Cu, Fe, Mg, Mn, Ni, Pb, Sn	distil with carbon powder as collector	[9]
Trichlorosilane, high-purity	$SiHCl_3$		Ag, Al, Bi, Ca, Cr, Cu, Fe, Ga, In, Mg, Mn, Mo, Ni, Pb, Sb, Sn, Ti, Zn, Zr	distil with triphenylchloromethane	[16]
Water (arctic glaciers)	H_2O		B and Al, As, Ca, Co, Cu, Fe, Mg, Mn, Sb, Ti, V		[49]
Zinc, high-purity metal	Zn	500	Ca, Fe, K, Mg, Mn, Na Ag, Bi, Co, Cu, Ga, Ni, Pb, Sb, Sn		[3]

Gallium oxide is used as a carrier to promote distillation of trace impurities from uranium dioxide into the excitation region. The impurity elements are distilled from a crater in a graphite electrode.

In 1956, Mandel'shtam *et al.* [55] and Zaidel *et al.* [56] adapted the Scribner–Mullin method for distillation of trace impurities from uranium dioxide in an electric resistance furnace, and collected them on an auxiliary cooled graphite electrode which was then submitted to excitation. In this modification, the method can also be used for isolating trace elements and determining them by any other procedure.

The carrier distillation method has been further developed [57,58] by introducing other types of carrier such as silver chloride, which also acts chemically as a chlorinating agent. In all the pertinent studies, a spectrographic finish was used, and carriers were considered in terms of their role in the distillation and in the subsequent excitation process. Distillation in a stream of air or other gas has frequently been used for separation of traces of mercury [59,60].

Elements which produce volatile covalent oxides can readily be isolated in this form, e.g. traces of osmium [61,62], ruthenium [63–65] and rhenium [66]. Strickland and Spicer [67] distilled traces of manganese as permanganic acid from $10M$ sulphuric acid containing potassium periodate.

Traces of sulphur and sulphur compounds can readily be distilled off after they have been converted into sulphur dioxide [68] or hydrogen sulphide [69–71]; the kinetics of the latter process have been studied [72].

Traces of nitrogen can be readily converted into, and distilled as, ammonia [73–76]. Nitrates and nitrites are first reduced to ammonia [77], and organic nitrogen is transformed into the ammonium ion by the Kjeldahl procedure [73,74,78].

Certain elements with volatile hydrides can be distilled off as such [79,80]. The classic example is the isolation of arsenic as arsine, used in the long-established Marsh test [81]. A modification of this method, known as the Gutzeit method [82], has been applied to the determination of arsenic in minerals [83], forensic specimens [84], foodstuffs, household articles [85], and other materials [86–89]. Antimony has also been distilled as its hydride [90,91], but the quantitativeness of the reaction has been questioned [92]. Recently, sodium borohydride and other reductants have been extensively used for producing arsine, stibine etc. followed by their measurement by atomic-absorption spectrophotometry [92a–g].

Any element forming a volatile halide can be distilled in trace amounts. Thus, traces of arsenic have been distilled as the bromide [93] or chloride [94], of germanium as the chloride [95], and of selenium in plant materials as the bromide [96]. The volatile oxychlorides have been used for the isolation of chromium and vanadium [97], and volatile fluorine compounds for the isolation of molybdenum [98] and silicon [99,100]. Vacuum evaporation has been used for separating lead as $PbCl_2$ and bismuth as BiOCl [101].

Boron can be separated in both major and trace amounts by distillation as methyl borate [102–104].

Distillation is also used to isolate traces of halide from solutions. Marczenko and Chołuj-Lenarczyk [105] have used this method to isolate trace amounts of fluorides and shown that depending on the conditions fluorine may distil as hydrogen fluoride rather than as fluorosilicic acid or silicon tetrafluoride, as claimed by many earlier investigators. Traces of chloride and bromide can be boiled off by heating, usually from sulphuric acid medium. Bromide and iodide can be distilled after oxidation to bromine and iodine. The differences between the redox potentials of chloride, bromide and iodide allow the halides to be oxidized selectively, starting with iodide and next bromide, each halide being removed after the particular oxidation step has been completed [106].

Finally, trace amounts of carbon can be volatilized and isolated. Typically, carbon is oxidized to carbon dioxide [107] or carbon disulphide [108] and expelled as such. Obviously, trace amounts of volatile carbon derivatives such as cyanides and thiocyanates can be distilled off as hydrogen cyanide [109] and thiocyanic acid, respectively.

Examples of distillative isolation of trace elements are given in Table 4.5.

4.5 ASHING OF ORGANIC MATRICES

Determination of inorganic constituents in organic materials requires removal of the organic matter, because it would either interfere with, or simply preclude, analytical reactions. The simplest way to remove organic matter is to ash or oxidize it. Carbon and hydrogen are oxidized to carbon dioxide (or monoxide) and water, and organic nitrogen is mainly liberated as free nitrogen. Since all the products are gases, ashing may be considered to be a method based on volatilization.

Ashing techniques have a very long tradition and, according to some authors, Fresenius and Babo [117] are due the credit for the first publication in the field. As early as 1844 they described the decomposition of animal tissue for the determination of inorganic trace constituents. It should be emphasised straight away that the very old ashing techniques have scarcely been modified to this day. Although the process is very simple and apparently offers no technical difficulties, in practice there is no one general

Table 4.5 Separation of trace impurities by distillation

Trace species	Distilled as	Sample or procedure	Ref.
Antimony	SbH_3	metals	[90]
Arsenic	AsH_3	rocks	[91]
		generally applicable	[79,80]
		minerals	[83]
		forensic	[84]
		foodstuffs, household articles high-purity zinc	[85]
		semiconductor-grade germanium	[86,87]
		others	[79,80, 88,89]
	$AsBr_3$	plant material	[93]
	$AsCl_3$	generally applicable	[111]
		generally applicable	[94]
Bismuth	Bi in hydrogen stream, 1000–1100°C	rocks, metals, oxides, etc.	[50]
	BiH_3	generally applicable	[79]
Boron	$B(OCH_3)_3$		[112,113]
Bromide	Br_2	generally applicable (separation from I^- and Cl^-)	[106]
Cadmium	Cd in hydrogen stream, 1000–1100°C	minerals, oxides, dust, metals	[50]
Carbon	CO_2	metals	[107]
	CS_2	semiconductor-grade silicon and germanium	[108]
Chloride	Cl_2	generally applicable (separation from I^- and Br^-)	[106]
Chromium	CrO_2Cl_2	separation from titanium	[97]
Cyanide	HCN	generally applicable	[109]
Fluoride	HF	generally applicable	[105]
Germanium	$GeCl_4$	bituminous coal	[95]
	Ge in hydrogen stream, 1000–1100°C	generally applicable	[50]
Indium	In in hydrogen stream, 1000–1100°C	minerals, oxides, dust, metals	[50]
Iodide	I_2	generally applicable (separation from Br^- and Cl^-)	[106]
Lead	Pb in hydrogen stream, 1000–1100°C	minerals, dust, metals, biological materials, salts	[50]

Table 4.5 (*continued*)

Trace species	Distilled as	Sample or procedure	Ref.
Manganese	$HMnO_4$	generally applicable	[67]
	Mn in hydrogen stream	high-purity aluminium	[115]
Mercury	Hg in hydrogen stream	generally applicable	[50]
Molybdenum			[98]
Nitrogen	NH_3	generally applicable	[73–76]
		ferro alloys	[114]
		after reduction of nitrates and nitrites	[77]
		after Kjeldahl digestion	[78]
Osmium	OsO_4	platinum metals	[61,62]
Rhenium	ReO_4	generally applicable	[66]
Ruthenium	RuO_4	generally applicable	[63,64]
	RuO_4	ores	[65]
Scandium	Sc in hydrogen stream, 1000–1100°C	copper	[52]
Selenium	$SeBr_4$	plant material	[96]
	SeH_4	generally applicable	[79,80]
Silicon	SiF_4	uranium, plutonium	[99]
		in solids	[100]
Sulphur	H_2S	generally applicable	[69–71]
	SO_2	copper	[68]
Tellurium	Te in hydrogen stream		[50]
Thallium	Tl in hydrogen stream	generally applicable	[50]
Vanadium	$VOCl_2$	separation from titanium	[97]
Zinc	Zn	generally applicable	[52]
	in hydrogen stream, 1000–1100°C	high-purity aluminium	[116]

procedure which is satisfactory for all kinds of organic and biological samples.

Like any other chemical process used in trace analysis, ashing should meet the following requirements.

(*i*) It should be quantitative: all the organic matter should be oxidized and volatilized, and the inorganic portion should remain quantitatively as the residue.

(*ii*) It should be rapid: in routine analyses of large numbers of samples the rate of ashing greatly affects the time and cost of the analysis.

(*iii*) It should be feasible in a simple and inexpensive apparatus.

(*iv*) No trace constituent must be lost from or introduced into (as an extraneous impurity) the sample examined.

No method developed so far meets all these requirements, nor is any new method likely to. Each modification of the existing methods must be regarded as a reasonable compromise between the requirements specified above. A vast number of papers have been published on this seemingly simple and apparently well-mastered process.

There are two fundamental techniques, viz., dry ashing and wet ashing (wet digestion).

Dry ashing, at its simplest, involves burning the organic portion of a sample in air. The sample is placed in a suitable vessel, e.g. a crucible, and heated either by a flame or in a muffle furnace. The organic matter is usually first charred and then the char is burned to leave an ash which consists of the inorganic constituents of the sample; these may retain some products of combustion of the organic matter, e.g. carbon dioxide in the form of carbonate. Combustion is sometimes facilitated by using oxygen-enriched air or pure oxygen. Even then, as is well known from organic elemental analysis, thermally stable compounds (e.g. between carbon and silicon or boron) may be formed from the matrix elements during combustion.

To overcome difficulties in burning the char formed in the first stage of combustion of organic substances, various oxidants are added to the sample (before or during combustion), e.g. nitric acid, sulphuric acid, ammonium nitrate, magnesium nitrate, to promote the process.

Special additives may be used to prevent losses of certain constituents on ashing. For example, calcium oxide prevents volatilization of boron by forming the borate. Ashing carried out in this way will be referred to as additive-modified dry ashing.

To avoid the problem of losses during dry ashing wet ashing is used instead. The most important liquid oxidants are conc. nitric acid, conc. sulphuric acid, and perchloric acid, used alone or in combination. Hydrogen peroxide has also been used alone, or together with sulphuric acid or nitric acid. Other oxidants are used only occasionally, usually to complement the major oxidants mentioned above.

Gorsuch's monograph on ashing of organic substances [118] and another relevant paper [119] report the following figures concerning the bibliography of ashing. Of some 250 papers, 49% were concerned with dry ashing and 51% with wet ashing.

More specifically, the distribution was: dry ashing—20%, wet ashing ($HNO_3 + H_2SO_4$)—14%, wet ashing ($HNO_3 + H_2SO_4 + HClO_4$)—12%, dry ashing with H_2SO_4 added—7%, dry ashing with HNO_3 added—6%, wet ashing ($HNO_3 + HClO_4$)—5%, dry ashing with Mg compounds added—5%, various methods with H_2O_2 added—5%.

None of the other ashing procedures formed more than 2% of the total.

As mentioned earlier, the volatility of elements and their compounds is the major cause of error in the application of ashing procedures in trace analysis. Substances may escape during simple or additive-modified dry ashing, which may require heating of the sample to 400–500°C, and also during wet ashing, which often involves heating to 250–270°C. Very often the composition of the sample favours the formation of volatile compounds, e.g. chlorine present in the organic matter may result in formation of a volatile chloride of the constituent to be determined, and carbon may reduce metal compounds to the free metals, many of which are rather volatile at the temperatures concerned.

Gorsuch [119] studied the recovery of traces of the 14 most frequently determined elements in the ashing of 2 g of cocoa beans by eight selected procedures. The procedures used included the first seven systems listed above, and the eighth was a hybrid procedure developed by Middleton and Stuckey [120].

Dry ashing was done at 550–560°C in a 500-ml quartz flask kept in an asbestos box which was heated by two gas burners. The flask was equipped with a quartz delivery tube to lead the volatile products into suitable absorption vessels. A gentle stream of air was allowed to pass through the whole apparatus during the ashing.

The additives used by Gorsuch included conc. nitric acid, added after the sample had charred, 5N sulphuric acid, and 7% magnesium nitrate solution added before ashing was started.

The Middleton–Stuckey ashing procedure was carried out in a 1-litre tall-form beaker; 10–20 ml of water and 5–10 ml of conc. nitric acid containing 5% of sulphuric acid were added to the sample. The mixture was evaporated to dryness on a hot-plate, the residue was moistened with conc. nitric acid and re-evaporated to dryness under a watch-glass, and the procedure was repeated until a white residue was obtained.

The wet ashing procedures with acid mixtures were carried out by adding the following volumes (ml) of the acids to 2 g of cocoa beans:

	HNO$_3$ (s.g. 1·42)	H$_2$SO$_4$ (s.g. 1·84)	HClO$_4$ (s.g. 1·54)
HNO$_3$ + HClO$_4$	15	—	10
HNO$_3$ + H$_2$SO$_4$ + HClO$_4$	15	5	10
HNO$_3$ + H$_2$SO$_4$	15	10	—

The cocoa beans and the acids were placed in a 500-ml conical flask fitted with a splash-head and adaptor. The distillate was collected for examination.

The flask was heated to start the reaction and then heated on a hot-plate until the ashing was complete. The flask was cooled, 5 ml of fuming nitric acid were added, and the flask was reheated until nitrogen oxides were no longer evolved.

A flask equipped with a reflux condenser (Fig. 4.1) was used in some of the studies.

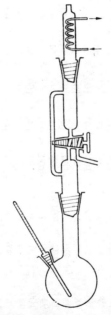

Fig. 4.1 Bethge's apparatus for distillation

Cocoa beans were chosen because they contain a wide variety of organic substances (proteins, fats, etc.). Recovery was studied by the radioactive tracer technique. The main results of the study are summarized in Table 4.6.

Table 4.6 Recovery of trace elements in cocoa analysis, after various ashing procedures
(from [119], by permission of the copyright holders, the Chemical Society)

Element	Percentage recovered by ashing procedure								Remarks
	HNO$_3$+HClO$_4$	HNO$_3$+HClO$_4$+H$_2$SO$_4$	HNO$_3$+H$_2$SO$_4$	Middleton-Stuckey	dry	dry +HNO$_3$	dry +H$_2$SO$_4$	dry +Mg(NO$_3$)$_2$	
Pb	100, 100	99, 93	90, 93	99, 101	100, 98 / 97, 99	99, 97	99, 99	98, 98	450°C
Hg*	78, 80	87, 90	93, 92		94, 95 / 71, 83	97, 99 / 83, 69	95, 96 / 96, 90	92, 94 / 91, 96	550°C 650°C
Zn	98, 100	101, 94	101, 99	97	0 / 96	97	100	99	no loss up to 900°C
Se	100, 101	100, 98	78, 80	1, 05	ND†	ND	ND	ND	
As	99, 99	98, 100	98, 97	91, 98	88	84	96	99	
Cu	98, 100	99, 99	101, 99	100, 99	86	94	96	98	
Co	98, 95	100, 98	99, 99	100, 99	99	99	99	101	
Ag	100, 101	99, 99	100, 100	94, 96	93, 99	87	97	100	
Cd	99, 94	102, 101	103, 100	99, 98	91	76	92	78	
Sb	100, 100	100, 100	100, 99	95, 98	96	92	94	97	
Cr	96, 98	101, 100	99, 101	100, 101	98	99	99	92	
Mo	96, 98	99, 97	101, 101	97, 99	99	98	100	98	
Sr	100, 98	97, 96	96, 100	98, 94	97	97	100	100	
Fe	97, 100	99, 99	102, 101	98, 95	99	101	100	100	

* Wet ashing carried out in Bethge's apparatus (Fig. 4.1).   † ND = not determined in view of negative results by the Middleton–Stuckey procedure.

Table 4.7 Recoveries of trace elements after dry and wet ashing
(from [121], by permission of the copyright holders, Pergamon Press Ltd.)

Ashing	Percentage recovered													
	Ag	As	Au	Co	Cr	Cu	Fe	Hg	Mn	Mo	Pb	Sb	V	Zn
Dry (24 hr at 400°C)	65	23	19	100	100	100	86	<1	100	100	100	67	100	100
Wet: HNO$_3$+HClO$_4$+H$_2$SO$_4$ (no reflux condenser)	100	94	65-100	100	100	100	85-98	24-84	100	100	100	94-99	100	100

Note: the tracer technique was used in the analyses.

The conclusions that may be drawn from these results, supplemented by our own observations [120a], are as follows.

(1) Dry ashing is most effective at 500–550°C (this temperature range has been recommended by numerous other investigators). Nevertheless, it is always necessary to check that none of the elements to be determined will escape under these conditions.

(2) Additives such as nitric acid and magnesium nitrate do not affect the results significantly.

(3) Addition of sulphuric acid prolongs the time of ashing.

(4) In wet ashing the best results are obtained with a mixture of nitric and perchloric acids; a mixture of nitric and sulphuric acids and the Middleton–Stuckey method give the worst results.

(5) Wet ashing is best performed in an apparatus equipped with a reflux condenser (Fig. 4.1). If mercury is present, this is the only technique which gives acceptable results.

For purposes of comparison, Table 4.7 lists results obtained by Pijck et al. [121] somewhat before Gorsuch's work. In general, the results are consistent with Gorsuch's, but more detailed study shows that Pijck et al. sometimes obtained rather variable results (e.g. recoveries of gold and mercury after wet ashing of 65–100% and 24–84%, respectively). Gorsuch obtained 99% recoveries of iron whereas Pijck et al. obtained only 85–98%. This illustrates the necessity for always checking that a proposed procedure will give complete recoveries of the traces to be determined.

Kowalczuk [122] demonstrated this in his study of dry and wet ashing procedures with six foodstuff samples, in which Ca, Fe, Mg, Mn, and Zn were determined. In the ten laboratories involved in the comparison test, ashings were carried out by two agreed procedures but highly disparate results were obtained.

One way of avoiding the losses by volatilization that occur in open-tube wet digestions is to use a closed system. Sealed glass tubes have long been used, but they have many practical disadvantages. More recently, the Teflon-lined steel or aluminium bombs originally devised for pressurized acid-digestion of geological samples [122a–d] have also been used for destruction of organic matter [122e–g].

4.5.1 Simple and Additive-Modified Dry Ashing

In dry ashing, whether additive-modified or not, in addition to the losses caused by volatility, there may be losses by reactions of the constituent to be determined, with the material of the vessel in which combustion is carried out. Numerous metals are known react with platinum vessels, and it should be borne in mind that, as mentioned earlier, heating with carbon provides conditions favourable to reduction to free metals. In quartz and porcelain vessels, the constituent to be determined can react with the vessel surface to yield a difficultly soluble silicate.

Gorsuch [119] experimented with quartz vessels and found that considerable amounts of gold and silver were retained in the presence of chloride and nitrate. Such losses have also been reported for copper [123] and iron [124]. In the dry ashing process, copper can readily be reduced to the metal and this will fail to pass into solution [125] if the ash is subsequently treated with hydrochloric acid (a very frequent procedure).

Ashing in a muffle furnace may result in contamination of the sample with volatile constituents of the furnace materials. In particular, possible contamination with boron must be borne in mind [126].

Boppel [127] has conducted extensive studies on the acceleration of dry ashing procedures as applied to routine analyses. He developed a special temperature regime which affords rapid and quantitative results. Although his procedures relate to the determination of trace amounts of radioactive isotopes, they are also applicable to other trace elements.

Combustion can be accelerated by the use of oxygen, in an oxygen bomb (as in the bomb calorimeter). Objections to this are based on the difficulties in quantitative removal of the traces from the bomb and possible contamination of the sample with the bomb material [128]. Better results are obtained by the Schöniger method [129–131] which involves burning of a sample in a quartz or glass flask filled with oxygen at atmospheric pressure [132].

As already indicated, ashing can be accelerated by various additives introduced before or during the process. According to Gorsuch [119], sulphuric acid as an additive protracts the ashing time; nevertheless, it continues to be used because of the non-volatility of sulphates. This is of particular importance for samples containing halogens and trace constituents convertible into a volatile form. Thus, for example, addition of sulphuric acid is recommended in ashing of plant and biological samples for the determination of cadmium [133].

The most common additives include nitric acid, usually added after the charring stage, and nitrates, added before ashing. With either of these additives, the sample often begins to burn or glow once a certain temperature has been attained. To avoid losses, the temperature must be raised slowly to prevent too vigorous a reaction. Nitric acid is usually added after the sample has charred and the char has partially burnt. Moistening the ash containing a residual char promotes oxidation of the carbon [134]. Tests have been made with nitric acid added before heating [135], and, as alternatives to nitric acid, ammonium nitrate [136], sodium nitrate–potassium nitrate mixtures [137], magnesium nitrate [138,139] (sometimes fortified with magnesia [140]), and calcium nitrate [141] have been suggested. Additives are intended to combine with some elements to form non-volatile salts stable at the ashing temperature. Such is the case, for example, in the determination of lead, which is readily reducible to the relatively volatile metal. To prevent losses, addition of sulphuric acid [142] or trisodium phosphate [143] has been suggested.

For extremely volatile metallic elements, e.g. mercury, special procedures have been developed [144]. When chlorides are present in considerable amounts, it is advisable to char the sample very slowly at temperatures lower than 420°C and to ash at that temperature overnight with magnesium oxide, magnesium nitrate or sodium carbonate added to avoid losses of iron, zinc, tin and antimony [145].

Dry and additive-modified dry ashings are lengthy procedures, but if a verified temperature regime is used, many samples can be readily ashed together in appropriate muffle furnaces with little labour. Thus, ashing continues to be broadly applied, as evidenced by Gorsuch's statistics.

4.5.2 Wet Ashing (Wet Digestion)

As already indicated in Section 4.1, wet ashing was introduced to avoid losses due to volatilization of trace constituents. The procedure is not completely safe, however. In determinations of mercury and of trace elements yielding volatile chlorides in chlorine-containing samples, considerable losses may occur if digestion is carried out in an open vessel. Losses may also be caused by precipitation of insoluble compounds as a result of reactions with the oxidant; e.g. tin [146] in a sample digested with nitric acid precipitates as insoluble metastannic acid. Similarly, from plumbiferous samples wet ashed in the presence of sulphuric acid, lead may be lost as sparingly soluble lead sulphate, especially when the sample contains large amounts of calcium (coprecipitation with calcium sulphate).

In wet ashing, it is important to maintain a suitable heating regime, especially if nitric acid is used, since it is much more volatile than sulphuric acid or perchloric acid. If the heating is too intense, nitric acid will distil off before it oxidizes the sample. On the other hand, sulphuric acid and perchloric acid work efficiently only at relatively high temperatures. The time required for the wet-ashing process depends on the temperature regime.

The classical wet-ashing procedures, viz., the Carius [147] and Kjeldahl [148] methods, are not used in trace analysis, the former because of the extremely small amounts of the substance that can be ashed and the latter because of the necessity of using catalysts which are likely to contaminate the sample.

The last important factor in wet ashing is the chemical composition of the sample. A variety of mechanisms of decomposition and oxidation are possible. Aromatic hydrocarbons, fats, proteins, and organic compounds containing heterocyclic nitrogen are difficult to ash. When treated with sulphuric acid and nitric acid, many of these compounds are sulphonated or nitrated and in this form they can withstand prolonged heating. Because of this, ashing procedures often recommend a preliminary charring step, followed by the actual wet ashing process. It may be generally stated that a clear solution obtained on wet ashing is by no means proof of complete ashing. Also, stable sulpho- or nitro-derivatives formed during ashing may have complexing properties, and thus hinder the determination of trace elements.

Nitric acid is relatively seldom used alone for ashing. It is used in hybrid treatments involving dry and wet ashing, e.g. in the determination of lead in tissue [149], bismuth in tissue [150], and trace elements in bone [151]. Nitric acid containing ammonium nitrate has also been used [152]. Nitric acid containing potassium permanganate [153] has been used for determining traces of mercury in plants. However, nitric acid vapour has been found to oxidize organic plant material conveniently and rapidly [153a].

Most frequently, a *sulphuric acid–nitric acid mixture* is used for ashing. The usual procedure is to start with sulphuric acid alone and, after the sample has charred, to add nitric acid to obtain a colourless solution.

This procedure has been recommended for application to the destruction of most organic materials, by the British Analytical Methods Committee [145,154,154a], except for the determination of traces of mercury, in which case ashing must be carried out under a reflux condenser. An exact and relatively rapid procedure for this type of ashing has been reported by Whalley [155]. Potassium sulphate [156] has been added to the sulphuric acid–nitric acid mixture when an increased ashing temperature is desired. Because of the difficulties in the wet ashing of urine for determination of lead, sulphuric acid has been replaced by nitrosylsulphuric acid and nitric acid is used as additive [157].

Kahane [158] was the first to report on the use of *perchloric acid* for ashing. Perchloric acid is useful because its oxidation potential is higher than that of sulphuric acid. At the same time perchlorates are, in general, more soluble than sulphates, so losses of trace constituents as sparingly soluble compounds are less likely to occur.

A considerable disadvantage of perchloric acid is that it may decompose violently when brought into contact with organic matter at elevated temperatures. The explosion power is high and the resulting hazard is so serious that there are official reports [159] warning against the use of perchloric acid alone for ashing. The explosion hazard is eliminated or at least considerably reduced when perchloric acid is used in mixture with nitric or sulphuric acid.

Smith [160] has extensively reviewed ashing procedures involving *nitric acid–perchloric acid mixtures*. Smith—the well-known specialist in the application of perchloric acid for ashing—has also established suitable temperature and concentration conditions for ashing with perchloric acid on its own [161]; under these conditions, he claims, there is no chance of violent decomposition of perchloric acid. Smith used perchloric acid for ashing organic materials such as cellulose, proteins, sugars, etc.

To return to the application of nitric acid–perchloric acid mixtures, it is usually recommended that ashing be started with nitric acid (to decompose the more readily oxidizable constituents) and continued with perchloric acid at a temperature progressively increased to achieve complete decomposition and total oxidation of organic substances.

Smith recommends that the mixture be used in a conical flask equipped with a special splash-head delivery tube to lead the evolved gases and vapours through a water-containing absorption vessel into an exhaust hood (Fig. 4.2). For samples containing volatile elements (Hg) to be ashed with perchloric acid alone or with a nitric acid–perchloric acip mixture, Smith recommends the use of Bethge's apparatus, shown iu Fig. 4.1. This apparatus ensures retention of volatile substances and also enables a specified temperature to be maintained.

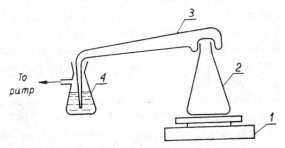

Fig. 4.2 Apparatus for wet-ashing with nitric–perchloric acid mixture: 1—hot plate, 2—conical flask, 3—splash head and delivery tube, 4—water-filled absorption vessel (from [160], by permission of the copyright holders, Elsevier Publ. Co.)

If nitric acid is replaced by *sulphuric acid* in the mixture with *perchloric acid*, the reaction temperature is considerably increased and a practically anhydrous medium can be achieved. Smith and Diehl [162,163] have recommended this mixture as a very powerful oxidant; it will rapidly ash materials such as alkaloids, plastics, bituminous coal, active carbon, etc., which are otherwise very difficult to ash. For further acceleration of the process, they advise vanadium (ammonium vanadate) as catalyst [164].

The oxidizing action of the mixture can be illustrated by the oxidation of a 3 g sample of anthracite in a Bethge apparatus. To a sample of anthracite (3 g) were added 18 ml of conc. sulphuric acid and 15 ml of 70% perchloric acid, and 2 mg of vanadium as ammonium vanadate. The temperature of the mixture was 42°C. The mixture was heated rapidly to 202°C in 5 min. The whole contents of the flask blackened and some froth appeared on the liquid surface. After 18 min the mixture had a temperature of 205°C and began to clear; after 23 min (still at 205°C), it turned greyish green. At this moment the reflux from the condenser was shut off and heating was continued as before. After 60 min, the temperature was 233°C and the transparent solution turned orange-brown, a colour characteristic of the resulting pervanadic acid: this indicated that ashing was complete. Evidently, this technique is quite efficient and lends itself to widespread application.

It is also worth emphasising that the sulphuric acid–perchloric acid mixture can be used for rapid and complete ashing of numerous heterocyclic nitrogen compounds such as quinoline, 8-hydroxyquinoline, phenanthroline derivatives and other reagents frequently used for separation of trace elements as their complexes (cf. Chapter 3) which are known to be extremely difficult to ash. Again, plastic materials, e.g. Nylon, Dacron, Orlon, rubber and the like, have also been ashed. For materials extremely difficult to ash, Smith and Diehl [165] recommend the use of sulphuric acid followed by nitric acid and finally by perchloric acid to complete the ashing.

An ashing procedure in which all three acids are used, nitric acid being added first followed by a sulphuric acid–perchloric acid mixture, is the best and most universal [166]. If results are to be achieved rapidly, a suitable temperature programme and suitable proportions of the acids must be established.

Hydrogen peroxide has long been known and applied as an additive to promote ashing. It has been used with nitric acid [167] and sulphuric acid [168]. Hydrogen peroxide often prevents frothing of the solution. Usually ashing is started by adding first a concentrated mineral acid and then aqueous 30% hydrogen peroxide progressively until the ashing is completed.

Attempts have been made to use more concentrated peroxide solutions, e.g. 50% or stronger. Such solutions do oxidize rapidly, but very often the reaction proceeds vigorously and loss of the substance to be ashed or even an explosion is likely to occur. However, 50% hydrogen peroxide does have certain advantages, and when used in mixtures with sulphuric acid it provides an effective means of destroying organic matter [169].

Sansoni [170] has described an interesting mild ashing procedure, involving the use of hydrogen peroxide along with iron(II) as catalyst for decomposition of the H_2O_2 into HO· radicals, which have a very high oxidation potential. Strongly oxidizing HO_2· radicals can also form. These radicals attack the organic substance present in the solution and give rise to chain oxidation reactions. In this manner Sansoni and Kracke [170] have ashed various samples of meat, bone, blood, vegetables, milk, sugar, cellulose, hay, wool, urine, and also organic reagents such as EDTA, 8-hydroxyquinoline, and glycerol. To a nearly boiling solution slightly acidified with nitric acid they added hydrogen peroxide dropwise to ash each of these materials completely within a period of time depending on the weight of sample. Several such ashings are listed in Table 4.8.

Table 4.8 Wet ashing with H_2O_2–Fe^{2+}
(from [170], by permission of the copyright holders, Springer-Verlag)

Material	Sample weight, g	Total consumption of 30% H_2O_2, ml	Ashing time, hr	Dry residue on ashing,		Residue on ignition at 800°C
				g	%*	%†
Venison	100	400	4	3·2	3·2	34
Dried blood	25	470	5	0·4	1·6	69
Calf bone	155	500	16	56·0	36·0	92
Powdered milk	50	500	4	3·8	7·6	72
Cellulose (filter paper)	45	1200	16	4·7	10·4	47
Hay	50	500	8	8·5	17·0	40
Wool	50	700	20	5·5	11·0	12
Urine	160	150	4–5	4·0	2·5	54
EDTA	25	250	6	0·05–0·1	0·2–0·4	—
8-Hydroxyquinoline	25	400	7	0·05	0·2	—
Glycerol	25	300	4	0·05	0·2	—

* Based on sample weight.
† Based on wet ashed dry-residue weight.

Table 4.8 shows that the procedure is rather slow, and that a lot of hydrogen peroxide is required. However, the mild ashing conditions and

Table 4.9 Ashing procedures applicable to individual trace elements

Element	Ashing	Ashing aid* (when added)	Temperature, °C	Recovery	Reference
Ag	dry	M (before ashing)	550	complete	[119]
	wet	N+S+P	ca. 200	complete	[119,121]
		N+S		complete	[119]
As	dry	M (before ashing)	550	complete	[119,140]
	wet	N+S+P, N+P		complete	[119]
				94%	[121]
				losses	[145]
Au	wet	N+S+P		complete only with refluxing	[121]
Bi	dry	N (during ashing)		complete	[150]
Ca	dry	0	550		[122]
	wet	N+P			[122]
Cd	dry	S (before ashing)	550		[119,133]
	wet	N+S, N+P, N+S+P		complete	[119]
Co	dry	0, N, S, M	550	complete	[119,121]
	wet	N+S, N+P, N+S+P		complete	[119,121]
Cr	dry	0, N, S, M	550	complete	[119,121]
	wet	N+S, N+P, N+S+P		complete	[119,121]
Cu	dry	0, N, S, M	400–550		[119,120, 123,125, 171]
	wet	N+S, N+P, N+S+P		complete	[119,121]
Fe	dry	0, N, S, M	400–550		[119,121, 122,129]
	wet	N+S, N+P, N+S+P			[145,171, 119,121]
Ge	dry		560–600	complete	[172]
Hg	dry			losses	[119,121]
	wet	N+S, N+P, N+S+P		78–92% (with reflux condenser)	[119,121, 153,154]
	combined				[144]
Mg	dry		550		[122]
	wet	N+P			[122]
Mn	dry	0	400–550	no loss	[121,122, 171,173]
	wet	N+P, N+S+P		no loss	[121,122]
Mo	dry	0, N, S, M	400–550	complete	[119,121]
	wet	N+S, N+S+P		complete	[119,121]
Pb	dry	0, N, S, M	400–450	complete	[119,121, 139,142]
				(data given) losses vary with temperature	[119,139, 174]

Table 4.9 (continued)

Element	Ashing	Ashing aid* (when added)	Temperature, °C	Recovery	Reference
	dry	Ca(NO$_3$)$_2$	500	complete	[141]
	dry	Na$_3$PO$_4$		complete	[143]
	wet	N+P, N+S+P		complete	[119,121]
		HNO$_3$+dry			[149]
	wet	nitrosylsulphuric acid +HNO$_3$		complete	[157]
b	dry	0, N, S, M	400–500	67–97%	[119,122, 140,145]
	wet	N+S, N+P, N+S+P		94–100%	[119,121]
e	dry		550	losses	[119]
	wet	N+P, N+S+P		complete	[119]
n	dry			prevent losses caused by Cl$^-$	[138,145]
	wet	HNO$_3$		losses	[146]
r	dry	S, M		complete	[119]
	wet	N+S, N+P, N+S+P		complete	[119]
	dry	0	400	complete	[121]
	wet	N+S+P		complete	[121]
n	dry	S, M	550	complete	[119,122]
	dry	0	400	complete	[121]
	dry			prevent losses caused by Cl$^-$	[145]
	wet	N+S, N+P, N+S+P		complete	[119,121, 122]

* Key: 0—no acid added, N—nitric acid, S—sulphuric acid, P—perchloric acid, M—Mg(NO$_3$)$_2$.

the high purity of reagent-grade hydrogen peroxide are advantages, particularly for large biological and foodstuff samples.

4.5.3 Conclusion

It must be stressed again that only general considerations indicating how the process can be performed have been mentioned here. Table 4.9 lists some references to wet ashing procedures applied to trace analysis.

In every application of ashing as a preliminary step in the separation and preconcentration of trace constituents, it is necessary to check that the recovery of trace elements is complete, and a blank test must be done, because the large amounts of acids and other reagents used may introduce significant amounts of impurities.

References

[1] Hoffman, J. I. and Lundell, G. E. F., *J. Res. Natl. Bur. Stand.*, **22**, 465 (1939).
[2] Wąsowicz, S. and Rutkowski, S., *Chem. Anal. (Warsaw)*, **11**, 603 (1966).
[3] Shvarts, D. M. and Kaporskii, L. N., *Zavodsk. Lab.*, **23**, 11 (1957).
[4] Gerken, E. B., *Nauch Tr. Gos. Nauchno-Issled. Inst. Tsvetn. Metal.*, (19), 800 (1962), cf. *Metody Analiza Veshchestv Vysokoi Chistoty (Methods for Analysing High-Purity Substances)*, Goskhimizdat, Moscow, 1965.
[5] Bogoyavlenskaya, A. N., Gerken, E. B. and Polyakova, V. V., *Nauch Gos. Nauchno-Issled. Inst. Tsvetn. Metal.*, (28), 138 (1968); *Chem. Abstr.*, **70**, 63982n (1969).
[6] Takahashi, M., Matsuda, Y., Toita, Y., Ouchi, M. and Komori, T., *Bunseki Kagaku*, **20**, 1085 (1971).
[7] Tiptsova, V. G., Dvortsan, A. G. and Malkina, E. I., *Zh. Analit. Khim.*, **23**, 1863 (1968).
[8] Pevtsov, G. A. and Krasil'shchik, V. Z., *Zh. Analit. Khim.*, **18**, 1314 (1963).
[9] Pevtsov, G. A. and Manova, T. G., *Sb. Tr. IREA*, **5**, 303 (1963).
[10] Pevtsov, G. A., Shirokova, M. Yu. and Mikheeva, I. P., *Metody Anal. Reaktivov Prep.* (12), 34 (1966); *Chem. Abstr.*, **67**, 70336 (1967).
[11] Gallus-Olender, J., *Chem. Anal. (Warsaw)*, **10**, 1039 (1965).
[12] Winslow, F. H., *U.S. Pat.* 2,812,235 (1957).
[13] Vecsernyés, L., *Z. Anal. Chem.*, **182**, 429 (1961).
[14] Pchelnitseva, A. F., Rakov, N. A. and Slyusareva, L. P., *Zavodsk. Lab.*, **28**, 677 (1962).
[15] Vecsernyés, L. and Hangos, I., *Z. Anal. Chem.*, **208**, 407 (1965).
[16] Vecsernyés, L., in *Reinststoffprobleme, II. Reinststoffanalytik*, Akademie-Verlag, Berlin 1966, p. 277.
[17] Notkina, M. A. and Solobodnik, S. M., *Zavodsk. Lab.*, **28**, 176 (1962).
[18] Brophy, V. A., Strock, L. W. and Peters, T., *Spectrochim. Acta*, **6**, 246 (1954).
[19] Willard, H. H. and Fogg, H. C., *J. Am. Chem. Soc.*, **59**, 40 (1937).
[20] Minczewski, J. and Chwastowska, J., *Chem. Anal. (Warsaw)*, **6**, 715 (1961).
[21] Chwastowska, J., *Chem. Anal. (Warsaw)*, **7**, 731 (1962).
[22] Nazarenko, V. A., Shustova, M. B., Rabitskaya, R. V. and Nikonova, M. P., *Zavodsk. Lab.*, **28**, 537 (1962).
[23] Pohl, F. A., Kokes, K. and Bonsels, W., *Z. Anal. Chem.*, **174**, 6 (1960).
[24] Fogg, D. N. and Wilkinson, N. T., *Analyst (London)*, **81**, 525 (1956).
[25] Shvarts, D. M. and Portnova, V. V., *Zavodsk. Lab.*, **24**, 731 (1958).
[26] Karabash, A. G., Peizulyaev, Sh. I., Sotnikova, N. P. and Sazanova, S. K., *Tr. Kom. Anal. Khim., Akad. Nauk SSSR*, **12**, 108 (1960).
[27] Fratkin, E. G., *Zavodsk. Lab.*, **30**, 170 (1964).
[28] Fratkin, E. G. and Shebunin, V. S., *Tr. Kom. Anal. Khim., Akad. Nauk SSSR*, **15**, 127 (1965).
[29] Ehrlich, P. and Keil, T., *Z. Anal. Chem.*, **165**, 188 (1959).
[30] Tarasevich, N. I. and Zheleznova, A. A., *Zh. Analit. Khim.*, **18**, 1345 (1963).
[31] Nazarenko, V. A. and Flyantikova, G. V., *Zavodsk. Lab.*, **24**, 663 (1958).
[32] Zilbershtein, Kh. I., *Zh. Tekh. Fiz.*, **25**, 1491 (1955).
[33] Fedyashina, A. F., Yudelevich, I. G. and Strokina, T. G., *Zh. Analit. Khim.*, **21**, 1232 (1966).

[34] Brodskaya, B. D., Notkina, M. A., Korneeva, S. A. and Men'shova, N. P., *Zh. Analit. Khim.*, **21**, 1447 (1966).
[35] Shvarts, D. M., *Zavodsk. Lab.*, **26**, 966 (1960).
[35a] Rodriguez-Vasquez, J. A., *Anal. Chim. Acta*, **73**, 1 (1974).
[35b] Kutal, C., *J. Chem. Educ.*, **52**, 319 (1975).
[35c] Komarov, V. A., *Zh. Analit. Khim.*, **34**, 366 (1976).
[35d] Uden, P. C. and Henderson, D. E., *Analyst (London)*, **102**, 889 (1977).
[36] Baudin, G., Lorrain, S. and Platzer, R., *Travaux XXXV Congrès Intern. Chimie Ind.*, Varsovie, 1965.
[37] Wang, M. S., *Appl. Spectrosc.*, **26**, 364 (1972).
[38] Kuźnia, Z., Ołdak, M., Rzeszotarska, J. and Zawadzki, B., *Chem. Anal. (Warsaw)*, **18**, 447 (1973).
[39] Karabash, A. G., Peizulyaev, Sh. I., Morozova, G. G. and Shirenkina, I. I., *Tr. Kom. Anal. Khim. Akad. Nauk SSSR*, **12**, 25 (1960).
[40] Vasilevskaya, L. S., Notkina, M. A., Sadof'eva, S. A. and Kondrashina, A. I., *Zavodsk. Lab.*, **28**, 678 (1962).
[41] Galkov, N. Ya., Ustinov, A. M. and Yudelevich, J. G., *Zh. Analit. Khim.*, **28**, 678 (1973).
[42] Fratkin, E. G., Volokhova, M. I. and Polivanova, N. G., *Zavodsk. Lab.*, **27**, 846 (1961).
[43] Galkov, N. Ya. and Ustinov, A. M., *Zavodsk. Lab.*, **37**, 149 (1971).
[44] Galkov, N. Ya., Yudelevich, I. G. and Ustinov, A. M., *Izv. Sib. Otd. Akad. Nauk SSSR, Ser. Khim. Nauk*, **1972**, 161.
[45] Zilbershtein, Kh. I., Nikitina, O. N., Nenarokov, A. V. and Panteleev, E. S., *Zh. Analit. Khim.*, **19**, 275 (1964).
[46] Ehrlich, P. and Keil, T., *Z. Anal. Chem.*, **166**, 254 (1959).
[47] Nozaki, T., Kawashima, T., Baba, H. and Araki, H., *Bull. Chem. Soc. Japan*, **33**, 1428 (1961); *Chem. Abstr.*, **55**, 18431 (1961).
[48] Onishi, H. and Sandell, E. B., *Anal. Chim. Acta*, **14**, 153 (1956).
[49] Boutron, C., *Anal. Chim. Acta*, **61**, 140 (1972).
[50] Geilmann, W. and Neeb, K., *Z. Anal. Chem.*, **165**, 251 (1959).
[51] Geilmann, W., Neeb, K. and Eschnauer, H., *Z. Anal. Chem.*, **154**, 418 (1957).
[52] Geilmann, W. and Neeb, R., *Angew. Chem.*, **67**, 26 (1955).
[53] Geilmann, W., *Z. Anal. Chem.*, **160**, 410 (1958).
[54] Scribner, B. F. and Mullin, H. R., *J. Res. Natl. Bur. Stand.*, **37**, 378 (1946).
[55] Mandel'shtam, S. L., Semenov, I. I. and Turovtseva, Z. M., *Zh. Analit. Khim.*, **11**, 9 (1956).
[56] Zaidel', A. V., Kaliteevskii, N. S., Lipis, L. V., Chaika, M. B. and Belyaso, I., *Zh. Analit. Khim.*, **11**, 21 (1956).
[57] Pszonicki, L., *Chem. Anal. (Warsaw)*, **5**, 261 (1960).
[58] Pszonicki, L., *Chem. Anal. (Warsaw)*, **7**, 947 (1962).
[59] Rains, T. C. and Menis, O., *J. Assoc. Off. Anal. Chem.*, **55**, 1339 (1972).
[60] Head, P. C. and Nicholson, R. A., *Analyst (London)*, **98**, 53 (1973).
[61] Sandell, E. B., *Ind. Eng. Chem., Anal. Ed.*, **16**, 342 (1944).
[62] McBryde, W. A. E. and Yoe, J. H., *Anal. Chem.*, **20**, 1094 (1948).
[63] Rogers, W. J., Beamish, F. E. and Russell, D. S., *Ind. Eng. Chem., Anal. Ed.*, **12**, 561 (1940).
[64] Hara, T. and Sandell, E. B., *Anal. Chim. Acta*, **23**, 65 (1960).
[65] Kalinina, V. E. and Bolgyreva, O. L., *Zh. Analit. Khim.*, **28**, 720 (1973).

[66] Geilmann, W. and Bode, H., *Z. Anal. Chem.*, **130**, 323 (1949).
[67] Strickland, J. D. H. and Spicer, G., *Anal. Chim. Acta*, **3**, 543 (1949).
[68] Pugh, H. and Waterman, W. R., *Anal. Chim. Acta*, **55**, 97 (1971).
[69] Pomeroy, R., *Anal. Chem.*, **26**, 570 (1954).
[70] Gustafsson, L., *Talanta*, **4**, 227, 237 (1960).
[71] Marczenko, Z. and Chołuj-Lenarczyk, Ł., *Chem. Anal. (Warsaw)*, 10, 729 (1965).
[72] Grintsaig, E. L., Nadezhina, L. S. and Bespalenkova. E. K., *Tr. Khim. Khim. Tekhnol. (Gorkii)*, **3**, (24), 120 (1969).
[73] Kirk, P. L., *Anal. Chem.*, **22**, 354, 611 (1950).
[74] Kuck, J. A., Kingsley, A., Kinsey, D., Sheenan, F. and Swigert, G. F., *Anal. Chem.*, **22**, 604 (1950).
[75] Committee for the Standardization of Microchemical Apparatus, A.C.S., *Anal. Chem.*, **23**, 523 (1951).
[76] Bremner, J. M. and Keeney, D. R., *Anal. Chim. Acta*, **32**, 485 (1965).
[77] McDonald, I. G. and Lench, A., *Analyst (London)*, **85**, 564 (1960).
[78] Baker, P. R., *Talanta*, **8**, 57 (1961).
[79] Schmidt, F. J. and Royer, J. L., *Anal. Lett.*, **6**, 17 (1973).
[80] Braman, R. S., Justen, L. L. and Foreback, C. C., *Anal. Chem.*, **44**, 2195 (1972).
[81] Marsh, J., *Edinburgh New Phil. J.*, **21**, 229 (1836).
[82] Gutzeit, M., *Pharm. Ztg.*, **24**, 263 (1879).
[83] Onishi, H. and Sandell, E. B., *Mikrochim. Acta*, **1953**, 34.
[84] Kubalski, J., *Acta Pol. Pharm.*, **14**, 221 (1957).
[85] Szymczak, J. and Żechałko, A., *Rocz. Państw. Zakł. Hig.*, **14**, 239 (1963).
[86] Fołdzińska, A. and Malinowski, J., *Nukleonika*, **7**, 153 (1962).
[87] Fołdzińska, A. and Malinowski, J., *Nukleonika*, **8**, 233 (1963).
[88] Oliver, W. T. and Funnell, H. S., *Anal. Chem.*, **31**, 259 (1959).
[89] Liederman, D., Bowen, J. E. and Milner, O. J., *Anal. Chem.*, **31**, 2052 (1959).
[90] Ciuhandu, G. and Rocsin, M., *Z. Anal. Chem.*, **174**, 118 (1960).
[91] Schnepfe, M. M., *Talanta*, **20**, 175 (1973).
[92] Rooney, R. C., *Analyst (London)*, **82**, 619 (1957).
[92a] Fernandez, F. J. and Manning, D. C., *At. Absorpt. Newsl.*, **10**, 86 (1971).
[92b] Fernandez, F. J., *At. Absorpt. Newsl.*, **12**, 93 (1973).
[92c] Pollock, E. N. and West, S. J., *At. Absorpt. Newsl.*, **11**, 504 (1972).
[92d] Pollock, E. N. and West, S. J., *At. Absorpt. Newsl.*, **12**, 6 (1973).
[92e] Schmidt, F. J. and Royer, J. L., *Anal. Lett.*, **6**, 17 (1973).
[92f] Smith, A. E., *Analyst (London)*, **100**, 300 (1975).
[92g] *Annual Reports on Analytical Atomic Spectroscopy*, The Chemical Society, London, 1975–1980, Vols. 5–10, etc. and references therein.
[93] Bartlet, J. C., Wood, M. and Chapman, R. A., *Anal. Chem.*, **24**, 1821 (1952).
[94] Liederman, D., Bowen, J. E. and Milner, O. J., *Anal. Chem.*, **30**, 1543 (1958).
[95] Cluley, H. J., *Analyst (London)*, **76**, 523 (1951).
[96] Handley, R. and Johnson, C. M., *Anal. Chem.*, **31**, 2105 (1959).
[97] Heczko, T., *Mikrochem. Mikrochim. Acta*, **36/37**, 825 (1951).
[98] Pometun, N. P., *Zh. Analit. Khim.*, **22**, 440 (1967).
[99] Holt, B. D., *Anal. Chem.*, **32**, 124 (1960).
[100] Szabó, Z. G., Zapp, E. É. and Perczel, S., *Mikrochim. Acta*, **1974**, 167.
[101] Gries, W. H., *Mikrochim. Acta*, **1974**, 249.
[102] Marczenko, Z. and Kasiura, K., *Chem. Anal. (Warsaw)*, **8**, 185 (1963).
[103] Spicer, G. S. and Strickland, J. D. H., *Anal. Chim. Acta*, **18**, 523 (1958).

[104]   Pohl, F. A., *Z. Anal. Chem.*, **157**, 6 (1957).
[105]   Marczenko, Z. and Chołuj-Lenarczyk, Ł., *Chem. Anal. (Warsaw)*, **13**, 405 (1968).
[106]   Kahane, E. and Kahane, M., *Bull. Soc. Chim. Fr.*, **1954**, 396.
[107]   Nazarenko, V. A., Birynk, E. A., Antonovich, V. P., *Zavodsk. Lab.*, **33**, 22 (1967).
[108]   Ducret, L. and Cornet, C., *Anal. Chim. Acta*, **25**, 542 (1961).
[109]   Bark, L. S. and Higson, H. G., *Analyst (London)*, **88**, 751, (1963).
[110]   Sakamato, T., Nakasima, T., Fukuda, K. and Mizuike, A., *Bunseki Kagaku*, **19**, 1218 (1970).
[111]   Magnuson, H. J. and Watson, E. B., *Ind. Eng. Chem., Anal. Ed.*, **16**, 339 (1944).
[112]   Karpen, W. L., *Anal. Chem.*, **33**, 738 (1961).
[113]   Mroziński, J., *Chem. Anal. (Warsaw)*, **12**, 93 (1967).
[114]   Maekava, S., Kato, K. and Kamada, T., *Bunseki Kagaku*, **19**, 1277 (1970).
[115]   Youngdahl, C. A. and Deboer, F. E., *Nature (London)*, **184**, 54 (1959).
[116]   Neeb, K. H., *Z. Anal. Chem.*, **194**, 54 (1959).
[117]   Fresenius, R. and Babo, L., *Ann. Chim. Pharm.*, **49**, 287 (1844).
[118]   Gorsuch, T. T., *The Destruction of Organic Matter*, Pergamon, Oxford, 1970.
[119]   Gorsuch, T. T., *Analyst (London)*, **84**, 135 (1959).
[120]   Middleton, G. and Stuckey, R. E., *Analyst (London)*, **79**, 138 (1954).
[120a] Minczewski, J., Chwastowska, J. and Hong Mai, P., *Chem. Anal. (Warsaw)*, **18**, 1189 (1973).
[121]   Pijck, J., Hoste, J. and Gillis, J., *Proceedings of the International Symposium on Microchemistry, Birmingham* 1958, Pergamon, London, 1960, p. 48.
[122]   Kowalczuk, J., *J. Assoc. Off. Anal. Chem.*, **53**, 926 (1970).
[122a] Ito, J., *Bull. Chem. Soc. Japan*, **35**, 225 (1962).
[122b] Langmyhr, F. J. and Sveen, S., *Anal. Chim. Acta*, **32**, 1 (1965).
[122c] Bernas, B., *Anal. Chem.*, **40**, 1682 (1968).
[122d] Langmyhr, F. J. and Paus, P. E., *Anal. Chim. Acta*, **49**, 358 (1970).
[122e] Kotz, L., Kaiser, G., Tschöpel, P. and Tölg, G., *Z. Anal. Chem.*, **260**, 207 (1972).
[122f] Šinko, I. and Gomišček, S., *Mikrochim. Acta*, **1972**, 163.
[122g] Van Eenbergen, A., and Bruninx, E., *Anal. Chim. Acta*, **98**, 405 (1978).
[123]   Comrie, A. A. D., *Analyst (London)*, **60**, 532 (1935).
[124]   Hamm, R., *Mikrochim. Acta*, **1956**, 268.
[125]   High, J. H., *Analyst (London)*, **72**, 60 (1947).
[126]   Williams, D. E. and Vlamis, J., *Anal. Chem.*, **33**, 967 (1961).
[127]   Boppel, B., *Proc. Int. Symp. Rapid Methods for Measuring Radioactivity in the Environment*, IAEA, Vienna 1971, p. 71.
[128]   Elvidge, D. A. and Garratt, D. C., *Analyst (London)*, **79**, 146 (1954).
[129]   Schöniger, W., *Mikrochim. Acta*, **1955**, 123.
[130]   Schöniger, W., *Mikrochim. Acta*, **1956**, 869.
[131]   Schöniger, W., *Z. Anal. Chem.*, **181**, 28 (1961).
[132]   Belisle, J., Green, C. B. and Winter, L. D., *Anal. Chem.*, **40**, 1006 (1968).
[133]   Shirley, R. L., Benne, E. J. and Miller, E. J., *Anal. Chem.*, **21**, 300 (1949).
[134]   Ansbacher, S., Remington, R. E. and Culp, F. B., *Ind. Eng. Chem., Anal. Ed.*, **3**, 314 (1931).
[135]   Cholak, J., *Ind. Eng. Chem., Anal. Ed.*, **9**, 26 (1937).
[136]   Muller-Mangold, D., *Stärke*, **19**, 251 (1967); *Anal. Abstr.*, **15**, 7565 (1968).
[137]   Bowen, J., *Anal. Chem.*, **40**, 969 (1968).

[138] Misk, E., *Compt. Rend.*, **176**, 138 (1923).
[139] Neumann, F., *Z. Anal. Chem.*, **155**, 340 (1957).
[140] Bamford, F., *Analyst (London)*, **59**, 101 (1934).
[141] Lockwood, H. C., *Analyst (London)*, **79**, 143 (1954).
[142] Abson, D. and Lipscomb, A. G., *Analyst (London)*, **82**, 152 (1957).
[143] Tompsett, S. L., *Analyst (London)*, **81**, 330 (1956).
[144] Gorbach, G. and Pohl, F., *Mikrochem. Mikrochim. Acta*, **38**, 328 (1951).
[145] Analytical Methods Committee, *Analyst (London)*, **85**, 643 (1960).
[146] de Ciacomi, R., *Analyst (London)*, **65**, 216 (1940).
[147] Carius, L., *Ann. Chem.*, **116**, 1 (1860); **136**, 129 (1865).
[148] Kirk, P. L., *Anal. Chem.*, **22**, 354 (1950).
[149] Danckworth, P. W. and Ude, W., *Arch. Pharm.*, **264**, 712 (1926).
[150] Cholak, J., *Ind. Eng. Chem., Anal. Ed.*, **9**, 26 (1937).
[151] Danckworth, P. W. and Jürgens, E., *Arch. Pharm.*, **266**, 492 (1928).
[152] Milton, R. F., Hoskins, J. L. and Jackman, W. K. F., *Analyst (London)*, **69**, 299 (1944).
[153] Waterman, R. E., Koch, F. C. and McMahon, W., *Ind. Eng. Chem., Anal. Ed.*, **6**, 409 (1934).
[154] Hanson, N. W., *Official, Standardised and Recommended Methods of Analysis*, Society for Analytical Chemistry, London, 1973.
[154a] Analytical Methods Committee, *Analyst (London)*, **86**, 608 (1961), **90**, 515 (1965).
[155] Whalley, C., *Analyst (London)*, **79**, 148 (1954).
[156] Lampitt, L. H., and Rooke, H. S., *Analyst (London)*, **58**, 733 (1933).
[157] Francis, A. G., Harvey, G. O. and Buchan, J. L., *Analyst (London)*, **54**, 725 (1929).
[158] Kahane, E., *L'action de l'acide perchlorique sur les matières organiques, I. Généralités, II. Applications*, Herman et Cie, Paris 1934.
[159] Analytical Methods Committee, *Analyst (London)*, **84**, 214 (1959).
[160] Smith, G. F., *Anal. Chim. Acta*, **8**, 397 (1953).
[161] Smith, G. F., *Anal. Chim. Acta*, **17**, 175 (1957).
[162] Diehl, H. and Smith, G. F., *Talanta*, **2**, 209 (1959).
[163] Smith, G. F. and Diehl, H., *Talanta*, **3**, 41 (1959).
[164] Smith, G. F., *Anal. Chim. Acta*, **13**, 115 (1955).
[165] Smith, G. F. and Diehl, H., *Talanta*, **4**, 185 (1960).
[166] Andrus, S., *Analyst (London)*, **80**, 514 (1955).
[167] Stock, A., Cucuel, F. and Köhle, H., *Angew. Chem.*, **46**, 187 (1933).
[168] Strafford, N. and Wyatt, P. F., *Analyst (London)*, **61**, 528 (1936).
[169] Analytical Methods Committee, *Analyst (London)*, **92**, 403 (1967).
[170] Sansoni, B. and Kracke, W., *Z. Anal. Chem.* **243**, 209 (1968).
[171] Warburg, O., *Biochem. Z.*, **187**, 255 (1927).
[172] Campbell, W. J., Carl, H. F. and White, C. E., *Anal. Chem.*, **29**, 1009 (1957).
[173] Richards, M. B., *Analyst (London)*, **55**, 554 (1930).
[174] Lynch, G. R., Slater, R. H. and Osler, T. G., *Analyst (London)*, **59**, 787 (1934).

Chapter 5

LIQUID–LIQUID EXTRACTION

5.1 FUNDAMENTAL PHYSICAL CHEMISTRY OF EXTRACTION

5.1.1 The Parameters Characteristic of Extraction

Liquid–liquid extraction is a process of transferring a chemical compound from one liquid phase to a second liquid phase, immiscible with the first. In trace analysis applications, one phase is usually water and the other a suitable organic solvent, so this discussion refers only to such systems.

Extraction is a process occurring in a two-phase ternary system. It follows from the Gibbs phase rule that such a system has, at constant temperature and pressure, one degree of freedom. This means that the concentrations of the solute in the two phases are mutually related, according to the distribution law proposed by Berthelot and Jungfleisch [1] and developed by Nernst [2]. This states that a solute is distributed between two immiscible solvents in such a way that at equilibrium the ratio of the concentrations of the solute in the two phases is constant at a given temperature, provided that the solute has the same molecular weight in each phase. This may be expressed by the formula

$$k = c_o/c_w \tag{5.1}$$

where k is the distribution constant, c_o is the concentration of the solute in the organic phase, and c_w is the concentration of the solute in the aqueous phase.

Not infrequently the experimental data fail to confirm the constancy of the distribution constant. This is due to the fact that in this form, the law is not strictly correct thermodynamically. The thermodynamic distribution law, as derived by Morrison and Freiser [3], is based on the observation that at constant temperature and pressure, equilibrium is reached when the chemical potentials (partial molar Gibbs free energies) of the solute in the two phases are equal.

The distribution constant then takes the form

$$K = (\gamma_w/\gamma_o) \exp\left[-(\phi_o^0 - \phi_w^0)/RT\right] \tag{5.2}$$

where ϕ_o^0 and ϕ_w^0 are the chemical potentials of the solute (equal to unity in the organic and the aqueous solution, respectively), and γ_o and γ_w are the corresponding molar activity coefficients in the organic and aqueous phases.

If the presence of the solute does not significantly affect the mutual solubility of the two phases, ϕ_o^0 and ϕ_w^0 can be assumed to be constant and the formula simplifies to the form

$$k = K \cdot \gamma_w / \gamma_o \tag{5.3}$$

where K is a constant for a given system.

Thus, the validity of the Nernst distribution law depends on the constancy of the ratio of the activity coefficients.

In very dilute solutions, activity coefficients are equal to unity, so the distribution constant is independent of solute concentration. In solutions of higher concentrations the distribution law may be followed only if the activity coefficients change in the same way in both solvents, which is very unlikely in practice.

Besides this approach to the extraction equilibrium (i.e. that based on an assumption of equality of chemical potentials in the two phases) attempts have been made to relate extractive properties to the properties of the individual components of the ternary system. These attempts include the electrostatic approach to the extraction system, application of the findings of studies on solubility of gases in liquids to two-phase ternary systems, or the Hildebrand theory of regular solutions.

Bell [4] and Kirkwood [5] derived a formula for the energy of displacement of a dipole of radius r and dipole moment μ from vacuum to a medium of dielectric constant ε

$$F = \frac{\mu^2}{3r^3}\left(\frac{1}{\varepsilon} - 1\right) \tag{5.4}$$

If the extracted molecule is assumed to be a dipole of definite dimensions, the difference between the standard chemical potentials that determines the distribution constant is approximately equal to the difference between the energy of displacement of the dipole from vacuum to the organic medium, and the energy of its displacement from vacuum to the aqueous phase. This is because the difference in energy of displacement of a dipole molecule from a medium of dielectric constant ε_1 to a medium of dielectric constant ε_2 is equal to the difference between the free energies of displacement of the molecule from vacuum to these media, and is expressed by the formula

$$F_{\varepsilon_1 - \varepsilon_2} = -\frac{\mu^2}{3r^3}\left(\frac{1}{\varepsilon_2} - \frac{1}{\varepsilon_1}\right) \tag{5.5}$$

The experimental data used by Banks [6] in his derivation of the empirical formulae that describe these relations showed considerable discrepancies from the expected results. According to Banks the theoretical formulae do not allow for chemical interactions of the solute molecules with solvents and water.

The approach which aims to correlate the extractive properties of a given compound with the properties of the pure components of the extraction system involves application of the data on solubility of gases in liquids from the work of Sisskind et al. [7], Uhlig [8] and Eiley [9] to two-phase ternary systems by Butler [10], McGowan [11] and Collander [12].

The third approach is based on the Hildebrand–Scatchard theory o solutions. The formulae derived by Hildebrand and Scatchard [13-15] for the free energy of mixing were used by Buchowski [16,17] in formulating the relation between the distribution constant k_z of a compound (z) in a system of two immiscible solvents (w, i), and the molar volumes V and the Hildebrand solubility parameters δ^* [13] of the pure compounds. For systems in which water is one of the components the expression for the distribution constant is fairly simple:

$$2\cdot 3\, RT \log (k_z)_{wi} = 2V_z(\delta_w - \delta_i)(\delta_z - \delta_{wi}) \tag{5.6}$$

where $\delta_{wi} = (\delta_w + \delta_i)/2$, δ_w being the solubility parameter of water, and δ the solubility parameter of the solvent.

From the data obtained, Buchowski reached a number of conclusions about the properties of extraction systems. He suggested that the miscibility series of liquids should be arranged according to increasing values of the solubility parameter. The value of the distribution constant is then determined by the relative positions of the solvents and the compound in the miscibility series.

The equations above for the distribution constant hold only when the solute is in the same form in both phases. Since in practice this is seldom the case, a more useful parameter is the distribution coefficient D

$$D = \sum(C_A)_o / \sum(C_A)_w \tag{5.7}$$

where $\sum(C_A)_o$ and $\sum(C_A)_w$ are the total concentrations of the compound A in the organic solvent and in water, respectively, at equilibrium, irrespective of the form in which A occurs.

---

* $\delta = \left(\dfrac{\Delta E^V}{V}\right)^{\frac{1}{2}} = $ (energy of vaporization per ml)$^{\frac{1}{2}}$.

In an ideal extraction process, when the extracted compound does not react further in either phase, the distribution coefficient is equal to the distribution constant. For a quantitative description of the extraction process in practice a quantity referred to as the extraction efficiency is used. This is equal to the weight fraction of compound extracted:

$$E = 100 \sum(A)_o / (\sum(A)_w + \sum(A)_o) \% \qquad (5.8)$$

where $\sum(A)_o$ and $\sum(A)_w$ are the weights of compound A in all possible forms dissolved in the organic and aqueous phases respectively at equilibrium.

The relation between extraction efficiency E and distribution coefficient D is

$$E = 100D/[D + (V_w/V_o)] \qquad (5.9)$$

where V_o and V_w are the volumes of the organic and the aqueous phases. This expression is illustrated in Fig. 5.1.

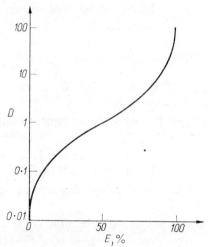

Fig. 5.1   Distribution coefficient *vs.* extraction efficiency

Some values of D along with the corresponding values of E, for a 1:1 volume ratio of phases, are given below.

D	1000	100	10	1	0·1	0·01	0·0001
$E, \%$	99·9	99	90	50	10	1	0·1

For an extraction efficiency of 99–100% the distribution coefficient varies from 100 to infinity. Thus there may be very large differences in the distribution coefficient, even though the extraction is practically complete. Thus a determination of E is often more meaningful than that of D.

For analytical purposes it may be taken that extraction is complete when $E = 99\%$ and that extraction does not occur when $E \leqslant 1\%$. Hence the limiting D values are $-2 < \log D < +2$.

If D is known, it is possible to calculate how many extractions will be required to achieve an effectively quantitative removal of a compound in an extraction process. The concentration c_{w1} of an extracted compound in the aqueous phase after a single extraction is

$$c_{w1} = c_{w0} V_w/(V_w + DV_o) \tag{5.10}$$

where V_w, V_o are the volumes of the aqueous and organic phases and c_{w0} is the initial concentration of the compound in the aqueous phase.

After n extractions, the final concentration of the compound in the aqueous phase will be

$$c_{wn} = c_{w0}[V_w/(V_w + DV_o)]^n \tag{5.11}$$

It follows from this expression that an extraction process is more efficient if it is done with several small portions of solvent, rather than with the whole volume of the solvent at once. Table 5.1 illustrates this for three D-values.

Table 5.1 Percentage of solute remaining in the aqueous phase, for various values of the distribution coefficient and different number of extractions (equal volumes of phases)

c_{wn}	$D = 1$	$D = 10$	$D = 1000$
c_{w0}	100	100	100
c_{w1}	50	9	0·1
c_{w2}	25	0·8	
c_{w3}	12·5	0·07	
.	.		
.	.		
.	.		
c_{w9}	0·195		
c_{w10}	0·098		

For a compound which is equally soluble in both phases ($D = 1$) the extraction process nears completion only after the 10th operation. Such a system is of no practical utility for batch extraction. For $D = 10$ only 3 extractions are necessary, and for $D = 1000$ a single extraction is sufficient.

No less important in analytical practice is the degree of separation of the compound of interest from accompanying compounds, since usually the systems to be handled are complex.

Separation of two metals which yield extractable species with a specific reagent is commonly described in terms of the separation factor α

$$\alpha_{1,2} = D_1/D_2 \qquad (5.12)$$

If the minimum extraction of metal 1 is 99% and the maximum extraction of metal 2 is 1%, then

$$\alpha_{1,2} = 10^2/10^{-2} = 10^4$$

Buchowski [16] defines the separation factor as the difference between the logarithms of the distribution constants. Thus, the separation factor for two compounds A and B in solvents w and i is given by the following formula, provided that the molar volumes of the compounds are nearly equal

$$\alpha_{A,B} = (\log k_A - \log k_B) = \frac{2V_A}{2 \cdot 3\, RT} (\delta_A - \delta_B)(\delta_w - \delta_i) \qquad (5.13)$$

Since the separation factor is proportional to the term $(\delta_w - \delta_i)$, the farther apart the solvents are in the miscibility series, the better the separation.

In studies primarily concerned with extraction, the separation factor is commonly defined as the ratio of either the distribution coefficients or the distribution constants.

5.1.2 Inorganic Extraction Systems

In terms of extractability, inorganic compounds may be tentatively classed into the following two groups.

(i) Compounds which occur in the aqueous phase as undissociated covalent species, e.g. I_2, Br_2, halides of some metals, e.g. $AsCl_3$, $AsBr_3$, $GeCl_4$, InI_3, $HgCl_2$, HgI_2, and some oxides, e.g. OsO_4.

(ii) Ionic compounds.

The extraction of covalent compounds is essentially the same as extraction of organic compounds. The interaction of a covalent compound with water as solvent depends on how polar the compound is. For non-polar compounds the interaction is due to dispersive forces and is comparable with the forces operative between these compounds and any organic solvent. Water as a solvent has peculiar properties; it has an exceptionally high value of the Hildebrand parameter δ because it is highly associated. For most solvents δ ranges from 6 to 15, but the value for water is 23·4. The Hildebrand parameter is equal to the square root of the specific cohesion energy (vaporization energy, see footnote on p. 99) of a liquid. Since the specific cohesion energy of a liquid is a direct measure of the energy required to remove a molecule from the liquid, it follows that in a two-phase water–organic solvent system a covalent compound will tend to pass into the phase of lower cohesion energy, thus into the organic phase. The gain in energy from this transfer is higher for large molecules. Berthelot and Jungfleisch noted in their early work on the distribution law that in the water–carbon disulphide system the distribution coefficient for iodine (I_2) is 400, and for bromine (Br_2) it is 80.

In principle ionic compounds should not be extracted from the aqueous phase into the organic phase, as can be shown by application of the Born charging equation to the extraction process. Born [18] derived an expression for the electrostatic energy of transfer of an ion of definite radius and charge from vacuum to a liquid phase of definite dielectric constant. Taking into account the value of that energy and the value of the association energy calculated from the theory of Bjerrum [19] and Fuoss-Kraus [20], it can be said that a necessary condition for a good extraction of an ionic species is the lowering of the electrostatic energy of the transfer by partial or complete neutralization of the charge, increase in the size of the neutral species formed and conferment of hydrophobic and organophilic properties on it. Three main mechanisms for extraction of inorganic cations may be distinguished, namely extraction of chelates, of ion-pairs, and of co-ordinatively solvated salts.

5.1.3 Classification of Extraction Systems

The three extraction mechanisms mentioned do not cover, of course, all possible systems. The most common basis for classification of extraction systems is the form in which the element extracted passes into the organic phase. Classification of extraction systems has been discussed by a number of authors [3,21–28]. As extraction theory and the chemistry of complex compounds develop, classification of extraction systems becomes more complicated. It is difficult to choose a classification which is correct and comprehensive, and also practically useful.

In trace analysis for inorganic species, two main types of system can be distinguished. In the first, the species extracted are uncharged covalent compounds (e.g. chelates), in the second, electrovalent compounds (e.g. anionic complexes, ion-pairs with amines, and co-ordinatively solvated salts).

These systems will be discussed in more detail later. Chelate systems are covered in detail, and a number of factors that affect the extraction process are discussed. A number of the relations derived for chelate systems may also be used for other extraction systems.

5.2 CHELATE SYSTEMS

5.2.1 Complex Compounds

Chelates are complexes of a metal ion with a multidentate ligand which occupies two or more co-ordination sites, and in which rings are formed. A ring may contain two electrovalent bonds, or one covalent and one electrovalent, or two covalent bonds with the metal. Inner complexes are chelate compounds in which the number of ligands is such that the charge on the central ion of the complex is exactly neutralized by the charges of the multidentate ligands. A stable ring is often formed when the ligand contains a charged group, which can form an electrovalent bond with the metal, and also an electron-donating group which can form a covalent bond with the metal. Also, the two groups must be positioned in the ligand molecule so that the ring formed can contain not less than 4 and not more than 7 or 8 members. The most stable are five-membered rings.

Typical charged groups are —OH, —COOH, —SO_3H, =NOH, =NH, —NH_2, —SH, —AsO_3H_2 and —PO_3H_2; and electron-donating groups usually contain oxygen, nitrogen or sulphur atoms, but selenium, tellurium, etc. are also possible. Typical examples are =O, —O—, —N=, —NH_2, =NOH, =S, —S—.

Sidgwick [29] and Ahrland et al. [30] have proposed a classification of metals on the basis of the stabilities of the chelates formed with given ions and on the nature, charge and structure of the ions.

Complexes may be co-ordinatively saturated or co-ordinatively unsaturated [31,32]. If the complexing agent used can simultaneously satisfy all the co-ordination sites of the metal ion and produce a complex with zero overall charge, the chelate compound is said to be co-ordinatively saturated. If, however, the charge on the central atom is neutralized but some co-ordination sites remain unsatisfied, the complex is said to be co-ordinatively unsaturated.

Bidentate singly charged ligands form co-ordinatively unsaturated chelate compounds with alkaline earth metals, Li^+, Tl^+, NpO_2^+, and its analogues, Mn^{2+}, Co^{2+}, Fe^{2+}, Zn^{2+}, Cd^{2+}, UO_2^{2+} and its analogues and tervalent lanthanide and actinide ions, and co-ordinatively saturated chelates with ter- and quadrivalent ions, e.g. Al^{3+}, Ga^{3+}, In^{3+}, Fe^{3+}, Cr^{3+}, Zr^{4+}, Hf^{4+}, Th^{4+}, U^{4+}, Pu^{4+}, and Np^{4+}.

The free co-ordination sites in co-ordinatively unsaturated chelates are usually occupied by water molecules. Hydrated chelates are only poorly extractable, so for a good extraction the solvating water molecules should be removed.

The hydrophilic character of an ion depends on the charge density, which is the ratio of the charge on the ion to its radius. The lower the charge density, the weaker is the hydration of the ion. Since the radius of a complex ion or organic molecule cannot be determined, the concept of specific charge has been introduced. This parameter is the ratio of the charge to the total number of atoms of all elements in the given compound. Ions with specific charges in the range 0·02–0·03 are usually assumed to be suitable for extraction (e.g. the n-octylamine cation has a specific charge of 0·03).

In some complexes, all the donor atoms are identical; i.e. the co-ordination sphere of the central ion is homogeneous but in others ('mixed complexes') more than one type of donor atom is involved. A distinction is made between mixed-donor complexes, in which the different donor atoms are incorporated into the same ligand molecule, and mixed-ligand complexes, in which the donor atoms belong to different ligand molecules.

If several ligands are present in solution, the number of mixed ligand complexes may be very high [33]. Beck [34] has discussed the equilibria between complexes of various kinds.

Mixed-ligand complexes are often kinetically inert and for that reason they are increasingly used in chemical analysis. Also, mixed-ligand complexes may increase the selectivity of some analytically important reactions. For example, Co(II) can be separated from Ni(II), and Pd(II) from Ni(II) as their dimethylglyoxime complexes by utilizing the formation of anionic mixed complexes [35]. Belcher, Leonard and West [36,37] have found that the fluoride ion produces a soluble coloured ternary ccmplex with the chelate of Ce(III) with alizarin fluorine blue. The reaction is specific for the fluoride ion.

The properties of such complexes and their application in extraction are discussed at greater length in Sections 5.2.4 and 5.2.5.

Outer-sphere complexes may also be useful in the extraction process. These are compounds in which anionic or neutral ligands are associated with a co-ordinatively saturated complex [34], and in which there is no direct bond between the central ion and the ligand. With respect to the nature of the bonding of the ligand to the central ion, outer-sphere and inner-sphere complexes represent two extremes. In between come the anisotropic complexes, in which the strengths of the linkages between the ligand atoms and the central atom differ (some of the donor atoms are held close to the central ion, to which they are strongly bound, others, more distant, are weakly bound). No clear-cut borderlines can be drawn, and occasionally an unambiguous distinction between outer- and inner-sphere species cannot be made.

Co-ordinatively saturated complexes generally bind additional ligands electrostatically. Bjerrum found that higher ionic charge and lower dielectric constant of the medium promote formation of complexes of that type. However, a charge-transfer mechanism may also be involved, with the resultant stabilization of the complexes.

Studies on the analytical applications of such complexes show that they can often be used to enhance both the extractability and the selectivity of the extraction process [38–40]. Particular attention has been given to the outer-sphere complexes of tris(phenanthroline) iron(II) which are used for the selective determination of iron following an extractive separation [41,42] or for determination of a variety of anions [43,1316].

Attention should also be paid to the existence of ligand-bridged heteronuclear complexes [43a–43e].

5.2.2 Extraction as a Chemical Reaction

The extraction reaction may be described by the following general equation which assumes that no side-reactions occur in either the aqueous or the organic phase:

$$M^{n+} + nHA_{(o)} \rightleftharpoons MA_{n(o)} + nH^+ \qquad (5.14)$$

where M^{n+} is the metal ion, HA is the organic chelate-forming agent and MA_n is the chelate. The subscript (o) refers to the organic phase: aqueous phase components are not given subscripts.

The equilibrium constant of this extraction reaction is

$$K_{ex} = \frac{[MA_n]_o [H^+]^n}{[M^n] [HA]_o^n} \qquad (5.15)$$

This equation describes the overall equilibrium. There are a number of intermediate equilibria, namely

$HA_{(o)} \rightleftharpoons HA$, distribution of the reagent HA between the phases,

$HA \rightleftharpoons A^- + H^+$, dissociation of HA in the aqueous phase,

$M^{n+} + nA^- \rightleftharpoons MA_n$, the reaction of the metal ions with the ions of the organic reagent in the aqueous phase,

$MA_n \rightleftharpoons MA_{n(o)}$, distribution of the chelate between the phases.

If no intermediate complexes of the metal with the organic reagent are formed, and there are no other reagents in the aqueous phase which can react with the metal ions, the ratio $[MA_{(o)}]/[M^{n+}]$ is equal to the distribution coefficient D. Then

$$K_{ex} = D \frac{[H^+]^n}{[HA]_o^n}$$

hence

$$D = K_{ex} \frac{[HA]_o^n}{[H^+]^n} \quad (5.16)$$

or $\quad \log D = \log K_{ex} + n\,\mathrm{pH} + n \log [HA]_o.$

If follows from this simplified model of an extraction reaction that for a given system the degree of extraction increases with the concentration of ligand in the organic phase and with the pH of the aqueous phase. If the concentration of complexing agent is constant, the extraction process depends on pH. In the pH range where the extraction equation (5.14) is followed, a graph of log D vs. pH is a straight line of slope n.

In practice, however, the slope of the linear portion of the log D vs. pH curve is often less than n. This is largely due to the side-reactions such as hydrolysis, reactions with other complexing agents, or the formation of intermediate complexes which occur in real solutions. Such side-reactions are taken into account by using a conditional extraction coefficient K'_{ex}, the value of which is also a function of pH [43a].

Most real systems are more involved than the one described by Eq. (5.14). A correct mathematical description of complex extraction reactions has for many years been the subject of research by a great many investigators, including Kolthoff and Sandell [44,45], Morrison and Freiser [46], Oosting [47,48], Starý [49], Dyrssen [50,51], Rydberg [52], Irving et al. [21,53,54], Schweitzer [55], Zolotov [56], and Siekierski [56a,56b].

Let us consider a model extraction system that is close to reality. The aqueous phase of such a system contains a metal ion M, a masking agent X, hydrogen ions, hydroxide ions, and background electrolyte of suitable concentration to ensure constant ionic strength of the solution. The organic phase is an organic solvent S containing a complex-forming reagent HA of known concentration and a reagent B that can form addition products with the extracted complexes. Both HA and B are usually sparingly soluble in water. When equilibrium between the phases is reached, a number of reactions will have occurred between the individual species of the system. Reagents B and HA partially pass into the aqueous phase, where HA may dissociate. Ions X, OH$^-$, A, and molecules HA, S, and B may react with metal ions to yield complexes. Each uncharged product formed as a result of the reaction partly passes into the organic phase.

For many systems, a general formula for a complex which will pass into the organic phase is $MA_n(HA)_a B_b$, where n is the charge on the metal ion. A general formula for a complex present in the aqueous phase is $MA_r(OH)_h X_x$. The subscripts n,a,b,r,h,x can take any integral values.

The distribution coefficient of the metal is given by

$$D = \frac{\sum [MA_n(HA)_a B_b]_o}{\sum [MA_r(OH)_h X_x]} \quad (5.17)$$

The constants for the individual reactions may be substituted into Eq. (5.17). These constants are the stability constant β_{nab} of the complex $MA_n(HA)_a B_b$, the dissociation constant K_{HA} of reagent HA, the distribution constant P_c of the chelate, the distribution constant P_{HA} of reagent HA, the distribution constant P_B of reagent B, the stability constant β_{rhx} of the complex $MA_r(OH)_h X_x$, and the ionic product of water, K_w. This leads to

$$D = \frac{\sum \beta_{nab} P_c K_{HA}^{n-r} [HA]_o^{n-r+a} [B]_o^b}{\sum P_{HA}^{n-r+b} P_B^b \beta_{rhx} K_w^h [H]^{n-r-h} [X]^x} \quad (5.18)$$

Thus the distribution coefficient is a function of pH, the concentration of HA in the organic phase, the concentration of reagent B in the organic phase, and the concentration of the complexing agent X in the aqueous phase. The equation will be further complicated if X and B can themselves react with other species in the system. Thus for a thorough understanding of an extraction process the dependence of the distribution coefficient on each of these parameters should be examined. A typical set of curves for these relationships is illustrated in Fig. 5.2. All the extraction curves shown have portions with different slopes, which usually means that complexes of several types are present in both phases. Curves of constant slope, which correspond to conditions under which only one type of complex is prevalent in each phase, have been disregarded.

It can be established from Eq. (5.18) that the slope of the graph of $\log D$ vs. pH at constant $[HA]_o$, $[X]$, $[B]_o$ is equal to $(n-r-h)$, for the graph of $\log D$ vs. $\log [HA]_o$ at constant pH, $[X]$, $[B]_o$, it is $(n-r+a)$, for the graph of $\log D$ vs. $\log [X]$, at constant $[HA]_o$, pH, $[B]_o$ the value is $(-x)$, whereas for the graph of $\log D$ vs. $\log [B]_o$ at constant pH, $[HA]_o$ and $[X]$ the slope is b.

Thus from the extraction curves, it is possible to predict the kind of complexes which predominate in the ranges corresponding to individual sections of constant slope on the curve. For instance on the curve of Fig. 5.2a $(n-r-h)$ assumes successively the values 2, 0, and -1. This suggests that the equilibria $[MA_2]_o/[M]$, $[MA_2]_o/[MA_2]$, and $[MA_2]_o/[MA_3]$ exist. The curve gives a good description of a system with:

$$D = \frac{[MA_2]_o}{[M] + [MA] + [MA_2] + [MA_3]}$$

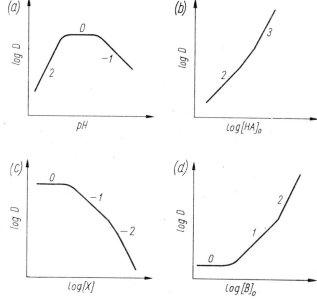

Fig. 5.2 Typical extraction curves

For the curve in Fig. 5.2b, a suitable expression is

$$D = \frac{[MA_2]_o + [MA_2HA]_o}{[M]}$$

for the curve in Fig. 5.2c

$$D = \frac{[MA_2]_o}{[M] + [MX] + [MX_2]}$$

and for the curve in Fig. 5.2d

$$D = \frac{[MA_2]_o + [MA_2B]_o + [MA_2B_2]_o}{[M]}$$

An even more general formula valid for all possible reactions occurring in both phases of an extraction system (including formation of polymeric species) is

$$D = \frac{\sum_{0}^{m}\sum_{0}^{a}\sum_{0}^{b} m\,[M_m A_n(HA)_a B_b]_o}{\sum_{0}^{m}\sum_{0}^{r}\sum_{0}^{h}\sum_{0}^{x} m\,[M_m A_r(OH)_h X_x]} \qquad (5.19)$$

5.2.3 Factors Influencing the Value of the Distribution Coefficient

Extraction is most often used in analytical practice to achieve a rapid, quantitative, and selective isolation of a definite element or group of elements. Thus, the distribution coefficient should be as high as possible for elements to be extracted into the organic phase and as low as possible for the elements to remain in the aqueous phase.

Analysis of Eq. (5.18) provides information on the effect of individual factors on the distribution coefficient.

Thus the distribution coefficient is increased if the stability constant of the complex to be extracted is increased, if the distribution constant of the complex is increased, if the reagent concentration is increased, and if the stability constant of complex $MA_r(OH)_h X_x$, the dissociation constant of reagent HA, the distribution constant of reagent HA, the hydrogen ion concentration, and the concentration of reagent X are decreased. Unfortunately, some of these desiderata are mutually exclusive, as discussed below.

5.2.3.1 THE EFFECT OF THE METAL ION

The nature of the metal ion in an extraction system is of prime importance, as it affects the value of the stability constant, the distribution constant of the complex which is extracted, and the stability constants of complexes of the metal in the aqueous phase.

Experimental data for the extraction of a number of metals with a variety of complex-forming reagents [57–60] show that often the most important factor associated with the metal is the stability constant, β_{nab}, of the complex extracted. Stability constants for some complexes and the $pH_{1/2}$ values* for the extraction of these complexes are listed in Table 5.2.

The stability series of the acetylacetonates is essentially the sequence of the extracted metals arranged according to decreasing $pH_{1/2}$ values for extraction of these compounds. Similar relationships are observed for the complexes of metals with other β-diketones, 8-hydroxyquinoline and cupferron.

The larger the stability constant of the complex, the more acidic the solutions from which extraction can be made, and the higher the distribution constant.

Another factor of some significance, associated with the metal, is its tendency to hydrolyse. The extent of hydrolysis depends on the hydrogen ion concentration $[H^+]$ and the stability constants β_{rhx} of the complexes

---

* $pH_{1/2}$ is the pH at which $D = 1$; cf.p.112

Table 5.2 Stability constants of complexes and corresponding $pH_{1/2}$ values

Complexing agent	Species	log β_{nab}	$pH_{1/2}$	Ref.
8-Hydroxyquinoline	Ag(I)	<2·0	8·5	[57]
	Be(II)	4·1	6·8	
	Co(II)	6·8	5·1	
	Ni(II)	7·9	3·2	
	In(III)	14·7	2·1	
	Fe(III)	20·5	1·5	
Dithizone	Pb(II)	18·6	8·0	[61]
	Zn(II)	20·1	5·0	
	In(III)	30·5	3·8	
	Bi(III)	36·9	2·0	
Acetylacetone	Mg(II)	6·1	9·4	[59, 62]
	Pb(II)	15·4	6·2	
	Al(III)	22·3	3·3	
	Fe(III)	26·2	1·6	
Thenoyltrifluoroacetone	Li(I)	−5·9	8·0	[58]
	La(III)	−2·6	4·2	
	Al(III)	−0·7	2·5	
	Th(IV)	1·9	0·5	
	Zr(IV)	3·0	−1·1	

$MA_r(OH)_h X_x$. The smaller the values of these parameters, the higher the value of the distribution coefficient. The effect of hydrolysis on the extraction process will be discussed in Section 5.2.3.3.

The concentration of the metal ion is also an important factor. The distribution constant of the metal should be independent of its concentration, if all the metal species concerned contain the same number of atoms of the metal, and the chelating agent is present in a suitable excess.

Numerous experimental data [63–65] support this theoretical proposition for metal concentrations not greater than $10^{-3}M$. Above this, increased metal concentration in the aqueous phase usually entails a decrease in the value of the distribution coefficient. The extraction efficiency of plutonium benzoylphenylhydroxamate as a function of the concentration of plutonium in the aqueous phase [66] is shown in Fig. 5.3, from which it can be seen that at concentrations of the metal exceeding $2 \times 10^{-3}M$ the extraction efficiency of plutonium decreases. This effect may be accounted for by the following:

(1) polymerization of hydroxy compounds in the aqueous phase, which occurs more easily with higher concentrations of metal;

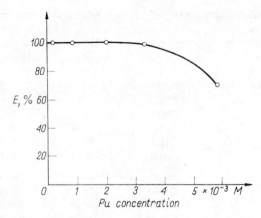

Fig. 5.3 Extraction efficiency for plutonium benzoylphenylhydroxamate *vs.* concentration of plutonium in the aqueous phase (from [66], by permission of the copyright holders Akad. Nauk SSSR)

(2) limited solubility of the extracted compound in the organic phase

(3) insufficient concentration of the organic reagent for the formation of the chelate in a quantity equivalent to the metal concentration.

On the other hand, if polynuclear metal species are extracted and only mononuclear species formed in the aqueous phase, the distribution coefficient should increase with metal concentration (e.g. [66a]).

5.2.3.2 The Effect of pH

When there are no side-reactions such as hydrolysis, the distribution coefficient increases as the pH of the aqueous phase increases.

The logarithmic form of Eq. (5.16) for the distribution coefficient is

$$\log D = \log K_{ex} + n\,\mathrm{pH} + n \log [\mathrm{HA}]_o \qquad (5.20)$$

or

$$-\frac{\log K_{ex}}{n} = \mathrm{pH} - \frac{\log D}{n} + \log [\mathrm{HA}]_o \qquad (5.21)$$

The plot of these equations is sigmoid in shape, and the position of the curves relative to the pH axis depends on the value of K_{ex}, and the slope on the value of n. It may be shown from Eq. (5.21) that if $D = 1$ ($E = 50\%$) and $[\mathrm{HA}]_o = 1$, the pH is constant and equal to $\log K_{ex}/n$. This value is termed $\mathrm{pH}_{1/2}$. This is a characteristic of the extraction process, and the difference between the $\mathrm{pH}_{1/2}$ values for two ions is a measure of the possibility of separating them.

As demonstrated, the slope of the extraction curves depends on the charge on the metal ion. It follows from Eq. (5.20) that an increase of one unit in pH results in an increase of a factor of 10 in the distribution coefficient for a singly charged metal ion, 100 for a doubly charged ion, and 1000 for a triply charged ion.

The dependence of $\log D$ on pH for three metal ions of different charges is illustrated graphically in Fig. 5.4. As seen from the curves, highly charged metal ions can be isolated from metal ions of lower charge by extraction at low pH. The theoretical curves of slope n sometimes agree well with experimental data. This is true, for instance, for extraction of the benzoylphenylhydroxamates of copper (for pH > 2) [67] and of thallium [68] (Figs. 5.5 and 5.6). In most cases, however, the slope of the extraction curves is smaller than n because of side-reactions in the aqueous phase.

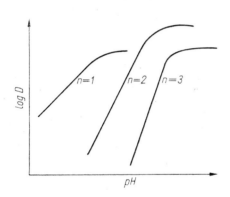
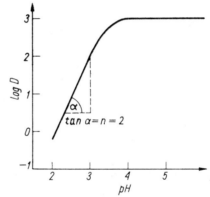

Fig. 5.4 Distribution coefficient vs. pH for ions of various charges

Fig. 5.5 Distribution coefficient of copper(II) benzoylphenylhydroxamate vs. pH; $n = 2$, $2 \times 10^{-4} M$ Cu, $0 \cdot 01 M$ BPHA in $CHCl_3$ (from [67], by permission of the author)

The curves for the extraction of zinc with a solution of 8-hydroxyquinoline in chloroform from a citrate medium of varied concentration, and based on the numerical data of Schweitzer [69], are shown in Fig. 5.7. The slope of the extraction curves is much smaller than n and depends on the concentration of citrate in the aqueous phase. Similar behaviour is observed in the extraction of uranium with benzoylacetone from a solution containing acetic acid at various concentrations [70] (Fig. 5.8).

8 Separation

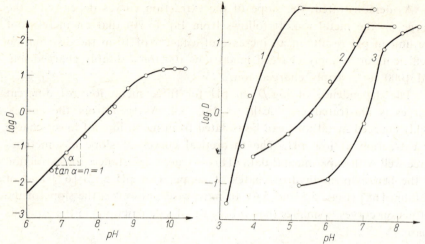

Fig. 5.6 Distribution coefficient *vs.* pH for thallium(I) benzoylphenylhydroxamate: $n = 1$, $1 \times 10^{-6} M$ Tl, $0{\cdot}1 M$ BPHA in $CHCl_3$ (after data from [68])

Fig. 5.7 Extraction curves for zinc oxinate from a solution containing citrate at various concentrations: $1 - 1 \times 10^{-4} M$, $2 - 1 \times 10^{-2} M$, $3 - 1 \times 10^{-1} M$ (after data from [69])

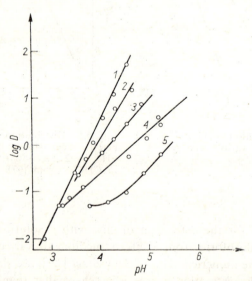

Fig. 5.8 Extraction curves for uranium(VI) with $0{\cdot}1 M$ benzoylacetone in benzene from a solution containing acetic acid: 1—no CH_3COOH, 2—$0{\cdot}01 M$ CH_3COOH, 3—$0{\cdot}02 M$ CH_3COOH, 4—$0{\cdot}05 M$ CH_3COOH, 5—$0{\cdot}10 M$ CH_3COOH (from [70], by permission of the copyright holders, North-Holland Publishers)

5.2.3.3 The Effect of Reactions Occurring in the Aqueous Phase

It has already been pointed out that real systems are extremely complex. Of necessity, they contain the reagents needed to bring the sample into solution, to adjust the pH of the aqueous phase to a suitable value and buffer it there, and to mask any elements that will interfere in the selective separation of the element or group of elements concerned. Some or all of these reagents may react with the ions of the metal to be extracted, and thus they may significantly influence the extraction process, particularly because their concentrations are usually high in comparison with the concentration of the metal. The magnitude of the effect of any competitive reactions depends, of course, on the stability of the complexes produced.

The most important interfering reactions are hydrolysis and polymerization of the metal ions, formation of complexes with reagents present in the aqueous phase, and formation of intermediate non-extractable complexes with the organic reagent. The effect of masking agents on the extraction of chelates of the elements has been studied by many authors. The most detailed investigations on the effect of the reactions occurring in the aqueous phase on the extraction process are those of Schweitzer [69, 71–73].

If the metal ions react with the anions in the aqueous phase, the equilibrium concentration of the metal in the aqueous phase which appears in the formula for the distribution coefficient must include the concentrations of all the forms in which the metal may occur, notably free metal ions, non-extractable complexes formed with the chelating agent, hydroxo complexes, and complexes with other ligands present in the aqueous phase. This can be written as follows:

$$C_M = [M^{n+}] + [MA^{n-1}] + \ldots + [MA_r^{n-r}] + [MOH^{n-1}] + \ldots$$
$$\ldots + [M(OH)_h^{n-h}] + [MX^{n-1}] + \ldots + [MX_x^{n-x}] + \ldots$$

or as

$$C_M' = [M^{n+}] + \sum_{i=1}^{I} \sum_{j=1}^{J} [M(L_i)_j]^{n-j} \tag{5.22}$$

where L_i denotes the ith complexing ligand present in the aqueous phase, I is the total number of ligands, j is the number of ligands in a particular complex and J is the maximum number of ligands L_i in the metal complex with that ligand.

Complex $M(L_i)_j^{n-j}$ is characterized by its stability constant

$$\beta_{ij} = \frac{[M(L_i)_j^{n-j}]}{[M^{n+}][L_i]^j}$$

Substitution of this in Eq. (5.22) leads to

$$C_M = [M^{n+}] + \sum_{i=1}^{I} \sum_{j=1}^{J} [M^{n+}] \beta_{i,j}[L_i]^j \qquad (5.23)$$

If F is defined as

$$F = 1 + \sum_{i=1}^{I} \sum_{j=1}^{J} \beta_{i,j}[L_i]^j \qquad (5.24)$$

then $C_M = [M^{n+}] F$ and

$$D = K_{ex} \frac{[HA]_o^n}{[H^+]^n F} \qquad (5.25)$$

It follows from Eq. (5.25) that reactions of the metal ions in the aqueous phase, such as hydrolysis, formation of complexes of the type MX_x and formation of intermediate, non-extractable complexes with the ions of the chelating agent of the type MA_r, all result in increased concentration of the metal in the aqueous phase under equilibrium conditions, that is, in a decreased distribution coefficient. The slope of the log D vs. pH curve is then smaller than n. As mentioned previously (Section 5.2.2), the slope is equal to the mean charge on the metal-containing species that occur in the aqueous phase. Hence, the larger the degree of hydrolysis of the metal ion, the lower the slope of its extraction curves.

Hydrolysis can be suppressed by increasing the acidity of the solution, or occasionally by raising the concentration of reagent HA in the organic phase. Hydrolysis may sometimes be avoided by reacting the metal with the organic reagent at low pH, then raising the pH to the optimum for the extraction process.

As shown previously, and as follows from Eq. (5.16), the log D–pH extraction curves are sigmoid in shape. Sometimes, however, minima are observed on the curves (Fig. 5.9). This is especially the case with metals such as niobium, tantalum, tungsten, zirconium, and iron. Minima have also been observed in the extraction of benzoylphenylhydroxamates of iron [74], bismuth [74], zirconium [75,76], hafnium [76], antimony [77], niobium and tantalum [78], of the oxinates of aluminium [79,80] and gallium [81], of the dinitro-oxinates of zirconium, niobium and tantalum [82], and of complexes of bismuth and hafnium with alkylphosphoric acids [83,84].

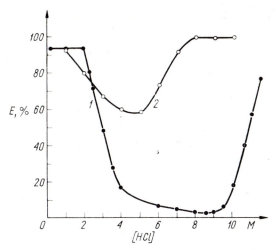

Fig. 5.9 Extraction curves for benzoylhydroxamates of antimony(III) (curve 1) (from [77], by permission of the Elsevier Publ. Co.) and zirconium(IV) (curve 2) (from [75], by permission of Polska Akad. Nauk)

Two kinds of minima occur; one is independent of the time of contact of the phases, and the other vanishes when the extraction time is increased. The minima arise from side-reactions such as hydrolysis, polymerization, and complex formation. Whether the minimum is permanent or temporary depends on which reaction is responsible. The concentration of the interfering (X) anions usually depends on pH. With increasing pH the concentrations of both these ions X and the ligand (A) ions rise, the increases being, of course, different and dependent on the dissociation constants and concentrations of the acids HA and $H_m X$. Thus, at certain pH values the extraction process may be suppressed because the interfering aqueous-phase reactions are enhanced. Also, at high pH the complexes of the metal with X may not be stable enough to prevent hydrolysis. The controlling factors here are the differences between the stability constants of the metal complexes with the ions A, X, and OH.

The temporary minima are usually connected with polymerization of the metal species or with formation of colloidal hydroxo compounds. Such minima are commonly observed in the extraction of compounds of zirconium, niobium and tantalum, which are notorious for their tendency to hydrolyse or polymerize. Polymeric species react much less readily than their parent monomers. Hence, at pH values at which polymers are likely to form, the reactions may be kinetically fairly complicated. Prolonged extraction times and elevated temperatures promote the reaction between the metal ions and the complexing agent and result in enhanced extraction efficiency and suppression of the minimum on the curve.

5.2.3.4 THE EFFECT OF THE CHELATING REAGENT

In every extraction system the nature of the chelating reagent determines the chelate stability constant, β_c, the chelate distribution constant, P_c, and the distribution and dissociation constants of the reagent, P_{HA} and K_{HA}, all of which influence the distribution coefficient.

Structural variations in the reagent molecule usually affect these constants, in particular P_{HA} and K_{HA}. In the series of reagents acetylacetone, benzylacetone, and dibenzoylmethane [52,73,85,86] the P_{HA} values are $10^{1\cdot4}$, $10^{3\cdot6}$ and $10^{5\cdot2}$, respectively, whereas the corresponding pK_{HA} values are 9·0, 8·7, and 9·2. In his studies on extraction with β-diketones Schweitzer [87] demonstrated that different substituents and chain lengths cause significant differences in the pK_{HA} values, e.g.

	pK_{HA}
2,4-pentanedione (acetylacetone)	9·03
3-ethyl-2,4-pentanedione	11·34
7-methyl-2,4-octanedione	9·72
1,1,1-trifluoro-7-methyl-2,4-octanedione	7·14
4,4,4-trifluoro-1-(2-thenoyl)-1,3-butanedione (thenoyltrifluoroacetone, TTA)	6·20

The effect of the structure of a reagent on its own properties and on the properties of the complexes formed has been demonstrated by Freiser [88] in his paper on dithizone derivatives.

From expressions involving the distribution coefficient and such quantities as β_c, P_c, K_{HA}, and P_{HA}, it can be seen that the more stable the extracted compound, the higher will be the distribution coefficient. Also, the larger the distribution constant of the chelate and the lower the distribution constant of the reagent, the lower can be the pH of the medium from which the extraction is made. It should be borne in mind, however, that the constants characteristic of the complexing agent are interdependent. It is well known that in general the stability of a complex is decreased by increased acidity of the reagent. Thus the effect on $pH_{1/2}$ of increasing the dissociation constant of the reagent is balanced by the decrease in the stability constant of the complex. However, it is possible to introduce into reagents substituents which, by their inductive effect, change the acidity of the reagent without greatly affecting the stability of its complexes, e.g. the trifluoromethyl group. Thus the net effect may be unpredictable and therefore the effect of the nature of the complexing reagent on the properties of the extraction system cannot and must not be discussed in terms of only one of the constants. Many studies have been concerned with comparison of reagents and their effect on $pH_{1/2}$;

in some cases a decrease in pK_{HA} is paralleled by a decrease in $pH_{1/2}$, while in others, the extraction conditions are similar despite differences in pK_{HA}.

For closely related compounds there is a relation between the chelate stability constant, β_c, and the reagent dissociation constant, K_{HA}:

$$(1/n) \log \beta_c = pK_{HA} + \text{const} \qquad (5.26)$$

Thus, if the value of pK_{HA} remains constant, the K_c values are also constant [59,89]. The two constants are linearly dependent. The greater the values of n, the less the effect of a decrease in pK_{HA} on the chelate stability, and the less the increased acidity of the reagent will be counterbalanced by lowered values of the complex stability constant.

Similar expressions can be derived for the relationship between the distribution constants of the complex and of the reagent [59,89,90]:

$$(1/n) \log P_c = \log P_{HA} + \text{const} \qquad (5.27)$$

Thus, the two quantities are directly proportional. If a suitable change is made in the structure of the reagent, e.g. by replacement of a hydrogen atom by an alkyl or aryl group, the distribution constant P_{HA} can be increased, and thus also P_c, which would result in a better extraction. Conversely, substitution of a CH_3 group for a hydrophilic group such as CF_3 results in decreased distribution constants. In that case, however, the effect of the substitution on pK_{HA} must also be taken into account.

It should be emphasised that the distribution of a reagent between the organic and aqueous phases often depends on the acidity of the aqueous phase. Figures 5.10 and 5.11 show the dependence of the distribution constants of 8-hydroxyquinoline [63] and of benzoylphenylhydroxylamine [91,92] on the acidity, for various extraction systems. In both cases, increased hydrogen ion concentration causes a decrease in the distribution coefficient of the reagent, which is due to its better solubility in water (caused by protonation) when the acidity is increased.

An important factor is the concentration of the chelate-forming reagent. Equation (5.18) shows that an increase in the concentration of the reagent in the organic phase causes the distribution coefficient to rise. Moreover, it may be seen from the expression for $pH_{1/2}$ [$pH_{1/2} = -(1/n) \log K_{ex} - \log [HA]_o$] that at higher $[HA]_o$ values, the $pH_{1/2}$ value is lowered, because the optimum extraction conditions are shifted to a higher hydrogen ion concentration range. If the compound extracted is of the type MA_n, a tenfold increase in the reagent concentration decreases $pH_{1/2}$ by one unit. These effects of varying reagent concentrations on extraction systems have been observed in practice. As an example, the extraction

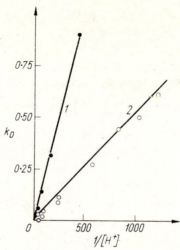

Fig. 5.10 Distribution coefficient of 8-hydroxyquinoline *vs.* $1/[H^+]$: 1 — benzene, 2 — chloroform, (from [63], by permission of the copyright holders, Akad. Nauk SSSR)

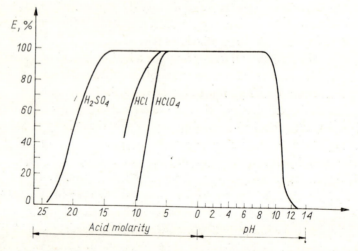

Fig. 5.11 Distribution of benzoylphenylhydroxylamine between chloroform and water as a function of the acidity of the aqueous phase (by permission of J. Chwastowska [91])

curves for thallium benzoylphenylhydroxamate [58] are shown in Fig. 5.12. A change in the concentration of BPHA (benzoylphenylhydroxylamine) from $0.01M$ to $0.1M$ results in a decrease in $pH_{1/2}$ from 9.2 to 8.2. Hence, for hydrolysable elements it is expedient to perform the extraction at higher concentrations of the complexing agent. The range is limited, of course, by the solubility of the reagent in the organic phase.

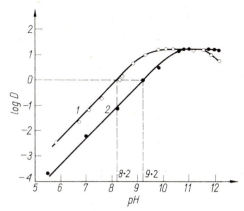

Fig. 5.12 Extraction curves for thallium(I) benzoylphenylhydroxamate for various reagent concentrations: 1—0·1M BPHA in CHCl₃, 2—0·01M BPHA in CHCl₃ (after data from [68])

The effects of varying the concentration of the chelating agent are more involved if the chelates extracted are of the type $MA_n(HA)_r$ or $MA_n(HA)_r(OH)_h$.

5.2.3.5 The Effect of Solvent

The suitability of a given solvent for an extraction depends on two distinct groups of properties, *viz.* the properties of practical significance, and the properties affecting the mechanism.

The first group includes the physical properties of solvents such as density, viscosity and surface tension, which are responsible for a good and fast phase separation; and also the boiling point and the vapour pressure at room temperature—solvents which are too volatile are troublesome in use and, whenever possible, a solvent with higher boiling point is preferred.

Among the properties that directly affect the thermodynamics of the process are heat of vaporization, heat of solidification, and the molar volume, which affect the miscibility of the solvents and the solubility of the extracted compounds. Dielectric constant, polarity and polarizability also belong in this group of properties. They influence strongly the solvent-solvent and solvent-solute interactions.

An important chemical property of a solvent is the basicity of the solvent molecules. Compounds with functional groups containing electronegative atoms such as oxygen or nitrogen increase in basicity with increasing electron density on these atoms. Mention should also be made of hydrogen-bonding ability, which is responsible for a tendency to produce polymers.

Thus the physical properties of solvents such as polarity, polarizability, molar volume, and also the chemical properties such as acid–base strength, and hydrogen-bonding ability, are responsible for the behaviour of a solvent, i.e. they influence the interactions with other organic liquids, with non-polar solutes, and with water and electrolytes. Pimentel and McClellan [93] introduced a classification of organic compounds that can be used for solvents (Table 5.3).

Table 5.3 Classification of solvents
(from [93], by permission of the copyright holders, G. C Pimental and A. L McClellan)

Class of compound	Characteristic groups		Examples
	proton-donor	electron-donor	
A	+	−	mainly halogen derivatives: $CHCl_3$, C_2HCl_5, etc.
B	−	+	ketones, aldehydes, ethers, tert. amines, esters, olefins
AB	+	+	H_2O, alcohols, carboxylic acids, prim. and sec. amines
N	−	−	saturated hydrocarbons, CS_2, CCl_4

Depending on its structure and on the kind of compound to be extracted, an organic solvent may be inert or active with respect to the compound. Some interesting work on this aspect has been done by Alimarin and Zolotov [32,94,95]. It was demonstrated that the nature of the solvent is of particular significance in the extraction of co-ordinatively unsaturated chelate complexes in which the free co-ordination sites are occupied by water molecules. Such hydrated compounds are poorly extracted into non-polar unreactive solvents by comparison with polar solvents capable of co-ordinating to the metal ion and replacing water molecules in the complex. Thus solvents such as alcohols, ketones, or esters are better extractants than hydrocarbons for these compounds. The work of Alimarin and Zolotov [94] on the extraction of neptunium thenoyltrifluoroacetonate indicates that this compound is not extractable by carbon tetrachloride, chloroform, or benzene; the best extraction solvent is isobutyl alcohol. Busev and Ivanov [96] have noticed that the compound of U(VI) with 4-(2-pyridylazo)resorcinol (PAR) is extracted better by alcohols and esters. Schweitzer showed that Tl(I) oxinate is readily extracted with isobutyl alcohol, but that it cannot be extracted with either alcohols or carbon tetrachloride.

Solvents can be classified, according to their suitability for extraction of co-ordinatively unsaturated chelates, as follows: alcohols > ketones > complex ethers > simple ethers > halogen derivatives of hydrocarbons > hydrocarbons.

Extraction of co-ordinatively saturated chelates is less dependent on the kind of the solvent used. Here the old rule of thumb that "like dissolves in like" is still useful. For chelates with no hydrophilic groups but containing hydrocarbon rings, the most suitable extractants will be low-polarity hydrocarbons and their halogen derivatives.

Dyrssen [50] has shown that co-ordinatively saturated complexes are extracted better with undissociated solvents (between the molecules of which only weak van der Waals forces are operative) such as benzene or carbon tetrachloride, than with ethers or alcohols.

The effect of solvent type on the extractability of complexes can be used to advantage to enhance the extraction selectivity [97,98].

Even though the techniques of extractive separation are widely used, the theoretical studies on the extraction process are still inadequate. Most investigations to date deal with reactions occurring in both phases, the composition of the extracted complex, etc., and little work has been done on the relationships between the physicochemical properties of the solvent and the distribution of the metal, despite the importance of these for understanding how to make a correct choice of solvent [1317,1318].

The factors controlling the course and the mechanism of extraction processes are currently being studied, and the effect of the solvent is one of them. It is now apparent that the empirical attempts to correlate the distribution coefficient with surface tension at the interface or dielectric constant are of little assistance. Any approach should involve consideration of the relationship between the thermodynamic functions of the solutions and the energy of molecular interaction of the solute in both phases. For the explanation of the solvent effect the commonest method is the Hildebrand–Scatchard theory of regular solutions. This was applied to extraction studies by Kemula and Buchowski [99,100], Siekierski et al. [101–106], Freiser and Mottola [107], Suzuki and co-workers [108–117], and Rydberg et al. [117a].

From the theory of regular solutions an expression can be derived which correlates the distribution constant with properties of the extractive system such as solubility parameters and molar volumes of individual components of the system:

$$2 \cdot 3 \, RT \log k_x = V_c \, [(\delta_c - \delta_w)^2 - (\delta_c - \delta_o)^2] \qquad (5.28)$$

where k_x is the distribution constant, in mole fraction units, δ_c is the solubility parameter and V_c the molar volume of the compound that is distributed, δ_w is the solubility parameter for the aqueous phase, and δ_o is the solubility parameter for the organic phase.

Freiser, in his studies on the distribution coefficient of 8-hydroxyquinoline in water–halogenated hydrocarbon systems, has shown that a plot of k_x against the solubility parameters of the solvents used is roughly a parabola described by Eq. (5.28) (Fig. 5.13).

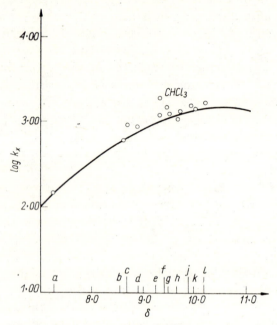

Fig. 5.13 Distribution constant *vs.* solubility parameter for 8-hydroxyquinoline in various solvents: a — n-hexane, b — carbon tetrachloride, c — bromobutane, d — toluene, e — chloroform, f — 2-nitropropane, g — 1-nitropropane, h — nitroethane with dichloromethane, j — 1,2-dichloromethane, k — dichlorobenzene, l — bromochloromethane with dibromomethane (from [107], by permission of the copyright holders, Pergamon Press Ltd.)

Wakayashi *et al.* verified the equations based on the theory of regular solutions in their studies on the extraction of β-diketones and their scandium chelates into inert solvents [108,109].

A relationship between the distribution coefficient of the chelate and the distribution constant of the chelating agent was also derived. For a given solvent this relation may be expressed as follows:

$$\log D_M = n \log P_{HA} + \text{const} \tag{5.29}$$

where n is the ratio of the molar volumes of chelate and chelating agent, and the constant is dependent on the type of solvent used.

Equation (5.2a) is approximately obeyed by the Sc–TTA system (cf. Fig. 5.14) [109].

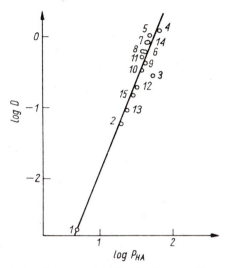

Fig. 5.14 Correlation between the distribution coefficient for TTA and its scandium chelate in: 1—n-hexane, 2—carbon tetrachloride, 3—chloroform, 4—methylene chloride, 5—ethyl bromide, 6—benzene, 7—chlorobenzene, 8—o-dichlorobenzene, 9—toluene, 10—m-xylene, 11—p-xylene, 12—mesitylene, 13—isopropyl benzene, 14—thiophene, 15—1,2,4-trichlorobenzene (from [109], by permission of the copyright holders, Pergamon Press Ltd.)

The studies were extended to extractions with oxygen-containing solvents. Despite the possibility of interaction of solvent and solute molecules, the theory of regular solutions was found to apply, at least qualitatively, even in systems where some kind of mutual interaction cannot be disregarded. Systems studied were Sc–TTA–alcohols [111,112], Sc–TTA–esters [113], and Sc–TTA–ethers [114], and also zinc [115] and uranium [116,117] systems.

Freiser has also shown that the magnitude of the distribution coefficients of 8-hydroxyquinoline and its complex with copper depend greatly on the polarity of the solvent. A plot of $\log k_x$ vs. dielectric constant (Fig. 5.15) is approximately a parabola with a maximum at a dielectric constant of 5–7.

The magnitude of the distribution constant is also considerably influenced by the mutual solubility of the two phases [107,118]. Figure 5.16 shows the dependence of the distribution coefficient of 8-hydroxyquinoline and its derivatives on the dielectric constant of the pure solvent and the molar solubility in water, S [107].

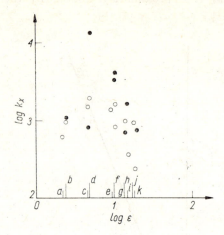

Fig. 5.15 Distribution constant for 8-hydroxyquinoline (open circles), and for copper 8-hydroxyquinoline (full circles) vs. solvent dielectric constant: a — carbon tetrachloride, b — toluene, c — isopentyl acetate, d — chloroform, e — dichloromethane, f — o-dichlorobenzene, g — octan-1-ol, h — methylpentan-2-one, i — 3-methylbutan-1-ol, j — pentan-3-one, k — butan-1-ol (from [107], by permission of the copyright holders, Pergamon Press Ltd.)

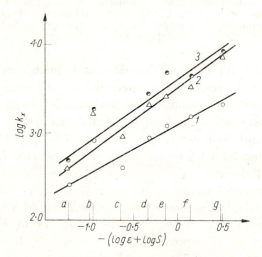

Fig. 5.16 Distribution constant for 8-hydroxyquinoline (1), 2-methyl-8-hydroxyquinoline (2), and 4-methyl-8-hydroxyquinoline (3) vs. dielectric constant (ε) and molar solubility in water (S) of pure solvents: a — butan-1-ol, b — diethylketone, c — isopentyl alcohol, d — methyl isobutyl ketone, e — dichloromethane, f — isopentyl acetate, g — chloroform (from [107], by permission of the copyright holders, Pergamon Press Ltd.)

The effects of the physicochemical properties of the solvent on the extraction process have been studied for other extraction systems [119–123,1319]. It has been found that, despite the correlation between dielectric constant and distribution coefficient, the dielectric constant is not the only factor controlling the extraction process.

The effect of the solvent on the extraction properties of a given system may also arise from interactions between the solvent molecules and the molecules of the complexing reagent. This is of particular significance with organic reagents which can dimerize or even polymerize, since these effects adversely affect the extraction process.

Among reagents which behave in this way are the organophosphorus acids [118,124–126]. The degree of dimerization was found to be very high in non-polar solvents, such as kerosene or carbon tetrachloride, whereas in solvents of higher polarity, the degree of dimerization decreases. In polar solvents such as alcohols, ketones or ethers a strong interaction takes place between the reagent molecules and the solvent molecules, and this counteracts dimerization. On this basis, solvents may be classified into several groups [127]. Alcohols with the highest association constants constitute the most basic group. Among these octyl alcohol exhibits the strongest basic properties, whereas cyclohexanol and hexanol are of lower basicity. Since alkyl groups repel electrons, whereas phenyl groups attract them, the basicity of aliphatic alcohols is higher than that of aromatic ones. In the second solvent group are ethers, esters and ketones with association constants of 10–50, the ketones being more basic than ethers or esters, from experimental evidence. The third group includes solvents of very low association constant, such as hydrocarbons and their halogen derivatives.

The following series of solvents [127,128] is arranged according to increasing ability to interact with solute molecules: kerosene < carbon tetrachloride < chloroform < ketones < alcohols.

As mentioned previously, no unambiguous criteria have yet been advanced for selection of a suitable solvent, and the lack of unanimity as to which properties are of prime significance in the extraction process is indicative of the fact that the suitability of a particular solvent depends on not just one property, but on the combination of several.

It is important that the solvent is of adequate purity, and that it is stable. Decomposition products, by-products of the synthesis, and other impurities present in the solvent may be detrimental, especially to the formation of the required complex. Thus, the solvent used should be thoroughly purified. Purification and stabilization of solvents have been studied by many workers [129].

5.2.3.6 THE EFFECT OF TEMPERATURE

Since extractions are usually done at room temperature, in most cases the effect of temperature is of little practical significance, but detailed investigations on the effect of temperature on the extraction process have been made by Schweitzer *et al.* [118,130,131] and other authors [1320-1323]. Their studies showed that as temperature increases, $pH_{1/2}$ increases and the distribution constant decreases. From Schweitzer's data [131] extraction curves have been plotted for cadmium oxinate at various temperatures (Fig. 5.17).

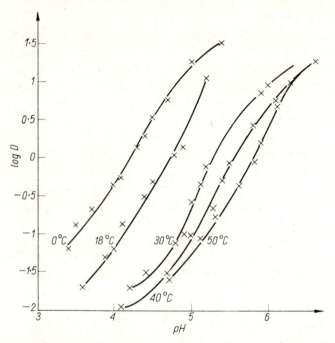

Fig. 5.17 Distribution coefficient *vs.* pH at various temperatures for the system: cadmium–chloroform–8-hydroxyquinoline solution (from [131], by permission of the copyright holders, Elsevier Publ. Co.)

Table 5.4 Effect of temperature on the constants for the extraction of cadmium oxinate with chloroform
(from [131], by permission of the copyright holders, Elsevier Publ. Co)

Temperature, °C	log P_{HA}	pK_{HA}	log K_{ex}	log P_c	log β_c
0	2·7	9·9	−4·4	4·9	21·2
18	2·7	9·8	−5·6	4·5	20·1
30	2·6	9·7	−6·4	3·9	19·4
40	2·5	9·6	−7·2	3·7	18·6
50	2·5	9·5	−7·6	3·0	18·1

The numerical values of the constants for the extraction are listed in Table 5.4. As seen, the increase in temperature is paralleled by a rise in $pH_{1/2}$ and in the reagent dissociation constant, whereas the distribution constant of 8-hydroxyquinoline and the complex stability constant decrease.

The effect of temperature is different if polymerized or hydrated molecules are present [132]. Then increased temperature favours dehydration and depolymerization, with resulting higher extraction efficiency.

5.2.4 Extraction Kinetics

In chelate extraction systems the problem of the extraction kinetics is involved, and equilibrium is often reached more slowly than with ion-association systems. Extraction processes involve several chemical reactions and distribution processes, and the overall rate is determined by the slowest step. The following chemical reactions should be taken into account [133–135].

1. Reactions related to the conversion of inactive forms of metal ions into reactive forms, such as decomposition of polymers or hydrolysis products.

2. Reactions related to the formation of an active form of the organic reagent: for instance the reagent may occur in enol or keto forms, only one of which can react with the metal ions.

3. The process of formation of the extractable complex, including reactions of the metal ion with the complexing reagent, and the filling of the co-ordination sphere with organic solvent molecules, etc.

The extraction kinetics are also significantly affected by mass transfer [1324,1325]. Three types of extraction can be distinguished: the kinetic type—where the extraction rate is controlled by the rate of a chemical reaction, the diffusion type—where the rate depends chiefly on the rate of mass transfer, and the combined type—where the rate of extraction depends on the rates of both processes. Recently Fomin has studied the kinetics of extraction [136,137,1326].

The extraction process may be depicted diagramatically as follows:

Organic phase

$$\xrightarrow[\text{etc.}]{\text{depolymerization}} HA \qquad\qquad\qquad\qquad\qquad MA_n$$

$$\updownarrow \qquad\qquad\qquad\qquad\qquad \updownarrow$$

Aqueous phase $HA \rightleftharpoons H^+ + A^-$

$$\xrightarrow[\text{dehydration etc.}]{\text{depolymerization}} M^{n+} + A^- \rightleftharpoons MA^{(n-1)} + \ldots MA_{n-1} + A^- \rightleftharpoons MA_n$$

The extraction rate may be expressed by the following formula

$$v = -d[M]/dt = k[M][A]^n = k\frac{K_{HA}^n}{P_r^n}\frac{[M][HA]_o^n}{[H^+]^n} \qquad (5.30)$$

In extraction processes where the compound to be extracted is formed and can exist in the aqueous phase, the reaction rate will not depend directly on the total concentration of this compound in the aqueous phase; it does depend on the magnitude of the distribution coefficient; the lower this is, the faster the reaction should proceed. Hence the choice of solvent is often crucial to the extraction speed. For example, in the extraction of dithizonates, the rate of extraction depends on whether the solvent used is chloroform or carbon tetrachloride. The higher extraction rate observed with carbon tetrachloride may result from the fact that the distribution constant is much larger for the chloroform–water system than for the carbon tetrachloride–water system [138].

For systems in which the compound to be extracted is formed at the interface, the extraction rate is related to the concentration of the reagent in the organic phase, and will be higher for solvents for which the reagent distribution constant is high [139,140].

As the reagent in such systems will be in the undissociated form in the organic phase, if other parameters are kept constant, the reaction rate is higher at high pH in the aqueous phase. This was found for complexes of dithizone [138,141,142] and of 8-hydroxyquinoline [57,143], for a number of elements.

The rate-determining step may be any one of the extraction stages. Usually, it is the chelate-formation step [144], but there are some systems in which the mass-transfer process is more important [145]. This process involves the diffusion in each phase, the transfer through the interface, and the reaction occurring at the interface [140,146].

An example of a study designed to establish the rate-determining steps in the extraction process is the work of Freiser and co-workers on the extraction kinetics of complexes of Cd, Zn, Co and Ni with dithizone and its derivatives [141,147,148]. In each case examined the rate-determining step of the extraction process was found to be the formation of a 1:1 complex in the aqueous phase. The rate of breaking the bond of the metal ion with water molecules is also a major factor. In the formation of zinc dithizonate, the rate-determining step is the removal of the water molecules from the hydrated species $Zn(H_2O)_6^{2+}$.

More recently Freiser [149] has studied the effect of some other complexing agents on the extraction kinetics of zinc and nickel dithizonates. If a complexing agent X is present in the aqueous phase, transitional complexes, MX_n, can form, and these then react with dithizone. The presence of ions of X may cause the rate of extraction of the dithizonate to increase if formation of MX_n is fast, and if this complex reacts with dithizone faster than do the hydrated metal ions.

Table 5.5 The rate constants k_1 for the reaction of metals with dithizone in the presence of ligand X (k_0 is the rate constant for perchlorate solution)
(from [149], by permission of the copyright holders, the American Chemical Chemistry)

Ligand X	Ni k_1/k_0	Zn k_1/k_0
Acetate	1·0	25·0
Thiocyanate	2·5	10·0
Mercaptoacetate	14·0	7·0
Oxalate	1·0	1·0
Tartrate	1·0	1·0

Freiser's data, shown in Table 5.5, indicate that ligands such as acetate, thiocyanate, and mercaptoacetate have a pronounced effect on the rate of formation of nickel and zinc dithizonates. Acetates accelerate the reaction of dithizone with zinc, and mercaptoacetates the reaction with nickel.

This effect has been examined for some other systems. It has been established that $Cu(OH)^+$ ions react with EDTA faster, and $CuAc^+$ ions slower, than do the hydrated copper ions [150,151]. On the other hand the liberation of co-ordinated water molecules is 13 times faster from the 1:1 complex of nickel with glycine than from hydrated nickel ions [152].

The various effects of individual complexing agents on the kinetics of the extraction process may be utilized in analytical practice. Specifically, the fact that the rate of formation of zinc dithizonate is 25 times that of nickel dithizonate in acetate media may serve as a basis for separation of the two elements. Attempts to separate certain ions on the basis of the differences in their extraction kinetics have been made [153,154]. McClellan and Sabel [155] have investigated the extractive separation of Cd(II), Zn(II), Co(II) and Ni(II) on the basis of observations that the rates of formation of the dithizonates, thus also the extraction rates, decrease from Cd to Ni.

The data on the kinetics of extraction of zinc, cobalt, and nickel dithizonates illustrated in Figs. 5.18–5.20 show that, at low pH, equilibrium in these systems is attained after a fairly long time, as long as 10 days for nickel. Optimum conditions (pH, [H$_2$Dz], time) were determined for separation of the following pairs of elements: Ni–Zn, Zn–Co, and Ni–Co (Table 5.6).

Table 5.6 Optimum conditions for separation of Ni, Zn, and Co as dithizonates
(from [155], by permission of the copyright holders, the American Chemical Society)

Ions	pH	H$_2$Dz concentration, M	Time, min	E, % Ni	E, % Zn	E, % Co
Ni, Zn	5·5	3·12 × 10^{-4}	30	2·5	99·0	
Zn, Co	5·0	1·00 × 10^{-4}	30		95·9	8·6
Ni, Co	6·1	1·00 × 10^{-3}	30	32·5		95·8

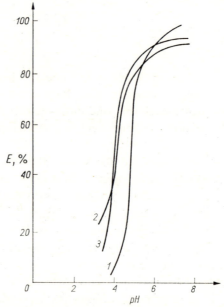

Fig. 5.18 Extraction curves for zinc dithizonate after various times of shaking: 1–15 min, 2–10 hr, 3–34 hr; [HDz] in CHCl$_3$ = 2·4 × 10^{-5}M (from [155], by permission of the copyright holders, the American Chemical Society)

The best results were achieved for separation of zinc from nickel, but quantitative separation of nickel and cobalt was impracticable. Niobium and tantalum have also been successfully separated by extraction, by utilizing differences in the kinetics of extraction of the 8-hydroxyquinoline complexes from tartrate media [156]. Zolotov *et al.* [157] have shown that the rate of extraction of a trace element as a chelate complex, in the presence

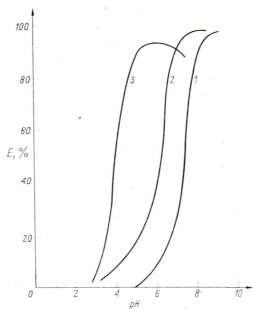

Fig. 5.19 Extraction curves for cobalt dithizonate after various times of shaking: 1—5 min, 2—1 hr, 3—20 hr; [HDz] in $CHCl_3 = 1.95 \times 10^{-4} M$ (from [155], by permission of the copyright holders, the American Chemical Society)

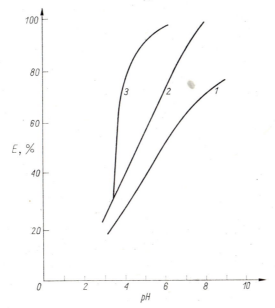

Fig. 5.20 Extraction curves for nickel dithizonate after various times of shaking: 1—15 min, 2—1 hr, 3—10 days; [HDz] in $CHCl_3 = 2.4 \times 10^{-4} M$ (from [155], by permission of the copyright holders, the American Chemical Society)

of large quantities of another element, is the higher, the lower the extraction constant of the complex of the macro component.

One method of increasing the rate of extraction is to form mixed complexes. It was known previously that, although quantitative extraction of nickel dithizonate from solutions of low pH (ca. 3), is possible, a high concentration of dithizone and a prolonged extraction time are required. On the other hand, in the presence of pyridine [158] or 1,10-phenthanthroline (phen) [159] a mixed complex is formed:

$$Ni^{2+} + 2H_2Dz_o + phen_o \rightleftharpoons Ni(Dz)_2phen_o + 2H^+$$

The rate of extraction of the mixed complex is very different from that of nickel dithizonate alone: the times required to attain equilibrium are 30 min and 10 days, respectively.

Mixed complexes may be used in extraction processes not only to increase extraction rates, but also to improve certain other properties of extraction systems. Thus, the complexes of dithiol with low-valency metal ions are not extractable, but in the presence of phenanthroline, the mixed complexes formed with zinc and with other bivalent metal ions are readily extractable [160].

Table 5.7 Extraction constants of mixed zinc complexes
(from [162], by permission of the copyright holders, the American Chemical Society)

Compound extracted	log K_{ex}
$Zn(Ox)_2 \cdot HOx$	-5.2
$Zn(Ox)_2 \cdot phen$	-2.2
$Zn(phen)_2 Ox^+, ClO_4^-$	6.8
$Zn(phen)_3^{2+}, 2ClO_4^-$	13.7

Ox = 8-hydroxyquinoline, phen = phenanthroline

It has often been shown that complexes of nickel with sulphur-containing ligands, such as dialkyldithiophosphates and alkylxanthates (with the exception of diethyldithiocarbamates), have a tendency to form mixed complexes with organic bases. The mixed nickel complexes containing pyridine, bipyridyl, phenanthroline, and ethylenediamine, are more useful analytically, as they are more stable, more soluble and more deeply coloured [161]. Mixed complexes of zinc with 8-hydroxyquinoline and 1,10-phenanthroline have higher extraction constants. As seen from Table 5.7, the extraction constant increases with the number of phenanthroline molecules in the compound extracted [162].

The mixed complexes of copper with 8-hydroxyquinoline and arylhydroxycarboxylic acids have been quite extensively investigated [163–165]. This problem will be discussed in more detail in Section 5.2.5.

Kawamoto and Akaiwa [166] recommend increasing the rate of attaining equilibrium by forming the compound to be extracted in a homogeneous phase by adding the reagent in solution in a solvent miscible with water. On completion of the reaction, phase separation is achieved by adding a salting-out agent.

5.2.5 Synergism

The term synergism was first applied to extraction systems by Blake *et al.* to describe the increase observed in the extraction of uranium(VI) with dialkylphosphoric acid on addition of an organophosphorus compound of the type R_3PO (where R is an alkyl or alkoxyl group). The synergic effect was originally believed to be operative for uranium alone [167], but later it was found to occur in the extraction of Pu(IV) and Pu(VI) [168], ter- and quadrivalent actinides and lanthanides ($Z > 64$) [169–171], bivalent elements [171,172], alkali metals [173] and other systems [174,175].

The occurrence of a synergic effect is usually connected with some processes that occur on account of the complexity of the organic phase. The organic phase may contain the following components: extractant (E), solvent (S), and diluent (D). The solvent and diluent form a phase immiscible with water. The extractant and the solvent may react with metal ions to yield compounds soluble in the organic phase. Siekierski [176] proposed that the following possible mixtures be distinguished: S_1+S_2, $S+E$, $S+D$, E_1+E_2, $E+D$, D_1+D_2, although in fact there are also ternary systems, such as: E_1+E_2+D, $E+S+D$, and E_1+E_2+S.

If the complex-forming reagent is denoted by HA, and the synergist as B, then an extractive system with synergic effect may be described as follows:

$M^{x+} + xHA_o \rightleftharpoons MA_{xo} + xH^+$,   extraction constant K_x

$MA_{xo} + yB_o \rightleftharpoons MA_x B_{yo}$,   stability constant β_{xy}

and the overall reaction is

$M^{x+} + xHA_o + yB_o \rightleftharpoons MA_x B_{yo} + xH^+$,   extraction constant $K_{x,y}$.

Assuming that two compounds MA_x and $MA_x B_y$ are extracted in the process, the distribution coefficient is given by

$$D = \frac{[MA_x]_o + [MA_x B_y]_o}{[M]} \qquad (5.31)$$

If the distribution constants P_x and $P_{x,y}$ and the stability constants, β_x and $\beta_{x,y}$, of the individual compounds, are introduced, this becomes

$$D = (P_x \beta_x + P_{x,y} \beta_{x,y})[B]^y[A]^x \qquad (5.32)$$

After substitution into Eq. (5.31) of the values for the extraction constants K_x and $K_{x,y}$, the expression for the distribution coefficient becomes

$$D = \frac{(K_x + K_{x,y})[B]_o^y[HA]_o^x}{[H^+]^x} \qquad (5.33)$$

or

$$D = \frac{K_x(1+\beta_{x,y})[B]_o^y[HA]_o^x}{[H^+]^x} \qquad (5.34)$$

According to Siekierski's definition the synergic effect occurs when the experimental distribution coefficient ($D_{x,y\,\text{exp}}$) for a mixture of two extractants is higher or lower than the distribution coefficient calculated on the principle of simple additivity ($D_{x,y\,\text{add}}$):

$$D_{x,y\,\text{exp}} \neq D_{x,y\,\text{add}}$$

To quantify the synergic effect, the synergic coefficient (s.c.) has been defined as

$$\text{s.c.} = \log \frac{D_{x,y\,\text{exp}}}{D_{x,y\,\text{add}}} \qquad (5.35)$$

or, after substitution for D

$$\text{s.c.} = \log(1 + \beta_{x,y}[B]_o^y) \qquad (5.36)$$

If s.c. > 0, a positive synergic effect is said to occur. If there is an antagonistic effect (negative synergism) s.c. < 0. If there is no synergic effect, s.c. = 0.

A great deal of work has been done on the synergic effect and its mechanism. Even the simplest systems, with alkylphosphoric acids as extractants, are complicated by dimerization of the acids in most organic solvents. The resulting HA_2^- species can react as such with the metal ions, because $O=P(OR)_2 \cdot OH \ldots OP(OR)_2-O^-$ can form an eight-membered chelate ring. The data of Zangen [177], and Hahn and Vander Wall [178] suggest that the synergist B combines directly with the uranium atoms in 1:1 or 1:2 ratio; in some of the possible mechanisms advanced by Zangen the co-ordination number of uranium changes from 6 to 7 and 8.

The nature and extent of the synergic effect depend primarily on the properties of the synergist itself, but the properties of the metal, and the diluent, and the nature of the chelate-forming reagent, can also contribute to it.

The formation of mixed complexes is believed to be the main cause of the synergic effect. It can be demonstrated that if the molecules of a neutral donor (synergist) have removed water molecules from an extractable chelate, the complex becomes less hydrophilic, so the extraction is improved.

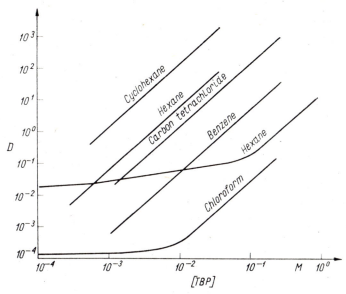

Fig. 5.21 Effect of diluent on the distribution coefficient of triply charged metal ions in a synergic system TTA/TBP/diluent (from [179], by permission of the copyright holders, Pergamon Press Ltd.)

The extent of the synergic effect was found to depend on the type of inert diluent used in the extraction process. For the system M/HA/B/D (HA is chelating reagent, B the synergist, and D the diluent, e.g, cyclohexane, hexane, carbon tetrachloride, benzene, chloroform) the distribution coefficient of compound $MA_x B_y$ depends appreciably on the solvent chosen. The distribution coefficient has been found to increase with solvent in the order: $CHCl_3 < C_6H_6 < CCl_4 < C_6H_8 < C_6H_{12}$, which is the order of decreasing water solubility [179–181] and decreasing solvent polarity [182]. In solvents of higher polarity mutual interactions of the diluent and the synergist molecules may be involved, and this would tend to inhibit formation of mixed complexes. Figure 5.21 shows the dependence of the synergic effect on the diluent used, in the extraction of americium–TTA and promethium–TTA (thenoyltrifluoroacetone) complexes with tri-n-butyl phosphate (TBP) [179]. The values of the distribution constants and of the stability constants $\beta_{x,y}$ shown in Table 5.8 for the synergic systems Zn–PMBP*–TBP and Zn–TTA–TBP with chloroform or carbon tetrachloride as diluents also support this conclusion.

* 1-Phenyl-3-methyl-4-benzoyl-5-pyrazolone.

Table 5.8 Extraction constants of ZnA_2 ($K_{ex(x)}$), and ZnA_2B ($K_{ex(x,y)}$) and the stability constants ($\beta_{x,y}$)
(from [181], by permission of the copyright holders, Pergamon Press Ltd.)

Diluent	log $K_{ex(x)}$		log $K_{ex(xy)}$(TBP)		log $\beta_{x,y}$(TBP)	
	PMBP	TTA	PMBP	TTA	PMBP	TTA
Chloroform	−6·18	−8·13	−4·47	−5·44	1·75	2·69
Carbon tetrachloride	−5·75	−8·04	−3·26	−3·70	2·49	4·34

Organophosphorus compounds are commonly used as synergists. The extent of the synergic effect depends on the donor ability of the oxygen in the compound used. Thus, the distribution coefficient is increased more when a more polar phosphine oxide is employed instead of tributyl phosphate. Studies on synergism in the U–TTA system indicate that the highest distribution coefficient is observed in mixtures of TTA and TBPO (tributylphosphine oxide), and the distribution coefficient is enhanced by a factor of not less than 10^4 compared with that for TTA used alone (Fig. 5.22). The synergic ability of organophosphorus compounds increases

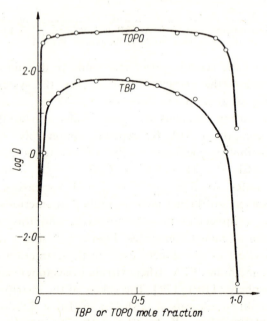

Fig. 5.22 Distribution coefficient of uranium(VI) in the systems: HNO_3/TTA/TBP or TBPO/cyclohexane: $[HNO_3] = 0·01M$; [TTA]+[TBP] or [TOPO] = $0·02M$ (from [183], by permission of the copyright holders, North-Holland publishers)

in the following order: phosphates, $(RO)_3PO$ < phosphonates, $R(RO)_2PO$ < phosphinates, $R_2(RO)PO$ < phosphine oxides, R_3PO [183]. In a study of synergism in the extraction of europium–salicylic acid complexes [184] it was found that the substitution of a phenyl for a butyl group in organophosphorus compounds results in increased donor ability of the oxygen. Similarly, substitution of arsenic for phosphorus results in the formation of more stable addition compounds, since $\equiv As \to O$ is a better co-ordinating group than $\equiv P \to O$. The sequence of efficiency for these synergists is: TBP < TPPO < TPAsO (triphenylarsine oxide).

Other common synergists are heterocyclic bases [185–190], sulphoxides [191,192] carboxylic acids [193], phenols [194], and amines [194a]. Figure 5.23 shows the synergic effect of pyridine and its analogues in the extraction of the cobalt–thenoyltrifluoroacetone complex. The synergic effect of individual bases increases with the basicity of the reagent. Reactive

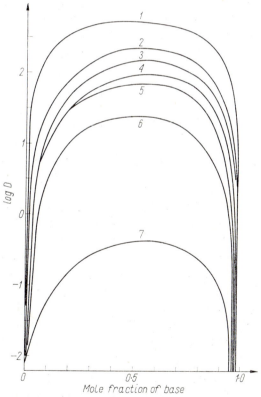

Fig. 5.23 Distribution coefficient for cobalt in the system: acetate buffer pH 4·93/TTA + heterocyclic bases in cyclohexane: (TTA + base) = 0·02M: 1 – isoquinoline, 2 – 3-methylpyridine, 3 – 4-methylpyridine, 4 – pyridine, 5 – 3-chloropyridine, 6 – quinoline, 7 – 2-methylpyridine (from [183], by permission of the copyright holders, North-Holland publishers)

oxygen-containing organic solvents, especially alcohols are also synergically active [184,195].

In most studies the influence of the chelating agent on synergism has been considered only within a particular group of reagents [180,183, 183a,196,197]. The results for a number of diketones (Table 5.9) [183] suggest that the more stable the complex formed by the chelating agent itself, the less pronounced is the tendency to yield mixed complexes, and so the less the part played by the synergic effect in the extraction process.

Table 5.9 Stability constants (log $\beta_{x,y}$) of the addition compounds of pyridine bases with 1,3-diketone copper chelates (in $CHCl_3$)
(from [183], by permission of the copyright holders, North-Holland publishers)

Base	Chelating agent				
	TTA	Trifluoro-acetylacetone	Ethyl acetoacetate	Benzoyl-acetone	Acetyl-acetone
4-Methylpyridine	372	233	8	4	3
Pyridine	267	135	3·3	3	2
2-Methylpyridine	46	26	1·6	0·74	0·66
Isoquinoline	291	157	4	—	2
Quinoline	16	11	0·7	—	0·4
log β_x (75% dioxan, 30°C)	13	17	18	23	23·5

The influence of the chelating agent was examined in more detail by Zolotov and Gavrilova [198], who studied synergic systems with a given synergist and diluent but with various complex-forming reagents that contained different ligand atoms. They found that in the extraction of zinc with an extractant containing sulphur as ligand atom (e.g. dithizone, sodium diethyldithiocarbamate) no synergic effect occurs on addition of TBP, although an effect is observed when TTA, PMBP or diphenyl-carbazone is the complex-forming reagent (Fig. 5.24). Many cations, especially those of transition metals, have variable co-ordination number. The co-ordination number depends on the electronic structure of the cation, and for a particular cation it depends on the nature of the bonding between the central atom and the ligand, which in turn is determined to a large extent by the nature of the ligand atoms directly linked to the metal. For instance, for ions such as Zn^{2+}, Ni^{2+}, Cu^{2+}, or Tl^+ M–S bonds are usually stronger than M–O bonds. Metal ions in complexes with sulphur-containing reagents have a low co-ordination number and because of that their extraction is independent of the composition of the organic phase. For complexes with ligand atoms O,O and O,N, high co-ordination numbers should be expected (for Zn^{2+}, 5 and 6): thus mixed complexes are more likely to form, and a synergic effect to occur.

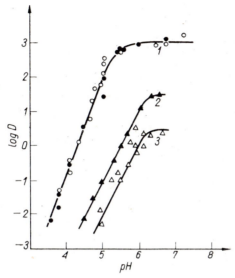

Fig 5.24 Extraction curves for zinc complexes: 1—(open circles) with Na diethyldithiocarbamate, (full circles) with sodium diethyldithiocarbamate +TBP, 2—with diphenylcarbazone + TBP, 3—with diphenylcarbazone (from [198] by permission of the copyright holders, Akad. Nauk SSSR)

The synergic effect may allow reagents to be used for the extraction of metals for which they would not otherwise be suitable [199]. For silver, the usual complexes extracted are chelates with reagents containing sulphur as ligand atoms. Reagents containing oxygen as the ligand atoms, e.g. diketones, are not used. However, in the presence of triphenylphosphine oxide a mixed complex of silver with β-diketones is produced. This is possible because of decreased electron density on the metal atom and stabilization of the Ag–O bond by formation of a π–bond between the silver and the phosphorus atoms. Thus, triphenylphosphine oxide is a synergist in the extraction of such silver chelates [200].

Jordanov [201–204] has shown that some carboxylic acids may also act as synergists. Extraction of thorium and uranium with trioctylphosphine oxide (TOPO) was found to be more efficient and more selective in the presence of such acids, although the acids alone do not extract thorium.

It was also found that a mixed complex is formed between thorium nitrate and benzoate, solvated with TOPO molecules, $ThB(NO_3)_3 \cdot TOPO$.

The mechanism of an extraction which involves a synergist may by determined in part from the composition of the compound extracted. In a number of systems, the composition of the complexes was studied with the help of a variety of primarily physico-chemical techniques [55,180, 187,205–209]. Although the composition of the extracted compounds can be determined fairly accurately, their structure has not been elucidated.

Newman [210] suggests that the synergic effect is largely due to the reaction occurring in the organic phase. Thus the measure of that effect is mainly the stability constant, $\beta_{x,y}$, of the mixed complex.

Table 5.10 lists $\beta_{x,y}$ values for a typical synergic system (TTA+TBP or TBPO) determined by Irving [183]. For most of the systems listed, the $\beta_{x,y}$ value depends more on the number of molecules of synergist in the mixed complex, than on the nature of the metal. This suggests that in similar cases there is more likely to be a bond between the synergist and the complex-forming reagent, rather than the metal atom. If a bond did exist between the synergist and the metal, difference $\beta_{x,y}$-values should be observed for different metals.

However, later work on synergism in the extraction of uranium with 1,3-diketones and TBP indicates that TBP is most probably bonded directly to the uranium, following displacement of water molecules from the uranium co-ordination sphere. In mixed complexes of cobalt with PMBP and TOPO it has been shown that the TOPO molecules add directly to the central atom [211].

The question of whether or not the chelate ring opens when the mixed complex is formed is connected with the co-ordination number

Table 5.10 Equilibrium constants for the extraction of TTA and TBP or TBPO complexes
(from [183], by permission of the copyright holders, North-Holland publishers)

Extracted compound	Solvent	log K_{ex}	B = TBP		B = TBPO	
			log $K_{ex(x,y)}$	log $\beta_{x,y}$	log $K_{ex(x,y)}$	log $\beta_{x,y}$
ThA$_4$B$_1$	benzene	1·00	5·70	4·7	7·70	6·7
UO$_2$A$_2$B$_1$		−2·26	2·48	4·7	—	—
ZnA$_2$B$_1$		−8·64	−4·26	4·38	—	—
UO$_2$A$_2$B$_2$	benzene	−2·26	2·48	4·74	8·61	10·87
TmA$_3$B$_2$		−6·96	−0·34	6·6	—	—
AmA$_3$B$_2$		−7·46	−0·96	6·5	2·51	9·97
PmA$_3$B$_2$		−7·77	−1·04	6·7	2·43	10·2
CmA$_3$B$_2$		−7·10	−0·70	6·4	2·83	9·9
CaA$_2$B$_2$		−12·0	−5·27	6·7	−3·27	8·7

of the metal. In lanthanide–TTA complexes the co-ordination number would have to increase from six, in $M(TTA)_3$, to eight, in $M(TTA)_3 B_2$, unless some TTA molecules act as unidentate ligands. There is a considerable body of evidence supporting a co-ordination number of eight for lanthanides and actinides. On the other hand, the spectral analysis suggests that TTA is present partly as a unidentate ligand [212].

The kinetic properties of the chelate are also relevant, as pointed out by Aggett [213]. Thus, the synergic effect is not observed for the kinetically inert cobalt acetylacetonate $Co(AA)_3$, but it is observed for the similar complex, $Fe(AA)_3$, which is kinetically labile. This behaviour may be ascribed to the cleavage of the chelate ring in the labile complex, followed by direct bond formation between TOPO and the iron.

According to Aggett [214] the synergic effect may arise from four district types of reaction, as follows.

1. Co-ordination of the synergist molecules to the metal ion, with the ligand remaining bidentate.

2. Interaction between the synergist and the co-ordinated ligand.

3. Co-ordination of the synergist to the metal ion so that it occupies the co-ordination positions made available as a result of dissociation of one end of the ligand. The ligand becomes unidentate instead of bidentate.

4. Co-ordination of the synergist to the metal ion as a result of complete dissociation of the ligand. In such cases the compound extracted must contain inorganic anions in its structure.

In addition to the various synergic systems discussed so far, there is another type involving two metals, which was first described by Healy [215]. In his investigations on the synergism of alkali metals Healy found that one metal can be extracted synergically in the presence of another metal. To give an example, caesium thenoyltrifluoroacetonate is extracted to a very small extent, and only from solutions of high pH, but in the presence of such ions as Na^+, K^+, Li^+, and especially Ca^{2+}, the extraction of caesium is greatly increased (Table 5.11). A reciprocal effect is also observed; the distribution coefficient of calcium increases at fairly high concentrations of caesium (Fig. 5.25).

Table 5.11 Effect of metal ions on the distribution coefficient of caesium when extracted with $0.2M$ TTA in benzene
(from [215], by permission of the copyright holders, North-Holland publishers)

Ion added	—	K^+	Na^+	Li^+	UO_2^{2+}	Ce^{3+}	Ca^{2+}
D_{Cs}	0.0015	0.006	0.07	0.2	0.8	4	10

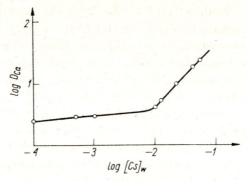

Fig. 5.25 The effect of addition of caesium on the extraction of calcium in the system Ca^{2+}/TTA in benzene: $[Cs^+]+[Ca^{2+}] = 0.5M$, $[TTA] = 0.2M$ (from [215], by permission of the copyright holders, North-Holland publishers)

There are two possible mechanisms:

$$Na^+ + Cs^+ + 2TTA \rightleftharpoons Na(TTA) \cdot Cs(TTA) + 2H^+ \quad (1)$$

$$Na^+ + Cs^+ + NO_3^- + TTA \rightleftharpoons Na(TTA) \cdot CsNO_3 + H^+ \quad (2)$$

Detailed studies suggest that mechanism (1) applies for some systems, and mechanism (2) for others.

Similar behaviour has been observed in the extraction of lanthanides with diethylhexylphosphoric acid (HDEHP) [216,217]. The distribution coefficient is substantially increased if the extracting agent is a mixture of HDEHP and Na(DEHP). The organic compound produced, Na(DEHP)·3HDEHP, extracts lanthanides by exchange of the sodium ion. With a greater excess of sodium ions, hydrated polymeric compounds are formed $Na(DEHP)_x \cdot zH_2O$, and this causes a fall in the distribution coefficient.

Not infrequently an excess of the synergist B results in a decreased synergic effect. This behaviour has been termed the antisynergic effect by Healy [218] and the antagonistic effect by Irving [219]. The antisynergism phenomenon is caused by all the factors that impair the formation of the mixed complex MA_xB_y. In the work by Healy [218] and Wang [172] it has been found that antisynergism in β-diketone extraction systems is caused by formation of ketohydrates on addition of excess of synergist. Decomposition of ternary complexes and formation of ketohydrates is favoured by the presence of an acid in the aqueous phase. Wang et al. [172,220] advanced the following general reaction to explain the antisynergism in the Zn–HPA–TOPO system (HPA is hexaphenylacetone):

$$Zn(HPA)_2 \cdot (TOPO)_2 + 2TOPO + 2H^+ + 4H_2O \rightleftharpoons Zn^{2+} + 2[HFA \cdot 2H_2O \cdot (TOPO)_2]$$

The authors gave the following structure for the ketohydrate, on the basis of infrared evidence:

$$\begin{array}{c} \quad\quad CF_3 \quad\quad OH\cdots OP(C_8H_{17})_3 \\ \quad\quad\ |\quad\quad\quad\quad\ \vdots \\ HO-C-CH_2-C-OH \\ \ \ \vdots\quad\quad\quad\quad\ | \\ (C_8H_{17})_3\, PO\cdots HO\quad\quad CF_3 \end{array}$$

Similar results were obtained for thenoyl- and benzoyltrifluoroacetone.

In the extraction of copper, with 4-methylpyridine as synergist [219], complex cations CuB_j^{2+} ($j = 1, 2, 3, ...$) were found to form, and this decreases the concentration of copper in solution, thereby hindering the formation of complexes CuA_2 or CuA_2B_x.

Subramanian and Pai [221] investigated the effect of temperature and concentration of salt in the aqueous phase on the synergic effect in the extraction of uranium.

The foregoing examples illustrate that the mechanism of the synergic effect is not at all simple, that it depends on a wide range of factors, and that it may be different for different systems [1327]. At the present time, no general conclusions can be drawn. Nevertheless, the synergic effect is extremely useful in practice, and it finds application in spectrophotometric analysis and quantitative analytical and manufacturing separation processes [221a].

Numerous studies deal with the question of how to increase the distribution coefficient for such elements as cobalt [222–224a,1328], iron [225,1329], nickel [226], actinides [227–230a,1330–1332], lanthanides [231, 233–233c], manganese, nickel, and cadmium [61,234], zinc [181,235], europium [227,228,232,236], or calcium [237,238]. The synergic effect has been taken advantage of for substoichiometric determination of thorium [239], and for photometric determination of manganese and cadmium with diphenylcarbazone [234] and europium and samarium with 2-naphthyltrifluoroacetone [240]. A synergic effect was also found in the extraction of uranium(VI) with a mixture of amines and organophosphorus acids. The extraction mechanism for systems of the type UO_2^{2+}, H_2SO_4, H_2O/amine, alkylphosphoric acid and diluent, and the composition and structure of the compounds extracted have been studied by Deptuła and Minc [241–243]. Newman and Klotz have described the synergic effect of tri-n-octylamine in the extraction of thorium and americium thenoyltrifluoroacetonates [244].

5.2.6 Co-extraction

Alimarin, Zolotov and Shakhova [245,246] have observed the effect of co-extraction in the extractive isolation of elements as chelates. Co-ex-

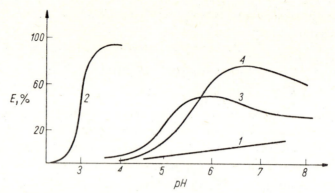

Fig. 5.26 Extraction of ruthenium with $1M$ naphthenic acids in benzene: 1 − Ru, 2 − Ru+Fe (10^{-2} gram-ion/l), 3 − Ru +Cu (10^{-2} gram-ion/l), 4 − Ru+Co (10^{-2} gram-ion/l) (from [247], by permission of the copyright holders, Akad. Nauk SSSR)

traction has been shown in a variety of extractive systems. Such ions as Ru(III), Cr(VI), Nb(V), and V(V) for instance were found not to be extracted with naphthenic acid, whereas in the presence of ferric or other salts they pass quantitatively into the organic phase (Fig. 5.26) [247]. Co-extraction also occurs in the isolation of the complexes of lanthanides and other metals with o,o'-dihydroxyazo compounds [248–250].

From the work done it is apparent that co-extraction during the isolation of the chelates may occur for a variety of reasons. It has been assumed to result from sorption of the co-extracted element on the micelles of the compound extracted [251]. It was suggested that the co-extraction of calcium with lanthanide oxinates results from formation of a compound of the type $CaA_2 \cdot 2MA_3$ [252]. Most frequently, however, the co-extraction has been explained by postulating the formation of ion-association compounds $M_I(M_{II} A_{n+1})$ (M_I is the co-extracted metal, M_{II} is the extracted metal, and A is the chelating reagent anion) [245,253]. Such ion-association compounds are formed if the metal extracted can form an anionic chelate. This property is characteristic of elements which give co-ordinatively unsaturated neutral chelates. Examples are the alkaline earth metals, lanthanides, cobalt(II) and nickel(II). Ions such as Fe^{3+}, Ca^{3+}, or Al^{3+} usually cannot form chelates of this type.

The findings of Zolotov et al. [254] on the co-extraction of calcium with the β-diketonates of some elements supports the 'ion-association' mechanism. In the system examined co-extraction of calcium occurs only in the extraction of β-diketonates of Sc, Nd, or Th, which are the elements that can form anionic chelates, whereas co-extraction does not occur with Al and Fe. The authors suggest that this mechanism also explains the observed effect of the reagent employed on the extent of co-extraction.

5.3 ION-ASSOCIATION EXTRACTION SYSTEMS

5.3.1 Fundamentals

Ion-association extraction systems are systems in which the compounds extracted may be a variety of species; non-solvated co-ordination salts, solvated co-ordination salts and anion complexes are the most important in separation processes.

Despite numerous attempts [255–257] it has not so far been possible to derive an equation to cover all possible systems. As already pointed out by Morrison and Freiser [3] the difficulties arise from the fact that extraction systems usually contain fairly high concentrations of electrolytes, and because a large number of equilibria are involved in the formation of an extractable species.

Two attempts to derive expressions to describe the extraction process for ion-association compounds will be discussed, namely the studies by Diamond and Tuck [256] and Irving and Lewis [258].

An important group of ion-association compounds for separation processes are the anionic complexes. Such complexes are formed by certain transition metals with halide or pseudohalide ions (e.g. CN^-, SCN^-). The earliest and best-known example of such extractions is the $FeCl_3$–HCl–$(C_2H_5)_2O$ system [259–261]. Diamond and Tuck [256] derived expressions for the extraction of anionic complexes, based on this system. One of the main reasons for the extraction of such anions is their bulkiness and weak interaction with water. The metal ion is surrounded with halide ions which interact weakly with water. Another factor of similar significance in the extraction process is the cation of the ion-association complex. Small and highly-charged cations are known to impede the extraction of the complex into the organic phase. Hydrogen ions, however, have properties which make them particularly suited to co-ordinative solvation. The peculiar behaviour of hydrogen ions in the aqueous phase is due to their high charge density. There is ample evidence [262,263] that the hydronium ion H_3O^+ adds three water molecules in the first hydration shell to yield a symmetrical species (I)

Since the charge on the hydronium ion is not distributed over its surface but is located on the three hydrogen atoms, the hydrogen-bond between the ion and water molecules is strong. The ion formed may be solvated by more water molecules or by basic organic molecules, because the positive charge localized on the hydrogen atoms of the hydronium ion may be displaced onto the hydrogen atoms of the water molecules of the hydration shell (species II). Thus, the hydrogen-bonding with the basic oxygen atoms of the solvent molecules is stronger for hydrogen ions than for other positive ions [264] (e.g. Li^+) for which the water molecules in the hydration shell are of low polarity. These special properties of the hydrogen ion are responsible for the excellent extractability of these protonated anionic complexes into basic organic solvents. The extraction is the better, the more basic is the solvent, and the more sterically available is the donor atom in the solvent molecule.

The properties of an ion-association extraction system depend to a considerable extent on the solvent used, because the form in which the extracted compound occurs in the organic phase depends on the nature of the solvent. In solvents of very low dielectric constant the ion-associates combine into larger aggregates—trimers, tetramers, etc. [265–267]. In solvents of medium dielectric constant the dominant species in the organic phase are ion-pairs. In solvents of high dielectric constant, the complexes largely dissociate into solvated hydrogen ions and complex anions.

Formation of an extractable compound in these systems necessitates the presence of a large excess of suitable acid which is also extracted into the organic phase.

Consider a two-phase system with organic solvent S, and metal ions M^{n+} which form a mononuclear complex HA_m^{n-m} with ligand A^- provided by acid HA, all in the aqueous phase. The complex that passes into the organic phase may be assumed to be $HS^+ \cdot M(S)_s A_{n+1}^-$, where s is the solvation number of the metal ions solvated by solvent S. The metal reacts with the ligand in stepwise fashion to form a series of complexes, so the concentration of metal in the aqueous phase is $C_M = \sum_{m=0}^{n+1} [MA_m^{n-m}]$.
The concentration of metal in the organic phase is equal to the sum of the metal concentrations of the various extractable species, namely the ion-pair $[H^+, MA_{n+1}^-]$, the dissociated species $[MA_{n+1}^-]$, and any polymers $(HMA_{n+1})_q$ (in this case the polymer concentration is multiplied by q, the degree of polymerization); hence, the distribution coefficient is given by

$$D = \frac{\{[MA_{n+1}^-] + [H^+, MA_{n+1}^-] + q[(HMA_{n+1})_q]\}_o}{\sum_{m=0}^{n+1}(MA_m^{n-m})} \qquad (5.37)$$

In solvents of low dielectric constant, such as aliphatic ethers, the concentration of the undissociated form is usually negligible and can be disregarded in the expression for the distribution coefficient. In such a system the controlling factor is the association reaction in the organic phase

$$q(HMA_{n+1}) \rightleftharpoons (HMA_{n+1})_q \qquad K_q = \frac{[(HMA_{n+1})_q]_o}{[HMA_{n+1}]_o^q}$$

Substituting K_q into Eq. (5.37) gives

$$D = \frac{[H^+, MA_{n+1}^-]_o + qK_q[HMA_{n+1}^-]_o^q}{\sum_{m=0}^{n+1}(MA_m^{n-m})} \quad (5.38)$$

In such systems at high aqueous-phase concentrations of acid HA and metal ion the ratio A^-/M in the organic phase is sometimes found to differ from the value obtained from the composition of the complex. The reason for that may be the formation of mixed ion-association complexes, e.g. $H^+FeCl_4^- \cdot H^+Cl^-$.

In solvents of higher dielectric constant the expression for the distribution coefficient must be changed, because in such a medium dissociation of the ion-pair is significant. Also, the extracted acid HA may dissociate [268,269].

$$D = \frac{[(MA_{n+1}^-) + (H^+, MA_{n+1}^-)]_o}{\sum_{m=0}^{n+1}(MA_m^{n-m})} \quad (5.39)$$

On substituting the equilibrium constants for the individual reactions involved in the extraction process, and applying the electroneutrality condition to the organic phase ($[MA_{n+1}^-]_o + [A^-]_o = [H^+]_o$), Diamond and Tuck [256] obtained the following equation for the extraction coefficient:

$$D = \{P_k \cdot [MA_{n+1}^-]([H^+]/P_r[A^-] + P_k[MA_{n+1}^-])^{1/2} +$$

$$P_k K_k[H^+][MA_{n+1}^-]\} / \sum_{m=0}^{n+1}(MA_m^{n-m}) \quad (5.40)$$

where P_k is the distribution constant for the complex, P_r is the distribution constant for the complexing agent HA, and K_k is the stability constant of the complex.

In ion-association extraction systems the value of the distribution coefficient may depend on the metal ion concentration, as discussed by Saldick [268] and Diamond [256,269].

The concentration of the acid HA is also important. The curves for the relationship between the distribution coefficient and the hydrochloric acid concentration for iron exhibit sharp maxima at various acid concentrations for several solvents (Fig. 5.27). The increase in the distribution coefficient with aqueous phase hydrochloric acid concentration is largely due to the increase in concentration of the extractable complex. Moreover, with increasing hydrochloric acid concentration the activity of water diminishes, and this facilitates solvation of ions by organic molecules, thereby increasing the extraction efficiency for ionic compounds. The drop in extraction efficiency observed on further increasing the concentration of the acid cannot be accounted for by the formation of non-extractable complexes with Cl^- ions, because chloro complexes of iron higher than $FeCl_4^-$ are not known [270].

Instead, the higher hydrochloric acid concentration increases the solubility of diethyl ether in the aqueous phase, and this results in an increased volume of the aqueous phase and decreased hydrochloric acid concentration by dilution. Thus, the final hydrochloric acid concentration passes through a maximum at an initial HCl concentration of $7M$. Solubility of higher ethers in water is lower (e.g. isopropyl ether) so the change in volume is less, and the maximum on the extraction curve occurs at higher hydrochloric acid concentrations. Butyl and β-chloroethyl ethers give no maxima on the extraction curves, however.

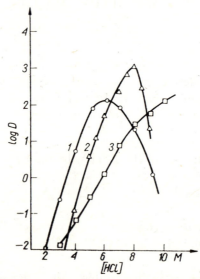

Fig. 5.27 Distribution coefficient for iron(III) *vs*. HCl concentration for various solvents: 1 — diethyl ether, 2 — di-isopropyl ether, 3 — β-chloroethyl ether (from [256], by permission of the copyright holders, Interscience Publ. Inc.)

The effect of the presence of additional ionic compounds was investigated in detail by Diamond [269]. He studied the effect of addition of nitric or perchloric acids on the distribution of the indium chloride complex for various solvents. In systems with a solvent of low dielectric constant the addition of a strong extractable acid HB results in a large distribution coefficient, because mixed ion-aggregates, e.g. $H^+ClO_4^- \cdot H^+MCl_4^-$, are formed in the organic phase.

In systems with solvents of high dielectric constant (such as β-chloroethyl ether) partial substitution of hydrochloric acid by perchloric or nitric acid results in a considerable decrease in the distribution constant.

Most studies of extraction systems are mainly concerned with the relationship between the distribution coefficient and the concentrations of the substances that can occur in the system. Occasionally authors deal with more general aspects of the extraction process, in particular the factors involved in the change in free energy associated with the partition of the solvated ion-pair [101, 108–110, 270–274]. Irving and Lewis [258a] have proposed another way of classifying ion-association complexes on the basis of the finding that the distribution constant of an ion-pair depends on the difference between the solvation energies in the aqueous and organic phases.

The solvation energy can be found from the physical properties of the solute and the solvent; the following three terms contribute.

1. Electrostatic energy of the ion treated as a point charge placed in the medium.

2. Internal solvation energy of the ion.

3. Energy of formation of the 'hole' in the solvent related to the work that has to be done against the cohesive forces between the solvent molecules.

For a two-phase system composed of aqueous phase W, and organic phase S, in which the mean solvation numbers of the cation by molecules of W and S in the phase rich in solvent W are denoted $\bar{w}$ and $\bar{s}$, and those in the phase rich in solvent S are designated $\overline{W}$ and $\overline{S}$, the authors derived the following equation for the distribution coefficient:

$$\log D = \frac{1}{2 \cdot 3RT} \left\{ \frac{(Ze)^2 N}{2\varepsilon_w} \left(\frac{1}{r_{+w}} + \frac{1}{r_{-w}} \right) - [\bar{w}I_{+w} + \bar{s}I_{+s}] \right.$$
$$\left. + p_w V_\pm + (\Delta G^o_\pm)_w \right\} - \frac{1}{2 \cdot 3RT} \left\{ \frac{(Ze)^2 N}{2\varepsilon_s} \left(\frac{1}{r_{+s}} + \frac{1}{r_{-s}} \right) \right.$$
$$\left. - [\overline{W}I_{+w} + \overline{S}I_{+s}] + p_s + V_\pm + (\Delta G^o_\pm)_s \right\} \qquad (5.41)$$

where r is the ionic radius in a given solvent, I is the energy of interaction between the ion and the solvent molecule, $(\triangle G_{\pm}^{\circ})$ is the standard energy of solvation of the ion-pair by a given solvent, $V_{\pm}$ is the sum of the molar volumes of the ions, p is the internal pressure of the solvent used, ε is the dielectric constant of the solvent, Z is the charge on the ion, e is the charge on an electron, N is Avogadro's number and the subscripts $+$ and $-$ refer to the cation and anion respectively.

This equation applies to partition of all ion-association complexes, and also other types of extraction system.

Irving [258b] was able to simplify this expression for the distribution coefficient which, based on the physicochemical studies [275–280] related to the parameters of this equation, he subsequently uses for the case when the organic phase consists of a mixture of solvents. From study of the extraction of In–HCl with various mixtures of solvents, he derived an equation for the distribution coefficient in terms of the parameters related to the physical and chemical properties of the components of the system, viz

$$\log D = k_1 - k_2 \{E/\varepsilon_m - X_d[B(\Delta\lambda_d - \Delta\lambda_h) - V(p_d - p_h)]\} \quad (5.42)$$

where h and d stand for the components of the organic solvent mixture which constitutes the organic phase, X_d is the mole fraction of component d in the mixture, k_1 and k_2 are constants for a specified temperature, E, B and V are constants calculated for the system concerned, ε_m is the dielectric constant of the solvent mixture, $\Delta\lambda_d$ and $\Delta\lambda_h$ are the component and solvent basicities and p is the internal pressure of the solvents.

Gordy and Stanford [280] have defined solvent basicity by

$$\Delta\lambda_s = 0\cdot 0147 \log(K_s/K_w) + 0\cdot 194$$

where K_w is the ionic product of water, and K_s is the equilibrium constant for the reaction $S + H_2O \rightleftharpoons SH^+ + OH^-$.

Equation (5.39) was found to be fairly valid for the distribution of the indium chloride complex between the aqueous phase and a mixture of methyl isobutyl ketone with a variety of solvents, naturally with simplifying assumptions and insufficient knowledge of some of the parameters of the equation.

The correlation of the distribution constant of the ion-pair with the difference between the solvation energies in the aqueous and the organic phases found by Irving and Lewis leads to an equation which is useful in the selection of the extraction conditions and, particularly, in choosing solvents.

Irving and Lewis made some simplifying assumptions, and these may be more or less valid in real systems. For instance the assumption that mutual solubility of phases is low enough not to affect the thermodynamic properties of the system at equilibrium is not valid when ethers are used as solvents.

5.3.2 Interactions Between Elements in the Extraction Process (Co-Extraction and Suppression of Extraction)

Interactions between elements in the extraction process have been extensively studied in ion-association extraction systems.

When protonated anionic complexes are extracted with solvents of fairly high dielectric constant or of high basicity, the extraction of trace elements is suppressed in the presence of a major component that is also extracted. Conversely, when low-polarity solvents (e.g. simple ethers) are used, trace components are co-extracted with the major component.

The co-extraction effect was investigated in detail by Zolotov et al. for the extraction of chloride [281-283], bromide [284], iodide [285], and thiocyanate [286] complexes.

Two possible mechanisms have been suggested, but in both, the processes responsible for the co-extraction effect are supposed to occur in the organic phase. According to Denisov and Khalonin [287], and McCorkell and Irvine [288] co-extraction is due to increase in the dielectric constant of the organic phase as a result of the extraction of the complex of the major component. The solvation energy of the trace component should then increase, and therefore its extractability. On the other hand, Zolotov et al. [289] are of the opinion that the mechanism is related to the formation of mixed ion-association complexes.

Zolotov et al. have found that maxima occur on the plots of log D for the trace component vs. the log of the concentration of the major component (Fig. 5.28). The plot of log D vs. the logarithm of the concentration of the major component is similar for the extraction of the

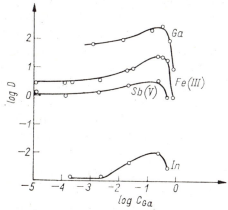

Fig. 5.28 The effect of gallium concentration on the extraction of trace elements and of gallium: $5 \cdot 8M$ LiCl $+1 \cdot 2M$ HCl $-$di-isopropyl ether (from [289], by permission of the copyright holders, Akad. Nauk SSSR)

major component itself. It is therefore likely that the cause of the maximum observed is the same in both cases, and is the formation in the organic phase of ion-association complexes or the association of the complex acids; the degree of association (the number of ionic dipoles in the aggregate) passes through a maximum with increasing electrolyte concentration.

For different extraction systems the maxima on the plots of log D vs. log C occur at different major-component concentrations. This may be because the degree of association depends not only on the dielectric constant of the solvent and on the electrolyte concentration, but also on the specific properties of the ions and ion-pairs involved. Co-extraction in iodide systems probably occurs at low major-component concentration because the iodide anion is bigger than the chloride or bromide anions.

Interactions between elements have been observed in amine extraction systems [290-290b] and in the extraction of solvated co-ordination salts [291].

These interactions are of great practical importance, because the extraction of halide acid complexes is widely used for removal of a major component before determination of impurities in the solution after the extraction.

5.3.3 The Extraction of Co-ordinatively Solvated Salts

One of the methods of extraction of ionic compounds is the formation of co-ordinatively solvated salts by the use of neutral extractants which have some electronegative groups capable of competing with water for a place in the first co-ordination shell of the cation [292,293].

In such systems, extractability depends primarily on the electronegativity of the extractant. Marcus [257] has divided oxygen-containing organic solvents into two classes. Class I includes ethers, ketones, alcohols and the like, whereas class II includes inert organophosphorus esters. Comparison of the data on extraction of salts formed with the solvents of the two groups shows that there are appreciable differences in the extractive power, because of the strongly polar nature of the class II solvents. Also, there are differences related to the behaviour of water. In systems with organophosphorus extractants water is generally expelled from the metal complexes, but in the solvates formed with class I solvents, some water molecules may remain to form a bond between the solvent and the salt.

The solvent classes also differ substantially in the solvation number of the compounds extracted. For class II extractants the degree of solvation can be strictly determined, whereas diluting class I extractants with inert diluents leads to ambiguous results. This points to the presence of various mixed hydro-solvates at equilibrium.

Systems involving class I extractants, (ethers, ketones, alcohols and esters) do not exhibit thermodynamically ideal behaviour, because of their miscibility with water, and as a result of dissociation of the extracted compounds in the organic phase.

As pointed out in Section 5.3.1, the hydrogen ion has a peculiar solvation mechanism. Other cations do not have the charge distributed over the hydrated ion surface, and this impairs the solvating power, so the solubility of the compound in a weakly basic solvent is low.

Garwin and Hixon [294] showed that, in the extraction of nickel and cobalt chlorides, ethers have a lower solvating power than ketones, esters and acids. This is due to steric factors, as the basic oxygen atom in ketones, esters or acids is more available (facilitating solvation) than in ethers of the same or higher basicity. Class I solvents are often used to extract salts, despite their basicity being lower than that of class II solvents [295]. The weak donor properties of the oxygen of these solvents accounts for the presence of water molecules in the solvates. The solvent molecules are usually hydrogen-bonded to the water molecules in the first hydration shell.

Specker et al. [296] suggest that complexes extracted into such solvents as cyclohexane (and phosphoryl compounds) have fewer molecules of inorganic ligands than compounds extracted into low-basicity solvents such as ethers or alcohols, but the total number of inorganic ligands and of solvating organic molecules is constant. With thiocyanate complexes, the solvate found in cyclohexane and TBP is $Fe(SCN)_3 \cdot 3S$ but in diethyl ether, despite similar experimental conditions, it is $HFe(SCN)_4 \cdot 2S$. Since a large number of complexes are likely to form in the organic phase, the description of the process of distribution of metal salts in such systems is difficult.

The most frequently used and studied extraction systems of this type are the nitrate systems [296a]. In addition to the much-studied uranyl nitrate system [297–303] this type of extraction has been examined and applied to the separation of nitrates of a number of elements, such as cerium, plutonium, palladium, thorium and zirconium. It is noteworthy that the distribution coefficient is higher for uranyl nitrate than for other elements such as manganese, cobalt and copper.

Of other metal salts possibly useful, sulphates are virtually non-extractable, perchlorates are extractable for just a few elements [304,305], and chlorides and thiocyanates are better typified as protonated anionic complexes [306].

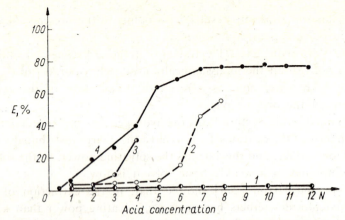

Fig. 5.29 Extraction of uranium(VI) with methyl isobutyl ketone from solutions of various acids: $1-H_2SO_4$, $2-HCl$, $3-HClO_4$, $4-HNO_3$ (from [308], by permission of the copyright holders, Pergamon Press Ltd.)

Methyl isobutyl ketone (MIBK) is a common extractant, and Ichinose [307,308] has discussed its use for extraction of the salts of metals. He found that methyl isobutyl ketone extracts uranium best from a nitric acid solution (Fig. 5.29). In the presence of a salting-out agent uranium can also be separated quantitatively and with satisfactory selectivity from hydrochloric acid solution [308].

As a group of solvents the higher alcohols can be differentiated from ethers, esters, and ketones. Their extracting ability is due to the presence of the hydroxyl group, which exhibits both donor and acceptor properties. The solvation of salts by aliphatic alcohols was the subject of studies by Katzin, who examined the composition of the solvates produced in the extraction of cobalt from solutions containing chlorides and nitrates [309]. Interesting data on the solvation mechanism for alcohols have been published [310–312].

The class II extractants are represented by the esters of orthophosphoric acid which have a $P \rightarrow O$ bond with a basic oxygen atom in a sterically favourable arrangement. These solvents have high dipole moments connected with the presence of the highly electronegative phosphoryl oxygen atom which, combined with the donor properties of that atom, give exceptionally good solvating properties, and thus extractive properties.

Solvents of this type which are important for extraction are, besides phosphates $(RO)_3PO$, phosphinates $R_2(RO)PO$, phosphonates $R(RO)_2PO$, and phosphine oxides R_3PO. Figure 5.30 illustrates the properties of the various types of compound. The gradual elimination of the ester oxygen atoms on passing from phosphates to phosphine oxides results in enhanced basicity and therefore in better extractive properties [167,313].

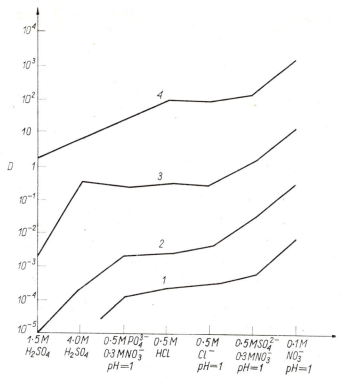

Fig. 5.30 Extractive properties of various organophosphorus compounds in the system U (VI) — solutions of various mineral acids; 1–TBP (tri-n-butylphosphate), 2 – DHHP (dihexylhexylphosphonate), 3 – BDHP (butyldihexylphosphinate), 4 – TOPO (trioctylphosphine oxide) (from [352])

The most typical and the most widely used solvent of this group is tri-n-butyl phosphate (TBP). It has found extensive application in industry, and this has resulted in a number of properties such as mutual solubility with water, dielectric constant, dipole moment, viscosity, and chemical resistance being investigated. An important chemical property is the hydrolysis to phosphoric acid, which can take place in both the organic and aqueous phases. The rate of hydrolysis increases with acid concentration in the aqueous and the organic phases. In the presence of bases, hydrolysis occurs only in the aqueous phase; the product is dibutyl phosphate.

The P → O group may participate in hydrogen bonding or it may occupy free sites in the first co-ordination shell of the cation. Thus, extraction with tri-n-butyl phosphate may proceed by various mechanisms, notably by hydrogen-bonding, whenever possible, and by the formation of a co-ordinatively unsaturated neutral solvate.

The simplest examples of solvate formation by hydrogen-bonding are dissolution of water in TBP, extraction of hydrogen peroxide [314], of weak organic acids [315], and nitric acid [315,316]. The hydrogen-bond can be produced not only by the hydrogen atoms of the extracted compound, but also by the water hydrating the hydrogen ion.

The extraction of the co-ordinatively solvated salts is of significance in the extraction of inorganic compounds. A typical system for such an extraction is that for uranyl nitrate (cf. Fig. 5.31). The extraction of chlorides, bromides, thiocyanates, sulphates, perchlorates and acetates was also studied.

The extraction mechanism and the distribution law for nitrates into TBP have been formulated by Moore [317] and extended by McKay et al. [318]. The mechanism advanced is of general applicability for any co-ordinatively saturated solvate. The extraction of nitrates with TBP may be described by the following equation:

$$M^{n+} + nNO_3^- + sTBP_{(o)} \rightleftharpoons M(NO_3)_n \cdot sTBP_{(o)}$$

The equilibrium constant for this reaction is

$$K = \frac{[M(NO_3)_n \cdot sTBP]_o}{[M^{n+}][NO_3^-]^n [TBP]_o^s} \quad (5.43)$$

The distribution coefficient of the extracted cation is given by the following formula

$$D = K[NO_3^-]^n [TBP]_o^s \quad (5.44)$$

It follows that for constant composition of the aqueous phase and low TBP concentration (i.e. when the activity coefficient in the organic phase may be neglected) a simple correlation exists between D and the TBP concentration

$$D \propto [TBP]_o^s$$

$$\frac{\partial \log D}{\partial \log [TBP]_o} = s \quad (5.45)$$

which allows the solvation number to be found.

The following formulae for the solvated salts have been established:

Trisolvates: $M(NO_3)_3 \cdot 3S$, where $M = Y, Ce, Eu, Tb, Tm, Lu, Am$.
Disolvates: $M(NO_3)_4 \cdot 2S$, where $M = Zr, Th, Np, Pu$.
$MO_2(NO_3)_2 \cdot 2S$, where $M = U, Np, Pu$.

By substituting activities into the equations a relation is obtained between the distribution constant k of the cation extracted and the concentrations and activity coefficients of individual components of the system

$$k = K \frac{[NO_3]^n \gamma_\pm^{n+1} [TBP]_{(o)}^s \gamma_{TBP(o)}^s}{\gamma_{k(o)}} \quad (5.46)$$

where $\gamma_\pm$ stands for a mean molar stoichiometric activity coefficient of the salt in the aqueous phase, and $\gamma_{k(o)}$ is the molar activity coefficient of the extracted complex in the organic phase.

It follows from Eq. (5.46) that the distribution coefficient is proportional to the activity of electrolyte in the aqueous phase and to the activity of TBP in the organic phase. Thus, increased electrolyte concentration in the aqueous phase increases the distribution coefficient. On the other hand, increased solvate concentration in the organic phase causes a drop in the TBP activity and an increase in the activity coefficient of the complex, which lowers the distribution coefficient.

The extractability of metal salts is affected by a number of factors such as the extractant structure, the diluent used, temperature, and the nature of the salting-out agent.

Although the extractability of salts solvated co-ordinatively with organophosphorus esters depends largely on the composition of the aqueous phase and on the equilibrium of the formation of the extracted compounds, the structure of the esters is also important. It is generally recognized that in the series of compounds $(RO)_n R_{3-n}-P=O$ the distribution coefficient increases with decreasing n. On the basis of numerous studies on the correlation between the extractive power of esters and their structure it can be said that basicity of an ester increases with chain-branching: extension of the alkyl chain may not affect the distribution coefficient or may cause it to increase. The presence of a doublebond in an ester decreases the distribution coefficient, as evidenced by a poorer extraction efficiency observed for phenyl derivatives of phosphoric acid. All the effects discussed are the result of the correlation between the electronegativity of the phosphoryl oxygen atom and the extractive power of the esters.

Because of its high viscosity, and because its density is close to that of water, pure tri-n-butyl phosphate is normally dissolved in a suitable diluent to facilitate separation of the phases. Many authors [319–326] have tried to establish a relation between the distribution coefficient and physico-chemical properties of solvents. Taube [327] has demonstrated that the extraction efficiency depends on the electrostatic interaction between the molecules of the complex and the permanent or induced dipole of the diluent molecules, and on the work required to form a 'hole' in the organic phase. When weakly polar TBP complexes are involved, the second factor is the more important and extraction into non-polar solvents is greater than into polar ones.

The differences between the distribution coefficients for various diluents are clearly less than expected from the activities of TBP. A natural corollary therefore is that another factor must be operative, which acts to the contrary. Detailed studies on this question have been carried out by Siekierski [101,103].

According to Eq. (5.46), for a constant composition of the aqueous phase the distribution constant k is given by

$$k = A \frac{a^s_{TBP(o)}}{\gamma_{k(o)}} \qquad (5.47)$$

where A is a constant and $a^s_{TBP(o)}$ is the activity of the extractant in the organic phase.

The distribution coefficient is therefore a function of the ratio $a^s_{TBP(o)}/\gamma_{k(o)}$. If the interactions between the diluent and the extractant and the complex are alike, both terms of Eq. (5.47) change analogously when the diluent is changed.

The changes thus caused in the distribution coefficient are generally small, but the less the self-compensation of the activity changes, the larger the differences in behaviour between the extractant and the complex molecules, and the more significant the solvent effect. A strong solvent effect can appear only when either the extractant or the extracted complex is strongly polar, or else when either combines with diluent molecules as a result of specific chemical forces, such as the formation of hydrogen-bonds between TBP molecules and chloroform or bromoform.

As demonstrated previously, a plot of the distribution coefficient of a solvated salt against the acid concentration in the aqueous phase exhibits a maximum (Fig. 5.31). The shape of the extraction curves observed results from the competitive interactions between the extractant and the acid molecules, accompanied by a decreased concentration of free extractant [328–330]. The effect can be eliminated by partial replacement of the acid by its alkali metal salt. Typical salting-out agents are alkali metal nitrates and alkaline-earth metal nitrates, but ammonium, aluminium and iron nitrates are also used. The selectivity of the salting-out effect has been found to depend on the properties of the cation of the salting-out agent, in particular its charge, radius, and hydration capacity [331–334]. Solovkin [331] has established a relation between the efficiency of a salting-out agent and the surface density of distribution of water molecules in the first hydration shell of the ions. The effects of the nature and quantity of the salting-out agents on the extraction of uranyl

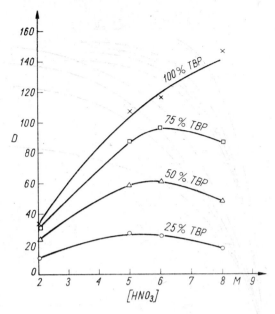

Fig. 5.31 Distribution coefficient of uranyl nitrate *vs.* nitric acid concentration for various percentage of TBP (from [352])

nitrate [335] are shown in Fig. 5.32. The salting-out effect of nitrates was said by Sato to decrease in the order Al > Fe > Li > Zn > Cu > Mg > Na > Ca > NH_4, but this order does not agree with other experimental evidence [336]. The efficiency of individual salts was observed to depend on the nitric acid concentration, and for different concentrations of acid the sequence of cations may be different (Fig. 5.33). This is because the change in the concentration of TBP depends on the amount of nitric acid extracted, and this in turn depends on the salting-out agent. This question has been studied by Ionov and Tikhomirov [336].

Temperature is an important factor in the attainment of equilibrium in the solvation extraction process. The temperature effect depends on the composition of the aqueous phase. Most data available refer to increasing the distribution coefficient during commercial nuclear-fuel processing, and not to detailed explanation of the nature of the effect [320,334,337–340]. In the extraction system $UO_2(NO_3)_2$–TBP the distribution coefficient decreases with temperature because the extraction is exothermic, but also depends on acid concentration. For a temperature change from 10° to 50°C, the distribution coefficient is halved for dilute solutions but scarcely affected for concentrated systems. The opposite effect is found for plutonium. Fletcher and Hardy [341] have suggested that the tem-

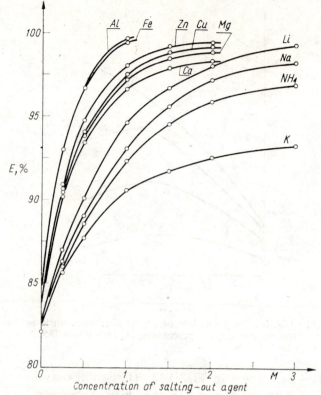

Fig. 5.32 The effect of the nature and concentration of salting-out agent on the extraction of uranyl nitrate, $[HNO_3] = 1M$ (from [335], by permission of the copyright holders Pergamon Press Ltd.)

cerature coefficient is greater than unity for metals having a distribution poefficient which does not decrease above a nitric acid concentration of $7M$. The stability of the solvated complexes may also affect the value of the temperature coefficient.

The organophosphorus reagents are typical group reagents. To enhance their selectivity, masking agents are used, and the concentration and nature of the acid and diluent can be optimized. For instance, uranium can be separated from thorium and cerium, and U(IV) from U(VI) [342–345] by extracting from solutions of chlorides instead of nitrates. Also, uranium can be isolated with a solution of TBP in methyl isobutyl ketone or other solvents [346–351].

A number of variants of the selective extraction of uranium have been developed, with various diluents [352], EDTA as a masking agent [351] or in the presence of thiocyanates [351]. The feasibility and the mechanism of the extraction of TBP from perchloric acid medium has also been closely examined [353,354].

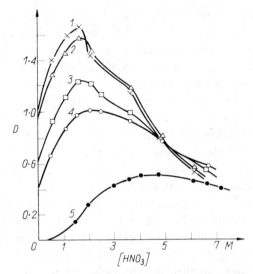

Fig. 5.33 The effect of the nitric acid concentration on the distribution coefficient of thorium nitrate for the extraction with $0\cdot 5M$ TBP in benzene in the presence of various salting-out agents at a concentration of $1M$: $1-Mg(NO_3)_2$, $2-Zn(NO_3)_2$, $3-Cd(NO_3)_2$, $4-Ca(NO_3)_2$, $5-$ no agent added (from [336], by permission of the copyright holders, Akad. Nauk SSSR)

5.3.4 Extraction with High Molecular-Weight Amines

High molecular-weight amines are excellent extractants of mineral and organic acids, and of ion-association complexes of metals [355–361]. Extraction with amines is an example of an ion-pair extraction. The process of extraction with amines in organic diluents can be fairly simply described in the following way [362]:

$$R_3N_{(o)} + H_3O^+ + X^- \rightleftharpoons (R_3NH^+, X^-)_{(o)} + H_2O$$

where R_3N stands for a tertiary amine, and X^- for an anion.

The amine salt anion in the organic phase can readily be exchanged for the anion from the aqueous phase. In this case amines behave like ion-exchangers:

$$(R_3NH^+ \cdot X_1^-)_{(o)} + X_2^- \rightleftharpoons (R_3NH^+ \cdot X_2^-)_{(o)} + X_1^-$$

The order for the exchange of anions is generally the same as the sequence established for ion-exchangers, i.e. $ClO_4^- > NO_3^- > Cl^- > HSO_4^- > F^-$.

It is not unimportant that the ion-pairs produced are much bulkier than simple ions. The extractive properties of large ion-pairs result primarily from the weakness of the interactions of the ions with water molecules; Thus, the ion-pair cation is not hydrated, the free energy of transferring a dipole from a medium of high dielectric constant to a medium of lower dielectric constant is decreased by the energy of the ion-pair formation.

Analytically, amines are used for extraction of elements that yield anionic complexes such as $FeCl_4^-$, $AuCl_4^-$, $InCl_4^-$. The distribution coefficient is affected by such properties of the extraction system as the structure of the amine, its concentration in the organic phase, the diluent used, and the anion involved. It follows from Born's charging equation that the larger the cation and anion of the ion-pair, the better the extraction. Hence the chain-length, degree of substitution and structure of the amine are all relevant to the extraction efficiency [363–363b]. It can generally be said that the extraction is better with long-chain amines and low solubility of the amine salt in the aqueous phase. On the other hand, the type of amine also has an effect, though this is not unequivocal. Extraction of iron and cobalt from hydrochloric acid solution is better with more highly substituted amines [364]. Rare-earth sulphates can be extracted only with primary amines, but their nitrates only with tertiary amines [365,366]. For simple acids the extraction efficiency is usually primary > secondary > tertiary [367,368]. For the simplest cases it can be assumed that the increase in the number of alkyl groups attached to the amine's nitrogen atom sterically hinders the access of the anion to the substituted quaternary ammonium ion produced by protonation and this weakens the interaction between the cation and the anion, and thus impairs extraction efficiency. For the same reason the selectivity of amines has the reverse order i.e. from primary through tertiary amines to quaternary ammonium salts. Chain-branching sometimes offsets the differences in the properties of amines of different degrees of substitution, e.g. the effect of branched-chain secondary amines may be similar to that of straight-chain tertiary amines.

In many cases chemical and electrostatic interactions are also important [369]. The ultimate effect is therefore the resultant of numerous factors, and a theoretical explanation of the variation in the distribution coefficient with degree of substitution has not so far been possible, despite extensive studies. The basic strength of the amine is also important. Grinsted investigated the effect of various structural factors of the amine on its basicity [368].

For many extraction systems with amines, the nature of the solvent used is also relevant. Coleman [362] argues that the amine–diluent system rather than the amine alone should be considered as the extractant. The correlation of the distribution constant of an ion-pair with particular physical properties of the diluent has been studied [370,371]. For certain systems a linear relationship has been established between the distribution coefficient and the dielectric constant of the diluent [272,273], but this has not been confirmed for all diluents studied. The chemical interaction connected with the basic or acidic character of the diluent is probably also relevant. If the diluent exhibits basic properties, it can interact with the slightly acidic substituted ammonium ion. Thus the good extraction with methoxybenzene can be explained by the basic character of the methoxy group. Likewise, benzene is a better extractant than cyclohexane because the π-electrons give benzene some weakly basic properties. If the diluent molecules are weakly acidic, they react with the molecules of a basic extractant, which decreases its actual concentration and impairs the extraction process. Thus, extraction with chloroform is often worse than with carbon tetrachloride [373,374]. The chloroform molecules, however, may also bond with the ammonium salt anion, to result in a more efficient extraction than could be expected from the diluent properties.

The diluent properties also decide to a large extent the form in which the compound to be extracted occurs in the organic phase. In a solution of high dielectric constant, such as nitrobenzene, the ammonium salt may be completely dissociated, or may also occur as an ion-pair, whereas in a solvent of very low dielectric constant the ammonium salt may associate to produce large aggregates [375].

The effect of the diluent in similar extraction systems has been elucidated in the work of Smułek and Siekierski [376—378] who took the Fe(III)–HCl–amine system as a model. The two factors found to be operative were: (1) interaction between the diluent and the extractant molecules; (2) interaction between the diluent and the molecules of the extracted complex. If the extractant and the complex interact with diluents in a similar manner, the interactions counteract one another, and the distribution coefficient is little affected by the diluent used. If, on the other hand, the interaction between the extractant and the diluent molecules differs from that between the complex and the diluent molecules, the diluent used strongly affects the value of the distribution coefficient. Then such physical properties of the diluents as dielectric constant, dipole moment, and the Hildebrand solubility parameter become important. A separate group of diluents is formed by chloroform and bromoform, for which the distribution coefficient is often small compared with other diluents of similar physical properties. This is because it is the formation of hydrogen-bonds rather than dispersion forces or dipole–dipole interactions that is responsible for the extractant–diluent interaction. Smułek has suggested that a hydrogen-bond is formed through a chlorine atom $R_3N:H-Cl \ldots H-CCl_3$.

In amine extraction systems the distribution coefficient of the metal depends in a complicated way on the concentration of the amine salt in the organic phase. It has often been shown that the distribution coefficient is proportional to the first power of the amine concentration, although there are two or three amine-salt molecules for one metal atom. According to Coleman [362] this results from the amine salt having an activity that remains constant, independent of its concentration in the diluent.

It has been suggested that the amine salt forms aggregates in the form of a stable colloidal suspension which is a separate phase of constant composition. The experimental evidence has proved the existence of such aggregates, but not in all systems investigated [379]. Smułek [378] has shown that over the amine concentration range of 10^{-3}–$10^{-5}M$ the amine activity is proportional to its concentration, so neither association nor dissociation of the amine in the organic phase is involved. At the higher concentrations ($\sim 0\cdot 1M$) which are of practical interest, the amine activity tends to a constant value.

The factors to be taken into account in the detailed consideration of an extraction system are the pH of the aqueous phase and the presence in solution of anions that compete with the formation of extractable complex ions. These two factors control the yield of metal-containing anions [362,365,380,381].

Comprehensive studies of the amine extraction mechanism for citrate and tartrate complexes of certain elements were made by Pyatnitskii et al. [382,383] and by Irving et al. [384]. It was pointed out that such extraction systems can be used for photometric determination of anions. Semenov et al. investigated the extraction properties of primary amines from sulphate solutions [388].

In the extraction of protonated anionic complexes the salts of high molecular-weight tertiary amines (and quaternary ammonium salts) which act on the ion-exchange principle have been widely used. Irving and Damodaran [385] investigated the extraction of cyanide complexes. Depending on the kind of metal involved, the cyanide complexes may occur as linear and singly charged anions, (e.g. $Ag(CN)_2^-$), tetrahedral and doubly charged, (e.g. $Zn(CN)_4^{2-}$), or octahedral and triply or quadruply charged ($Fe(CN)_6^{3-}$ and $Fe(CN)_6^{4-}$). For ions of the same geometry and charge the extractability increases with the atomic number of the element: $Ag(CN)_2^- \ll Au(CN)_2^-$; $Zn(CN)_4^{2-} < Cd(CN)_4^{2-} \ll Hg(CN)_4^{2-} < Ni(CN)_4^{2-} < Pd(CN)_4^{2-} < Pt(CN)_4^{2-}$

Generally, singly charged anions are extracted better from cyanide solution than doubly charged ones, and triply charged anions are not extracted at all.

Amines as extractants are best known and most commonly used for uranium [365,386,387].

The high selectivity of some high molecular-weight amines is extremely valuable for a number of difficult separations which arise in analytical practice and industrial applications, and this explains the exceptional interest in such extraction systems. Basic data on amine extraction processes are given by Diamond *et al.* [375], and the review by Olenkovich *et al.* [355] covers most of the information relating to the use of amines for extraction of acids [358,389,390] and protonated anionic complexes. Certain amines are powerful masking agents for certain elements [390].

5.4 APPLICATIONS OF EXTRACTIONS IN SEPARATION PROCESSES

The first part of this chapter covered the theory of the extraction process. Familiarity with theory is helpful when an extraction system suitable for a particular separation has to be found, and when the system has been selected, it helps in the choice of appropriate extraction conditions (e.g. solvent and concentration of complexing agent).

This section discusses optimum conditions for extraction of various elements with a variety of complexing reagents. Only main groups of compounds will be discussed in detail: an exhaustive discussion is impracticable because of the large number of reagents that can be used in extractions.

Every reagent is discussed in the following manner. If sufficient literature data are available, the characteristics of the aqueous phase (pH, concentration and type of acid used, presence of other complexing agents) and the organic phase (solvent, concentration of complexing reagent) are given. For the specified aqueous and organic phases the maximum extraction efficiency is reported for a given element, along with such quantities as $pH_{1/2}$, extraction constant, or distribution coefficient.

If the data for individual elements are compared, the selectivity of a given complexing reagent, and its usefulness for separation of single elements or groups of elements may be estimated.

An extractive separation in trace analysis will have one of the following three aims.

1. Extraction of the major component, to allow the impurities left in the aqueous phase to be determined.

2. Isolation of a group of elements which are to be determined.

3. Selective isolation of a single element from the material to be analysed.

Typical systems which can be used for isolation of major components and for isolation of groups of micro components, are listed in Tables 5.12 and 5.13. Halide complexes are particularly useful for isolation of major

Table 5.12 Extractive isolation of major components

Species	Sample	Aqueous phase	Organic phase	Elements isolated	Method of determination	Concentration level %	Ref.
Ag	silver	5M HNO$_3$	60% tri-iso-octyl-thiophosphate in CCl$_4$	Cu, Fe	flame photometry	10^{-4}	[1166]
As	arsenic chloride	conc. HCl	benzene	Ag, Al, Ba, Bi, Ca, Co, Cr, Cu, Fe, Ga, In, Mg, Mn, Ni, Pb, Sb, Sn, Te, Ti, Tl	spectrography	10^{-7}–10^{-5}	[1167]
As		HBr, H$_2$SO$_4$	benzene	Au, Cr, Cu, Fe, In, Ga, Mn, Ni, Zn	activation analysis	5×10^{-7}–10^{-4}	[1168]
As		2·5M HCl	diethyl ether	Ag, Cr, Fe, Hg, In, Mn, Ni, Os, Pb, Pd, Pt	spectrography	10^{-5}–10^{-3}	[1169]
Au	gold	6M HCl	MIBK + isoamyl acetate	Ag, Bi, Cu, Mn, Ni, Pb, Pd	atomic absorption		[1170]
Au	gold	10% HCl	ethyl acetate	Ag, Cu, Fe	colorimetry		[1171]
Au	gold	2M HCl	isoamyl acetate	Fe, Ag, Cu, Al, Bi, Cd, Zn, Co, Ni, Pb, Mn	colorimetry	10^{-5}–10^{-3}	[1172]
Au	gold	2M HCl	isoamyl acetate	Pd, Pt	colorimetry	10^{-5}–10^{-2}	[1173]
Au	gold	3M HBr	isoamyl acetate	Bi, Cd, Cu, Ni, Pb, Zn	polarography	10^{-6}–10^{-5}	[1174]
Au	gold	1–5M HBr	di-isopropyl ether	Ag, Bi, Cu, Fe, Ni, Pb	colorimetry	10^{-4}–10^{-3}	[1175]
Bi	bismuth	HI	cyclohexane	As, Cu, Fe, Ni, Pb, Zn	polarography and colorimetry	10^{-5}–10^{-4}	[1176]

Cd	cadmium chloride	3·4M HI	diethyl ether	Zn	colorimetry		[1178]
				Al, Ca, Cd, Cr, Fe, Mg, Mn, Ni, Ti, Zn	spectrography	10^{-5}–10^{-4}	[1179]
Cd	cadmium sulphide	3·4M HI	diethyl ether + isoamyl alcohol	Al, Co, Cr, Fe, Mg, Mn, Ni, Ti, Zn	spectrography	10^{-5}–10^{-4}	
Ce	cerium dioxide	ca. 10M HNO_3	HDEHP in CCl_4	La, Pr, Nd	spectrography	3×10^{-4}–2×10^{-3}	[1180]
Fe	iron, high purity	conc. HCl	bis-β-chloroethyl ether	Ag, Al, As, Ba, Bi, Ca, Cu, Co, Cd, Cr, In, Mg, Mn, Ni, Pb, Pt, Te, Ti	spectrography	10^{-7}–10^{-4}	[1181]
Fe	iron, steel	6·5M HCl	diethyl ether	Ag, Al, Be, Bi, Ca, Cd, Co, Cr, Cu, In, Mn, Mo, Os, Pb, Pd, Sn, Ti, V, U, W, Zn, Zr	spectrography		[1182]
Fe	iron, steel	6M HCl	MIBK	Ag, Al, As, Bi, Pb	spectrography	10^{-4}–10^{-3}	[1183]
Fe	iron	9M HCl	butyl acetate	Ag, Au, As, Co, Cu, Cr, Ga, Mn, Ni, Mo, Sb, W, Zn	activation analysis		[1184], [1185]
In	indium, high purity	8M HBr	bis-β-chloroethyl ether	Al, Bi, Cd, Ca, Co, Cu, Mn, Mg, Ni, Pb, Ag, Cr, Zn, Be	spectrography		[1186]

Table 5.12 (continued)

Species	Sample	Aqueous phase	Organic phase	Elements isolated	Method of determination	Concentration level %	Ref.
In	indium arsenide	conc. HBr	bis-β-chloroethyl ether	Al, Bi, Cd, Ca, Co, Cu, Mn, Mg, Ni, Pb, Ag, Cr, Zn, Be	spectrography	10^{-7}–10^{-4}	[1187]
In	indium and its compounds	5M HBr	isopropyl ether	Cd, Cu, Pb, Zn	polarography	10^{-6}	[1188]
In	indium	5M HBr	diethyl ether	Al, Bi, Cd, Cu, Mg, Mn, Ni, Pb, Zn	spectrography	10^{-5}–10^{-4}	[1189, 1190]
In	indium, high purity	4·5M HBr	di-isopropyl ether	As, Au, Cu, Sb, Zn	activation analysis	10^{-7}–10^{-5}	[1191]
In	indium antimonide	5M HBr	diethyl ether	Al, Bi, Cd, Cu, Mg, Mn, Ni, Pb, Zn	spectrography	10^{-5}–10^{-4}	[1192]
In	indium, high purity	ca. 5M HBr	di-isopropyl ether	Bi, Ca, Cd, Co, Cu, Hg, Mg, Ni, Pb, Zn	colorimetry		[1193]
		0·5M HNO$_3$	mono- and di(2-ethylhexyl) phosphates in CCl$_4$	As, Cd, Co, Cr, Cu, Mn, Ni, Se, Te, Zn	spectrography or activation analysis		[1194]
In-Sb	indium antimonide	8M HBr	di(2-ethylhexyl) phosphoric acid	Ag, Ba, Ca, Cd, Co, Cr, Cu, Bi, Al, Mg, Mn, Ni, Pt, Pb, Te, Zn	spectrography	1×10^{-7}–5×10^{-5}	[1195]
Ga	gallium antimonide	8M HBr	bis-β-chloroethyl ether	Ag, Al, Bi, Ca, Cd, Co, Cr, Cu, Mn, Mg, Ni, Pb,	spectrography		[1196]

Applications

Ga	gallium, gallium chloride, gallium arsenide	6–8M HCl	di-isopropyl ether	Ag, Al, Ca, Cd, Co, Cr, Cu, Mg, Mn, Mo, Ni, Pb, Ti, V, Zn, Zr	spectrography	10^{-6}–10^{-3}	[1197–1199]
Ga	gallium, gallium arsenide	conc. HCl	bis-β-chloroethyl ether	Ag, Co, Cu, In, Se, Te, Zn	activation analysis	10^{-7}–10^{-5}	[1200]
Ga	gallium, gallium arsenide	6M HCl	butyl acetate	Al, Bi, Cd, Co, Cr, Cu, In, Mg, Mn, Ni, Pb, Ti	spectrography	10^{-7}–10^{-6}	[1201, 1202]
Hg	mercury	2–3M HCl	isoamyl alcohol	Ag, Al, Ca, Cd, Cu	spectrography	10^{-7}–10^{-6}	[1203]
Hg	mercury	HI	cyclohexanone	Cd, Cu, Fe, Mn, Ni, Pb, Zn	spectrography	10^{-6}–10^{-5}	[1204]
Mo	molybdenum and its compounds	6M HCl	diethyl ether	Ag, Al, Ba, Bi, Ca, Cd, Co, Cr, Cu, Fe, Mg, Mn, Ni, Pb, Sb	spectrography	10^{-6}–10^{-4}	[1205]
Mo	molybdenum oxide	pH 2	50% acetylacetone in CHCl$_3$	Cr	colorimetry	10^{-3}	[1206]
Nb	niobium	1M HCl	amyl acetate	Al, Ca, Cd, Co, Cr, Cu, In, Mn, Pb, Sn, Hf, V, Zn	spectrography	10^{-5}–10^{-4}	[1207]
Nb+Ta	tantalum ores	3M HF + 4M H$_2$SO$_4$	MIBK	Fe, Mn, Al, Sn, Ti	spectrography		[1148]
Nb	potassium, lithium, rubidium niobates	H$_2$SO$_4$	TBP in benzene	Contaminants	spectrography	10^{-6}–10^{-5}	[1208]
Pd	palladium	0·3–0·4M HBr + 5–6M H$_2$SO$_4$	cyclohexanone	Cu, Ni, Pb, Cd, Zn	spectrography		[1209]

172 Liquid–Liquid Extraction [Ch. 5

Table 5.12 (continued)

Species	Sample	Aqueous phase	Organic phase	Elements isolated	Method of determination	Concentration level %	Ref.
Pu	plutonium	$4M$ HNO_3	20% trilaurylamine in Shell-Sol. T	Ce, Dy, Er, Gd, Sm	spectrography	5×10^{-4}–5×10^{-3}	[1210]
Pu+U	plutonium, uranium	$7M$ HNO_3	TBP	Al, Cd, Co, Cr, Cu, Fe, Mn, Mo, Ni, Am	spectrography	10^{-5}–10^{-2}	[1211]
Pu	plutonium	HNO_3	TBP	Al, Fe	spectrophotometry		[1212]
Re	rhenium	HNO_3 weakly acidic	$0\cdot 4M$ 2-n-nonyl-pyridine oxide in toluene	Al, Ba, Be, Ca, Cd, Co, Cr, Cu, In, Mn, Ni, Pb, Tl, Zn	spectrography	1×10^{-7}–2×10^{-5}	[1213]
Re	rhenium	pH 1 (HNO_3) H_2SO_4, HCl	$0\cdot 3M$ TOA in toluene	As, Al, Ba, Be, Bi, Ca, Cd, Co, Cr, Fe, In, Mg, Mn, Ni, Pb, Sb, Si, Te, Tl, Zn	spectrography	1×10^{-7}–2×10^{-4}	[1214]
Sb	antimony	$10M$ HCl	butyl acetate	Ag, Cd, Co, Cu, Mg, Ni, Pb, Zn, and others	spectrography	10^{-6}–10^{-4}	[1215]
Sb	antimony	$12M$ HCl	bis-β-chloroethyl ether	As, Co, Cu, In, Sn, Te, Zn	activation analysis	10^{-7}–10^{-4}	[1216]
Sb	antimony			Ag, Al, As, Bi, Ca, Cd, Co, In, Mg, Mn, Ni, Pb, Te, Zn	spectrography	10^{-7}–10^{-4}	[1217, 1218]
Sb	antimony	10–11M HCl	bis-β-chloroethyl ether	As	spectrophotometry	10^{-6}–10^{-4}	[1219]

§ 5.4] Applications 173

Tl	thallium, thallium chloride	6M HCl	diethyl ether	Cu, Fe, Mo, Ni, Pb, Sb, Ti, Zr	spectrography	2×10^{-6}–5×10^{-5}	[1220]
Tl	thallium	7M HCl	MIBK	Ag, Al, Cd, Cu, In, Mg, Mn, Ni, Pb, Zn	spectrography	10^{-6}–10^{-4}	[1221]
Tl	thallium	3M HCl	bis-β-chloroethyl ether	Cd, Cu, Ni, Pb, Zn	polarography		[1217]
Tl	thallium	1M HBr	bis-β-chloroethyl ether	As, Co, Cu, Fe, Ga, In, Sb, Sn, Fe, Zn	activation analysis		[1222]
Tl	thallium			Ag, Al, Ba, Bi, Co, Cd, Cr, Ca, Cu, Fe, Ga, In, Mg, Mn, Ni, Pb, Pt, Te, Zn	spectrography	10^{-7}–10^{-4}	[1223]
U	uranium	5–6M HNO$_3$ + NH$_4$F	TBP	Ag, Be, Bi, Ca, Cd, Co, Cr, Cu, Fe, In, Mg, Mn, Mo, Nb, Ni, Pb, Sb, Sn, Ti, V, Zr	spectrography	10^{-6}–10^{-4}	[1224]
Y	yttrium, yttrium oxide	0·1M HNO$_3$	tri-isoamylphos-phine oxide in CCl$_4$	Al, Ca, Co, Cr, Cu, Fe, Mg, Ni, Pb, Si	spectrography	10^{-6}–1×10^{-4}	[1225]
Y	yttrium, yttrium oxide	15M HNO$_3$	TBP	Ba, Bi, Ca, Cd, Co, Cr, Cu, In, Mg, Mn, Ni, Pb, Sb, Zn	spectrography	10^{-7}–10^{-3}	[1226]
Y and heavy rare earth elements	yttrium and heavy rare earth element compounds	0·5M HCl	1M 2-ethylhexyl-phosphoric acid	La, Ce, Nd, Pr	spectrography		[1227]

Table 5.13 Group isolation of elements

Reagent	Aqueous phase	Organic phase	Elements isolated	Method of determination	Concentration level %	Materials used	Ref.
HOx	pH 5·5–6·0	CHCl$_3$ + isoamyl alcohol	Al, Co, Cr, Cu, Mn, Mo, Ni, Pb, V	spectrography	10^{-8}–10^{-6}	hydrogen iodide	[1228]
HOx	pH 5·5–6·0	CHCl$_3$	Al, Cu, Fe, Mn, Ni, Pb, Sn, Ti, V	spectrography	10^{-7}–10^{-6}	bases	[1229]
HOx	pH 5·0–5·5	CHCl$_3$	Al, Fe	spectrography	10^{-4}	magnesium, high purity	[1230]
HOx	pH 5·0–6·0	CCl$_4$ + isoamyl alcohol	Al, Co, Cu, Fe, Ge, Mn, Ni, Pb, Zn	spectrography	10^{-7}–10^{-5}	calcium oxide, high purity	[1231]
HOx	pH 4·5–11·3	CHCl$_3$	Zr, Hf	spectrophotometry	10^{-5}–10^{-4}	uranium	[1232]
HOx		CHCl$_3$	Dy, Eu, Gd, Sm	spectrography		thorium	[1233]
HOx	pH 11	MIBK	Ca, Mg	atomic absorption	10^{-4}–10^{-3}	aluminium salts	[1234]
HOx	pH 8·0–10·7	CHCl$_3$	Al, Fe	spectrophotometry	10^{-3}–10^{-2}	titanium, vanadium	[1235]
HOx	pH 8·0	CCl$_4$ + isoamyl alcohol	Ag, Al, Bi, Co, Cu, Fe, Ga, Mn, Ni, Pb, Ti, Zn	spectrography	10^{-7}–10^{-5}	phosphorus pentoxide	[1236]
HOx	pH 3·2	CHCl$_3$	Ga, In	spectrography		ores and minerals	[81]
NaDDTC	pH 9·0	CCl$_4$	Ag, Bi, Cd, Cu, Mn, Pb, Zn	spectrography	5×10^{-5}–5×10^{-3}	silicate minerals	[1237]
NaDDTC		CHCl$_3$	Au, Cu, Co, Bi, In, Ni, Pb, Pd, Tl, Cd, Hg	spectrography		sodium chloride	[1239]
NaDDTC	pH 7·5	CHCl$_3$	Bi, Cd, Cu, Fe, Mn, Pb, Zn	spectrophotometry	10^{-7}–10^{-6}	alkali metals	[1240]

§ 5.4] Applications 175

Reagent	pH	Solvent	Elements	Method	Concentration	Matrix	Ref.
NaDDTC	pH 5·5–6·0	$CHCl_3$	Ag, Cu, Fe, Mn, Ni, Pb, Sb, Sn, Zn	spectrography	10^{-7}–10^{-6}	water	[1241]
NaDDTC	10–12M HCl, 1M HCl, 3M HCl	$CHCl_3$	As, An, Sn, Mn, Cd, Cu, Ni, Mo, Sb	activation analysis		beryllium	[1242]
NaDDTC	pH 4·5	ethyl acetate	Cd, Co, Cu, Fe, Mn, Ni, Pb, Zn	polarography	2×10^{-5}	beryllium	[1243]
NaDDTC	pH 6·0–6·5	$CHCl_3$	Cu, Mn, Ni	flame photometry	20 µg/ml	aluminium alloys	[1244]
NaDDTC	pH 6	$CHCl_3$	Cd, Cu, Fe, Mn, Ni	spectrography	10^{-6}–10^{-5}	gallium phosphide	[1245]
NaDDTC	pH 6, HF, tartaric acid	$CHCl_3$	Bi, Cd, Co, In, Ni, Mn, Pb, Sb, V, Zn	spectrography		niobium and tantalum, high purity	[1246]
NaDDTC	pH 6·0–6·5	$CHCl_3$ or CCl_4	Ag, Au, Bi, Cd, Co, Cu, Fe, In, Mn, Ni, Pb, Se, Tl, Zn	spectrography	10^{-5}–10^{-4}	tantalum niobium	[1247]
NaDDTC	pH 6·4–6·6	$CHCl_3$	Co, Cu, Fe, Ni, Zn	spectrophotometry	10^{-5}	niobium	[1248]
NaDDTC	pH 4–5	$CHCl_3$	Bi, Cd, Co, Cu, Fe, Mn, Ni, Pb, Zn	spectrography	10^{-5}–10^{-4}	chromium	[1249]
NaDDTC	pH 5·5	butane + 4-methyl-pentan-2-ol	Co, Cr, Cu, Fe, Mn, Zr	atomic fluorescence		sea water	[1250]
NaDDTC	pH 3·6	MIBK	Co, Cu, Fe, Cr, Mn, Ni, Pb, Zn	atomic absorption	10^{-6}–10^{-5}	natural waters	[1251]
NaDDTC	pH 11	$CHCl_3$ or CCl_4	Bi, Pb, Cu	polarography	1×10^{-4}	iron	[1252, 1253]
NaDDTC		isoamyl acetate or MIBK	Co, Cu, Fe, Mn, Ni, Zn	atomic absorption	0·05–0·2 µg/l	sea water	[1254]

Table 5.13 (continued)

Reagent	Aqueous phase	Organic phase	Elements isolated	Method of determination	Concentration level %	Materials used	Ref.
NaDDTC	pH 8–9	CCl_4	Ag, Bi, Cd, Co, Cu, Hg, Mn, Ni, Pb, Zn	spectrography	10^{-6}–10^{-5}	silicate minerals	[1255]
NaDDTC	0.1M HCl	CCl_4	Co, Cu, Fe, Ni, V	flame photometry		pure materials	[1256]
NaDDTC	pH 6–9 (masking agents)	CCl_4, $CHCl_3$	Cu, In, Mn, Zn	activation analysis		glass, biological materials	[1257]
PDTC	pH 4–5	MIBK	Cd, Cu, Pb, Zn	atomic absorption	10^{-5}–10^{-4}	waters	[1258]
DTC derivative	pH 8	MIBK, amyl acetate	Cu, Co, Ni	atomic absorption		rocks	[1259]
CuDDTC		$CHCl_3$	Ag, Au, Cu, Pd	spectrography	10^{-5}	minerals, alloys	[1260]
H_2Dz	pH 9	CCl_4	Cu, Zn	polarography	10^{-4}	sodium chloride	[1261]
H_2Dz	pH 8.5	CCl_4	Bi, Cd, Co, Cu, Ni, Pb, Zn	spectrophotometry		high-purity bases	[1262]
H_2Dz	pH 6–7, 8–9	CCl_4	Bi, Cd, Co, Cu, Ni, Pb, Zn	extractive titration	1.5×10^{-4}	various reagents	[1263]
H_2Dz	pH 8–10	CCl_4	Ag, Bi, Cd, Co, Cu, Hg, Zn	extractive titration		gallium arsenide	[1264]
H_2Dz	pH 2.0	CCl_4	Ag, Cd, Cu, Bi, Zn	spectrophotometry	10^{-6}–10^{-5}	high-purity lead	[1265]
H_2Dz	pH 2.0	CCl_4	Bi, Cd, Cu, Pb, Zn	polarography		high-purity vanadium	[1266]
H_2Dz	pH 7.5	CCl_4	Ag, Au, Cd, Co, Cu, Fe, Hg, Mn, Ni, Pb, Pt, Zn	extractive titration	10^{-5}–10^{-6}	high-purity antimony	[1267]
H_2Dz	pH 3–9	CCl_4	Bi, Cd, Co, Cu, Ni, Pb, Zn	spectrophotometry	10^{-4}–10^{-5}	chromium	[1268]

§ 5.4] Applications 177

Reagent	Conditions	Solvent	Metals	Method	Concentration	Application	Ref.
H_2Dz	pH 10	$CHCl_3$	Bi, Pb	spectrophotometry		sten and its oxides high-purity tellurium	[1270]
H_2Dz	pH 8.0–8.5	CCl_4	Ag, Cu, Pb, Zn, Cd, Co, Cr, Mn, Mo, Ni, Sn, V	spectrophotometry		natural waters	[1271, 1272]
H_2Dz	pH 9	$CHCl_3$, sulphosalicylic acid	Ag, Bi, Cd, Cu, Hg, Co, Ni, Pb, Zn	spectrography		silicate minerals	[1273]
H_2Dz	pH, various values	$CHCl_3$, CCl_4	Bi, Cd, Cu, Pb, Tl, Zn	spectrophotometry		rhenium	[1274]
Cupferron	20% HCl $H_2C_2O_4$	$CHCl_3$ + butyl alcohol	Nb, Ta, Zr, Ti, Sn	spectrography		alkali metals	[1238]
BPHA	pH 6–9	$CHCl_3$	Al, Bi, Co, Cu, Fe, Mn, Ni, Pb, Zn	spectrography	10^{-4}–10^{-2}	silver, lithium fluoride	[674]
BPHA	pH 2.0–2.5 7.0–7.5	$CHCl_3$ + isoamyl alcohol	Cd, Co, Cu, Fe, Mn, Mo, Ni, Pb, Ti, V	spectrography		aqueous solutions	[520]
Diethyldithiophosphate	6M HCl	$CHCl_3$	As, Bi, Sb	spectrography	10^{-6}–10^{-5}	nickel sulphate	[1275]
1,5-Di(β-naphthyl)thiocarbazone	pH 8–9	$CHCl_3$	heavy metals	spectrography	10^{-5}	hydrochloric and nitric acids, high purity	[1276]
Trifluoroacetyl acetone	pH 6.3–10 pH 2–10	toluene TIBA in toluene	Fe, In Co, Zn	chromatography		sea water	[1277]

12 Separation

Table 5.13 (*continued*)

Reagent	Aqueous phase	Organic phase	Elements isolated	Method of determination	Concentration level %	Materials used	Ref.
Tri-iso-octyl-thio-phosphate	4–8M HNO_3	MIBK	Ag	atomic absorption	5×10^{-6}	rocks, sediments and soils	[1278]
PMBP	pH 7.0–7.5	$CHCl_3$ + isoamyl alcohol	Cd, Co, Cu, Mn, Mo, Ni, Fe, Pb, Ti, V	spectrography	10^{-7}–10^{-6}	alkali metals, halides	[1279]
Batho-phenan-throline	pH 3	MIBK	Cu, Fe, Ni	atomic absorption	1×10^{-4}	uranium compounds	[1280]
PAN	pH 4–5	MIBK	Co, Cu, Fe, Ni, Zn	atomic absorption		waters	[1281]
Aliphatic mono-carboxylic acids	pH 8–9 pH 3–4.5		Ag, Al, As, Bi, Cd, Cr, Cu, Fe, Ga, In, Pb, Sb, Sn, Ti, Zn	spectrography	10^{-7}–10^{-5}	alkali metal salts	[1282]
TTA	pH 5.5	MIBK	La, Nd, Sc, Yb	atomic absorption	0.006–0.33 μg/ml	aqueous solutions	[459]
Tri-fluoro-acetyl-acetone	pH 4.5	$CHCl_3$ or benzene	Al, Cu, Fe	gas chromatography		alloys	[1283–1285]
Dithiol		n-butyl acetate cyclohexane	Mo, W	spectrophotometry		steel	[1286]
HDEHP	1–11M HCl, HNO_3, $HClO_4$		51 elements	activation analysis		rocks and metallurgical raw materials	[1287]

Applications

phosphoric acid							
Cl⁻	4M HCl	MIBK	Te	atomic absorption spectrophotometry		conductor copper	[1289]
Cl⁻	7M HCl	MIBK	Ga, Fe	spectrophotometry		high-purity aluminium	[1290]
Cl⁻	7M HCl	isoamyl acetate	Au, Fe, Ga, Mo, Tl			high-purity indium	[1291]
Cl⁻	6M HCl	diethyl ether	Ga, Fe, Tl	spectrophotometry	10^{-6}–10^{-5}	zinc sulphide	[1292]
Cl⁻	2.5–3M HCl 0.2–0.3M HNO$_3$	CHCl$_3$	Bi, Cd, Sb, Sn, Zn	spectrography		high-purity aluminium	[1293]
Cl⁻	6M HCl	isopropyl or diethyl ether	Fe, Tl	polarography	10^{-6}	indium	[1127, 1294]
Cl⁻	6M HCl	diethyl ether	Fe, Ga, Tl	spectrography	10^{-6}–10^{-4}	high-purity indium	[1295]
Cl⁻	10M HCl	bis-β-chloroethyl ether	Au, Fe, Ga, Sb	activation analysis	10^{-8}–10^{-5}	silicon	[1222]
Cl⁻	5% HCl, 2% NaI	MIBK	Pb, Cd, In, Bi, Cu, Sb	spectrophotometry		high-purity iron	[1296]
Cl⁻	2.3M HCl, 0.2M KI, 0.1M ascorbic acid	MIBK	Bi, Cu, Sb, Te	spectrophotometry		iron alloys	[1297]
I⁻	1M HI	diethyl ether	In, Tl	spectrography		silicates	[1298]
I⁻	HI	CHCl$_3$, C$_6$H$_6$, diantipyrylmethane	Cd, Bi, Sc	spectrography		steels, magnesium and aluminium alloys	[1299]
I⁻, TOA	S2–2.5M H$_2$O$_4$ 1M KI	0.2M TOA in MIBK	Cu, Ag, Au, Zn, Cd, Hg, In, Sn, As, Sb, Bi, Te, Tl	atomic absorption	10^{-5}–10^{-4}	steel	

Table 5.13 (continued)

Reagent	Aqueous phase	Organic phase	Elements isolated	Method of determination	Concentration level %	Materials used	Ref.
SCN⁻	pH 1, EDTA, 2·4M NaSCN	TBP	alkaline earth metals	complexometric analysis		rare earth elements	[1300]
SCN⁻	HCl, NH$_4$SCN	CHCl$_3$, diantipyryl-methane	Cu, Fe, Zn	spectrography	10^{-4}	chromium	[1301]
SCN⁻	NH$_4$SCN	MIBK or TBP	Co, Fe	spectrophotometry		nickel and its salts	[1302]
HOx + DDTC	pH 5·5-6·0	CHCl$_3$ + isoamyl alcohol	Ag, Al, Au, Bi, Cd, Cu, Ga, In, Mn, Ni, Pb, Ti, Tl, Zn	spectrography	10^{-9}–10^{-6}	sodium and potassium nitrate	[1303]
HOx + DDTC	pH 5·5-6·0 NaI	CHCl$_3$ + isoamyl alcohol	Al, Co, Cu, Fe, Ni, Mn, In, Sn, Ti, Zn, Pb	spectrography		sodium iodide	[1304]
HOx + DDTC	pH 7·5-8·0	CHCl$_3$ or CCl$_4$	Ag, Al, Au, Bi, Cd, Co, Cu, Fe, In, Mn, Ni, Pb, Tl, Zn	spectrography	10^{5}–10^{-4}	vanadium	[1207]
HOx + cupferron	pH 6-7	CHCl$_3$ + butyl alcohol	Fe, Mn, Mo, Ni, Nb, Sn, Ta, W, Zr	spectrography	3×10^{-6}–10^{-4}	alkali metals	[1238]
HOx + DDTC + H$_2$Dz	various pH values	CHCl$_3$	Ag, Al, As, Co, Cd, Cr, Cu, Fe, Mg, Mn, Ni, Pb, Sb, Sn, Ti, V, Zn	spectrography	10^{-6}–10^{-5}	sodium hydroxide	[1305]
HOx + DDTC	pH 5, 7, 9	CHCl$_3$	Ag, Cu, Mo, Ni, Pb, Sn, V	spectrography		mineral waters	[1306]

Reagents	Conditions	Solvent	Metals determined	Method of analysis	Concentrations	Materials	Ref.
HOx + PMBP	pH ca. 8	MIBK, CCl_4 + isoamyl alcohol	Al, Bi, Ca, Co, Cu, Fe, Ga, Mg, Mn, Ni, Pb, Ti, Zn	spectrography	10^{-7}–10^{-5}	arsenic chloride	[1167]
HOx + PMBP + DDTC	pH 8	CCl_4 + isoamyl alcohol	Ag, Al, Bi, Ca, Co, Cu, Fe, Ga, Mn, Ni, Pb, Ti, Zn	spectrography	10^{-6}–10^{-5}	phosphorus pent-oxide	[1236]
HOx + H_2Dz	pH 8	$CHCl_3$, CCl_4	Al, Cd, Cu, Co, Fe, Mn, Ni, Zn	spectrophotometry	ca. 10^{-4}	silver salts	[1308]
HOx + H_2Dz	2·5M HCl, pH 10	cyclohexane	Ag, Al, Au, Bi, Cd, Co, Cu, Fe, Ga, Hf, Hg, In, La, Mn, Mo, Ni, Pb, Pt, Sb, Sc, Sn, Th, Ti, Tl, U, V, Y, Zn, Zr	spectrography	10^{-7}–10^{-6}	high-purity selenium	[1309]
HOx + H_2Dz	pH 5–6, 7, 9	$CHCl_3$	Ag, Al, Au, Bi, Cd, Co, Cu, Fe, Ga, Hg, In, Mn, Mo, Ni, Pb, Pd, Pt, Sb, Se, Sn, Ta, Th, Ti, Y, Zn, Zr, lanthanides	spectrography		acids	[1310]
PDTC + H_2Dz	pH 3, 5, 7, 9	$CHCl_3$	Ag, As, Au, Bi, Cd, Co, Cr, Cu, Fe, Ga, Hg, In, Mo, Ni, Pb, Pd, Pt, Sb, Se, Sn, Te, Tl, U, V, Zn	spectrography	1×10^{-6}	textile fibres	[1311]

Table 5.13 (continued)

Reagent	Aqueous phase	Organic phase	Elements isolated	Method of determination	Concentration level %	Materials used	Ref.
DDTC+ TOPO cupferron	pH 8–9	$CHCl_3$	18 elements	spectrography	10^{-7}–10^{-5}	alkali-metal halides	[1312]
HOx+ DDTC+ H_2Dz	pH 3, 5, 7, 9	$CHCl_3$	Ag, Al, Au, Bi, Cd, Co, Cr, Cu, Fe, Ga, Hg, In, Mn, Mo, Ni, Pb, Pd, Sn, Ti, Th, Tl, V, Zn, rare earth elements	spectrography	10^{-9}–10^{-7}	waters	[1313, 1314]
					10^{-4}–10^{-1}	soils	[1315]
H_2Dz+ DDTC	pH 3·5–4·0, 8–9	$CHCl_3$	Ag, Bi, Cd, Co, Cr, Cu, Mn, Mo, Ni, Pb, Pd, Sn, V, Zn	spectrography	10^{-5}–10^{-4}	rocks, soils	[1314]
PDTC+ cupferron	pH 2·0, 4·8	$CHCl_3$	Ag, Bi, Cd, Co, Cu, Ga, In, Mn, Mo, Ni, Pb, Sn, Ti, V, Zn, Zr	spectrography	10^{-4}	ores, steels, minerals, soils	[1100]
acetyl-acetone +DDTC	pH, various values	$CHCl_3$	Co, Cu, Fe, Mn, Mo, Pb, Zn	spectrography		biological materials	[409]
BPHA+ DDTC	pH, various values	$CHCl_3$	60 elements (5 groups depending on acidity)				[669]

components, and for isolation of groups of trace elements the most commonly used systems involve chelating extractants such as 8-hydroxyquinoline, sodium diethyldithiocarbamate, dithizone, and their mixtures.

It is not possible to prescribe a brief, ready-to-use procedure for the selective isolation of a single chemical element. In each different analytical system, the procedure for isolation of a specified element will differ: a choice must be made from the wide range of complex-forming reagents, media (including the pH and anion used), and masking agents. Evidently, a description of one of the many possible extraction systems would not be representative of all the possible procedures for a given element, and an exhaustive treatment of every chemical element is beyond the scope of this book.

Detailed information about the selective isolation of individual elements is to be found in the books by Sandell [391,391a] and Marczenko [392] on colorimetric methods of analysis, besides those on solvent extraction.

Cresser's recent monograph [392a] gives a concise account of the role of solvent extraction (for preconcentration) in flame spectrometric analysis, discusses theoretical and practical aspects, and surveys ~ 800 literature procedures for 58 elements. A brief review is given by Fritz [392b] and Nishimura [1333].

5.4.1 Extraction with β-Diketones

The active grouping in β-diketones is

$$-\underset{\underset{O}{\|}}{C}-CH_2-\underset{\underset{O}{\|}}{C}- \rightleftharpoons -\underset{\underset{OH}{|}}{C}=CH-\underset{\underset{O}{\|}}{C}-$$

$$\text{I} \qquad\qquad \text{II}$$

They can occur in both the enolic (II) and keto (I) forms. In their reactions with metals they yield six-membered chelate rings. The stability of the chelates and the optimum pH for the extraction process depend on the substituents attached to the diketo group and on the solvents used [87,393,394]. Aromatic derivatives yield more stable complexes than their aliphatic analogues. The presence of a $-CF_3$ group in the diketone molecule lowers the basicity of the ligand and favours extraction from a more acidic medium.

The most important β-diketones are acetylacetone and thenoyltrifluoroacetone; others are benzoylacetone, dibenzoylmethane, furoyltrifluoroacetone and thiothenoyltrifluoroacetone.

5.4.1.1 Acetylacetone (AA)

$$CH_3-\underset{\underset{O}{\|}}{C}-CH_2-\underset{\underset{O}{\|}}{C}-CH_3$$

In aqueous solution, acetylacetone behaves as a weak acid. It is unstable in alkaline solutions, and is miscible with chloroform, carbon tetrachloride, and benzene.

Acetylacetone can be used as both the chelate-forming reagent and as the solvent, but usually a solution in an organic solvent is used for extractions.

Acetylacetone reacts with about 50 metals. The compounds are thermally stable, and they dissolve well in organic solvents [395]. Thus, both major components and traces can be isolated as acetylacetonates. Comprehensive studies on acetylacetonate extractions were conducted by Freiser *et al.* [396–399], Shigematsu and Tabushi [400], and Starý and Hladký [59].

The optimum extraction conditions for metals as acetylacetonates are collected in Table 5.14.

5.4.1.2 Thenoyltrifluoroacetone (TTA)

$$\underset{I}{\underset{S}{\bigcirc}}-\underset{O}{\overset{}{C}}-CH_2-\underset{O}{\overset{}{C}}-CF_3 \rightleftharpoons \underset{II}{\underset{S}{\bigcirc}}-\underset{O}{\overset{}{C}}-CH=\underset{OH}{\overset{}{C}}-CF_3$$

Thenoyltrifluoroacetone reacts with metal ions to form mainly chelate compounds of the type MA_n, soluble in organic solvents.

Compared with other β-diketones TTA is a better extractant because of the acidity conferred by the —CF_3 group ($pK_{HA} = 6 \cdot 2$) and because of the considerable stability of the chelates formed in acidic solutions.

The properties and the application of TTA in extraction processes have been extensively studied [21,58,437–438a].

The optimum conditions for the extraction of metals as TTA complexes are listed in Table 5.15.

5.4.1.3 Dibenzoylmethane (DBM)

$$C_6H_5-\underset{O}{\overset{}{C}}-CH_2-\underset{O}{\overset{}{C}}-C_6H_5$$

Dibenzoylmethane has properties close to those of acetylacetone. It reacts with many metals, but its use is restricted for kinetic reasons. For certain elements equilibrium is reached only after shaking for a few hours or even a few days [Mo(VI), Ni(II), etc.].

The optimum extraction conditions for metals as DBM complexes are listed in Table 5.16.

Table 5.14 Extraction of metals with acetylacetone

Species	Aqueous phase	Organic phase	E, % max.	$pH_{1/2}$	$\log K_{ex}$	Reference
Al(III)	pH 3–6	100% AA	90	1.75		[399]
	pH 5–9	0.1M AA in benzene	90	3.3	−6.5	[59]
	pH 5–10	AA in CCl$_4$, AA in CHCl$_3$	100			[401, 402]
Be(II)	pH 1.5–3	AA in ethyl ether 100% AA	98	0.67	−3.3	[398]
	pH 3.5–8	0.1M AA in benzene	100	2.45	−2.8	[59, 403]
		other solvents				[404, 405]
Ce(III)	pH 8–9	0.1M AA in benzene	80			[32]
	pH 5–6 + 0.65M NaBrO$_3$	0.6M AA in benzene	ca. 95			[406]
Co(II)	pH 7–10	0.1M AA in benzene	<30			[59]
	pH 8.1–8.4	AA in 2-butanol or cyclohexane	60			[32, 407]
Cu(II)	pH 2–5	100% AA	80	1.10	−4.2	[396, 398]
	pH 4–10	0.1M AA in benzene	90	2.9	−3.9	[59, 408]
		other solvents				[400]
Dy(III)	pH 6.5	100% AA	ca. 52	5.8		[395]
Er(III)	pH 6	100% AA	ca. 68	4.9		[395]
Fe(III)	pH ≈ 1	100% AA	100	0.07	−3.2	[409]
	pH 2.5–7	0.1M AA in benzene	100	1.60	−1.4	[59, 410]
		other solvents				[400, 411, 414]
Ga(III)	pH 2–6	100% AA	97	1		[399]
	pH 3.5–8	0.1M AA in benzene	ca. 100	2.9	−5.5	[59]
	6M HCl	100% AA	100			[415]
Gd(III)	pH 6	100% AA	40			[416]

Table 5.14 (continued)

Species	Aqueous phase	Organic phase	E, % max.	pH$_{1/2}$	log K_{ex}	Reference
Hf (IV)	pH > 3	100% AA	ca. 80	1.75		[397]
	pH > 3	2M AA in benzene	ca. 80	1.6		[417]
	pH ≈ 7	0.05M AA in CHCl$_3$	> 95	4.7		[418]
Hg (II)	pH 4-10	0.1M AA in benzene	< 25			[59]
Ho (III)	pH 6.5	100% AA	ca. 62	5.1		[395]
In (III)	pH 3-6	100% AA	ca. 100	1.7	−8.1	[399]
	pH > 5.5	0.1M AA in CCl$_4$	ca. 100	4.15	−7.2	[419, 420]
	pH > 5.5	0.1M AA in benzene	100	3.95	−7.2	[419, 420]
	pH > 5.5	0.1M AA in CHCl$_3$	100	4.55	−9.1	[419, 420]
La (III)	pH 6-10	0.1M AA in benzene	20			[59]
Mg (II)	pH 9-12	0.01M AA in benzene	< 60	9.4		[59]
Mn (II)	pH 5.5-6.5	100% AA	10-20			[154]
Mn (III)	pH 8-9.5		ca. 100			[421]
Mo (VI)	0.005-3M H$_2$SO$_4$	100% AA or 50% AA in CHCl$_3$	96-98			[154, 422]
	2M HCl	100% AA	100			[154]
Nb (V)	pH 2-5	20% AA in CHCl$_3$	90	0.8		[423]
	pH 2-5	2M AA in benzene	90			[423]
Nd (III)	pH 6	100% AA	28			[395]
Ni (II)	pH 5-6	0.1M AA in benzene	> 20			[59]
Pa (V)	acetate buffer	50% AA in benzene	ca. 40			[424]
Pb (II)	pH 6-8	100% AA	ca. 80	5.65	−13.3	[396]

Applications

Element	Conditions	Reagent	% Extraction			References
Pu (IV)	pH 4–7	1M AA in benzene	ca. 100	2·5		[418, 425]
Ru (III)	pH 4–7	30% AA in CHCl$_3$	partially	2·95	−5·8	[426]
Sc (III)	pH 3·5–9	0·1M AA in benzene	ca. 100			[59]
Sm (III)	pH ≈ 3	100% AA	33	1		[395]
Sn (II)	pH 3–9	0·1M AA in benzene	>75			[59]
Tb (III)	pH ≈ 6	100% AA	50			[395]
Tc (VII)	pH ≈ 4	100% AA	ca. 55			[427]
Th (IV)	pH 5–9	0·1M AA in benzene	100	4·1	−12·1	[427, 428]
Ti (IV)	pH 3–5	0·1M AA in benzene	35			[59]
	pH 0–2	100% AA	75			[429, 430]
Tl (III)	pH 2–10	0·1M AA in benzene	100	1·3		[59]
U (IV)	pH > 3	0·5M AA in benzene	100	2·0		[418, 431]
U (VI)	pH 3–7	100% AA	ca. 95	1·7		[396]
V (III)	pH 2·3–3·0	50% AA in CHCl$_3$	100	0		[432]
V (IV)	pH 2–4	50% AA in CHCl$_3$	80	1·4		[432]
V (V)	pH ≈ 2·1	50% AA in CHCl$_3$	<70	1·2		[433]
	2M NaCl	AA in butanol	ca. 94			[434]
	6M HCl	50% AA in benzene	100			
Y (III)	pH 5·5–10	100% AA	>50	5·1		[395, 435]
Yb (III)	pH 6	100% AA	ca. 85	4·5		[395]
Zn (II)	pH 5·5–7	100% AA	>50			[397, 398]
	pH 8–10	0·1M AA in benzene				[59, 145, 436]
Zr (IV)	pH ≈ 2	100% AA	70	1·5		[430, 432]
	pH 3–8	2M AA in CHCl$_3$	98			[423]

Table 5.15 Extraction of metals with thenoyltrifluoroacetone

Species	Aqueous phase	Organic phase	E, % max.	$pH_{1/2}$	$\log K_{ex}$	Reference
Ac (III)	pH 5·4	0·25M TTA in benzene	ca. 100	4·6		[439]
Al (III)	pH 5·5	0·02M TTA in hexane	ca. 100			[440]
	pH > 3·5	0·2M TTA in benzene	ca. 100	2·4	−5·2	[58]
Am (III)		0·2M TTA in benzene in other solvents	> 95	3·2	−7·5	[58] [21, 58, 441]
Au (III)		0·015M TTA in xylene	ca. 100			[442]
Ba (II)	pH 4·0 + 10M LiCl	0·2M TTA in benzene	ca. 100	8·0	−14·4	[58, 443]
Be (II)	pH ≈ 7	0·1M TTA in benzene	ca. 100			[152]
	pH < 4	0·5–1M TTA in benzene	ca. 100	ca. 2·3		[58]
Bi (III)	pH > 2·5	0·25M TTA in benzene	ca. 100	1·7		[439]
	pH 5·8–6·1	0·01M TTA in CCl$_4$	ca. 100			[444]
Bk (III)	pH ≈ 3·4	0·2M TTA in benzene	80	3·0	−7·5	[58, 445]
Ca (II)	pH 8·5	0·2M TTA in benzene in other solvents	ca. 100	6·7		[58] [446–448]
Cd		0·1M TTA in CHCl$_3$	ca. 100	6·7	−11·4	[449]
		0·2M TTA in benzene	ca. 100	> 8		[58]
Ce (IV)	pH 5·4	0·15M TTA in benzene	ca. 100	2·9		[450]
	0·5M H$_2$SO$_4$	0·5M TTA in xylene	ca. 100			[451]
Ce (III)		0·2M TTA in benzene	ca. 100	3·9	−9·3	[452]
		0·5M TTA in xylene	ca. 100			[453]
Cf (III)	pH 4·3 ± 0·2	0·2M TTA in benzene	ca. 100	3·1		[58, 441]
Cm (III)	pH 3·4	0·2M TTA in toluene	ca. 100			[58, 441]
	pH 3·4	0·2M TTA in benzene	ca. 100	3·4	−7·3	[58]
Co (II)	pH 7·6–8·8	0·1M TTA in ethyl methyl ketone	ca. 100	4·1		[32, 58, 407, 454]
Cr (III)	pH 6	0·15M TTA in benzene	> 80	ca. 4		[455, 456]
		0·15M TTA in benzene	ca. 100	0·83		[456]

§5.4] Applications

	Aqueous phase	Organic phase	% extraction			References
Cs (I)	1M HCl	0.15M TTA in benzene	100	1.79		[456]
	pH 8-9.5 + 0.3-0.17M citrate	0.25M TTA in mixture of nitromethane and xylene				[457]
Cu (II)	pH 3.4	0.2M TTA in benzene	ca. 90	6.5		[58]
Dy (III)	pH 3	0.5M TTA in MIBK	ca. 100	1.35	−1.2	[458, 459]
	pH ≈ 3	0.2M TTA in benzene	ca. 85	2.7		[21, 58]
Er (III)	pH 5.5	0.1M TTA in MIBK	ca. 85	3.1	−7.0	[459]
Es (III)	pH ≈ 3.4	0.2M TTA in benzene	ca. 60	3.1		[58, 441]
Eu (III)	pH 3.3	0.5M TTA in toluene	ca. 50	2.9	−7.6	[458]
	pH 5.5	0.1M TTA in MIBK in other solvents	100			[459] [21, 205, 460]
Fc (III)	pH 2.5	0.15M TTA in benzene	>99	1.0		[461, 462]
	2M HNO₃ + 9M NH₄NO₃	10% TTA in xylene	ca. 100			[463]
	10M HNO₃					
Fm (III)	pH 3.4	0.5M TTA in xylene	ca. 90	0.95	−7.7	[464-467]
Gd (III)	pH 3.0	0.2M TTA in toluene	70		−7.6	[441]
	pH 5.5	0.5M TTA	ca. 75	2.9		[437, 458]
Hf (IV)	0.35-3.5M HNO₃	0.1M TTA in MIBK	100			[459]
Ho (III)	pH > 0	0.2M TTA in benzene	ca. 100	3.1	−7.2	[468, 469]
In (III)	pH > 4	0.2M TTA in benzene	ca. 100	2.8		[470]
	pH 2.5-3.5	0.005M TTA in benzene	100	1.9	−4.3	[471]
Ir (III)	pH > 7	0.5M TTA in benzene TTA in benzene	100			[58, 472] [58]
K (I)	pH 9	0.5M TTA in benzene	ca. 82			[457]
La (III)	pH > 3.5	0.5M TTA in benzene	ca. 100	3.7	−10.5	[58, 473]
	pH ≈ 5	0.1M TTA in MIBK	100	−1.7		[459]
Mn (III)	pH 5.5	0.2M TTA in MIBK	100			[459]
	0.5M H₂SO₄	0.5M TTA in xylene	100			[474]
Mn (II)	pH 6.7-8.0	0.15M TTA in mixture of acetone and benzene (3:1) in other solvents	100			[253, 453] [473]

Table 5.15 (continued)

Species	Aqueous phase	Organic phase	E, % max.	$pH_{1/2}$	$\log K_{ex}$	Reference
Mo (VI)	9M HCl	0·15M TTA in mixture of butyl alcohol and acetophenone (5 : 8)	ca. 100			[475]
Na (I)	0·5M HCl	0·15M TTA in benzene	100			[456]
Nb (V)	pH ≈ 9	0·5M TTA in nitrobenzene	ca. 45			[457]
	10M HNO₃	0·5M TTA in xylene	ca. 95			[464]
Nd (III)	pH > 3	0·5M TTA in benzene	ca. 100	3·1	−8·6	[58, 437]
Ni (II)	pH 5·5–8	0·15M TTA in mixture of benzene and acetone (1 : 3)	ca. 100	3·8		[476]
Np (IV)	pH 6·5	0·001M TTA in CCl₄	ca. 100			[477]
Pa (IV)	1M HCl	0·5M TTA in xylene	ca. 100			[478, 479]
Pd (II)	6M HCl	0·3–1·0M TTA in benzene	90			[480]
Pb (II)	HClO₄	TTA in benzene	ca. 100			[454, 481]
Pm (III)	pH ≈ 5	0·25M TTA in benzene	100	3·2	−5·2	[439]
		0·5M TTA in benzene	ca. 100	3·0	−8·0	[58, 437]
Po (V)	pH ≈ 2	0·25M TTA in benzene	100	0·9		[439]
Pr (III)	pH > 4	0·1–0·5M TTA in benzene	100			[437, 482]
Pu (III)	pH ≈ 4·2	0·4M TTA in benzene	99		8·5	[58, 205, 478]
Pu (IV)	2–6M HCl	0·5–1·0M TTA in benzene	100		−4·4	[58, 464]
	10M HNO₃	0·05M TTA in benzene	100		5·3	[483]
	0·5M, 1M HNO₃	TTA in CCl₄	100		5·0	[484, 485]
Pu (VI)		TTA in benzene	ca. 100		−1·8	[58, 205]
Pt (IV)	3·5M HCl	0·1M TTA in mixture of butyl alcohol and acetophenone (2 : 1)	100			[454, 486]
	0·5M H₂SO₄	0·15M TTA in mixture of butyl alcohol and acetophenone (2 : 1)	100			[454]
Rb (I)	pH ≈ 9	0·5M TTA in nitrobenzene	92			[457]

Applications

Element	Aqueous phase	Organic phase	% Extraction			References
Se(IV)	pH 3–4	0.2M TTA in xylene	>95	0.5	−0.8	[58, 437]
	pH 0.5–4.5	0.03M TTA in xylene	ca. 100			[487]
Sm(III)						[488]
Sn(IV)	>0.5M HCl	0.5M TTA in benzene	ca. 100	2.9	−7.7	[58, 437]
		0.5M TTA in MIBK	100			[489]
Sr(II)	pH 10–12	0.2M TTA in benzene	ca. 80	8.0	−14.0	[153, 490]
	pH 10–12	0.2M TTA in MIBK	>99	6.0		[490]
Th(IV)	pH >1	0.15–0.45M TTA in benzene	>98		−0.9	[439, 493, 494]
Ti(IV)	10M HCl	0.1M TTA in mixture of isobutyl alcohol and benzene (2 : 1)	ca. 100		—	[495]
Tl(I)	pH ≈ 7	0.25M TTA in benzene	>95	5.8	−5.2	[439]
Tl(III)	pH ≈ 4	0.25M TTA in benzene	100	2.6		[439]
Tu(III)		0.5M TTA in benzene	ca. 100	3.1	−6.7	[58, 437]
U(IV)	pH ≈ 4.2	0.4M TTA in benzene	98.5		5.3	[58, 478, 496]
U(VI)	pH 3.5–8	0.15M TTA in benzene in other solvents	100			[450, 496–498]
V(IV)	pH ≈ 4	0.25M TTA in benzene	ca. 70	1.2–4.0		[499, 500]
V(V)	pH 2.5–4.1	0.3M TTA in n-butyl alcohol	100			[437, 500]
W(VI)	9M HCl	0.15M TTA in mixture of n-butyl alcohol and acetophenone (5 : 8)	100			[501]
						[475]
Y(III)	5M HCl	0.15M TTA in benzene	100		8.7	[453]
	pH > 5.4	0.3M TTA in benzene in other solvents	99	3.8	−6.8	[58, 437, 486]
Zn(II)	pH > 5	0.1M TTA in MIBK in other solvents				[446, 502, 503]
		0.5M TTA in benzene	99	3.85	−5.7	[504]
Zr(IV)	2M HNO$_3$	0.5M TTA in xylene	ca. 100			[504]
	2M HCl, 3M HClO$_4$		100			[505]
	or 1.5–10M HNO$_3$					[506–509]

Table 5.16 Extraction of metals with dibenzoylmethane

Species	Aqueous phase	Organic phase	E, % max.	$pH_{1/2}$	$\log K_{ex}$	Reference
Ag (I)	pH 10–11	0.1M DBM in benzene	60	9.9	−8.6	[59]
Al (III)	pH 5–10	0.1M DBM in benzene	80	4.0		[59]
Ba (II)	pH > 12	0.1M DBM in benzene	> 50	ca. 12.0		[59]
Be (II)	pH 4.5–8	0.1M DBM in benzene	ca. 100	−2.7	−3.5	[59]
Bi (III)	pH 9–12	0.1M DBM in benzene	< 60	10.5		[59]
Ca (II)	pH > 10.5	0.1M DBM in benzene	100	9.9	−18.0	[59]
Cd (II)	pH 9–11	0.1M DBM in benzene	100	8.0	−14.0	[59]
Co (III)	pH 7.5–10.0	0.1M DBM in benzene	100	6.4	−10.8	[59]
Cu (II)	pH 4–9	0.1M DBM in benzene	100	2.9	−3.8	[59]
Fe (III)	pH 2–4	0.1M DBM in benzene	100	1.7	−1.9	[59]
Ga (III)	pH 4.0–5.5	0.1M DBM in benzene	100	3.9	−5.8	[59]
Hg (II)	pH 5.5–7.5	0.1M DBM in benzene	100	3.9		[59]
In (III)	pH 4.5–5.5	0.1M DBM in benzene	100	3.6	−7.6	[59]
La (III)	pH ≈ 9.0	0.1M DBM in benzene	100	8.5	−19.5	[59]
Mg (II)	pH > 9.5	0.1M DBM in benzene	100	8.5	−14.7	[59]
Mn (II)	pH 9–12	0.1M DBM in benzene	ca. 100	7.8	−13.7	[59]
Mo (VI)	pH 1–4	0.1M DBM in benzene	< 10			[59]
Ni (II)	pH 7.5–11.0	0.1M DBM in benzene	100	6.4	−11.0	[59]
Pb (II)	pH > 7.5	0.1M DBM in benzene	ca. 100	5.6	−9.4	[59]
Pd (II)	pH 3–10	0.1M DBM in benzene	> 90	1.8		[59]
Sc (III)	pH 4.8	0.1M DBM in benzene	100	3.1	−6.0	[59]
Sr (II)	pH ≈ 12	0.1M DBM in benzene	ca. 80	11.1	−20.9	[59]
Th (IV)	pH 3.5–8.0	0.1M DBM in benzene	ca. 100	2.6	−6.4	[59]
Ti (IV)	pH > 3	0.1M DBM in benzene	ca. 100	2.6		[59]
Tl (III)	pH 5–9	0.1M DBM in benzene	ca. 80	3.8		[59]
U (VI)	pH 5–6.5	0.1M DBM in benzene in other solvents	100	3.6	−4.1	[59] [85, 510, 511]
Zn (II)	pH 8–11	0.1M DBM in benzene	100	6.4	−10.7	[59]

5.4.1.4 1-PHENYL-3-METHYL-4-BENZOYL-5-PYRAZOLONE (PMBP)

$$\underset{\substack{C_6H_5-N \\ }}{\overset{N=C-CH_3}{\underset{OO}{C-CH-C-C_6H_5}}} \rightleftharpoons \underset{\substack{C_6H_5-N \\ }}{\overset{N=C-CH_3}{\underset{OOH}{C-C=C-C_6H_5}}}$$

4-Acyl derivatives of 1-phenyl-3-methyl-5-pyrazolone were first used as chelating agents by Jensen [512]. These reagents are β-diketones, and various derivatives have been investigated by Navrátil et al. [513–517a]. PMBP is decomposed by oxidants, and occasionally it can itself reduce metal ions, such as V(V). It is readily soluble in most organic solvents.

PMBP can be used for extraction of many elements [522, 1334]. It has been established that in many cases quantitative extraction with PMBP is possible from more acidic solutions than for TTA.

The reagent can be used for some group separations. The optimum extraction conditions for metals as PMBP complexes are given in Table 5.17.

5.4.2 Extraction with 8-Hydroxyquinoline and Its Derivatives

5.4.2.1 8-HYDROXYQUINOLINE (HOx, OXINE)

This compound is poorly soluble in water, but easily soluble in numerou organic solvents, in particular chloroform, benzene and carbon tetrachloride. In aqueous solutions it displays amphoteric properties: in acidic solutions it occurs as positive oxinium ions, and in alkaline solutions as oxinate ions.

8-Hydroxyquinoline is a typical group reagent: it reacts with many metals to form chelate compounds which are mostly readily soluble in organic solvents and can be extracted.

Sometimes the extraction of a metal is done by first precipitating the oxinate with an alcoholic or acetic acid solution of oxine, then dissolving the precipitate in a suitable solvent.

Much basic research on the properties of oxine and the extraction of oxinates has been done [57,79,80,542–546].

The extraction mechanism for many elements is not clear. Most elements yield oxine complexes of type MA_n, but with bivalent elements, complexes of type $MA_n \cdot HA$ and $MA_n \cdot 2HA$ have been shown to be extracted, and hydrated complexes [547,548] and ion-association complexes [548,549] have also been observed.

The optimum conditions for extraction of metals as 8-hydroxyquinolinates are reported in Table 5.18.

Table 5.17 Extraction of metals with 1-phenyl-3-methyl-4-benzoyl-5-pyrazolone (PMBP)

Species	Aqueous phase	Organic phase	E, % max.	pH$_{1/2}$	log D	Reference
Ac (III)	pH 2·6	0·25M PMBP in octyl alcohol + benzene	99			[523]
Al (III)	pH 4–5 or 11	0·01M PMBP in various solvents	88–95			[522]
Ca (II)	pH 6·5–7·0	0·1M PMBP in isoamyl alcohol + CHCl$_3$(1 : 1)		4·70		[524]
Cd (II)	pH > 7·2	0·01M PMBP in isoamyl alcohol		5·15	2	[525]
Cm (III)	pH > 8	PMBP in benzene	ca. 100			[526]
Ce (III)	pH 2	0·05M PMBP in isoamyl alcohol	ca. 100			[227]
Cr (III)	pH 3–4	0·01–0·1M PMBP in CHCl$_3$	100	3·05		[527]
	pH 3–6 CH$_3$COO– elevated temperature	CHCl$_3$	90–100			[528]
Co (II)	pH > 3·5	0·05M PMBP in isoamyl alcohol	100			[223, 529, 522]
		0·01M PMBP in various solvents				
Cu (II)	pH ⩾ 3	0·1M PMBP in various solvents	100	0·4	3·2	[522, 526, 530, 531]
Er (III)	pH 3–4	0·01–0·1M PMBP in CHCl$_3$	100	2·45		[527]
Eu (III)	pH 2	0·05M PMBP in isoamyl alcohol	ca. 100			[227]
Fe (III)	1–6·5M HNO$_3$	0·1M PMBP in benzene	ca. 100			[532]
	10–11M HCl	0·1M PMBP in benzene	ca. 100			[532]
	9M HCl	0·1M PMBP in dichloroethane	ca. 100			[533]
	pH 3–7	0·01M PMBP in various solvents	100			[522]
Gd (III)	pH 3–4	0·01–0·1M PMBP in CHCl$_3$	ca. 100	2·7		[527]

Element	Aqueous phase	Organic phase	% extraction			References
Mn (II)	pH > 4.6	0.01M PMBP in isoamyl alcohol			2.4	[525]
Nb (V)	8M HCl	0.05M PMBP in CHCl$_3$	95–98			[535, 536]
	5M HNO$_3$	0.1M PMBP in benzene	99.3	3.2		[513, 514, 537]
Nd (III)	pH 3–4	0.01–0.1M PMBP in CHCl$_3$	90–95			[527]
Ni (II)	pH ⩾ 3	0.005M PMBP in isoamyl alcohol		2.8	1.5	[531]
Pa (V)	0.1–0.5M HCl, HNO$_3$, 0.05–0.25M H$_2$SO$_4$	0.1M PMBP in benzene	98			[538]
Pu (III)	7M HNO$_3$	0.1M PMBP in benzene	76.7			[537]
Pu (IV)	3M HNO$_3$	0.1M PMBP in benzene	99.9		2.9	[539]
	1M HCl	0.1M PMBP in benzene	93.7		1.2	[539]
	0.5–2M H$_2$SO$_4$	0.1M PMBP in benzene	98–99			[539]
Sc (III)	pH > 1.4	0.01M PMBP in isoamyl alcohol	100		3.9	[525]
Sr (II)	pH 6.7–9.5	0.01M PMBP in isoamyl alcohol			1.4	[525]
Ta (V)	9M HCl	0.1M PMBP in benzene			1.5	[536]
Ti (IV)	pH 2.5–3.0 10^{-4}–10^{-1} M tartaric acid	PMBP in CHCl$_3$	96	5.8		[540]
Th (IV)	pH 0–4 (NO$_3^-$)	0.01M PMBP in various solvents	100			[522]
	0.25–1M HCl or HNO$_3$	0.1M PMBP in benzene	98			[541]
U (VI)	0.1M HNO$_3$	0.1M PMBP in benzene	85.5			[537]
Zn (II)	pH ⩾ 6	0.01M PMBP in isoamyl alcohol, other solvents			ca. 3	[530]
Zr (IV)	6M HNO$_3$	0.1M PMBP in benzene	100		> 3	[537]
	7M HNO$_3$	0.1M PMBP in benzene	100		> 3	[537]
	3.5M H$_2$SO$_4$	0.1M PMBP in benzene	98.7		1.9	[537]
Rare earth elements	pH > 2.5	0.05M PMBP in various solvents				[541a, 541b, 541c]

196 Liquid–Liquid Extraction [Ch. 5

Table 5.18 Extraction of metals with 8-hydroxyquinoline

Species	Aqueous phase	Organic phase	E, % max.	pH$_{1/2}$	log K$_{ex}$	Reference
Ag (I)	pH 8–9	0·1M HOx in CHCl$_3$	90	6·5–6·9	−4·5, −4·8,	[57, 130]
Al (III)	pH 4·5–11	0·01M HOx in CHCl$_3$	100	3·8	−5·2	[57, 79, 80, 543, 544, 550–552]
		in other solvents				[553–555]
Ba (II)	pH > 10	0·5–1M HOx in CHCl$_3$	partial		−20·9	[57, 544, 557]
Be (II)	pH 6–10	0·5M HOx in CHCl$_3$	ca. 87	5·1	−9·5	[57, 544, 557]
Bi (III)	pH 2·5–11	0·1M HOx in CHCl$_3$	100	2·1	−1·2	[57]
	pH 4·5–2	0·01M HOx in CHCl$_3$	100			[544]
Ca (II)	pH > 10·7	0·5M HOx in CHCl$_3$	100			[57, 544, 558]
		in other solvents				[559]
Cd (II)	pH 5·5–9·5	0·1M HOx in CHCl$_3$	100	4·7	−5·3	[57, 560–562, 544]
Ce (III) and (IV)	pH 9·9–10·6 (citrates, tartrates)	1·5M HOx in CHCl$_3$	ca. 100			[563]
Co (II)	pH 4·5–10·5	0·1M HOx in CHCl$_3$	100	3·2	−2·2	[57, 544, 564]
		in other solvents				[565]
Cr (III)	pH 6–8	CHCl$_3$ at elevated temperature	100			[544, 566–568]
Cu (II)	pH 2–12	0·1M HOx in CHCl$_3$	ca. 100	1·5	1·8	[57, 79, 80, 544, 569]
		in other solvents				[568]
Fe (III)	pH 2–10	0·01M HOx in CHCl$_3$	100	1·5	4·1	[57, 79, 80, 544, 570]
		in other solvents				[571]
Ga (III)	pH 2·2–12	0·01M HOx in CHCl$_3$		1·1	3·7	[57, 544]
Ge (IV)	pH 4	CHCl$_3$			−2	[572]

§ 5.4] Applications 197

Hg (II)	pH > 3	0.1M HOx in CHCl$_3$	ca. 100		[57]
In (III)	pH 3.0–11.5	0.01M HOx in CHCl$_3$	100	2.1	0.9–1.3	[57, 544, 64, 574]
		in other solvents				[64]
La (III)	pH 7–10	0.1M HOx in CHCl$_3$	100	6.5	−15.7	[57, 673]
Mg (II)	pH 9	0.1M HOx in CHCl$_3$	100	8.6	−15.1	[57, 544]
		in other solvents				[575]
Mn (II)	pH 6.5–10	0.1M HOx in CHCl$_3$	100	5.7	−9.3	[57, 79, 80, 544]
Mo (VI)	pH 1.0–5.5	0.01M HOx in CHCl$_3$	100	0.5		[57, 544]
Nb (V)	pH 6–9	2.5% HOx in CHCl$_3$	100			[544, 576–579]
	pH 5	0.1M HOx in CHCl$_3$	⩾95			[156]
	0.1M tartaric acid					
Ni (II)	pH 4.0–10.0	0.01M HOx in CHCl$_3$	100	3.2	−2.2	[57, 79, 80, 543, 580]
Pa (V)	saturated (NH$_4$)$_2$CO$_3$	amyl acetate	67			[424]
Pb (II)	pH 6–10	0.1M HOx in CHCl$_3$	100	5.0	−8.0	[57, 79, 80, 543, 544]
Pd (II)	pH 0–10	0.01M HOx in CHCl$_3$	<100			[57, 544]
Pu (IV)	pH < 8	amyl acetate	ca. 100			[581]
Pu (VI)	pH 4–6	amyl acetate	100			[581]
Rh (III)	pH 6–9 temp. 90–100°	CHCl$_3$	ca. 100			[49]
Ru (III)	pH 6.4	5–15% HOx in CHCl$_3$	ca. 100			[49]
Sc (III)	pH 4.5–10	0.1M HOx in CHCl$_3$	100	3.6	−6.6	[57, 544, 582, 583]
Sm (III)	pH 6–8.5	0.5M HOx in CHCl$_3$	100	5.0	−13.3	[573]
Sn (IV)	pH 2.5–5.5	0.1M HOx in CHCl$_3$	100			[580]
	pH 0.85±0.1 (H$_2$SO$_4$)	CHCl$_3$	100			[584]
	3M KCl	0.001M HOx in isoamyl alcohol	100	ca. 0.0		[565]
		0.001M HOx in CHCl$_3$	100	ca. 1.5		[565]

Table 5.18 (continued)

Species	Aqueous phase	Organic phase	E, % max.	$pH_{1/2}$	$\log K_{ex}$	Reference
Sr(II)	pH > 11·5	0·5M HOx in CHCl$_3$	100	10·5	−19·6	[57, 544, 557, 585]
Ta(V)	pH 0–7	isoamyl alcohol and other oxygen-containing solvents	partial			[576]
Th(IV)	pH 4–10	0·1M HOx in CHCl$_3$	100	3·1	−7·1	[57, 544, 585]
Ti(IV)	pH 2·5–9·0	0·1M HOx in CHCl$_3$	100	0·9		[57, 544, 586]
	pH 1·0–2·0 0·1M tartaric acid	0·1M HOx in CHCl$_3$	97			[540]
Tl(I)	pH 12	0·05M HOx in CHCl$_3$	60	11·5	−9·5	[587]
		0·05M HOx in isobutyl alcohol	85			[68, 587]
Tl(III)	pH 3·5–11·5	0·01M HOx in CHCl$_3$	100	2·0		[57, 544, 588]
U(VI)	pH 5–9	0·1M HOx in CHCl$_3$ in other solvents	100	2·6	−1·6	[57, 543, 544, 573] [78, 589, 590–592]
V(V)	pH 2–6	0·1M HOx in CHCl$_3$				[57]
	pH 2·5–3·5	0·1M HOx in CHCl$_3$	100	0·9	1·7	[593]
W(VI)	pH 1–5	0·1M HOx in CHCl$_3$	>99			[402, 544]
Y(III)	pH 7–10	0·2M HOx in CHCl$_3$	ca. 100	ca. 5		[544, 594, 595]
Yb(III)	pH > 8	0·1M HOx in CHCl$_3$	ca. 100			[49]
Zn(II)	pH 4–5	0·1M HOx in CHCl$_3$	ca 100	3·3	−2·4	[57, 69, 544, 547, 549]
Zr(IV)	pH 1·5–4·0	0·1M HOx in CHCl$_3$	ca. 100	1·0	2·7	[57, 544]
	pH 8·9–9·0 (2M HF + 0·5M tartaric acid + HOx in acetone)	CHCl$_3$	ca. 100			[597]

5.4.2.2 Derivatives of 8-Hydroxyquinoline

Numerous derivatives of 8-hydroxyquinoline are used for extraction of metals [556]. Hydroxyquinaldine (2-methyl-8-hydroxyquinoline) is a reagent of similar properties to the parent compound, but it is more selective than 8-hydroxyquinoline. It was originally reported not to yield an extractable chelate with aluminium [598], but later evidence shows that it does [598a] and the species can be extracted as ion-association complexes [199]. The extraction of methylhydroxyquinolinates has been studied by Motojima and Hashitani [599].

Other derivatives include 5,7-dichloro-8-hydroxyquinoline [600–605, 1335], 5,7-dinitro-8-hydroxyquinoline [606] and 8-mercaptoquinoline, the sulphur analogue of 8-hydroxyquinoline [607–615].

The optimum conditions for extraction of metals as methyl-8-hydroxyquinolinates are listed in Table 5.19.

Table 5.19 Extraction of metals with methylhydroxyquinoline (MHOx)

Species	Aqueous phase	Organic phase*	E, % max	Reference
Al(III)	pH 6.5 2M NaClO$_4$	1.85M TBP in CHCl$_3$	37	[199]
	pH 6	0.01M MHOx+1.85M TBP+0.32M caproic acid in CHCl$_3$	100	[199]
Be(II)	pH 7.5–8.5	CHCl$_3$	100	[599, 616, 617]
Bi(III)	pH ≈ 10	1% MHOx in CHCl$_3$	100	[618]
Cd(II)	pH 10	1% MHOx in CHCl$_3$	100	[618]
Ce(III)	pH ⩾ 10	CCl$_4$	ca. 100	[619]
Co(II)	pH 10	1% MHOx in CHCl$_3$	ca. 100	[618]
Cr(III)	pH 5.3–9.5	CHCl$_3$	ca. 100	[620]
Cu(II)	pH 4.2–12.5	CHCl$_3$	100	[599, 619]
Fe(III)	pH 4.5–12.2	CHCl$_3$	ca. 100	[599]
Ga(III)	pH 5.5–9	CHCl$_3$	ca. 100	[621, 622]
In(III)	pH 4.6–13	CHCl$_3$ or benzene	ca. 100	[599]
Mn(II)	pH 10	CHCl$_3$	ca. 100	[623, 624]
Mo(VI)	pH 3.5–4.5	CHCl$_3$	ca. 100	[599]
Ni(II)	pH 8.5–10.7	CHCl$_3$	ca. 100	[599]
Pb(II)	pH 8.2–11.8	CHCl$_3$	ca. 100	[599]
Ti(IV)	pH 5.0–9.3	CHCl$_3$	100	[623, 625, 626]
Tl(III)	pH ⩾ 4	CHCl$_3$	ca. 100	[599]
V(V)	pH 4–4.8	CHCl$_3$	100	[599]
W(VI)	pH 10	CHCl$_3$	ca. 100	[402, 618]
Y(III)	pH 10	1% MHOx in CHCl$_3$	ca. 100	[618]
Zn(II)	pH 10	1% MHOx in CHCl$_3$	ca. 100	[618]

* It should be noted that most cases MHOx is added as an acetone or acetic acid solution to the aqueous phase.

Table 5.20 Extraction of metals with cupferron

Species	Aqueous phase	Organic phase	E, % max.	pH$_{1/2}$	log K_{ex}	Reference
Al(III)	pH 3·5-9	CHCl$_3$, MIBK	100	2·5	−3·5	[60, 440, 632]
Au(III)	0·5-8N H$_2$SO$_4$	CHCl$_3$	100			[633]
Be(II)	pH > 3	CHCl$_3$	100	2·1	−1·5	[60]
Bi(III)	pH 2-12	CHCl$_3$	100	0·6	5·1	[60]
Ca(II)	pH 6-8	TBP, other solvents	95	4·7	−8·7	[645]
Ce(III)	pH 4-5	CHCl$_3$	ca. 100		4·6	[634]
Ce(IV)	0·1-0·15M H$_2$SO$_4$	butyl acetate, amyl acetate	ca. 100			[49]
Co(II)	pH > 4·5	CHCl$_3$	100	3·2	−3·6	[60]
Cu(II)	pH 2-10	CHCl$_3$	100	0·03	2·7	[60]
Fe(III)	pH 2-10	CHCl$_3$	100	−2	9·8	[60, 635]
		other solvents				[635, 636]
Ga(III)	pH 1·5-12	CHCl$_3$	100	0·7	4·9	[60]
	H$_2$SO$_4$, HCl	various solvents				[637]
Hf(IV)	dilute acids	CHCl$_3$	100		> 8	[573]
Hg(II)	pH 2-5	CHCl$_3$	ca. 98	0·85	0·9	[60]
In(III)	pH 3-8	CHCl$_3$	100	1·5	2·4	[60]
La(III)	pH 4-10	CHCl$_3$	ca. 90	3·4	−6·2	[60]
Mg(II)	pH 6-8	TBP	99	4·5		[645]
Mn(II)	pH 4·5-9·5	CHCl$_3$	ca. 15			[60]
Mo(VI)	pH 0-1·5	CHCl$_3$	ca. 100			[60]
	3M H$_2$SO$_4$	isoamyl alcohol	100			[638]
Nb(V)	pH 0·5+2% ammonium tartrate	CHCl$_3$	ca. 90			[639]
Ni(II)	pH 9-12	CHCl$_3$	ca. 50			[60]
	pH 6-8	mixture of isoamyl alcohol and benzene (1:1)	ca. 100			[640]

§ 5.4] Applications 201

Element	Conditions	Solvent	%			Ref.
Pa (V)	0.1–6M HCl	various solvents	> 90			[641]
Pb (II)	pH 3–9	CHCl$_3$	100	2.1	−1.5	[60]
		other solvents				[640]
Pd (II)	pH 0–12	CHCl$_3$	100			[60]
Pu (IV)	pH 0.3–2	CHCl$_3$	ca. 100		7.0	[642]
Sb (III)	pH 0–9	CHCl$_3$	100			[60]
Sc (III)	pH 3–12	CHCl$_3$	ca. 95	0.2	3.3	[60, 643]
Sn (II)	dilute acids	CHCl$_3$ or benzene	ca. 100			[644]
Sn (IV)	dilute acids	CHCl$_3$ or butyl acetate	100			[644]
Sr (II)	pH 8	TBP	80			[645]
Ta (V)	pH ≈ 0 + 0.5% tartaric acid	isoamyl alcohol	ca. 100	6.2		[639]
Th (IV)	pH 2.8–8.5	CHCl$_3$	100	1.2	4.4	[573]
		other solvents				[573]
Tl (IV)	pH 0–4	CHCl$_3$	100			[60]
		other solvents				[639, 646]
Tl (I)	pH 7–11.5	CHCl$_3$	ca. 50			[60]
Tl (III)	pH 1.5–1.0	CHCl$_3$	ca. 50		ca. 3	[68]
U (IV)	acid solutions	ethyl ether or ethyl acetate	ca. 100		8.0	[60]
						[647–649]
U (VI)	pH 3.5–6	CHCl$_3$	ca. 30			[60]
V (V)	pH 0–2.5	CHCl$_3$	100			[60]
W (VI)	pH 0–3	CHCl$_3$	< 25			[60]
Y (III)	pH > 5	CHCl$_3$	> 75			[60]
Zn (II)	pH 9–10.5	CHCl$_3$	ca. 82			[60]
		other solvents				[640]
Zr (IV)	pH 0–3	CHCl$_3$	100			[60, 650]
						[651]

5.4.3 Extraction with Nitrosoarylhydroxylamines

The characteristic nitrosohydroxyamino group occurs in two tautomeric forms

$$\begin{array}{c} N=O \\ | \\ -N-OH \end{array} \rightleftharpoons \begin{array}{c} N-OH \\ \| \\ -\overset{\oplus}{N}-O^{\ominus} \end{array}$$

These compounds form five-membered chelate rings.

The most important members of this group are cupferron and neocupferron.

5.4.3.1 CUPFERRON

$$\begin{array}{c} N \overset{O}{\nearrow} \\ | \\ C_6H_5 \overset{N}{\diagdown} O^{\ominus}-NH_4^{\oplus} \end{array}$$

Cupferron, the ammonium salt of N-nitroso-N-phenylhydroxylamine, forms chelates with numerous metals, and is used for separation, by precipitation or extraction, of many elements. The cupferronates of many metals are water-insoluble: they dissolve in organic solvents such as ethyl acetate, ethyl ether, benzene, isoamyl alcohol and chloroform.

A serious disadvantage of cupferron as an analytical reagent is its instability. It decomposes at elevated temperatures and when treated with oxidants.

The properties of cupferron and its analytical applications have been widely studied [60,627–631,644].

The optimum conditions for extraction of metals as cupferronates are listed in Table 5.20.

5.4.4 Extraction with Hydroxamic Acids

The hydroxamic acids, with their characteristic grouping

$$O=C\diagup_{NH-OH}$$

react with a large number of elements to give extractable chelate compounds. Benzo-, salicylo-, anthranilo-, and thiophenylhydroxamic acids and some others fall into this category [652–660a]. N-Benzoyl-N-phenylhydroxylamine is of most analytical importance.

In recent years, thiohydroxamic acids have come to be used for the extraction and spectrophotometric determination of some elements [661–665].

5.4.4.1 N-Benzoyl-N-Phenylhydroxylamine (BPHA)

$$\begin{array}{c} C_6H_5 \diagdown \diagup O \\ C \\ | \\ N \\ C_6H_5 \diagup \diagdown OH \end{array}$$

BPHA is relatively stable to oxidizing and reducing agents, and to hydrochloric, sulphuric and perchloric acids. Only nitric acid ($> 3M$) decomposes it. It is sparingly soluble in water, and readily soluble in numerous organic solvents. In aqueous solutions of high pH and in acidic solutions its solubility is increased because of the dissociation and protonation reactions.

With metals, it forms chelate compounds by replacement of the hydrogen of the hydroxamic group and a dative bond of the carbonyl oxygen. Such compounds are formed in slightly acidic, neutral, or slightly alkaline solutions with numerous elements.

The reaction mechanism in highly acidic solution is not fully understood. One possibility is that an ion-association compound composed of the protonated BPHA cation and an anionic metal complex is formed, but recent studies on mechanisms of formation of ion-association complexes lend no support to this assumption. Studies on the reaction of BPHA with Hf, Zr, and Ge suggest that the inorganic acid anions present may enter the complex formed to yield a mixed complex [666–668]. Fouché [666] proposed the formula $M(BPHA)_i X_{4-i}$ (X is the acid anion) for the complex with Zr or Hf. In highly acidic solution the BPHA anion may be completely displaced to yield another complex, $MX_4 \cdot 2BPHA$, in which two neutral BPHA molecules are co-ordinatively bonded to the metal cation.

BPHA reacts with most of the elements of the periodic table, but there are also many possibilities for group separations [75,92,669].

The general properties of BPHA have been described by several authors [670–678a].

The optimum conditions for extraction of metals as their BPHA complexes are listed in Table 5.21.

5.4.5 Extraction with Dithizone (H_2Dz)

$$\begin{array}{cc} C_6H_5\text{—NH—NH} & C_6H_5\text{—NH—N} \\ \diagdown & \diagdown \\ C=S \rightleftarrows & C\text{—SH} \\ \diagup & \diagup \\ C_6H_5\text{—N}=N & C_6H_5\text{—N}=N \\ \text{I} & \text{II} \end{array}$$

Table 5.21 Extraction of metals with benzoylphenylhydroxylamine

Species	Aqueous phase	Organic phase	E, % max.	$pH_{1/2}$	Reference
Al (III)	pH 6	0.01M BPHA in $CHCl_3$	100	4.3	[74]
Ba (II)	1M NH_4OH, pH > 9 (NaOH)	0.3M BPHA in $CHCl_3$	ca. 100		[669, 679]
Be (II)	pH ≈ 5	0.01M BPHA in $CHCl_3$	100		[673]
Bi (III)	pH 2–13	0.01M BPHA in $CHCl_3$	99.5	0.3	[74]
	0.1M HCl	0.6% BPHA in $CHCl_3$			[680]
Ca (II)	1M NH_4OH, pH > 9 (NaOH)	0.3M BPHA in $CHCl_3$	ca. 100		[669, 681]
Cd (II)	pH 10–11	0.01M BPHA in $CHCl_3$	78		[74]
Ce (III, IV)	pH ≈ 7	0.01M BPHA in $CHCl_3$	100	9.0	[673, 682]
	pH 8–10	$CHCl_3$	100		[683]
Co (II)	pH 10–11	0.01M BPHA in $CHCl_3$	100	7.6	[74]
Cr (III)	pH 3–4	0.01M BPHA in $CHCl_3$	ca. 25		[74]
Cu (II)	pH 4–11	0.01M BPHA in $CHCl_3$	100	2.3	[74]
Fe (II)	pH ~ 5	$CHCl_3$			[673]
Fe (III)	pH 2–3 (HCl)	0.01M BPHA in $CHCl_3$	100	1.0	[74]
	12M HCl	0.1M BPHA in $CHCl_3$	ca. 80		[75]
Ga (III)	7M H_2SO_4 to pH 8	0.1M BPHA in $CHCl_3$	ca. 100	2.5	[684]
	pH > 3, 0.05M acetate	BPHA in $CHCl_3$	100		[77]
Ge (IV)	> 8M HCl	BPHA in $CHCl_3$	100		[680, 685]
	2–6M $HClO_4$	0.1M BPHA in benzene	90		[686]
Hg (II)	pH 7.5–9	0.01M BPHA in $CHCl_3$	88		[74]
Hf (IV)	3M HCl	BPHA in $CHCl_3$ and in benzene in other solvents	> 90		[687, 688]
					[689]
In (III)	> 2M $HClO_4$	BPHA in benzene	100		[687]
	pH 5–12	BPHA in $CHCl_3$	100		[680]
	pH > 5 ($NaClO_4$)	0.001–0.1M BPHA in $CHCl_3$	100		[690]
La (III)	pH ≈ 7	0.1M BPHA in $CHCl_3$	ca. 99		[671, 682]
Mg (II)	pH 9–10	0.2–0.3M BPHA in $CHCl_3$	ca. 100		[682a]

§ 5.4] Applications 205

Nb (V)	2·5–10M HCl	0·2M BPHA in CHCl$_3$ and benzene	100		[75, 688]
	1·5–3M H$_2$SO$_4$	0·2% BPHA in CHCl$_3$	>98		[691]
	pH 4–6, H$_2$SO$_4$, (tartaric acid)	1% BPHA in CHCl$_3$	100		[691, 693]
	2–5M HClO$_4$ + HF	0·025M BPHA in CHCl$_3$	100		[694]
	1M HCl + 0·05M HF; 1–11M HCl	0·5% BPHA in CHCl$_3$	100		[695]
Nd (III)	pH > 6	CHCl$_3$	100		[675]
Ni (II)	pH 9–11	0·01M BPHA in CHCl$_3$	97	7·6	[74]
Np (IV)	2–5M HNO$_3$	0·4M BPHA in CHCl$_3$	>90		[66]
Pa (V)	2–7M H$_2$SO$_4$; 1–10M HCl	0·1M BPHA in benzene	>99		[696, 697]
	> 10M HCl	0·5% BPHA in CHCl$_3$	>90		[695]
Pb (II)	pH 7–11	0·01M BPHA in CHCl$_3$	97	6·1	[74]
Pd (II)	pH ≈ 3	CHCl$_3$	ca. 100		[673]
Pr (III)	pH > 3	CHCl$_3$	ca. 100		[673]
Pu (IV)	1–6M HNO$_3$; pH 3	0·4M BPHA in CHCl$_3$	>97		[66]
Re (VII)	9M HClO$_4$, 9M HCl	5 × 10^{-3}M in C$_6$H$_6$	ca. 100		[698]
Sb (III)	0·3–17M H$_2$SO$_4$	0·05M BPHA in CHCl$_3$	80–99		[699]
	0–2M HCl	0·05M BPHA in CHCl$_3$	93		[700]
Sb (V)	>11M HCl	0·5% BPHA in CHCl$_3$	100		[77]
	>9M HClO$_4$	1% BPHA in CHCl$_3$	98·5		[75, 680]
Sc (III)	pH 4–6	1% BPHA in CHCl$_3$	100		[77, 680]
Sn (IV)	5M H$_2$SO$_4$	0·5% BPHA in isoamyl alcohol	100		[701]
	0·5–0·8M HCl	0·4M BPHA in CHCl$_3$	90		[700]
	4M HClO$_4$	1% BPHA in CHCl$_3$	95		[75, 77, 702]
Sr (II)	1M NH$_4$OH, NaOH	1% BPHA in CHCl$_3$ or in benzene	100		[680, 686]
	10–28N H$_2$SO$_4$	0·3M BPHA in CHCl$_3$	ca. 100		[669, 679]
Ta (V)	9–10M HCl	1% BPHA in CHCl$_3$	100		[693]
	2–5M HClO$_4$ + HF	0·2M BPHA in CHCl$_3$ and benzene	100		[75, 688]
	1M HCl + 0·05M HF	0·025M BPHA in CHCl$_3$	100		[694]
		0·5% BPHA in CHCl$_3$	100		[695]

Table 5.21 (continued)

Species	Aqueous phase	Organic phase	E, % max.	$pH_{1/2}$	Reference
Tc (VII)	$6M$ HClO$_4$,	0.05M BPHA in CHCl$_3$	100		[699]
	$6M$ HCl	0.05M BPHA in CHCl$_3$	100		[699]
Th (IV)	pH 3–9	0.1M BPHA in CHCl$_3$ or in other solvents	100		[671] [676]
	pH 3.5–7	3% BPHA in isoamyl alcohol	100	2.6	[703]
Ti (IV)	1–10M HCl	0.2M BPHA in CHCl$_3$	100		[75, 701, 704]
	3–4M HClO$_4$	0.1M BPHA in benzene	100		[686]
	>6M HCl, HClO$_4$, 3M H$_2$SO$_4$	0.1% BPHA in CHCl$_3$	100		[705]
Tl (I)	pH 9–12	0.1M BPHA in CHCl$_3$	ca. 90		[65]
Tl (III)	pH ≈ 4	CHCl$_3$	ca. 100	8.3	[673]
U (VI)	pH > 3.5	0.1M BPHA in CHCl$_3$	>90		[671]
V (IV)	pH 3–4	CHCl$_3$	100		[703]
V (V)	2M H$_2$SO$_4$	CHCl$_3$	100		[703]
	1–11M HCl	BPHA in CHCl$_3$	100		[92]
	3N HCl	BPHA in benzene in other solvents	100		[706] [707, 708]
W (VI)	4M HCl, 2M H$_2$SO$_4$	BPHA in CHCl$_3$	100		[709]
	1–11M HCl	BPHA in CHCl$_3$	90		[92]
Y (III)	pH ≈ 6	CHCl$_3$	ca. 100		[673]
Zn (II)	pH 9–11	0.01M BPHA in CHCl$_3$	100		[74]
Zr (IV)	pH 0–2	0.2% BPHA in CHCl$_3$ in benzene	98		[66, 691]
	pH 7.2–8.0, H$_2$O$_2$, citrate		100		[710]
	8–10M HCl	0.2M BPHA in CHCl$_3$ and benzene	100		[75, 687, 690]
	>4M HClO$_4$	0.1M BPHA in benzene	100		[6586, 687]

Dithizone, or diphenylthiocarbazone, is a typical chelate-forming group reagent primarily used for colorimetric determination of traces of numerous elements, but it is also important as an extractant for separation of traces of metals.

Dithizone solutions are unstable and decompose when exposed to light or treated with oxidants. In organic solutions dithizone occur as a mixture of the two tautomers, keto (**I**) and enol (**II**).

From the reaction of dithizone with metal ions, primary or secondary dithizonates can be produced, depending on whether dithizone reacts as a monobasic or dibasic acid (to yield HDz^-, or Dz^{2-}). Some elements are capable of forming both primary and secondary dithizonates. Primary dithizonates are typically formed in acidic solutions with an excess of dithizone, whereas secondary ones are produced in alkaline solutions with insufficient reagent.

The structure of dithizonates, and in particular of secondary ones, is uncertain [711–718]. The structure of primary dithizonates is different for different metals.

The extractability of dithizonates depends primarily on the pH of the aqueous phase and on the dithizone concentration. Highly stable dithizonates (Pt, Pd, Au, Ag, Hg, Cu) can be extracted from highly acidic solutions: dithizonates of Bi, In, Zn, and Sn, from slightly acidic solutions, and dithizonates of Co, Ni, Pb, Tl, and Cd, from neutral or alkaline solutions. The extraction process for dithizonates is not very selective, but the selectivity can be improved by choice of suitable pH, and use of masking agents such as EDTA, cyanides, thiosulphate, citrates, etc.

Notable work on dithizonates and their extraction has been done by Kolthoff and Sandell [44], Irving et al. [54,719,720,729,729a], Babko and Pilipenko [721,722], Iwantscheff [723], and others [147,724–726].

A particularly useful method is Irving's 'reversion' technique for successive stripping of metals extracted as dithizonates [719].

The optimum extraction conditions for metals as dithizonates are listed in Table 5.22.

5.4.6 Extraction with Dithiocarbamates

Dithiocarbamates are compounds containing sulphur atoms with donor properties, of general formula

$$\left[\begin{array}{c} R \\ R \end{array} \!\!\!\! N\!-\!C \!\!\!\! \begin{array}{c} S \\ S \end{array} \right]^{-}$$

Table 5.22 Extraction of metals with dithizone

Species	Aqueous phase	Organic phase	E,% max.	log K_{ex}	Reference
Ag (I)	from pH 7 to 4M H_2SO_4	CCl_4	100	7·1–8·9	[118]
	1M $HClO_4$	CCl_4 or $CHCl_3$	100		[118]
		other solvents			[118]
Au (III)	0·5–2M H_2SO_4	CCl_4	100		[723, 727, 728]
	3–7M H_2SO_4	benzene	100		[730]
	0·1M HCl	$CHCl_3$	30–70		[731]
	0·1M HCl + 0·1M HBr	$CHCl_3$	85–100		[731]
	0·1M KCN	$CHCl_3$	70–85		[731]
	H_2SO_4 + $NaClO_4$, NaCl, or CCl_3COOH	$CHCl_3$	100		[732]
Bi (III)	pH 3–10 ammonium acetate or potassium cyanide	CCl_4	100	9·7–10·7	[118]
	pH 5–11	$CHCl_3$	100		[723]
		other solvents			[733]
Cd (II)	pH 6·5–10	CCl_4	100	1·6–2·1	[723, 734, 735]
	pH 7–14	$CHCl_3$	100	0·5	[71, 723]
Co (II)	pH 5·5–8·5	CCl_4	100	1·6	[723]
	pH 8	$CHCl_3$	100	ca. 1·5	[723]
Cu (I)	pH 0–10	CCl_4	100		[723, 736]
Cu (II)	pH 1–4	CCl_4	100	9·5–10·5	[45, 722, 7
	pH 7	CCl_4	100		
Fe (II)	pH 7–9	CCl_4	100		[723, 737]
Ga (III)	pH 4·5–6	$CHCl_3$	90	−1·3	[738]
Hg (I)	ca. 1M H_2SO_4	$CHCl_3$	100		[723]
	pH 7	$CHCl_3$	100		[723]
Hg (II)	from 6M H_2SO_4 to pH 4	CCl_4 or $CHCl_3$	100	26·7	[723, 739]
In (III)	pH 8·2–9·5	$CHCl_3$	100	0·6	[72]
	other solvents				[72]
Mn (II)	pH 9·5–11	$CHCl_3$	60		[61]
Ni (II)	pH 6–9	CCl_4	100	−0·6–1·2	[722, 723]
	pH 8–11	$CHCl_3$	100	−2·9	[723]
Pb (II)	pH 8–10	CCl_4	100	0·4	[723, 740]
	pH 8·5–11	$CHCl_3$	100	−0·9	[723, 740–743]
		other solvents			[744, 745]
Pd (II)	15M acids	CCl_4	100		[723, 72
Po (V)	pH 0–5	$CHCl_3$	100		[746]
	pH 0·6–9	CCl_4	100		[746]

5.22 (continued)

ecies	Aqueous phase	Organic phase	$E,\%$max.	log K_{ex}	Reference
(II)	0·1–5·25M H$_2$SO$_4$	benzene	100		[747]
(IV)	6M HCl	CCl$_4$	100		[748, 749]
(II)	pH 5–9	CCl$_4$	100	−2	[723]
(IV)	0·1–1M acids	CCl$_4$	95		[750]
(I)	pH 11–14·5	CHCl$_3$	80	3·8	[68]
(II)	pH 6–9·5	CCl$_4$	100	2·0–2·3	[751]
	pH 7–10	CHCl$_3$	100	0·6–1·0	[751]

Dithiocarbamates react with metal ions that form sparingly soluble sulphides to form compounds that contain four-membered rings and are usually insoluble in water. They react with a wide range of elements, but their selectivity can be enhanced by the use of suitable masking agents.

The properties of dithiocarbamates have been studied by Hulanicki [752,753], and their possible analytical uses have been described by Podchainova et al. [754, 755].

The most commonly used are sodium diethyldithiocarbamate and diethylammonium diethyldithiocarbamate [756,759].

5.4.6.1 Sodium Diethyldithiocarbamate (Na-DDTC)

$$\begin{array}{c} C_6H_5 \\ \diagdown \\ N-C \\ \diagup \diagdown \\ C_6H_5 SNa \end{array} \begin{array}{c} S \\ \diagup \end{array}$$

This reagent is readily soluble in water but much less soluble in organic solvents. The aqueous solution slowly decomposes.

Na-DDTC decomposes in acidic solutions into diethylamine and carbon disulphide: hence, extraction from acidic solutions must be done immediately, preferably with a good excess of reagent present.

The reagent reacts with a range of elements, especially heavy metals. Its use in extractive separations requires suitable masking agents. Hulanicki [752] has classified ions according to their reactions with dithiocarbamates at various pH values and in the presence of various masking agents (Table 5.23).

The compounds of metals with Na-DDTC are generally insoluble in water and, since they are uncharged, weakly dissociated chelates, are readily extracted with organic solvents.

The stability of diethyldithiocarbamates of metals decreases in the following order [757,758]:

Hg > Pd > Ag > Co > Cu > Ni, Bi > Pb > Cd > Zn > In > Fe(III) > Mn

Table 5.23 Reactivity of the elements with dithiocarbamates at various pH values
(from [752], by permission of A. Hulanicki)

Masking agent	pH ≈ 5–6	pH ≈ 9	pH ≈ 11
Tartrate	Ag, As, Bi, Cd, Co, Cu, Fe, Hg, In, Mn, Ni, Pb, Pd, Sb, Se, Sn, Te, Tl, Zn	Ag, Bi, Cd, Co, Cu, Fe, Hg, In, Mn, Ni, Pb, Pd, Sb, Te, Tl, Zn	Ag, Bi, Cd, Co, Cu, Hg, Ni, Pb, Pd, Tl, (Zn)
EDTA	Ag, As, Bi, Cd, Co, Fe, Hg, In, Mn, Pb, Pd, Sb, Se, Sn, Te, Tl	Af, Bi, Cu, Fe, Hg, Pd, Sb, Te, Tl	Ag, Bi, Cu, Hg, Pd, Tl
KCN		Bi, Cd, Fe, In, Mn, Pb, Sb, Te, Tl	Bi, Cd, Pb, Tl
Citrate + EDTA		Ag, Bi, Cu, Hg, Sb, Te, Tl	

Extractions with diethyldithiocarbamates have been studied by Bode and Neumann [759,760] and by Malissa and Gomišček [761].

The optimum extraction conditions for various metals as diethyldithiocarbamates are given in Table 5.24.

5.4.7 Extraction with Organophosphorus Compounds

The organophosphorus compounds are important for the extractive separation of metals. Because of their chemical and physical properties they are capable of competing with water for the available sites in the first co-ordination shell of the cation, and thus capable of forming extractable co-ordination solvated salts.

The formation of metal–organophosphorus compound complexes, and their chemical, physical, and extraction properties have been extensively studied [787–789].

The most important compounds of this group are organophosphorus acids, mono- and dialkylphosphoric acids, neutral alkyl derivatives of phosphoric acid (esters) and organophosphorus oxides.

5.4.7.1 ORGANOPHOSPHORUS ACIDS

Monoalkylphosphoric acids

$$\underset{HO}{\overset{RO}{}}P\underset{OH}{\overset{O}{}}$$

Table 5.24 Extraction of metals with sodium diethyldithiocarbamate

Species	Aqueous phase	Organic phase	E,%max.	Reference
Ag (I)	pH 4–11	CCl_4	100	[760]
	pH 0–10	$CHCl_3$ + acetone (5 : 2)	100	[401]
As (III)	pH 0–8	$CHCl_3$ or $CHCl_3$ + acetone (5 : 2)	100	[760]
Au	0·5M HCl	isoamyl alcohol	93	[762]
Bi (III)	pH 4–11	CCl_4 or $CHCl_3$	100	[760, 763]
Cd (II)	pH 5–11	CCl_4	100	[760]
	pH 1–10	$CHCl_3$ + acetone (5 : 2)	100	[401]
Co (III)	pH 4–11	CCl_4	100	[760]
Co (II)	pH 0–10	$CHCl_3$ + acetone (5 : 2)	100	[401]
Cr (VI)	pH 6·0	$CHCl_3$	100	[764]
Cu (II)	pH 4–11	CCl_4	100	[760, 765–768]
	pH 0–10	$CHCl_3$ + acetone (5 : 2)	100	[401]
Fe (III)	pH 4–11	CCl_4	100	[760]
	pH 0–5	$CHCl_3$ or $CHCl_3$ + acetone (5 : 2)	100	[401]
Ga (III)	pH 1·5–5·0	ethyl acetate	100	[768, 769]
Hg (II)	pH 4–11	CCl_4	100	[401, 771, 772]
	pH 0–10	$CHCl_3$ or $CHCl_3$ + acetone (5 : 2)	100	[401]
In (III)	pH 4–10	CCl_4	100	[760]
	pH 3–10	ethyl acetate	100	[769, 770]
Ir (III, IV)	pH 7–9 elevated temperature	dichloroethane	99	[773]
Mn (III)	pH 6–9	CCl_4	100	[760, 774]
	pH 5·2	$CHCl_3$	100	[774]
	pH 1·7	$CHCl_3$ + acetone (5 : 2)		[401]
Ni (II)	pH 0–8	$CHCl_3$ or $CHCl_3$ + acetone (5 : 2)	100	[401]
	pH 5–11	CCl_4	100	[760]
Mo (VI)	slightly acidic	ethyl acetate	100	[770]
Nb (V)	pH < 6 tartrates	other solvents	partially	[760]
Pb (II)	pH 4–11	CCl_4	100	[760, 775]
		other solvents		[764, 776, 777]
Pd (II)	pH 4–11	CCl_4	100	[760]
	0·001–8M HCl	$CHCl_3$, isoamyl alcohol	100	[762, 778]
Pt (II)	pH 4–11	CCl_4	100	[401]
	7M HCl	CCl_4	100	[778]
Pu (IV)	pH ~ 3	amyl alcohol or amyl acetate	100	[581]

Table 5.24 (*continued*)

Species	Aqueous phase	Organic phase	E,%max.	Reference
Sb (III)	pH 4–9·5	CCl$_4$	100	[760]
	pH 0–7	CHCl$_3$ + acetone (5 : 2)	100	[401]
Se (IV)	pH 4–6·2	CCl$_4$	100	[760]
Sn (IV)	pH 4–6·2	CCl$_4$	100	[760]
	pH 1–8	CHCl$_3$ or CHCl$_3$ + acetone (5 : 2)	100	[401]
Te (IV)	pH 4–8·8	CCl$_4$	100	[760]
Tl (I)	pH 5–13	CCl$_4$	100	[760]
	pH 9–12	CHCl$_3$	100	[68]
Tl (III)	pH 4–11	CHCl$_3$ or CCl$_4$	100	[760]
U (VI)	pH 2·0–3·5	methyl ethyl ketone	100	[779]
	pH 1·5–3·0	butanol	100	[780]
		other solvents		[764, 781, 783]
V (V)	pH 3–6	CCl$_4$ or CHCl$_3$	100	[760, 770–784]
W (VI)	0·1–0·2N acids	CCl$_4$	100	[785]
Zn (II)	pH 1–3	ethyl acetate	100	[785]
	pH 4–11	CCl$_4$	100	[760]
	pH 1–7	CHCl$_3$	100	[760, 786]
	pH 1–10	CHCl$_3$ + acetone (5 : 2)	100	[401]

and dialkylphosphoric acids

$$\begin{array}{c} RO \diagdown \quad \diagup O \\ P \\ RO \diagup \quad \diagdown OH \end{array}$$

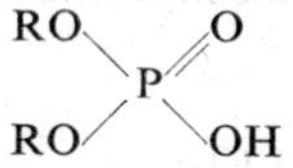

are produced together in the reaction of alcohols with phosphorus pentoxide and their preparation in pure form requires rather involved separation and purification procedures.

Both the mono- and dialkylphosphoric acids are strongly polymerized in organic solvents: the degree of polymerization differs for different acids and diluents [790–792].

Reactions of metal ions with organophosphorus acids can be written in general form as follows:

$$M^{n+} + (H_2A)_{x(o)} \rightleftharpoons M(H_{2x-n}A_x)_o + nH^+$$
$$M^{n+} + n(H_2A_2)_o \rightleftharpoons M(HA_2)_{n(o)} + nH^+$$

The properties of organophosphorus acids were mainly studied by Peppard, Mason, and others [170, 790–798b].

The most important organophosphorus acids are mono-2-ethylhexylphosphoric acid, dibutylphosphoric acid, and di(2-ethylhexyl)phosphoric acid.

Dialkyl and diaryldithiophosphoric acids have also been used as analytical reagents [799].

Mono-2-ethylhexylphosphoric acid (H_2MEHP)

$$C_4H_9-\underset{\underset{C_2H_5}{|}}{CH}-CH_2-O-\underset{\underset{OH}{|}}{\overset{\overset{OH}{|}}{P}}=O$$

exhibits, like other monoalkylphosphoric acids, a variable degree of polymerization in different organic solvents; for instance, it is 4 in chloroform, 14 in n-hexane, and 2 in acetone, whereas in methyl alcohol the acid occurs as the monomer. The applications of this acid in the extraction of metals are summarized in Table 5.25.

Table 5.25 Extraction of metals with mono-2-ethylhexylphosphoric acid

Species	Aqueous phase	Organic phase	log K_{ex}	log D	Reference
Ac (III)	1M HClO$_4$	toluene	ca. 1		[793]*
Am (III)	pH 0·7 (HCl+NaCl)	toluene		2·5	[793]*
	12M HCl	0·5M H$_2$MEHP in toluene		1·1	[795]
Ce (III)	acid solutions pH 1	0·1M H$_2$MEHP in chlorobenzene	2·5		[794]
Ga (III)	H$_2$SO$_4$, HBr	TBP, ketones, alcohols	⩾2		[800]
La (III)	HClO$_4$	toluene	1·0		[793]*
Np (IV)	12M HCl	toluene		3	[797]
Pm (III)	pH 1 (HClO$_4$)	0·1M H$_2$MEHP in chlorobenzene	2·8		[794]
Pu (III)	12M HCl	toluene		−1	[797]
Sc (III)	HClO$_4$	toluene	6·5		[793]
Th (IV)	12M HCl	0·25M H$_2$MEHP in toluene		4	[797]
Tl (III)	various acids				[798]
U (IV)	12M HCl	toluene		3·8	[798]
U (VI)	12M HCl	toluene		−1·5	[797]
	6M HCl	toluene		1	[797]
Y (III)	HClO$_4$	toluene	1·5		[793]

* The extraction of Sc, Y, La, Ac, Pm, Tm and Am from HClO$_4$ solutions was studied. The variation of the distribution coefficient with concentration of extractant in the organic phase, hydrogen ions in the aqueous phase and position of M(III) in the periodic table, was determined.

Table 5.26 Extraction of metals with dibutylphosphoric acid

Species	Aqueous phase	Organic phase	E,% max.	log D	Reference
Am (III)	$0.1M$ HNO$_3$	$0.3M$ HDBP in hexane	99		[801]
		other diluents			[801, 802]
Be (II)	$0.1M$ HNO$_3$	$0.5M$ HDBP in toluene	100	1.55	[803]
	other acids	other diluents			[803–805]
Ca (II)		CHCl$_3$		−2.5	[801]
Ce (III)	$0.1M$ HClO$_4$, $0.1M$ HNO$_3$	dibutyl ether	>99		[802]
Cf (III)		0.1–$0.6M$ HDBP in dibutyl ether		3.5	[801]
Cm (III)		dibutyl ether		1.9	[801]
Hf (IV)	$2M$ HCl, HClO$_4$	0.06–$0.08M$ HDBP in benzene	ca. 100		[808]
In (III)	$1M$ HNO$_3$	dibutyl ether	>95		[806]
Lanthanides:					
Eu (III)	$0.1M$ HClO$_4$ $0.1M$ HNO$_3$	hexane	>99	2.6	[801, 807]
Gd (III)	$0.1M$ HNO$_3$	dibutyl ether	100	3.3	[801]
Ho (III)	$0.1M$ HNO$_3$, $0.1M$ HClO$_4$	dibutyl ether	>99	4.9	[801]
La (III)	$0.1M$ HNO$_3$, $0.1M$ HClO$_4$	dibutyl ether	>99	1.3	[801]
Lu (III)	$0.1M$ HNO$_3$, $0.1M$ HClO$_4$	dibutyl ether	>99	6.8	[801]
Nd (III)	$0.1M$ HNO$_3$ $0.1M$ HClO$_4$	CCl$_4$	>99	1.6	[809]
Pm (III)	$0.1M$ HNO$_3$, $0.1M$ HClO$_4$	dibutyl ether	>99	2.4	[801]
Pr (III)	$0.1M$ HNO$_3$, $0.1M$ HClO$_4$	dibutyl ether	>99	2.0	[809]
Sm (III)	$0.1M$ HNO$_3$, $0.1M$ HClO$_4$	dibutyl ether	>99	2.7	[801]
Tb (III)	$0.1M$ HNO$_3$, $0.1M$ HClO$_4$	dibutyl ether	>99	4.1	[801]
Yb (III)	$0.1M$ HNO$_3$, $0.1M$ HClO$_4$	CCl$_4$	>99	4.8	[810]
Mo (VI)	$1M$ HNO$_3$	dibutyl ether	20		[806]
Nb (V)	$1M$ HNO$_3$	dibutyl ether	>95		[806]
	$0.05M$ H$_2$SO$_4$, H$_2$O$_2$, EDTA	$1M$ HDBP in benzene	95		[811]
Pa (V)	$2M$ HNO$_3$	isoamyl acetate and other diluents	99.8		[812, 836]

Table 5.26 (*continued*)

Species	Aqueous phase	Organic phase	E,%max.	log D	Reference
Re (VII)	3–4M HNO$_3$	neat HDBP	20		[813]
Sc (III)	0·1–10M HNO$_3$	chloroform	>95	10·8	[814]
Sn (IV)	1M HNO$_3$	dibutyl ether	50		[806]
Sr (II)		CHCl$_3$		−3·5	[801]
Ta (V)	1M HNO$_3$	dibutyl ether	85		[806]
	0·05M H$_2$SO$_4$	1M HDBP in benzene	89		[811]
Th (IV)	1M HNO$_3$	0·1M HDBP in	100		[814a]
Ti (IV)	pH 1	chloroform	100		[815]
U (VI)	0·1M HClO$_4$	chloroform and other diluents	>99·7	3·6	[804, 816]
Y (III)	0·1M HNO$_3$	chloroform		3·2	[817]
Zr (IV)	·2–4M HNO$_3$	toluene	>99		[818]

Di-n-butylphosphoric acid (HDBP)

$$\begin{array}{c} C_4H_9-O \\ C_4H_9-O \end{array} P \begin{array}{c} O \\ OH \end{array}$$

is a liquid sparingly soluble in water, but readily soluble in organic solvents. In aqueous solutions it is a strong acid (pK_{HA} = 1·00), and in most organic solvents it occur as dimers.

Applications of HDBP in the extraction of metals are listed in Table 5.26.

Di(2-ethylhexyl)phosphoric acid (HDEHP or D2EHP)

$$\begin{array}{c} C_2H_5 \\ | \\ C_4H_9-CH-CH_2-O \\ C_4H_9-CH-CH_2-O \\ | \\ C_2H_5 \end{array} P \begin{array}{c} O \\ OH \end{array}$$

is a liquid sparingly soluble in water, but easily soluble in organic solvents. It occurs as a dimer in most solvents.

The reagent has found wide application for extractive separation of a number of elements [1336, 1337]. Applications of HDEHP in the extraction of metals are listed in Table 5.27.

Table 5.27 Extraction of metals with di(2-ethylhexyl)phosphoric acid

Species	Aqueous phase	Organic phase	$E,\%$ max.	$\log K_{ex}$	$\log D$	Reference
Ac (III)	0.1M HCl	1.5M HDEHP in toluene	ca. 99			[820]
Ag (I)	1M HCl	1.5M HDEHP in toluene	50			[809]
Al (III)	slightly acidic or neutral	kerosene	100			[821]
Am (III)	pH > 2	0.15M HDEHP in toluene	> 90			[822]
	0.05–0.1M HCl	1.5M HDEHP in toluene	100			[822]
Be (II)	0.25M HNO₃	0.5M HDEHP in toluene	95			[217]
	pH 3 (HCl) + 2M KCNS	0.2M HDEHP in hexane			2	[822a]
	pH 1–3 (HCl)	0.4M HDEHP in kerosene			2	[822b]
Bi (III)	0.01M HCl, 0.005M H₂SO₄	1.5M HDEHP in toluene	100			[823]
	0.125–0.5M H₂SO₄	n-octane, other solvents		2.95		[824]
Bk (III)	0.13M HCl	0.5M HDEHP in toluene	75			[793, 795]
Ca (II)	pH 4–5 (HCl + NaCl)	0.6M HDEHP in toluene			−3.5	[825, 826]
	mineral acids	petrol	ca. 100			[827]
Cf (III)	< 0.2M HCl	1.5M HDEHP in toluene	> 90			[828]
Cm (III)	pH > 1	1.5M HDEHP in toluene	> 90			[829]
Cu (II)	0.01M HCl	1.5M HDEHP in toluene	50			[823]
Fe (III)	0.1M HCl	1.5M HDEHP in toluene	100		3.6	[830]
Ga (III)	0.1M HCl	1.5M HDEHP in toluene	99			[823]
Hf (IV)	0.1–1.0M HCl	1.5M HDEHP in toluene	100			[823]
Hg (II)	0.1M HCl	1.5M HDEHP in toluene	ca. 50			[823]
In (III)	0.1M HCl	1.5M HDEHP in toluene	100			[823, 831]
Lanthanides:						
Ce (III)	pH > 1	0.7M HDEHP in heptane	ca. 100		−2.0	[790, 832]
Ce (IV)	10M HNO₃				> 3.9	[820]
Dy	0.5M HCl	1.5M HDEHP in toluene	> 95			[832, 833]

Eu, Gd	pH > 0.5 (HCl)	1.5M HDEHP in toluene and n-heptane	> 95			[832–834]
Ho	pH > 0.3 (HCl)	1.5M HDEHP in toluene	> 95			[832, 833]
La	$HClO_4$, HNO_3	100% HDEHP	ca. 100	−3.5		[820]
Lu	2M HCl	1.5M HDEHP in toluene	> 95			[832, 833]
Nd	0.1M HCl	1.5M HDEHP in toluene	> 95			[832, 833]
Pm	pH > 1 (HCl)	1.5M HDEHP in toluene	> 95			[832, 833]
Pr	pH > 0 (HCl)	1.5M HDEHP in toluene	> 95			[832, 833]
Tb	0.3M HCl	1.5M HDEHP in toluene	> 95			[832, 833]
Tm	2M HCl	1.5M HDEHP in toluene	> 95			[832, 833]
Yb	2M HCl	1.5M HDEHP in toluene	> 95			[832, 833]
Mg (II)	0.01M HCl	1.5M HDEHP in toluene	ca. 8			[823]
Mn (II)	pH 4	0.6M HDEHP in toluene	100			[823]
Mo (VI)	0.1–1M HCl	1.5M HDEHP in toluene	90			[823]
Nb (V)	0.5M HCl	1.5M HDEHP in toluene	50			[823]
Np (IV)	0.5M HCl	0.2M HDEHP in toluene	100		5.3	[835]
Np (VI)	0.5M HCl	0.15M HDEHP in toluene	98			[835]
Os	0.01–1N $HClO_4$	1M HDEHP in toluene	90			[823]
Pa (V)	10M HNO_3	0.05M HDEHP in kerosene	100			[836]
Pb (II)	0.01M HCl	1.5M HDEHP in toluene	ca. 90			[823]
	0.2M HNO_3	0.3M HDEHP in n-heptane	ca. 100			[837]
Pu (IV)	1M $HClO_4$	decane	100			[802]
Sb	0.1M HCl	1.5M HDEHP in toluene	98			[823]
Sb (III)	pH 0 (HCl, $HClO_4$, HBr, H_2SO_4, HNO_3)	1M HDEHP in benzene, other solvents			ca. 1	[838, 838a]
Sc (III)	0.1–1M $HClO_4$	1.5M HDEHP in toluene	100		8.8	[790]
Sn (IV)	1M HCl, 1M HBr	0.5M HDEHP in n-heptane, other solvents	100			[790a]
Sr (II)	pH ≈ 5	0.6M HDEHP in toluene	99			[825, 826]

Table 5.27 (continued)

Species	Aqueous phase	Organic phase	$E,\%$ max.	log K_{ex}	log D	Reference
Ta (V)	0·5M HCl	0·01–0·1M HDEHP	ca. 80			[823]
Th (IV)	0·1M HCl	0·04M HDEHP in toluene	100		3	[839]
Ti (IV)	0·1M HCl	1·5M HDEHP in toluene	90			[832]
Tl (III)	0·1–0·5M, HClO$_4$, HNO$_3$, 0·5–0·25M H$_2$SO$_4$	0·3M HDEHP in heptane	ca. 100			[840, 841]
U (VI)	H$_2$SO$_4$, HNO$_3$	0·1M HDEHP in kerosene	100		4·5	[842–849]
U (VI)	0·33M H$_3$PO$_4$	0·1M HDEHP in hexane				[850]
U (IV)	2M HClO$_4$	5% HDEHP in kerosene			2·7	[851]
U (VI)	2M HClO$_4$	5% HDEHP in kerosene other solvents			1·8	[851]
V (IV)	pH 0·2–3 (HCl) H$_2$SO$_4$	4% HDEHP in kerosene kerosene	ca. 100			[852] [854]
V (V)	0·01M HCl, HNO$_3$	1·5M HDEHP in toluene in kerosene	50			[823, 853]
W (VI)	0·01M HCl	1·5M HDEHP in toluene	15			[823]
Zn (II)	0·2M HNO$_3$	0·3M HDEHP in n-heptane	ca. 100			[837]
Zr (IV)	0·1–1M HCl	1·5M HDEHP in toluene	100			[823, 855]
Y (III)	1–4M HNO$_3$	various solvents	100			[856]
Rare earth elements	3M HNO$_3$, HClO$_4$ various acids	various solvents				[857] [858–861]
Actinides	various acids	various solvents				[862, 862a, 862b]

5.4.7.2 Trialkyl Phosphates and Phosphine Oxides

As mentioned previously in the general discussion of extraction systems (Section 5.3.2) organophosphorus compounds of this type have good solvating properties, and thus are good extractants. Their properties are due to the high dipole moment and the presence of the electronegative —P=O oxygen atom with donor properties.

The extractive properties of phosphoric acid derivatives depend on the number of ester oxygen atoms and the nature of the substituents present in the molecule [863]. The removal of the ester oxygen atoms, as the R—O— groups are replaced by alkyl residues, results in the formation of compounds of increasing basicity and hence improved extraction properties.

Neutral organophosphorus compounds have been widely used for extraction separations.

The compounds of this kind most frequently used in analysis are tri-n-butyl phosphate and trioctylphosphine oxide.

Tri-n-butyl phosphate (TBP)

$$\begin{array}{c} C_4H_9O \\ C_4H_9O-P\rightarrow O \\ C_4H_9O \end{array}$$

is a liquid of large dipole moment (3·3 D) and high dielectric constant (7·97). It is miscible in any proportions with many organic solvents, but is almost insoluble in water, the solubility decreasing with increasing salt concentration in aqueous solution [864,865]. The reagent is resistant to acids, alkalis, elevated temperature and radiation. Extractions with TBP can proceed both by hydrogen-bond formation (extraction of organic or mineral acids and protonated anionic complexes [866–869, 1338]) and formation of co-ordinatively saturated uncharged solvates.

TBP is an unselective reagent that reacts with scores of elements. The optimum conditions for extraction of metals as TBP complexes are listed in Table 5.28.

Tri-n-octylphosphine oxide (TOPO)

$$(C_8H_{17})_3P\rightarrow O$$

has been used to extract many metals. Usually, elements in their highest oxidation states yield the most extractable TOPO complexes. For numerous elements extraction is quantitative from a hydrochloric acid solution of concentration $> 3M$.

The optimum conditions for extraction of metals as TOPO complexes are summarized in Table 5.29.

Table 5.28 Extraction of metals with tri-n-butyl phosphate

Species	Aqueous phase	Organic phase	E, % max.	log D	Reference
Ag (I)	0·1M HCl	100% TBP		1·1	[870–872]
	1M HNO$_3$ + 2M LiNO$_3$	40% TBP in butanol	ca. 100		[873]
	0·1M NaSCN + various acids	100% TBP	ca. 100		[873a]
Al (III)	pH 4 to 0·5M HCl + 1·6–3·2M NaClO$_4$	100% TBP	>90		[870]
Am (III)	5M NaSCN	5% TBP in hexane			[874]
Au (III)	3M HCl + 2M LiCl (MgCl$_2$)	50% TBP in toluene	ca. 100		[875]
Be (II)	0·1M HCl, LiCl	100% TBP			[874]
Bi (III)	0·1–0·6M HClO$_4$	100% TBP	100		[876, 877]
	dil. HNO$_3$ + KI	100% TBP	100		[878]
Ca (II)	1% NaOH	TBP in CCl$_4$	100		[879, 880]
Cd (II)	dil. HNO$_3$ + KI	100% TBP	100		[878]
	4M HCl	100% TBP	>90	3	[878a]
Co (II)	10M LiCl	100% TBP	100		[224, 871, 881–890]
	8·5M HCl	60% TBP in toluene	95		
	LiCl, HCl in other conc.	—			
Cr (VI)	0·01–0·04M H$_2$SO$_4$	25% TBP in benzene	100		[888, 889]
Cu (II)	8M HCl	100% TBP	100		[870, 871, 890]
	6M HCl	50% TBP in benzene	100		[891, 892]
	pH 2·8 + 2% KSCN	100% TBP	100		[887]

§ 5.4] Applications 221

	Aqueous	Organic	%		Refs.
	2–6M HCl	50% TBP in CCl$_4$	100		[893, 896]
		100% TBP	100		[296, 890–898]
Ga (III)	6M HCl	3% TBP in non-polar solvents	>90	1·1–2·4	[899]
	4M HCl	25% TBP in benzene or in toluene	100		[900, 901]
Ge (IV)	1·5M HCl	100% TBP	100		[901a]
Hf (IV)	>8M HNO$_3$	100% TBP	100		[902, 903]
	>7·5M HCl	1M TBP in benzene	ca. 100	>2	[904]
	≥12·5M H$_2$SO$_4$	30% TBP in CCl$_4$	100		[905]
Hg (II)	dil. HNO$_3$ + KI	100% TBP	100		[878, 906]
In (III)	6M HCl	100% TBP	100		[900]
	8M HCl	1M TBP in benzene		<−0·2	[907]
Ir (IV)	HCl, LiCl, CaCl$_2$	100% TBP			[908]
Lanthanides					
Ce (III)	0·01–0·001M HClO$_4$	100% TBP	100		[909, 910]
La (III)	HNO$_3$ + nitrate	100% TBP	100		[911–913]
other lanthanides					[328, 909, 914]
Mn (II)	1M HCl + 2·5M AlCl$_3$	40% TBP in xylene	100		[870, 882, 915]
Mo (VI)	6M HCl	100% TBP	100		[916, 917]
	3·5M H$_2$SO$_4$, H$_2$O$_2$	100% TBP	100		[918]
Nb (V)	10M HCl	2% TBP in CHCl$_3$	100		[919]
	1M HF	80% TBP in kerosene	99·7–100		[920, 921]
	10–12M HNO$_3$	50% TBP in kerosene	80–90		[922]
	7M HNO$_3$	100% TBP	100		[933]
Np (V)	10M HCl	32% TBP in CHCl$_3$	100		[919]
	0·5–1·3M HNO$_3$	0% TBP in CCl$_4$	100		[923–926]

Table 5.28 (continued)

Species	Aqueous phase	Organic phase	E, % max.	log D	Reference
Pa (V)	6M HCl	15% TBP in kerosene	85		[927, 928]
	5M HCl	100% TBP		1·6	[342]
Pb (II)	3M HCl + 2M LiCl	30% TBP in MIBK	ca. 100		[929]
Pd (II)	4M HCl + I$^-$	100% TBP	62	0·3	[887, 930]
		15% TBP in hexane	100		[887]
Pt (IV)	0·5M HNO$_3$	100% TBP	44·4		[887]
	0·5–2M HCl + SCN$^-$	100% TBP	100	2·14	[887, 931]
	pH 1 + KSCN	100% TBP		1·8	[931]
	I$^-$	15% TBP in hexane	100		[930]
	4M HCl, > 3M CaCl$_2$, > 6M LiCl	100% TBP		1·2	[932]
Pu (IV)	6M HNO$_3$	100% TBP	100		[329, 934–939]
	pH 3 (HNO$_3$)	30% TBP in kerosene	100		[940]
Sb (V)	⩾ 2M HCl + 1M MgCl$_2$	20% TBP in toluene	ca. 100		[941]
Sc (III)	8–12M HCl, chloride, mineral acids	100% TBP		2–3	[942, 943, 1340]
	1M NH$_4$SCN	15% TBP in CHCl$_3$	97·5		[944]
Se (IV)	4M HCl + 2M MgCl$_2$	60% TBP in toluene	ca. 100		[945]
Sn (II)	6·3M HCl	5% TBP in octanol		0·77	[946]
Sn (IV)	6·3M HCl	5% TBP in octanol		0·74	[946]
Ta (V)	0·5–2N HF + H$_2$SO$_4$	100% TBP	100		[920, 921, 947]
	1M HF	80% TBP in kerosene	100		[920]
Th (IV)	7M HCl	100% TBP		−1·1	[342]
	2M HNO$_3$	100% TBP		1	[330]

§ 5.4]   Applications   223

Ti (IV)	5M HCl + 4M MgCl$_2$	60% TBP in xylene	100	[948–950]
	11M H$_2$SO$_4$	30% TBP in CCl$_4$	>85	[949]
Tl (III)	H$_2$SO$_4$ + Cl$^-$	TBP in n-octane and n-decane	~100	[951, 952]
U (VI)	8–9M HNO$_3$	100% TBP	~100	[953]
	4–6M HNO$_3$ + Al, Ca, Na(NO$_3$)$_x$	50% TBP in MIBK	100	[343, 954–956]
	pH 3–6, Al(NO$_3$)$_3$	30% TBP in iso-octane	100	[346]
	6M HNO$_3$ + Al(NO$_3$)$_3$	20% TBP in kerosene	100	[957]
	HNO$_3$ saturated Al(NO$_3$)$_3$	10% TBP in hexane	100	[954]
	pH 0·5–1·4 (HNO$_3$)	9% TBP in CHCl$_3$	100	[958]
	0·5M HCl	100% TBP	>95	[344, 959]
	pH 2·5–3·0 (HCl)	20% TBP in CCl$_4$	ca. 100	[960]
	pH 3·0 (HCl) + 6M NaNO$_3$	25% TBP in iso-octane	100	[961]
	pH 1·5 + 0·25M NH$_4$SCN	10% TBP in CCl$_4$	100	[962]
	pH 3·5–3·9 + SCN$^-$	32% TBP in CHCl$_3$	100	[351]
	0·8M HNO$_3$ + salting-out agents	TBP in various diluents	100	[963–965]
V (V)	1–6M HCl	20% TBP in benzene	100	[966]
W (VI)	8M HCl	100% TBP	100	[967]
Y (III)	12M HNO$_3$	100% TBP	97	[328, 968]
	pH 2·5–3 (HNO$_3$ + HF)	100% TBP	100	[969–971]
Zr	8–10M HCl	1% TBP in toluene or 5% TBP in kerosene	100	[855, 972]
	HNO$_3$	100% TBP		[973, 974]
	SCN$^-$	100% TBP		[975]
Alkaline earth elements	pH 1, SCN$^-$, EDTA	100% TBP	100	[976]

Table 5.29 Extraction of metals with tri-octylphosphine oxide

Species	Aqueous phase	Organic phase	E, % max.	log D	Reference
As (III)	10M HCl	0·1M TOPO in cyclohexane	partial		[977]
Au (III)	HCl+SO$_4^{2-}$, NO$_3^-$, ClO$_4^-$	0·5M TOPO in cyclohexane	100		[977]
Bi (III)	< 1M HCl	"	100		[977]
	< 1M HNO$_3$	"	100		
	< 0·5M H$_2$SO$_4$	"	100		
Cr (VI)	< 1M HCl	0·1M TOPO in cyclohexane	99	1·9	[978]
	1M H$_2$SO$_4$	"		1·5	
	1M HNO$_3$	"		1·2	
	1M HClO$_4$	"		0·7	
Cu (II)	7M HCl	"	100		[977]
Fe (III)	7M HCl	"	100		[979]
Ga (III)	7M HCl	"	100		[977]
	7M H$_2$SO$_4$	"	100		
	7M HClO$_4$	"	100		
Hf (IV)	1–7M HCl, HNO$_3$, 0·5–3·5M H$_2$SO$_4$	0·01–0·05M TOPO in cyclohexane	100		[977, 980]
Ce (IV)	1M HNO$_3$+7M NaNO$_3$	0·1M TOPO in cyclohexane	97		[977]
	3–5M H$_2$SO$_4$	"	80		[981]
Mn (VII)	0·1M HCl, 0·5M H$_2$SO$_4$	0·1M TOPO in cyclohexane	100		[977]
Mo (VI)	pH ≈ 1 (HCl, HNO$_3$, H$_2$SO$_4$, HClO$_4$)	"	100	2	[977]
Nb (V)	4M HCl+0·3M tartaric acid	"	>99		[982]
Np (IV)	6M HNO$_3$+NaNO$_3$	0·1M TOPO in kerosene	100	2	[982]
Np (V)	2–8M HNO$_3$	0·3M TOPO in kerosene	50		[983]
Pu (IV)	9–10M HCl	0·1M TOPO in cyclohexane	$ca.$ 100	$ca.$ 2	[984]
	6–7M HNO$_3$	"	$ca.$ 100	$ca.$ 3	[985, 1339*]

§ 5.4] **Applications** 225

Element	Aqueous phase	Organic phase	%		Ref.
Sb (III)	2–4M HCl	"	100	ca. 2	[977]
Sn (II, IV)	1–6M HCl	"	100	⩾3	[977]
	1M HCl + 5M LiCl	"	100	>3	
Ta (V)	6M HCl + 0.2M lactic acid	"	60		[982, 986]
Tc (VII)	1M H$_3$PO$_4$	"	100	2.1	[987]
	1M H$_2$SO$_4$	"	100	1.7	
	1M HCl	"	100	1.7	
Th (IV)	5–7M HCl	"	100	2	[202, 988]
	0.5–1M HNO$_3$ + 7–7.5M NaNO$_3$	"	100		
	0.5–2M HNO$_3$	"	100	ca. 2	
	0.5M HNO$_3$ + 4M NaNO$_3$ + 2M H$_2$SO$_4$	"	100		
	0.5M H$_2$SO$_4$ + 0.1M H$_3$PO$_4$	"	100		
Ti (IV)	0.3M HNO$_3$	TOPO in kerosene	ca. 100		[989]
	>7M H$_2$SO$_4$, NH$_4$SCN	0.01M TOPO in cyclohexane	100	1.3	[990–992]
	8M HCl	cyclohexane			
U (IV)	0.3M HNO$_3$	TOPO in kerosene	ca. 100		[989]
U (VI)	2–10M HCl	0.1M TOPO in cyclohexane	100	⩾3	[977]
	1M HNO$_3$	"		2.7	[993, 994]
	1M H$_2$SO$_4$	"		0	
	1M HClO$_4$	"		<1	
Zn (II)	1M HCl	0.1M TOPO in ethyl ether	ca. 100		[995]
	3M HCl	0.1M TOPO in cyclohexane	100	⩾2	[977]
Zr (IV)	>5M HCl; 1M HCl + 4–5M NaCl; 0.5–10M MNO$_3$; 1M HNO$_3$ + 3–4M NaNO$_3$; 0.2–7M H$_2$SO$_4$	0.02M TOPO in cyclohexane	100	2–3	[996]

* Transplutonium elements

15 Separation

Table 5.30 Extraction of acids with a 5% solution of MDOA in $CHCl_3$
(from [355], by permission of Nauchno-Tekhnicheskoe Izdat., Moscow)

Acid	E, %	Acid	E, %
Hydrochloric	98·0	Picric	96·0
Nitric	98·0	Maleic	97·5
Sulphuric	98·0	Malic	90·8
Phosphoric	76·5	Oxalic	98·0
Acetic	75·8	Succinic	90·0
Trichloroacetic	98·7	Ascorbic	0·0
Hydrofluoric	87·1	Glutamic	0·0
Formic	89·9		

Table 5.31 Extraction of elements with high molecular-weight amines

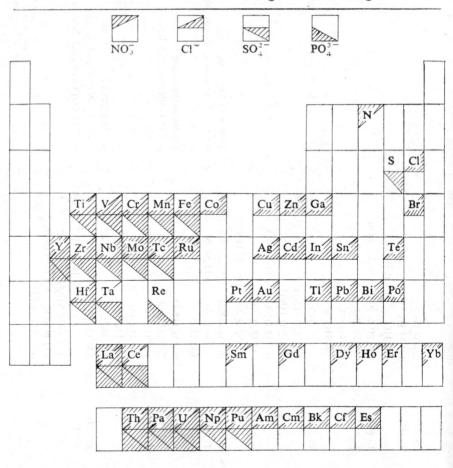

5.4.8 Extraction with High Molecular-Weight Amines

In the last few years high molecular-weight amines (M.W. 250–600) have found increasing application in extraction processes for both analytical and industrial purposes, in particular in the extraction and purification of compounds of uranium, neptunium and plutonium.

Extractions with amines are usually done from strongly acidic solutions. Only when quaternary ammonium salts are used can the process be effected from alkaline solutions.

As mentioned in Section 5.3.3 amines are suitable for the extraction of organic or mineral acids, and numerous elements.

Table 5.30 lists the acids that can be extracted with a chloroform solution of methyldioctylamine (MDOA).

Elements can be extracted with high molecular-weight amines from solutions of various acids, mainly hydrochloric, sulphuric, nitric and phosphoric (Table 5.31).

The data on the extraction of the elements with various amines are collected in Table 5.32.

5.4.9 Extraction with Halides and Pseudohalides

Metal ions react with halides and pseudohalides to form two types of product, the anionic complexes, such as $AuBr_4^-$ $FeCl_4^-$ and $SbCl_6^-$, and molecular (covalent) species such as $HgCl_2$, SnI_4, and $AsBr_3$.

The extraction properties of the two groups differ substantially so they must be considered separately. The covalent compounds are extractable with non-polar solvents, whereas the anionic complexes can be extracted only with solvents of high basicity, which contain atoms of donor properties, i.e. oxygen, sulphur or nitrogen.

The extraction of molecular complexes with non-polar solvents is very useful for separations, because of its selectivity. Many useful methods for the determination of metals depend on a preliminary extractive separation of major components that form covalent halide compounds.

Zolotov [1079] considered the formation of such compounds by various metals, and found that uncharged halide compounds can be produced by the elements at the bottom of groups IV, V, and VI of the periodic table, notably germanium ($GeCl_4$, $GeBr_4$), tin (SnI_4), arsenic ($AsCl_3$, $AsBr_3$, AsI_3), antimony ($SbBr_3$, SbI_3), bismuth (BiI_3), selenium ($SeCl_4$, $SeBr_4$), and mercury ($HgCl_2$, $HgBr_2$, HgI_2). In considering the relation of the co-ordination number of the element to its position in the periodic system and to the ligand ion involved, he concluded that such

Table 5.32 Extraction of elements with amines

Species	Aqueous phase	Organic phase	Reference
Ag (I)	HCl, LiCl, CsCl	10% MDOA in trichloroethylene	[997]
Ag (II)	pH > 2, (ClO_4^-, Cl^-)	tetrabutylammonium perchlorate in $CHCl_3$	[998]
	1·25M H_2SO_4, 3M HCl	5% TOA in $CHCl_3$	[999]
Bi (III)	0·5M HCl	TLA in xylene	[1000]
	0·1M HBr	TLA in xylene	[1000]
	0·5–2M HCl	0·1M TOA	[1001]
Ca (II)	pH 12·2–12·9	8-hydroxyquinoline, quat. ammonium salt in MIBK	[1002]
Cd (II)	pH 3 + KI	Amberlite LA-2 in xylene	[1003]
	pH 0·9–11, 0·2–2M KI	5% Aliquat 336-S in xylene	[1004]
	0·1–0·5M HBr	Amberlite LA-1 in xylene	[1005]
	5–6M HCl	0·1M TBA in benzene	[355]
	1–8M HCl	10% Amberlite LA-1 in xylene	[1006]
	Cl^-, Br^-, I^-	TOA	[1007, 1008]
Co (II)	8–10M HCl	8% MDOA in trichloroethylene	[1009]
	6–10M HCl	0·1M TOA(TIOA) in various solvents	[1010]
	8M HCl	Amberlite XE-204 in xylene	[1011]
	NH_4SCN	Amberlite LA-1 in CCl_4	[1012]
	KSCN	20% TOA in CCl_4	[1013]
	pH 6·2, 0·15M citrate	0·5M TBA in amyl alcohol	[1014]
Cr (III)	0·1M HCl, 4·75M KSCN	0·2M TOA in CCl_4	[1015]
	aqueous solution of $H_2C_2O_4$	5% MDOA in $CHCl_3$	[1016]
Cr (VI)	6M HCl	5% TBA in $CHCl_3$	[1016]
	0·1M H_2SO_4	0·01M TOA in benzene	[1017]
	0·02–0·5M H_2SO_4	5% Aliquat 336-S in $CHCl_3$	[1018]
	0·1M H_2SO_4	TBA in ethylene chloride	[1019]
	0·05M HCl + H_2O_2; 0·05M H_2SO_4	0·01M TOA in benzene	[1020]

§ 5.4] Applications 229

Cu (II)	7M HCl	8% MDOA in trichloroethylene	[1009]
	6.5M HCl	0.1M TOA in xylene	[1006]
	>3M HCl	10% Amberlite LA-1 in xylene and other amines	[1021]
			[1022]
Eu, transpluto-	0.1–0.2M HNO₃, 7.2M LiNO₃	0.4M TOA in cyclohexane	[1009]
nium elements	>4M HCl	8% MDOA in trichloroethylene	[1023]
	5M HCl	0.2M TBA in CHCl₃	[1010]
	>1M HCl	5–20% TIOA in xylene	[1024]
Fe (III)	pH 2.3–2.7, H₂SO₄	TOA	[1025]
	1M SO₄²⁻	prim. amines in hydrocarbons	[364]
	1–12M HCl	tert. amines in CHCl₃	[355]
Ga (III)	7M HCl	0.1M n-dodecylamine in CHCl₃	[1006]
	>4M HCl	10% Amberlite LA-1 in xylene	[1016]
Hf (IV)	11–12M HCl	5% MDOA in xylene	[1026]
	8M HCl	0.2M TOA in cyclohexane	[1027]
Hg (II)	0.1–1M H₂SO₃	0.1M TOA or TIOA in kerosene	[1028]
	sulphate, selenate and oxalate complexes from acid medium	MDOA in CHCl₃	
In (III)	NaOH, CN⁻	quat. ammonium salts in hexane	[1029]
	6–7M HCl	10% Amberlite LA-1 in xylene	[1006]
	Cl⁻, Br⁻, I⁻, SCN⁻	tetra-n-hexylammonium salts in ethylene dichloride	[1030]
Ir (IV)	12M HCl	TOA in benzene	[1030a]
Mn (II)	8M HCl	0.1M TOA in xylene	[1006]
	>4M HCl	10% Amberlite LA-1 in xylene	[1006]
MnO₄⁻ NO₃⁻, NO₂⁻	H₂O	0.1M N (C₇H₁₅)₄Cl in ethylene dichloride	[1031]
Mo (VI)	pH 1–2, 1M SO₄²⁻	0.2M prim. and sec. amines in aromatic hydrocarbons	[1016]

Table 5.32 (continued)

Species	Aqueous phase	Organic phase	Reference
Mo(IV)	pH 2–3 (SO_4^{2-})	TOA in kerosene or toluene	[1032, 1033]
	HNO_3	TIOA in toluene or ethylene dichloride	[1034]
	HCl	TIOA	[1035]
	1·5–2·5M HCl + 0·2–0·8M KSCN		[1036]
Nb(V)	>9M HCl	1% TBA in $CHCl_3$	[467, 1045a]
	11M HCl	5% MDOA in xylene	[1037]
	9M HCl	8% TBA in methylene chloride	[1010, 1038, 1039]
	1·5M H_2SO_4	5% TOA in xylene, benzene	[1040]
	1·85M H_3PO_4	5% MDOA in trichloroethylene	[1040]
	4·0–4·8M H_2SO_4	5% MDOA in trichloroethylene	[1037]
	HF + HNO_3	8% TBA in methylene chloride	[1041]
	pH 1–4 tartaric acid or oxalic acid	TIOA in kerosene	[1042]
Pb(II)	1·8M HCl	TOA in kerosene	[1043, 1043a]
	1·5M HCl	Aliquat 336 in benzene or xylene	[1006]
Po(IV)	1M HCl, HBr	10% Amberlite LA-1 in xylene	[1000]
Pt(IV)	2M HCl	TLA in xylene	[1044]
	1M HCl; 1M HBr	0·05M TOA in toluene	[1000]
		0·01M TLA in xylene	[1045]
		other amines	
Re(VII)	alkaline and acidic medium	quat. ammonium salts	[467, 1045a]
Ru(IV)	2M H_3PO_4	1% TBA in $CHCl_3$	[1046]
	0·1–9M HCl	5% TIOA in xylene	[1010]
	HNO_3	tert. amines	[1047]
Si(IV)	≤1M HF	TOA in xylene	[1048]
Sn(II)	7–8M HCl	Amberlite LA-2 in xylene	[238]
Sn(IV)	5–9M HCl	10% Amberlite LA-2 in xylene	[1049]
Ta(V)	4M HNO_3 + 1M HF	20% TOA in xylene	[1050]
	>4·8M H_2SO_4	8% TBA in methylene chloride	[1037]

§ 5.4] Applications 231

Tc (VII)	$6M$ Al$(NO_3)_3$	$0.3M$ TLA in Amsco 125-82, 5% tridecanol	[1051]
Te (IV)	$4-6M$ HCl	Amberlite LA-1 or Amberlite LA-2 in xylene	[1052]
Ti (IV)	pH 1.4–3, $1M$ SO$_4^{2-}$, $8M$ HCl	$0.1M$ Amine 21F-81 in kerosene	[1053]
		10% Amberlite LA-1 in kerosene	[1006]
	pH 4, $0.3M$ citric acid	$0.2-5M$ tributylamine	[1054]
Tl (III)	$0.1-0.5M$ HBr, HCl	Amberlite LA-1 in xylene	[1055]
V (V)	pH 2, $1M$ SO$_4^{2-}$	$0.1M$ prim., sec. and tert. amines in aromatic hydrocarbons	[1025, 1056]
W (VI)	$6-7M$ HCl	10% Amberlite LA-1 in xylene	[1006]
	pH 2 (HCl or HNO$_3$)	$0.15M$ TIOA in benzene, other solvents	[1056a]
Y (III)	$7-8M$ HN$_4$SCN	5% MDOA in xylene	[365]
	$0.5M$ NHO$_3$	$1M$ TOA or TIOA in benzene	[1025, 1027]
Zn (II)	$\geq 2M$ HCl	8% MDOA in trichloroethylene	[1009]
		5% TBA in CHCl$_3$	[1009]
		5% TIOA in hexane	[1057]
Zr (IV)	$2M$ HCl	MDOA or TIOA in MIBK	[1057]
	$6M$ HCl + ascorbic acid	MDOA in xylene	[1058]
	$11-12M$ HCl	5% MDOA in xylene	[1010, 1016, 1059]
	$>8M$ HCl	$0.2M$ TOA or TIOA in xylene	[1010]
	$10M$ HCl	$0.1M$ TOA in benzene, other amines	[1060, 1061]
	$0.1-1M$ H$_2$SO$_4$	5% MDOA in xylene	[1027]
		$0.1M$ TOA or TIOA in kerosene	[1027]
Lanthanides	$12M$ LiCl	TIOA in xylene	[1062]
	pH 8–12, $1.2 \times 10^{-3}M$ EDTA	$0.1M$ Capriquat in CCl$_4$	[1062a]
Eu, Ce (IV)	$0.1M$ HCl + $11.9M$ LiCl	20% TIOA in methylene chloride	[1062]
	pH 1, $1M$ SO$_4^{2-}$	prim. amines in hydrocarbons	[1025]
	dil. HNO$_3$ + BrO$_3^-$	Aliquat 336-S-NO$_3^-$ in xylene	[1063]

Table 5.32 (continued)

Species	Aqueous phase	Organic phase	Reference
Actinides			
Am, Cm (III)	11·6M LiCl+0·1M HCl	20% TIOA in xylene	[1062]
Am, Cm, Bk,	>10M (LiCl+AlCl₃)	tert. amines in isopropylbenzene	[386]
Np (IV)	1–10M HNO₃	10% TOA in xylene	[1064]
	2M HNO₃	0·3M TIOA in xylene	[983]
	0·1–10M HNO₃	0·1M N (C_7H_{15})$_4$NO₃ in xylene	[1065]
		other amines	[1066]
Pa (V)	1–5M H_2SO_4	0·001M Primene JM-T in xylene	[983]
	6–11M HCl	5% MDOA in xylene	[1067]
	8M HCl	5–10% Amberlite A-1 in kerosene	[1068]
	0·5M H_2SO_4	5% MDOA in trichloroethylene	[1040]
	1·8M H_3PO_4	5% MDOA in trichloroethylene	[1069]
Po (IV)	6M HCl	5% TBA in CHCl₃	[1016]
Pu (IV)	4M HNO₃+0·03M NaNO₂	0·15M TLA in Amsco + n-octyl alcohol	[1070]
	1–12M HNO₃+NaNO₂	10% TOA in xylene	[1064]
	2–10M HNO₃	10% TLA in xylene	[1064]
	3M HNO₃	0·01M N (C_7H_{15})$_4$NO₃ in benzene or toluene	[1071]
	H_2SO_4	Primene JM-T in xylene	[983]
Pu (VI)	4·8M HCl+0·01M $K_2Cr_2O_7$	5% TIOA in xylene	[1010]
	6–9M HCl	5% TBA in CHCl₃	[1016]
	1–12M HNO₃	10% TOA in xylene	[1064]
Th (IV)	5–6M HNO₃	0·4M TOA or TIOA in benzene or toluene	[1072]
	pH 0·05, 12% SO_4^{2-} + 3% PO_4^{3-}	0·1M Primene JM in kerosene	[1072]
U (IV)	1–10M HNO₃	10% TOA in xylene	[1064]
	1M (NH₄)₂ SO₄	0·1M Amine 21F-81 in benzene	[355]

§5.4] Liquid–Liquid Extraction

Element	Aqueous phase	Organic phase	Ref.
U (VI)	5M HCl	5% TIOA in xylene	[1010]
	9M HCl	5–20% TIOA in xylene	[1026]
	8M HCl	8% TBA in CHCl$_3$	[1073]
	3M HCl	Amberlite LA-1 in kerosene	[1068]
	>1M HCl	5% MDOA in xylene	[1016]
	6M HCl + 0.6M H$_3$PO$_4$	5% TOA in xylene	[1074]
	0.1–1M H$_2$SO$_4$	0.1M sec. and tert. amines in kerosene	[1016, 1025]
	pH 1.5	TOA in butylbenzene	[1072]
	pH 0.95 (H$_2$SO$_4$) + 0.01M Al(III)	5% MDOA in xylene	[1075]
	0.1–1M H$_2$SO$_4$ + Na$_2$SO$_4$	benzyldodecylamine in CHCl$_3$	[1076]
	2M CH$_3$COOH	other amines	[1016]
		5% MDOA in xylene	[1016, 1025, 1077]
	1–2M HF		[1016]
	pH 10–12.5 (CO$_3^{2-}$)	8-hydroxyquinoline + quat. ammonium salts in MIBK	[592]
Co, Zn, Fe, Hg	1M KSCN + 0.1M HCl	0.1M TOA in CCl$_4$	[1015]
54 elements	0.2M HNO$_3$ + 0.3–0.9M LiNO$_3$ or 0.1M H$_2$SO$_4$ + 0.1–2.4M Li$_2$SO$_4$	various amines	[1078]
63 elements	HCl, LiCl	various amines	[360]
16 elements heteroligand complexes	pH 2–11 + 4×10^{-3} M EDTA	0.2M Aliquat-336 in 1,2-dichloroethane	[1078a] [1341]

MDOA – methyldi-n-octylamine
TOA – tri-n-octylamine
TIOA – tri-iso-octylamine
TBA – tri-benzylamine
TLA – tri-n-laurylamine

Primene JM – primary amine mixture
Primene JM-T – mixture of primary amines with C$_{18}$–C$_{22}$ chains
Primene 81-R – mixture of primary amines with C$_{12}$–C$_{14}$ chains

Aliquat 336 or 336-S – tricaprylmonoethylammonium chloride
Amberlite LA-1-N-dodecyl(trialkylmethyl)amine
Amberlite LA-2-N-lauryl(trialkylmethyl)amine
Amine 9D-178-secondary amine mixture
Arquard 2C – R$_2$N(CH$_3$)$_2$Cl where R contains ca. 16 carbon atoms (75% solution in isopropyl alcohol)
Amine 21F-81-1-(3-ethylpenty)-4-ethyloctylamine
Capriquat – tri-n-octylmethylammonium chloride

compounds are likely to be formed by other elements as well, largely those from group III of the periodic system. The elements capable of forming uncharged halide complexes are listed in Table 5.33.

Table 5.33 The elements capable of forming covalent halide complexes: those useful for extractions are shaded

H																	He
Li	Be											B	C	N	O	F	Ne
Na	Mg											Al	Si	P	S	Cl	Ar
K	Ca	Sc	Ti	V	Cr	Mn	Fe	Co	Ni	Cu	Zn	Ga	*Ge*	*As*	*Se*	Br	Kr
Rb	Sr	Y	Zr	Nb	Mo	Tc	Ru	Rh	Pd	Ag	Cd	In	*Sn*	*Sb*	*Te*	I	Xe
Cs	Ba	La	Hf	Ta	W	Re	Os	Ir	Pt	Au	*Hg*	Tl	*Pb*	*Bi*	*Po*	At	Rn
Fr	Ra	Ac															

Ce	Pr	Nd	Pm	Sm	Eu	Gd	Tb	Dy	Ho	Er	Tm	Yb	Lu
Th	Pa	U	Np	Pu	Am	Cm	Bk	Cf	Es	Fm	Md	No	

The compounds that are important in extraction separations are the chlorides, bromides, iodides, fluorides, and thiocyanates. The best known extraction system (used since 1892) is the extraction of iron from chloride medium with ethyl ether.

The widespread use of the chloride system in analysis stems primarily from the fact that the extraction is done from hydrochloric acid solution, which is often of advantage, because this acid is used to dissolve a variety of materials, so an additional complex-forming agent need not be introduced. Moreover, the reaction conditions enable the process to be conducted in the presence of the elements which would precipitate in a slightly acidic or alkaline solution. The presence of chloride causes formation of sparingly soluble compounds in only a few cases.

The optimum extraction conditions for chloride–metal complexes are given in Table 5.34.

The remaining halide systems involve only a fairly small number of elements which, on the one hand, restricts their application, but on the other hand increases their selectivity. In many analytical problems this is an advantage, as such complexes are suitable for the extraction of major components because of their appreciable solubility in the organic phase.

Thiocyanate complexes have been useful in analysis; frequently they are coloured, and the extracts can be used for spectrophotometric determinations.

§ 5.4] Applications 235

Table 5.34 Extraction of metal chloride complexes

Species	Aqueous phase	Organic phase	$E,\%$max.	Reference
As (III)	$11M$ HClO$_4$ + $2M$ NaCl	CCl$_4$	96·2	[1090]
	$7\cdot5M$ H$_2$SO$_4$ + $2M$ KCl	CCl$_4$	70	[1090]
	$10M$ HCl	CCl$_4$	57·5	[1090]
	$6M$ HCl	diethyl ether	68	[3, 628]
	$7\cdot7M$ HCl	di-isopropyl ether	67	[628]
	$7M$ HCl	MIBK + amyl acetate (1 : 1)	80	[1100, 1101]
	$11M$ HCl	benzene	94	[1102]
	$7M$ HCl	MIBK	88	[1100, 1101]
	$7M$ LiCl	MIBK	33	[1100, 1101]
As (V)	$6M$ HCl	diethyl ether	2–4	[628]
	$7M$ HCl	MIBK + amyl acetate (1 : 1)	25	[1100, 1101]
Au (III)	$6M$ HCl	diethyl ether	95–100	[3, 628]
	$7M$ HCl	MIBK	>99	[1100, 1101]
	$7\cdot7M$ HCl	di-isopropyl ether	ca. 99	[628]
	$0\cdot5$–$8M$ HCl	MIBK	99	[308]
	10% HCl	ethyl acetate	100	[3]
	>$0\cdot1M$ HCl	butyl alcohol	>98	[1103]
Cd (II)	$7M$ HCl	MIBK	12–13	[1100, 1101]
Co (II)	$0\cdot85M$ CaCl$_2$	octan-2-ol	9·1	[628]
Co (III)	$4\cdot5M$ HCl	octan-2-ol	9·1	[628]
Cr (VI)	$1M$ HCl	MIBK	ca. 99	[343, 1104]
Fe (III)	$6M$ HCl	diethyl ether	100	[3, 628]
	$7\cdot7M$ HCl	di-isopropyl ether	100	[3, 628]
	$9M$ HCl	bis-β-chloroethyl ether	ca. 99	[3]
	$7M$ HCl	MIBK	100	[1100, 1101]
	$7M$ LiCl	MIBK		[1100, 1101]
	$7M$ HCl	MIBK + amyl acetate (1 : 1)	100	[1100, 1101]
	$10\cdot7M$ HCl + 4% H$_2$SO$_4$	amyl acetate		[1104]
	HCl	butyl acetate	100	
	HCl	methyl ethyl ketone	100	[1105]
	HCl	ethers and ketones	100	[1106, 1089]
	$8M$ HCl + $5M$ HF	amyl acetate	ca. 100	[1107]
Ga (III)	$6M$ HCl	diethyl ether	97	[3, 628]
	$6M$, $7M$ HCl	di-isopropyl ether	100	[3, 628, 1108]
	$3M$ HCl	TBP	100	[3]
	$7M$ HCl	MIBK	100	[1094, 1100, 1101]
	$7M$ LiCl	MIBK	100	[1100, 1101]
	$2M$ HCl + $1M$ NH$_3$Cl + $1M$ H$_2$SO$_4$	MIBK	99·9	[1109]

Table 5.34 (*continued*)

Species	Aqueous phase	Organic phase	E,%max.	Reference
Ge (IV)	6M HCl	diethyl ether	40–60	[3, 628]
	10·5M HCl	CCl₄	99·5	[1110–1112]
	11M HCl	benzene	99·6	[1102]
	7M HCl	MIBK + amyl acetate (1 : 1)	97	[1100, 1101]
Hg (II)	0·125M HCl	ethyl acetate	80	[3, 1113]
	0·1M HCl	n-butyl acetate + trichloro-acetic acid	82–89	[3]
	0·1–3M HCl	ketones and alcohols		[1114]
In (III)	8M HCl	diethyl ether	3	[628]
	7M HCl	MIBK	94	[1100, 110]
	7M HCl	MIBK + amyl acetate (1 : 1)	48	[1100, 110]
	7M LiCl or 8M HCl	MIBK	ca. 60	[1094, 110-1101]
Mo (VI)	6M HCl	diethyl ether	76–90	[3, 628]
	7·7M HCl	di-isopropyl ether	21	[3, 60]
	5M, 7M HCl	amyl acetate	99	[3, 1115]
	7M HCl	MIBK	ca. 96	[1100, 110]
	7M HCl	MIBK + amyl acetate (2 : 1)	92	[1100, 110]
	HCl	various solvents		[1088]
Nb (V)	12M HCl	di-isopropyl ketone	100	[3]
	HCl	solvent mixture		[1116]
	7·7–9·4M HCl	100% TBP	100	[1117]
Pa (V)	6M HCl	isobutyl alcohol	100	[1118]
	6M HCl	di-isopropyl ketone	99	[1119]
Pb (II)	HCl	oxygen-containing solvents		[1082]
	3M HCl + 2M LiCl	30% TBP in MIBK	100	[1120]
Po (IV)	6–8M HCl	oxygen-containing solvents	90–100	[1082]
Pt (II)	3M HCl (SnCl₂)	diethyl ether	>95	[1118]
Sb (III)	6M HCl	diethyl ether	6	[3, 628]
	6·5–8·5M HCl	di-isopropyl ether	ca. 2	[3, 628]
	7M HCl	MIBK	69	[1100, 11]
	7M HCl	MIBK + amyl acetate (2 : 1)	59	[1100,110]
Sb (V)	6M HCl	diethyl ether	81	[3, 628]
	7·7M HCl	di-isopropyl ether	100	[3, 628]
	7M HCl	MIBK + amyl acetate (1 : 1)	94	[1100, 11]
	7M LiCl	MIBK + amyl acetate (2 : 1)	ca. 100	[1100, 11]
	1–2M HCl	ethyl acetate	83	[1121]

§ 5.4] Applications 237

Table 5.34 (continued)

Species	Aqueous phase	Organic phase	$E,\%$ max.	Reference
Se (IV)	7M HCl	MIBK + amyl acetate (1 : 1)	ca. 6	[1100, 1101]
	HCl	other solvents		[1114, 1344]
	5–9M HCl	aliphatic and cyclic unsaturated hydrocarbons in CHCl$_3$	100	[1114a]
Sn (II)	6M HCl	diethyl ether	15–30	[3, 628]
Sn (IV)	7·7M HCl	diethyl ether	17	[3, 628]
	7M HCl	MIBK	ca. 93	[1100, 1101]
	7M HCl	MIBK + amyl acetate (2 : 1)	78	[1100, 1101]
	10·7M HCl + 4% H$_2$SO$_4$	amyl acetate	68	[1104]
		various solvents		[1122]
Te (IV)	4M HCl	MIBK, other ketones	95	[1123]
Tl (III)	6M HCl	diethyl ether	90–95	[3, 628]
	6M HCl	di-isopropyl ether	ca. 99	[628]
	7M HCl	MIBK + amyl acetate (1 : 1)	ca. 100	[1100, 1101]
	6M HCl	ethyl acetate	ca. 100	[1124]
	1·5–8M HCl	MIBK	> 99	[307]
U (VI)	7M HCl	MIBK	ca. 22	[1100, 1101]
	5–8M HCl + 10M LiCl	MIBK	> 99	[308]
	7–8M HCl + 1M MgCl$_2$	MIBK	> 99	[308]
V (V)	7·7M HCl	di-isopropyl ether	22	[3, 628]
	7M HCl	MIBK	81	[1100, 1101]
W (VI)	7M HCl	MIBK + amyl acetate (1 : 1)	49	[1100, 1101]
Zn (II)	7M HCl	MIBK	5–6	[1100, 1101]

Irrespective of the detailed studies on the application of halide extraction systems for individual analytical problems, investigations of a more general nature such as the dependence of the distribution coefficient on the acid concentration, the kind of the solvent used, ligand concentration, etc., have been made by Boswell and Brooks [1080,1081], by Iofa et al. [1082–1084] and by others [21,1085–1090,1342] (on halide extractive systems), by Študlar [1091] and others [1092] (on the extraction of bromide complexes), by Byrne [1343] (on the extraction of iodide complexes), by a Japanese group [307,308,1093–1095] (on the extraction of various complexes with methyl isobutyl ketone) and by Różycki [1096,1097], and others [1098,1099] (on the extraction of thiocyanates).

Tables 5.35–5.38 summarize the optimum conditions for extraction of bromide, iodide, fluoride and thiocyanate complexes, respectively.

Table 5.35 Extraction of metal bromide complexes

Species	Aqueous phase	Organic phase	E,%max.	Reference
As (III)	6M HBr	diethyl ether	73	[1125, 1126]
	5M HBr	di-isopropyl ether	31	[1127]
	8M H$_2$SO$_4$ + 5% KBr	CCl$_4$	100	[1091, 1128]
Au (III)	0·5–4M HBr	diethyl ether	ca. 100	[1125, 1126]
	6M HBr	di-isopropyl ether	ca. 99	[1127]
Bi (III)	9–10M HClO$_4$ + 0·005M NaBr	CCl$_4$	ca. 100	[1090]
Cd (II)	7M H$_2$SO$_4$ + 0·012M NaBr	isobutyl acetate	90	[1129]
Cu (I)	2M NH$_4$Br + 3M HBr	MIBK	50	[1130]
Cu (II)	6M HBr	diethyl ether	6	[1125, 1126]
	2M NH$_4$Br + 3M HBr	MIBK	37	[1130]
Fe (III)	4–5M HBr	diethyl ether	ca. 97	[1125, 1126]
	4M NH$_4$Br	MIBK	26	[1130]
	2M NH$_4$Br + 3–4M HBr	MIBK	100	[1130]
	5M HBr	di-isopropyl ether	35	[1127]
	5–7M HBr	ketones, ethers, alcohols	ca. 100	[1131]
Ga (III)	5–5·5M HBr	diethyl ether	97–96	[1125, 1126]
	5M HBr	di-isopropyl ether	29	[1127]
Ge (IV)	8M H$_2$SO$_4$ + 5% KBr	CCl$_4$	100	[1091]
Hg (II)	pH 1 (HBr)	diethyl ether	94	[1125, 1126]
In (III)	3–5M HBr	diethyl ether	98–100	[1125, 1126]
	5M HBr	di-isopropyl ether	99	[1127]
Mo (VI)	6M HBr	diethyl ether	54	[1125, 1126]
Os	5M HBr	di-isopropyl ether	25–33	[1127]
Sb (III)	2M HBr	diethyl ether	38	[1125, 1126]
Sb (V)	4·5M HBr	diethyl ether	97	[1125, 1126]
	5M HBr	di-isopropyl ether	6·5	[1127]
	8M H$_2$SO$_4$ + 5% KBr	CCl$_4$	100	[1091]
Se (IV)	6M HBr	diethyl ether	ca. 31	[1125, 1126]
	0·2–1M HBr	benzene	> 99	[1132]
	8M H$_2$SO$_4$ + 5% KBr	CCl$_4$	100	[1091]
Sn (II)	4M HBr	diethyl ether	84	[1125, 1126]
	0·75M HBr + 1M NH$_4$Br	MIBK	ca. 89	[1130]
	4M HBr + 2M NH$_4$Br	MIBK	97·5	[1130]
Sn (IV)	4M HBr	diethyl ether	85	[1125, 1126]
	0·75M HBr + 4M NH$_4$Br	MIBK	89	[1130]
	4M HBr + 2M NH$_4$Br	MIBK	99	[1130]
	8M H$_2$SO$_4$ + 5% KBr	CCl$_4$	100	[1091]
	5M HBr	di-isopropyl ether	ca. 6	[1127]
Te (IV)	5M HBr	di-isopropyl ether	ca. 1·5	[1127]
	0·05–0·1M HBr	nitrophenol	> 99	[1132]
Tl (I)	1M–3M HBr	diethyl ether	100	[1125, 1126]

§ 5.4] Applications 239

Table 5.35 (continued)

Species	Aqueous phase	Organic phase	E,%max.	Reference
(III)	0·1–5M HBr	diethyl ether	100	[1125, 1126]
	5M HBr	di-aminopropyl ether	99	[1127]
(II)	3M HBr	diethyl ether	10	[1125, 1126]
	2M NH$_4$Br	MIBK	7·5	[1130]
	0·75M HBr + 2M NH$_4$Br	MIBK	16	[1130]
	3·8M HBr + 2M NH$_4$Br	MIBK	50	[1130]
	7M H$_2$SO$_4$ + 0·01M NaBr	isobutyl acetate	80	[1129]

Table 5.36 Extraction of metal iodide complexes

Species	Aqueous phase	Organic phase	E,%max.	Reference
(I)	1M H$_3$PO$_4$ + 0·2M KI	MIBK	100	[1133]
(III)	3·5M H$_2$SO$_4$ + 2M KI	CCl$_4$	96·3	[1090]
	11M HClO$_4$ + 2M NaI	CCl$_4$	95·0	[1090]
	4M HI	CCl$_4$	87·5	[1090]
	0·9M HI	diethyl ether	62	[1134]
	H$_2$SO$_4$ + KI	CCl$_4$ or cyclohexane	ca. 100	[1085, 1086]
(III)	6·9M HI	diethyl ether	100	[1134]
(III)	6·9M HI	diethyl ether	34	[1134]
	0·75M H$_2$SO$_4$ + 1·5M KI	diethyl ether	< 10	[1134]
	HNO$_3$ + KI	cyclohexane or TBP	ca. 100	[1135]
	1M H$_3$PO$_4$ + 0·2M KI	MIBK	100	[1133]
(II)	6·9M HI	diethyl ether	100	[1134]
	0·75M H$_2$SO$_4$ + 1·5M KI	diethyl ether	100	[1126]
	HNO$_3$ + KI	cyclohexane or TBP	ca. 100	[1135]
	3M H$_2$SO$_4$ + 0·1M KI	MIBK	⩾ 99	[1136]
	4M H$_2$SO$_4$ + 0·001M KI	isobutyl acetate	98	[1129]
	1M H$_3$PO$_4$ + 0·2M KI	MIBK	100	[1133]
(II)	1M H$_3$PO$_4$ + 0·2M KI	MIBK	100	[1133]
(II)	6·9M HI	diethyl ether	100	[1134]
	0·75M H$_2$SO$_4$ + 1·5M KI	diethyl ether	33	[1126]
	HNO$_3$ + KI	cyclohexane or TBP	ca. 100	[1135]
(III)	6·9M HI	diethyl ether	ca. 8	[1134]
	0·5–2·5 HI + 0·75M H$_2$SO$_4$	diethyl ether	100	[1126]
	HNO$_3$ + KI	cyclohexane or TBP	100	[1135]
(VI)	6·9M HI	diethyl ether	6·5	[1134]
(II)	excess KI, 5% HCl	methyl isopropyl ketone	97	[1137]
	1M H$_3$PO$_4$ + 0·2M KI	MIBK	100	[1133]
(III)	2M NaI + 2·5M H$_2$SO$_4$	50% TBP in toluene	ca. 100	[1138]
	2M NaI + 2M HClO$_4$	50% TBP in toluene	ca. 94	[1138]

Table 5.36 (*continued*)

Species	Aqueous phase	Organic phase	E,%max.	Reference
Sb (III)	$5M\ H_2SO_4 + 10^{-5}M\ KI$	benzene	ca. 100	[1139]
	$3.5M\ H_2SO_4 + 0.05M\ KI$	CCl_4	96	[1090, 114
	$0.2-0.5M\ HI$	CCl_4	64	[1090]
	$5M\ H_2SO_4 + 0.02M\ KI$	diethyl ether	100	[1134]
	$5M\ HClO_4 + 0.02M\ KI$	diethyl ether	100	[1134]
	$0.75M\ H_2SO_4 + 1.5M\ KI$	diethyl ether	< 50	[1126]
Sb (V)	$5M\ H_2SO_4 + 0.05M\ KI$	CCl_4	96	[1090]
	$8M\ HClO_4 + 0.001M\ NaI$	CCl_4	98	[1090]
	$0.5-0.7M\ HI$	CCl_4	ca. 100	[1090]
Sn (II)	$6.9M\ HI$	diethyl ether	100	[1134]
	$0.75M\ H_2SO_4 + 1.5M\ KI$	diethyl ether	100	[1126]
Sn (IV)	$1.2M\ HCl + KI$	ethyl iodide, toluene	100	[1141]
Te (IV)	$6.9M\ HI$	diethyl ether	5.5	[1126]
Tl (I)	$0.5-2M\ HI$	diethyl ether	100	[1126]
Tl (III)	$0.05-2M\ HI$	diethyl ether	100	[1126]
Zn (II)	$6.9M\ HI$	diethyl ether	10	[1126]
	$0.75M\ H_2SO_4 + 1.5M\ KI$	diethyl ether	33	[1126]
	$HNO_3 + KI$	cyclohexane or TBP	ca. 100	[1135]
	$5M\ H_2SO_4 + 0.01M\ KI$	isobutyl acetate	98	[1129]

§ 5.4] Applications 241

ble 5.37 Extraction of metal fluoride complexes

ecies	Aqueous phase	Organic phase	$E,\%$max.	Reference
(III)	$4 \cdot 6M$ HF	diethyl ether	62	[1142]
(III)	$20M$ HF	diethyl ether	38	[1143]
(V)	$20M$ HF	diethyl ether	14	[1143]
(II)	$20M$ HF	diethyl ether	4	[1143]
(II)	$20M$ HF	diethyl ether	$1 \cdot 5$	[1143]
(II)	$20M$ HF	diethyl ether	$1 \cdot 7$	[1143]
(II)	$20M$ HF	diethyl ether	$1 \cdot 3$	[1143]
(IV)	$20M$ HF	diethyl ether	$ca.$ 7	[1143]
(II)	$20M$ HF	diethyl ether	$2 \cdot 7$	[1143]
(II)	$20M$ HF	diethyl ether	$1 \cdot 3$	[1143]
(VI)	$3 \cdot 5M$ HF	diethyl ether	9	[1142]
	$20M$ HF	diethyl ether	9	[1143]
	$10M$ HF $+ 6M$ H_2SO_4 $+ 2 \cdot 2M$ NH_4F	MIBK	$9 \cdot 7$	[1144]
(V)	$20M$ HF	diethyl ether	$65 \cdot 8$	[1144]
	$9M$ HF $+ 6M$ H_2SO_4	di-isopropyl ketone	90	[1145]
	$6M$ HF $+ 6M$ H_2SO_4	2,6-dimethylheptan-4-ol	98	[1067]
	$10M$ HF $+ 6M$ H_2SO_4 $+ 2 \cdot 2M$ NH_4F	MIBK	96	[1144] [1146]
	$1 \cdot 6M$ HF $+ 6 \cdot 3M$ H_2SO_4	MIBK	$ca.$ 100	[1147]
	$1M$ HF $+ 4M$ H_2SO_4	MIBK		[1148]
(VII)	$20M$ HF	diethyl ether	62	[1143]
(III)	$20M$ HF	diethyl ether	6	[1143]
(IV)	$20M$ HF	diethyl ether	13	[1143]
(II)	$4 \cdot 6M$ HF	diethyl ether	100	[1142]
(IV)	$1 \cdot 2 - 1 \cdot 6M$ HF	diethyl ether	100	[1142]
(V)	$10M$ HF $+ 6M$ H_2SO_4 $+ 2 \cdot 2M$ HF	MIBK	100	[1144]
	$0 \cdot 4M$ HF $+ 3 \cdot 7M$ HCl	di-isopropyl ketone	80	[1145]
	$0 \cdot 4M$ HF $+ 3 \cdot 9M$ HNO_3	di-isopropyl ketone	79	[1145]
	$0 \cdot 5M$ HF $+ 1M$ HNO_3	ethyl acetate	100	[1149]
	$0 \cdot 4M$ HF $+ 4 \cdot 5M$ H_2SO_4	di-isopropyl ketone	95	[1145]
	$0 \cdot 4M$ HF $+ 4 \cdot 6M$ $HClO_4$	di-isopropyl ketone	90	[1145]
	$20M$ HF	diethyl ether	79	[1143]
	$1M$ HF $+ 4M$ H_2SO_4	MIBK	100	[1148]
	HCl $+$ HF	MIBK	100	[1110]
(V)	$20M$ HF	diethyl ether	23	[1143]
(I)	$20M$ HF	diethyl ether	1	[1143]
)	$20M$ HF	diethyl ether	$8 \cdot 5$	[1143]
(I)	$20M$ HF	diethyl ether	$0 \cdot 9$	[1143]
(V)	$20M$ HF	diethyl ether	3	[1143]

16 Separation

Table 5.38 Extraction of metal thiocyanate complexes

Species	Aqueous phase	Organic phase	E,%max.	Reference
Al (III)	$0.5M$ HCl + $6M$ NH$_4$SCN	diethyl ether	19.4	[1150]
	pH 1.0–2.5 (HCl) + $8M$ NH$_4$SCN	MIBK	95	[1096]
	pH 1.0–3.7 (HCl) + $8M$ NH$_4$SCN	isoamyl alcohol	95	[1096]
	pH 1.0–2.5 (HCl) + $8M$ NH$_4$SCN	diethyl ether	48	[1096]
	pH 0.8–3.0 (HCl) + $8M$ NH$_4$SCN	isobutyl acetate	67	[1096]
Be (II)	$0.5M$ HCl + $7M$ NH$_4$SCN	diethyl ether	92.2	[1150]
Bi (III)	pH 1.5–4.2 (HNO$_3$) + $8.2M$ NH$_4$SCN	MIBK	40	[1096]
	pH 0.6–1.5 (HNO$_3$) + $8.2M$ NH$_4$SCN	isoamyl alcohol	31	[1096]
Ca (II)	pH 1.0–6.5 + $3.6M$ NH$_4$SCN	isoamyl alcohol	19.0	[1097]
Cd (II)	pH 3.0–5.0 (HCl)	MIBK	55	[1096]
	pH 2.0–5.3 (HCl)	isoamyl alcohol	19	[1096]
Co (II)	$0.5M$ HCl + $7M$ NH$_4$SCN	diethyl ether	75.2	[1150]
	$7M$ HCl + 1% NH$_4$SCN	MIBK	ca. 100	[1151]
	pH 1.0–7.4 (HCl)	MIBK	100	[1096]
	pH 2.5–5.3 (HCl)	isoamyl alcohol	84	[1096]
	pH 2.0–5.3 (HCl)	isobutyl acetate	70	[1096]
Cr (III)	pH 2.1 + $9M$ NH$_4$SCN	isoamyl alcohol	72.0	[1096]
	pH 1–3 + $0.03M$ NH$_4$SCN	MIBK, other solvents	99	[1152]
Cu (I)	pH 2.7–6.0 (HCl)	MIBK	49.5	[1097]
Fe (III)	$0.5M$ HCl + $1M$ NH$_4$SCN	diethyl ether	88.9	[1150]
	pH 0.8–3.0 (HCl) + $0.9M$ NH$_4$SCN	diethyl ether	78	[1097]
	pH 0.8–3.0 (HCl) + $0.9M$ NH$_4$SCN	MIBK	98	[1097]
	pH 0.8–3.0 (HCl) + $0.9M$ NH$_4$SCN	isoamyl alcohol	94	[1097]
	pH 1.5–3.0 (HCl) + $0.9M$ NH$_4$SCN	isobutyl acetate	82	[1097]
Ga (III)	$0.5M$ HCl + $7M$ NH$_4$SCN	diethyl ether	99.3	[1150]
	$1M$ HCl–pH 1.7 + $4.7M$ NH$_4$SCN	diethyl ether	94	[1097]
	$2M$ HCl, pH 2.2 + $4.7M$ NH$_4$SCN	MIBK	99	[1096]
	$1M$ HCl, pH 2.0 + $4.7M$ NH$_4$SCN	isoamyl alcohol	92	[1096]
	$1M$ HCl, pH 2.2 + $4.7M$ NH$_4$SCN	isobutyl acetate	98	[1096]

Table 5.38 (continued)

Species	Aqueous phase	Organic phase	$E,\%$ max.	Reference
n (III)	$0.5M$ HCl + $3M$ NH$_4$SCN	diethyl ether	75.3	[1150]
	pH 1.5–4.4 (HCl) + $0.9M$ NH$_4$SCN	diethyl ether	75	[1096]
	pH 1.0–4.4 (HCl) + $0.9M$ NH$_4$SCN	MIBK	99	[1096]
	pH 1.0–4.7 (HCl) + $0.9M$ NH$_4$SCN	isoamyl alcohol	97	[1096]
	pH 2.0–4.4 (HCl) + $0.9M$ NH$_4$SCN	isobutyl acetate	83	[1096]
a (III)	pH 1.0–7.1 + $4.7M$ NH$_4$SCN	isoamyl alcohol	28.0	[1150]
g (II)	pH 1.9–8.0 + $3.6M$ NH$_4$SCN	isoamyl alcohol	25.5	[1097]
	pH 4, $0.03M$ NaSCN	$3.6M$ TBP in CCl$_4$	80	[1153]
n (II)	pH 1.8–7.4 + $3.6M$ NH$_4$SCN	MIBK	40.0	[1097]
	pH 1.8–7.9 + $3.6M$ NH$_4$SCN	isoamyl alcohol	33.7	[1097]
o (V)	$0.5M$ HCl + $1M$ NH$_4$SCN	diethyl ether	99.3	[1150, 1154]
	$6M$ HCl, SCN$^-$	diethyl ether	100	[1155]
	$1M$ HCl, $0.5M$ H$_2$SO$_4$	MIBK, other solvents	66.7–80.5	[1156, 1157]
b (V)	$3M$ HCl + $1.5M$ KSCN + $0.3M$ SnCl$_2$	diethyl ether	98.0	[1158–1160]
	$2.4M$ HCl + 0.4–$1.7M$ KSCN	15% TBP in CHCl$_3$	ca. 95	[1161]
(III)	$0.5M$ HCl + $7M$ NH$_4$SCN	diethyl ether	89.0	[1150]
(IV)	$0.5M$ HCl + 1–$7M$ NH$_4$SCN	diethyl ether	99.5–99.9	[1150]
(II)	pH 1.8–7.0 + $3.6M$ NH$_4$SCN	isoamyl alcohol	16.0	[1097]
(IV)	$0.5M$ HCl + $3M$ NH$_4$SCN	diethyl ether	84.0	[1150]
(IV)	pH = 1–3, $3.6M$ KSCN	10% TBP in CCl$_4$	ca. 100	[1162]
(VI)	$0.5M$ HCl + $1M$ NH$_4$SCN	diethyl ether	45.1	[1150]
(IV)	$0.5M$ HCl + $1M$ NH$_4$SCN	diethyl ether	15.0	[1150]
(III)	pH 1.0–6.1 (HCl)	MIBK	5.5	[1097]
	pH 2.5–6.7 (HCl)	isoamyl alcohol	44.2	[1097]
(II)	$0.5M$ HCl + $3M$ NH$_4$SCN	diethyl ether	97.4	[1150]
	pH 2.0–5.0 (HCl)	diethyl ether	58.0	[1096]
	$3M$ HCl to pH 6.5	MIBK	100	[1096, 1163]
	pH = 1, $1M$ KSCN	MIBK	100	[1164]
	$1M$ H$_2$SO$_4$, SCN$^-$	isobutyl acetate	84	[1162]
	pH 0.6–5.5 (HCl)	isoamyl alcohol	97	[1096, 1165]
	pH 1.5–5.5 (HCl)	isobutyl acetate	93	[1096]

References

[1] Berthelot, M. and Jungfleisch, E., *Ann. Chem. Phys.*, **4**, 26, 400 (1872).
[2] Nernst, W., *Z. Phys. Chem.*, **8**, 110 (1891).
[3] Morrison, G. H. and Freiser, H., *Ekstrakcja w chemii analitycznej*, PWN, Warsaw, 1960; *Solvent Extraction in Analytical Chemistry*, Wiley, New York, 1957.
[4] Bell, R. P., *Trans. Faraday Soc.*, **27**, 797 (1937).
[5] Kirkwood, J. G., *J. Chem. Phys.*, **2**, 351 (1934).
[6] Banks, W. H., *Trans. Faraday Soc.*, **33**, 215 (1937).
[7] Sisskind, B. and Kasarnowskij, I., *Z. Anorg. Chem.*, **214**, 385 (1933).
[8] Uhlig, H. H., *J. Phys. Chem.*, **41**, 1215 (1937).
[9] Eiley, D. D., *Trans. Faraday Soc.*, **35**, 1281, 1421 (1939).
[10] Butler, J. A. V., *Trans. Faraday Soc.*, **33**, 171, 229 (1937).
[11] McGowan, J. C., *J. Appl. Chem.*, **2**, 323, 651 (1952); **4**, 41 (1954).
[12] Collander, R., *Acta Chem. Scand.*, **3**, 717 (1949); **4**, 1085 (1950); **5**, 774 (1951).
[13] Hildebrand, J. H. and Scott, R. L., *The Solubility of Nonelectrolytes*, 3rd Ed., Reinhold, New York, 1950.
[14] Hildebrand, J. H. and Wood, S. E., *J. Chem. Phys.*, **1**, 817 (1933).
[15] Scatchard, G., *Chem. Rev.*, **8**, 321 (1931).
[16] Buchowski, H., *Nature (London)*, **194**, 674 (1962).
[17] Buchowski, H., *Wpływ własności rozpuszczalników na współczynniki podziału nieelektrolitów w układach woda-rozpuszczalnik organiczny (Effect of the Properties of Solvents on the Distribution Coefficient of Nonelectrolytes in Water-Organic Solvent Systems)*, PWN, Warsaw, 1963.
[18] Born, M., *Z. Physik*, **1**, 45 (1920).
[19] Bjerrum, N., *Kgl. Danske Selskab.*, **9**, 7 (1926).
[20] Fuoss, C. A. and Kraus, C. A., *J. Am. Chem. Soc.*, **55**, 1019 (1933).
[21] Irving, H. M., *Quart. Rev., Chem. Soc.*, **5**, 200 (1951).
[22] Diamond, R. M. and Tuck, D. G., *Ekstraktsiya neorganicheskikh soedinenii*, (*Extraction of Inorganic Compounds*), Gosatomizdat, Moscow, 1962.
[23] Libuś, W., Siekierska M., and Libuś, Z., *Rocz. Chem. (Warsaw)*, **31**, 1293 (1957).
[24] Kuznetsov, V. I., *Usp. Khim.*, **23**, 659 (1954).
[25] Fomin, V. V., *Khimiya ékstraktsionnykh protsessov (Chemistry of Extraction Processes)*, Atomizdat, Moscow, 1960.
[26] Frolov, Yu. G. and Ochkin, A. V., *Zh. Neorgan. Khim.*, **7**, 1468 (1962).
[27] Alimarin, I. P. and Zolotov, Yu. A., *Chem. Anal. (Warsaw)*, **13**, 941 (1968).
[28] Morrison, G. H., *Anal. Chem.*, **36**, 93 R (1964).
[29] Sidgwick, N. V., *The Chemical Elements and their Compounds*, OUP, London, 1950.
[30] Ahrland, S., Chatt, J. and Davies, N., *Quart. Rev., Chem. Soc.*, **12**, 265 (1958).
[31] Zolotov, Yu. A. and Alimarin, I. P., *Dokl. Akad. Nauk SSSR*, **136**, 603 (1961).
[32] Zolotov, Yu. A. and Alimarin, I. P., *Radiokhimiya*, **4**, 272 (1962).
[33] Preetz, W. and Blasius, E., *Z. Anorg. Chem.*, **332**, 190 (1964).
[34] Beck, M. T., *Chemistry of Complex Equilibria*, Van Nostrand Reinhold, London, 1970.
[35] Burger, K. and Ruff, I., *Acta Chim. Acad. Sci. Hung.*, **45**, 77 (1965).
[36] Belcher, R., Leonard, M. A. and West, T. S., *Talanta*, **2**, 92 (1959).
[37] Leonard, M. A. and West, T. S., *J. Chem. Soc.*, **1960**, 4477.

[38] Archer, V. S. and Doolittle, F. G., *Anal. Chem.*, **39**, 371 (1967).
[39] Archer, V. S. and Doolittle, F. G., *Talanta*, **14**, 921 (1967).
[40] Nosouri, F. G., Shahine, S. A. and Magee, R. J., *Anal. Chim. Acta*, **36**, 346 (1966).
[41] Margerum, D. W. and Banks, C. V., *Anal. Chem.*, **26**, 200 (1954).
[42] Vydra, F. and Přibil, R., *Talanta*, **3**, 72 (1959).
[43] Yamamoto, Y., Kumamaru, T. and Uemura, Y., *Anal. Chim. Acta*, **39**, 51 (1967).
[43a] Schultz, W. W., Mendel, J. E. and Phillips, J. F., *J. Inorg. Nucl. Chem.*, **28**, 2399 (1966).
[43b] Crouse, D. J. and Horner, D. E., in *Solvent Extraction of Metals*, H.A.C. McKay (Ed.), Macmillan, London, 1965, p. 305.
[43c] Irving, H.M.N.H. and Tomlinson, W. R., *Talanta*, **15**, 1267 (1968); *Chem. Commun.*, **1968**, 497.
[43d] Chikryzova, E. G. and Topaly, E. E., *Zh. Neorgan. Khim.*, **16**, 708 (1971).
[43e] Gilbert, T. W., Newman, L. and Klotz, P., *Anal. Chem.*, **40**, 2123 (1968).
[44] Kolthoff, I. M. and Sandell, E. B., *J. Am. Chem. Soc.*, **63**, 1906 (1941).
[45] Geiger, R. W. and Sandell, E. B., *Anal. Chim. Acta*, **8**, 197 (1953).
[46] Morrison, G. H. and Freiser, H., *Anal. Chem.*, **30**, 632 (1958).
[47] Oosting, M., *Anal. Chim. Acta*, **21**, 301, 397, 505 (1959).
[48] Oosting, M., *Rec. Trav. Chim., Pays-Bas*, **79**, 627 (1960).
[49] Starý, J., *The Solvent Extraction of Metal Chelates*, Pergamon, Oxford, 1964.
[50] Dyrssen, D., *Sven. Kem. Tidskr.*, **68**, 212 (1956).
[51] Dyrssen, D., Dyrssen, M. and Johansson, E., *Acta Chem. Scand.*, **10**, 341 (1956).
[52] Rydberg, J., *Ark. Kemi*, **8**, 101 (1955).
[53] Irving, H., Rossotti, F. J. C. and Williams, R. J. P., *J. Chem. Soc.*, **1955**, 1906.
[54] Irving, H. and Williams, R. J., *J. Chem. Soc.*, **1949**, 1841.
[55] Schweitzer, G. K., *Anal. Chim. Acta*, **30**, 68 (1964).
[56] Zolotov, Yu. A., *Ekstraktsiya vnutrikompleksnykh soedinenii* (*Extraction of Internal Complexes*), Izd. Nauka, Moscow, 1968.
[56a] Siekierski, S., Inst. of Nucl. Research, *Report 'P'*, No. 1339/v/c, 1971.
[56b] Siekierski, S., *J. Radioanal. Chem.*, **31**, 335 (1976).
[57] Starý, J., *Anal. Chim. Acta*, **28**, 132 (1963).
[58] Poskanzer, A. M. and Foreman, B. M., *J. Inorg. Nucl. Chem.*, **16**, 323 (1961).
[59] Starý, J. and Hladký, E., *Anal. Chim. Acta*, **28**, 227 (1963).
[60] Starý, J. and Smižanska, J., *Anal. Chim. Acta*, **29**, 546 (1963).
[61] Marczenko, Z. and Mojski, M., *Anal. Chim. Acta*, **54**, 469 (1971).
[62] Sillén, L. G. and Martell, A.E., *Stability Constants*, Special Publication No. 17, Chemical Society, London, 1964, *Supplement No. 1*, S. P. No. 25, Chemical Society, London, 1971.
[63] Zolotov, Yu. A. and Kuzmin, N. M., *Zh. Analit. Khim.*, **20**, 476 (1965).
[64] Zolotov, Yu. A. and Lambrev, V. G., *Zh. Analit. Khim.*, **20**, 1153 (1965).
[65] Rudenko, N. P. and Starý, J., *Radiokhimiya*, **1**, 52 (1959).
[66] Chmutova, M. K., Petrukhin, O. M. and Zolotov, Yu. A., *Zh. Analit. Khim.*, **18**, 588 (1963).
[66a] Haffenden, W. T. and Lawson, G. T., *J. Inorg. Nucl. Chem.*, **29**, 1133 (1967).
[67] Chwastowska, J., *Doctoral Thesis*, Politechnika Warszawska (Warsaw Technical University). Warsaw, 1964.

[68] Schweitzer, G. K. and Norton, A. D., *Anal. Chim. Acta*, **30**, 119 (1964).
[69] Schweitzer, G. K. and van Willis, W., *Anal. Chim. Acta*, **30**, 114 (1964).
[70] Starý, J., *Proc. Intern. Conf., Gothenburg*, 1966, 1 (*Solvent Extraction Chemistry*, Dyrssen, D., Liljenzin, J.-O. and Rydberg, J., (Eds.), North-Holland, Amsterdam, 1967).
[71] Schweitzer, G. K. and Dyer, F. F., *Anal. Chim. Acta*, **22**, 172 (1960).
[72] Schweitzer, G. K. and Coe, G. R., *Anal. Chim. Acta*, **24**, 311 (1961).
[73] Schweitzer, G. K. and Mottern, J. L., *Anal. Chim. Acta*, **26**, 120 (1962).
[74] Chwastowska, J. and Minczewski, J., *Chem. Anal. (Warsaw)*, **8**, 157 (1963); **9**, 791 (1964).
[75] Chwastowska, J., Lissowska, K. and Sterlińska, E., *Chem. Anal. (Warsaw)*, **19**, 671 (1974).
[76] Hála, J., *Proc. Intern. Conf., Gothenburg*, 1966, 135 (see [70]).
[77] Lyle, S. J. and Shendrikar, A. D., *Anal. Chim. Acta*, **36**, 286 (1966).
[78] Alimarin, I. P., Petrukhin, O. M. and Zolotov, Yu. A., *Zh. Analit. Khim.*, **17**, 544 (1962).
[79] Gentry, G. H. and Sherrington, L. G., *Analyst (London)*, **71**, 432 (1946).
[80] Gentry, G. H. and Sherrington, L. G., *Analyst (London)*, **75**, 17 (1950).
[81] Minczewski, J., Maleszewska, H. and Steciak, T., *Acta Chim. Acad. Sci. Hung.*, **28**, 91 (1961).
[82] Kuznetsov, V. I. and Fan Min-E., *Zh. Neorgan. Khim.*, **5**, 1375 (1960).
[83] Levin, N. S., Yukin, Yu. M. and Vorosina, N. A., *Zh. Analit. Khim.*, **25**, 752 (1970).
[84] Navrátil, O., *J. Inorg. Nucl. Chem.*, **29**, 2007 (1967).
[85] Moučka, V. and Starý, J., *Collect. Czech. Chem. Commun.*, **26**, 763 (1961).
[86] Starý, J., *Collect. Czech. Chem. Commun.*, **25**, 890, 2630 (1960).
[87] Schweitzer, G. K. and van Willis, W., *Anal. Chim. Acta*, **36**, 77 (1966).
[88] Math, K. S., Fernando, Q. and Freiser, H., *Anal. Chem.*, **36**, 1763 (1964).
[89] Jensen, B. S., *Acta Chem. Scand.*, **13**, 1347 (1959).
[90] Starý, J., *Chem. Listy*, **53**, 556 (1959).
[91] Chwastowska, J., *Personal communication*.
[92] Vita, O. A., Levier, W. A. and Litteral, E., *Anal. Chim. Acta*, **42**, 87 (1968).
[93] Pimental, G. C., and McClellan, A. L., *The Hydrogen Bond*, Freeman, San Francisco, 1960.
[94] Alimarin, I. P. and Zolotov, Yu. A., *Talanta*, **9**, 891 (1962).
[95] Zolotov, Yu. A. and Alimarin, I. P., *Acta Chim. Acad. Sci. Hung.*, **32**, 327 (1962).
[96] Busev, A. I. and Ivanov, V. M., *Vestn. Moscow, Univ. Ser. II*, **15**, (3), 52 (1960).
[97] Begreev, V. V. and Zolotov, Yu. A., *Zh. Analit. Khim.*, **18**, 425 (1963).
[98] Zaharovskii, F. G. and Ryzhenko, V. L., *Zh. Analit. Khim.*, **22**, 1142 (1967).
[99] Kemula, W., Buchowski, H. and Lewandowski, R., *Bull. Acad. Polon. Sci., Ser. Sci. Chim.*, **12**, 267 (1964).
[100] Kemula, W., Buchowski, H. and Teperek, J., *Bull. Acad. Polon. Sci., Ser. Sci. Chim.*, **12**, 343, 347 (1964).
[101] Siekierski, S., *J. Inorg. Nucl. Chem.*, **24**, 205 (1962).
[102] Siekierski, S. and Olszer, R., *J. Inorg. Nucl. Chem.*, **25**, 1351 (1963).
[103] Siekierski, S., *Nukleonika*, **9**, 601 (1964).
[104] Olszer, R. and Siekierski, S., *J. Inorg. Nucl. Chem.*, **28**, 1991 (1966).

References

[105] Siekierski, S. and Narbutt, J., *Nukleonika*, **12**, 487 (1967).
[106] Narbutt, J., Dancewicz, D. and Halpern, A., *Radiochim. Acta*, **7**, 55 (1967).
[107] Mottola, H. A. and Freiser, H., *Talanta*, **13**, 55 (1966); **14**, 864 (1967).
[108] Wakahayashi, T., Oki, S., Omori, T. and Suzuki, N., *J. Inorg. Nucl. Chem.* **26**, 2255 (1964).
[109] Omori, T., Wakahayashi, T., Oki, S. and Suzuki, N., *J. Inorg. Nucl. Chem.*, **26**, 2265 (1964).
[110] Oki, S., Omori, T., Wakahayashi, T. and Suzuki, N., *J. Inorg. Nucl. Chem.*, **27**, 1141 (1965).
[111] Suzuki, N., Akiba, K., Kanno, T. and Wakahayashi, T., *J. Inorg. Nucl. Chem.*, **30**, 2521 (1968).
[112] Suzuki, N., Akiba, K., Kanno, T. Wakahayashi, T. and Takaizuni, K., *J. Inorg. Nucl. Chem.*, **30**, 3047 (1968).
[113] Suzuki, N., Akiba, K. and Kanno, T., *Anal. Chim. Acta*, **43**, 311 (1968).
[114] Suzuki, N., Akiba, K. and Asano, H., *Anal. Chim. Acta*, **52**, 115 (1970).
[115] Suzuki, N. and Akiba, K., *J. Inorg. Nucl. Chem.*, **33**, 1897 (1971).
[116] Akiba, K., Suzuki, N., Asano, H. and Kanno, T., *J. Radioanal. Chem.*, **7**, 203 (1971).
[117] Akiba, K., Suzuki, N. and Kanno, T., *Anal. Chim. Acta*, **58**, 379 (1972).
[117a] Allard, B., Johnsson, S., and Rydberg, J., *Proc. Int. Solv. Extr. Conf.*, *Lyon*, 1974, Vol. 2., London 1974, 1419.
[118] Dyer, F. F. and Schweitzer, G. K., *Anal. Chim. Acta*, **23**, 1 (1960).
[119] Kovtun L. V. and Rudenko, N. P., *Zh. Neorgan. Khim.*, **12**, 3123 (1967).
[120] Rudenko, N. P. and Kovtun, L. V., *Zh. Analit. Khim.*, **24**, 1390 (1969).
[121] Rudenko, N. P., Smirnov, N. N. and Kovtun, L. V., *Zh. Analit. Khim.*, **29**, 152 (1974).
[122] Jamil, M., Zur Nedden, P. and Duyckaerts, G., *Anal. Lett.*, **4**, 87 (1971).
[123] Kuznetsova, E. M., Panchenkov, G. M. and Klinovskaya, T. V., *Zh. Fiz. Khim.*, **44**, 2222 (1970).
[124] Hardy, C. J. and Scargill, D. J., *J. Inorg. Nucl. Chem.*, **11**, 128 (1959).
[125] Dyrssen, D. and Liem, D. H., *Acta Chem. Scand.*, **14**, 1091 (1960).
[126] Dyrssen, D., Ekberg, S. and Liem, D. H., *Acta Chem. Scand.*, **18**, 135 (1964).
[127] Krašovec, F., Ostanek, M. and Klofutar, C., *Anal. Chim. Acta*, **36**, 431 (1966).
[128] Dyrssen, D. and Liem, D. H., *Acta Chem. Scand.*, **18**, 224 (1964).
[129] Weissberger, A., Proskauer, E., Riddick, J. A. and Toops, E. E., *Organic Solvents*, Interscience, New York, 1955.
[130] Schweitzer, G. K. and Bramlitt, E. T., *Anal. Chim. Acta*, **23**, 419 (1960).
[131] Hellwege, H. E. and Schweitzer, G. K., *Anal. Chim. Acta*, **28**, 236 (1963).
[132] Bankovskii, Yu. A., Chera, L. M. and Ievin'sh, A. F., *Zh. Analit. Khim.*, **18**, 555 (1963).
[133] Zolotov, Yu. A., Alimarin, I. P. and Bodnya, V. A., *Zh. Analit. Khim.*, **30**, 750 (1964).
[134] Zolotov, Yu. A., Vorobeva, G. A. and Izosenkova, L. A., *Kinetika i mekhanizm élementarnogo akta ékstraktsii vnutrikompleksnykh soedinenii* (*The Kinetics and Mechanism of the Extraction of Internal Complexes*), GEOKhI, Akad. Nauk SSSR, Moscow, 1964.
[135] Alimarin, I. P., Zolotov, Yu. A. and Bodnya, V. A., *Pure Appl. Chem.*, **25**, 667 (1971).
[136] Fomin, V. V., *Radiokhimiya*, **17**, 744, 754 (1975).

[137] Fomin, V. V. and Leman, G. A., *Zh. Neorgan. Khim.*, **19**, 6, 1677 (1974).
[138] Irving, H., Bell, C. F. and Williams, R. J. P., *J. Chem. Soc.*, **1952**, 352.
[139] Kletenik, Yu. B. and Navrotskaya, V. A., *Zh. Neorgan. Khim.*, **12**, 3114 (1967).
[140] Navrotskaya, V. A., and Kletenik, Yu. B., *Zh. Neorgan. Khim.*, **14**, 1900 (1969).
[141] Honaker, C. B. and Freiser, H., *J. Phys. Chem.*, **66**, 127 (1962).
[142] Barnes, H., *Analyst (London)*, **72**, 469 (1947).
[143] Rudenko, N. P. and Kovtun, L. V., *Vestn. Mosk. Univ., Khim.* **24**, (4), 103 (1969); *Chem. Abstr.*, **71**, 95542f (1969).
[144] Schweitzer, G. K. and Benson, E. W., *Anal. Chim. Acta*, **30**, 79 (1964).
[145] Schweitzer, G. K. and Rimstidt, J. R., *Anal. Chim. Acta*, **27**, 389 (1962).
[146] Baumgärtner, F. and Finsterwalder, L., *Proc. 5th Intern. Conf. Solvent Extr. Chem.*, **1969**, 313, (*Solvent Extraction Research*, Kertes, A. S. and Marcus, Y., (Eds.), Wiley, New York, 1969).
[147] McClellan, B. E. and Freiser, H., *Anal. Chem.*, **36**, 2262 (1964).
[148] Oh, J. S. and Freiser, H., *Anal. Chem.*, **39**, 295 (1967).
[149] Subbaraman, P. R., Cordes, S. M. and Freiser, H., *Anal. Chem.*, **41**, 1878 (1969).
[150] Margerum, D. W., Zabin, B. A. and Jones, D. J., *Inorg. Chem.*, **5**, 250 (1966).
[151] Bydalek, T. G., *Inorg. Chem.*, **4**, 232 (1965).
[152] Hammes, G. H. and Stenifeld, V. I., *J. Am. Chem. Soc.*, **84**, 4639 (1964).
[153] Bolomey, R. A. and Wish, K., *J. Am. Chem. Soc.*, **72**, 4483 (1950).
[154] McKaveney, J. P. and Freiser, H., *Anal. Chem.*, **29**, 290 (1957).
[155] McClellan, B. E. and Sabel, P., *Anal. Chem.*, **41**, 1077 (1969).
[156] Pyatnitski, I. V. and Sereda, E. S., *Zh. Analit. Khim.*, **25**, 1552 (1970).
[157] Spivakov, B. Ya. and Zolotov, Yu. A., *Zh. Analit. Khim.*, **24**, 1773 (1969).
[158] Math, K. S., Bhatki, K. S. and Freiser, H., *Talanta*, **16**, 412 (1969).
[159] Freiser, B. S. and Freiser, H., *Talanta*, **17**, 540 (1970).
[160] Hamilton, H. G. and Freiser, H., *Anal. Chem.*, **41**, 1310 (1969).
[161] Math, K. S. and Freiser, H., *Anal. Chem.*, **41**, 1682 (1969).
[162] Woodward, C. and Freiser, H., *Anal. Chem.*, **40**, 345 (1968).
[163] Schulman, S. G. and Gershon, H., *Anal. Chim. Acta*, **50**, 348 (1970).
[164] Morpurgo, L. and Williams, R. J. P., *J. Chem. Soc. A*, **1966**, 73.
[165] Schulman, S. G. and Gershon, H., *J. Inorg. Nucl. Chem.*, **31**, 2467 (1969).
[166] Kawamoto, H. and Akaiwa, H., *Chem. Lett.*, **1973**, No. 3, 259.
[167] Blake, C. A., Baes, C. F., Brown, K. B., Coleman, C. F. and White, J. C., *Proc. 2nd Intern. Conf. Peaceful Uses At. Energy, Geneva*, **28**, 289 (1958).
[168] Blake, C. A., Horner, D. E. and Schmidt, J. M., *U.S. At. Energy, Comm. Rep.*, ORNL-2259 (1959).
[169] Zangen, M., *J. Inorg. Nucl. Chem.*, **25**, 1051 (1963).
[170] Mason, G. W., McCarty, S. and Peppard, D. F., *J. Inorg. Nucl. Chem.*, **24**, 967 (1962).
[171] Healy, T. V., *J. Inorg. Nucl. Chem.*, **19**, 314 (1961).
[172] Wang, S. M., Park, D. Y., and Li, N. C., *Proc. Intern. Conf., Gothenburg*, **1966**, 111 (see [70]).
[173] Healy, T. V., *Proc. Intern.. Conf., Gothenburg*, **1966**, 119 (see [70]).
[174] Upor, E., *Proc. 3rd Symp. Coord. Chem., Budapest*, **1**, 143 (1970).
[175] Dolgatova, N. V. and Fridman, Ya. D., *Zh. Analit. Khim.*, **27**, 1453 (1972).

References

[176] Siekierski, S. and Taube, M., *Nukleonika*, **6**, 489 (1961).
[177] Zangen, M., *J. Inorg. Nucl. Chem.*, **25**, 581 (1963).
[178] Hahn, H. T. and Vander Wall, E. M., *J. Inorg. Nucl. Chem.*, **26**, 191 (1964).
[179] Healy, T. V., *J. Inorg. Nucl. Chem.*, **19**, 328 (1961).
[180] Li, N. C., Wang, S. M. and Walker, W. R., *J. Inorg. Nucl. Chem.*, **27**, 2263 (1965).
[181] Zolotov, Yu. A. and Gavrilova, L. G., *J. Inorg. Nucl. Chem.*, **31**, 3613 (1969).
[182] Honjo, T., Horiuchi, M. and Kiba, T., *Bull. Chem. Soc. Japan*, **47**, 1176 (1974).
[183] Irving, H. M., *Proc. Intern. Conf.*, *Gothenburg*, **1966**, 91, (see [70]).
[183a] Sekine, T., Murai, R., Takahashi, K. K. and Iwahori, S., *Bull. Chem. Soc. Japan*, **50**, 3415 (1977).
[184] Irving, H. M. and Sinha, S. P., *Anal. Chim. Acta*, **51**, 39 (1970).
[185] Irving, H. M., *Proc. Symp. Coord. Chem.*, *Tihany*, **1965**, 219.
[186] Akaiwa, H. and Kawamoto, H., *Anal. Chim. Acta*, **48**, 438 (1969).
[187] Dyrssen, D., *Nucl. Sci. Eng.*, **16**, 448 (1963).
[188] Rao, A. P. and Dubey, S. P., *Talanta*, **18**, 1076 (1971).
[189] Rane, A. T., *J. Radioanal. Chem.*, **8**, 117 (1971).
[190] Rocca, J. L. and Porthault, M., *Anal. Chim. Acta*, **61**, 457 (1972).
[191] Subramanian, M. S. and Viswanathe, A., *J. Inorg. Nucl. Chem.*, **31**, 2575 (1969).
[192] Burgett, Ch. A., *Anal. Chim. Acta*, **67**, 325 (1973).
[193] Shigematsu, T., Tabushi, M., Matsui, M. and Honyo, T., *Bull. Chem. Soc. Japan*, **42**, 976 (1969).
[194] Bray, L. A., *Nucl. Sci. Eng.*, **20**, 362 (1964).
[194a] Noriki, S. and Nishimura, M., *Anal. Chim. Acta*, **94**, 57 (1977).
[195] Kurmaiah, N., Satyanarayana, D. and Rao, V. P. R., *Anal. Chim. Acta*, **35**, 484 (1966).
[196] Walker, W. R. and Li, N. C., *J. Inorg. Nucl. Chem.*, **27**, 2255 (1965).
[197] Sekine, T. and Dyrssen, D., *J. Inorg. Nucl. Chem.*, **26**, 1727 (1964).
[198] Zolotov, Yu. A. and Gavrilova, L. G., *Zh. Analit. Khim.*, **25**, 813 (1970).
[199] Zolotov, Yu. A., Demina, L. A. and Petrukhin, O. M., *Zh. Analit. Khim.*, **25**, 1487 (1970).
[200] Petrukhin, O. M., Zolotov, Yu. A. and Izosienkova, P. A., *Proc. 3rd Anal. Chem. Conf.*, *Budapest*, **1970**, 165.
[201] Jordanov, N., Mareva, St., Borisov, G. and Jordanov, B., *Talanta*, **15**, 221 (1968).
[202] Jordanov, N., Mareva, St. and Nguyen duc Thach, *Proc. 3rd Anal. Chem. Conf.*, *Budapest*, **1**, 159 (1970).
[203] Mareva, St., Jordanov, N. and Konstantinova, M., *Anal. Chim. Acta*, **59** 319 (1972).
[204] Konstantinova, M., Mareva, St. and Jordanov, N., *Anal. Chim. Acta*, **68**, 237 (1974).
[205] Irving, H. and Edgington, D. N., *J. Inorg. Nucl. Chem.*, **15**, 158 (1960); **20**, 314, 321 (1961); **21**, 169 (1961).
[206] Healy, T. V. and Ferraro, J. R., *J. Inorg. Nucl. Chem.*, **24**, 1449 (1962).
[207] Walker, W. R. and Li, N. C., *J. Inorg. Nucl. Chem.*, **27**, 411 (1965).
[208] Ferraro, J. R., *J. Inorg. Nucl. Chem.*, **26**, 225 (1964).
[209] Gal, I. J. and Nikolić, R. M., *J. Inorg. Nucl. Chem.*, **28**, 563 (1966).
[210] Newman, L., *J. Inorg. Nucl. Chem.*, **25**, 304 (1963).

[211] Noskova, M. P., Zolotov, Yu. A. and Gribov, L. A., *Zh. Analit. Khim.*, **25**, 220 (1970).
[212] Ferraro, R. and Healy, T. V., *J. Inorg. Nucl. Chem.*, **24**, 1463 (1962).
[213] Aggett, J., *Chem. Ind. (London)*, **1966**, 27.
[214] Agget, J., *J. Inorg. Nucl. Chem.*, **32**, 2767 (1970).
[215] Healy, T. V., *Proc. Intern. Conf.*, Gothenburg, **1966**, 119, (see [70]).
[216] McDowell, W. J. and Coleman, C. F., *J. Inorg. Nucl. Chem.*, **27**, 1117 (1965).
[217] McDowell, W. J. and Coleman, C. F., *J. Inorg. Nucl. Chem.*, **28**, 1083 (1966).
[218] Healy, T. V., Peppard, D. F. and Mason, G. W., *J. Inorg. Nucl. Chem.*, **24**, 1429 (1962).
[219] Irving, H. M. and Al-Niaimi, N. S., *J. Inorg. Nucl. Chem.*, **27**, 717 (1965).
[220] Wang, S. M., Walker, W. R. and Li, N. C., *J. Inorg. Nucl. Chem.*, **28**, 875 (1966).
[221] Subramanian, M. S. and Pai, S. A., *J. Inorg. Nucl. Chem.*, **32**, 3677 (1970).
[221a] Umland, F., *Zh. Analit. Khim.*, **33**, 612 (1978).
[222] Rahaman, M. S. and Finston, M. L., *Anal. Chem.*, **41**, 2023 (1969).
[223] Zolotov, Yu. A. and Gavrilova, L. G., *Radiokhimiya*, **11**, 389 (1969).
[224] Irving, H. M. and Edgington, D. N., *J. Inorg. Nucl. Chem.*, **27**, 1359 (1965).
[224a] Aly, H. F., Raieh, M., Mohamed, S. and Abdel-Rassoul, A. A., *J. Inorg. Nucl. Chem.*, **40**, 567 (1978).
[225] Tomažič, B. B. and O'Laughlin, J., *Anal. Chem.*, **45**, 1519 (1973).
[226] Akaiwa, H., *Kagaku To Koge*, **26**, 200 (1973).
[227] Chmutova, M. K. and Kochetkova, N. E., *Zh. Analit. Khim.*, **24**, 1757 (1969).
[228] Chmutova, M. K. and Kochetkova, N. E., *Zh. Analit. Khim.*, **25**, 710 (1970).
[229] Goffart, J. and Duyckaerts, G., *Anal. Chim. Acta*, **48**, 99 (1969).
[230] Bykhovtsov, V. L., *Radiokhimiya*, **12**, 539 (1970).
[230a] Khopkar, P. K. and Mathur, J. N., *J. Inorg. Nucl. Chem.*, **39**, 2063 (1977).
[231] Taketatsu, T. and Banks, Ch. V., *Anal. Chem.*, **38**, 1524 (1966).
[232] Sinha, S. T. and Irving, H. M., *Anal. Chim. Acta*, **52**, 193 (1970).
[233] Kassierer, F. and Kertes, A. S., *Proc. 3rd Symp. Coord. Chem.*, **1**, 119 (1970).
[233a] Gerow, I. H., Hayden, J. G., Gaggar, K. and Davis, M. W., *Sep. Sci.*, **12**, 511 (1977).
[233b] Honjo, T., Matsui, M. and Shigematsu, T., *Bull. Inst. Chem. Res. Kyoto Univ.*, **55**, 423 (1977).
[233c] Tashimori, S. and Nakamura, H., *J. Radioanal. Chem.*, **44**, 37 (1978).
[234] Gavrilova, L. G. and Zolotov, Yu. A., *Zh. Analit. Khim.*, **25**, 1054 (1970).
[235] Zolotov, Yu. A., Petrukhin, O. M. and Gavrilova, L. G., *J. Inorg. Nucl. Chem.*, **32**, 1679 (1970).
[236] Irving, H. M. and Edgington, D. N., *J. Inorg. Nucl. Chem.*, **21**, 168 (1961).
[237] Gorbenko, F. P. and Sachko, V. V., *Zh. Analit. Khim.*, **25**, 1884 (1970).
[238] Boguszewska, Z., *Doctoral Thesis*, Politechnika Warszawska (Warsaw Technical University), 1971.
[239] Remov, A. G. and Sovotovich, E. V., *Radiokhimiya*, **12**, 182 (1970).
[240] Shigematsu, T., Matsui, M. and Wake, R., *Anal. Chim. Acta*, **46**, 100 (1969).
[241] Deptuła, C. and Minc, S., *Nukleonika*, **10**, 343, 421 (1965).
[242] Deptuła, C. and Minc, S., *J. Inorg. Nucl. Chem.*, **29**, 159, 221, 229, 1097 (1967).
[243] Deptuła, C., *J. Inorg. Nucl. Chem.*, **32**, 277 (1970).
[244] Newman, L. and Klotz, P., *J. Phys. Chem.*, **67**, 205 (1963); **70**, 461 (1966).

References

[245] Shakhova, N. V., Alimarin, I. P. and Zolotov, Yu. A., *Dokl. Akad. Nauk SSSR*, **152**, 884 (1963).
[246] Alimarin, I. P., Zolotov, Yu. A. and Shakhova, N. V., *Tr. Kom. Anal. Khim., Akad. Nauk SSSR*, **14**, 24 (1963).
[247] Alekperov, R. A. and Markov, N. N., *Zh. Analit. Khim.*, **23**, 460 (1968).
[248] Alimarin, I.P., Gibalo, I. M. and Pigaga, A. K., *Zh. Analit. Khim.*, **25**, 2336 (1970).
[249] Alimarin, I. P., Gibalo, I. M. and Pigaga, A. K., *Dokl. Akad. Nauk SSSR* **186**, 1323 (1969).
[250] Alimarin, I. P., Pigaga, A. K. and Gibalo, I. M., *Vestn. Mosk. Univ., Khim.*, **24**, 94 (1969).
[251] Kysh, M., Slutski, P. and Pishtek, P., *Zh. Neorgan. Khim.*, **10**, 2764 (1965).
[252] Keil, R., *Z. Anal. Chem.*, **245**, 362 (1969).
[253] Dyrssen, D., *J. Inorg. Nucl. Chem.*, **8**, 291 (1958).
[254] Shakhova, N. V., Ryabakova, E. V. and Zolotov, Yu. A., *Zh. Analit. Khim.*, **29**, 682 (1974).
[255] Irving, H., Rossotti, F. and Williams, R., *J. Chem Soc.*, **1955**, 1906.
[256] Diamond, R. R. and Tuck, D. G., *Prog. Inorg. Chem.*, **2**, 158 (1960).
[257] Marcus, Y. and Kertes, A. S., *Ion Exchange and Solvent Extraction of Metal Complexes*, Wiley-Interscience, New York, 1969.
[258] Irving, H. and Lewis, D., *Ark. Kem.*, **28**, 131 (1967).
[258a] Irving, H. M. N. H. and Lewis, D., *Ark. Kem.*, **32**, 121 (1970).
[258b] Irving, H. M. N. H. and Lewis, D., *Ark. Kem.*, **32**, 131 (1970).
[259] White, J. M., Kelly, P. and Li, N. C., *J. Inorg. Nucl. Chem.*, **16**, 337 (1961).
[260] Campbell, D. E., Lawrence, A. H. and Clark, H. M., *J. Am. Chem. Soc.*, **74**, 6193 (1952).
[261] Nekrasov, B. V. and Ovsyankina, V. V., *Zh. Obshch. Khim.*, **11**, 573 (1941).
[262] Wicke, F., Eigen, M. and Ackermann, T., *Z. Phys. Chem.*, (*Frankfurt*), **1**, 340 (1954).
[263] Vdovenko, V. M., Lipovskii, A. A. and Kuzina, M. G., *Zh. Neorgan. Khim.*, **2**, 975 (1957).
[264] Diamond, R. M., *J. Phys. Chem.*, **62**, 659 (1959).
[265] Fuoss, R. M. and Kraus, C. A., *J. Am. Chem. Soc.*, **55**, 2387 (1933).
[266] Meyers, R. J. and Metzler, D. E., *J. Am. Chem. Soc.*, **72**, 3772 (1950).
[267] Nachtrieb, N. H. and Fryxell, R. E., *J. Am. Chem. Soc.*, **74**, 897 (1952).
[268] Saldick, J., *J. Phys. Chem.*, **60**, 500 (1956).
[269] Diamond, R. M., *J. Phys. Chem.*, **61**, 69, 75, 1552 (1957).
[270] Friedman, H. L., *J. Am. Chem. Soc.*, **74**, 5 (1952).
[271] Friedman, H. L. and Haugen, G. R., *J. Am. Chem. Soc.*, **76**, 2060 (1954).
[272] Goble, A. G. and Maddock, A., *Trans. Faraday Soc.*, **55**, 591 (1959).
[273] Goble, A. G. and Maddock, A., *J. Inorg. Nucl. Chem.*, **7**, 94 (1958).
[274] Pierotti, R. A., *J. Phys. Chem.*, **67**, 1840 (1963).
[275] Millen, W. A. and Watts, D. W., *J. Am. Chem. Soc.*, **89**, 6051 (1967).
[276] Ritson, D. M. and Hasted, J. B., *J. Chem. Phys.*, **16**, 11 (1948).
[277] Schellman, J. A., *J. Chem. Phys.*, **26**, 1225 (1957).
[278] Buckingham, A. D., *Discuss. Faraday Soc.*, **24**, 151 (1957).
[279] Feakins, D., *Physico-chemical Processes in Mixed Aqueous Solvents*, Heinemann, London, 1967.
[280] Gordy, W. and Stanford, S. C., *J. Chem. Phys.*, **9**, 204 (1941).

[281] Zolotov, Yu. A. and Golovanov, V. I., *Dokl. Akad. Nauk SSSR*, **191**, 92 (1970); **193**, 626 (1970).
[282] Zolotov, Yu. A. and Golovanov, V. I., *Zh. Analit. Khim.*, **25**, 610 (1970).
[283] Zolotov, Yu. A. and Golovanov, V. I., *Zh. Neorgan. Khim.*, **17**, 1118 (1972).
[284] Sokolov, A. B. and Zolotov, Yu. A., *Zh. Neorgan. Khim.*, **17**, 1123 (1972).
[285] Zolotov, Yu. A. and Prokoshev, A. A., *Zh. Analit. Khim.*, **26**, 2307 (1971); **28**, 629 (1973).
[286] Zolotov, Yu. A. and Sultanova, Z. Kh., *Chem. Anal.*, (*Warsaw*), **17**, 1113 (1972).
[287] Denisov, E. G. and Khanovin, A. E., *Zh. Prikl. Khim.*, **38**, 296 (1965).
[288] McCorkell, R. H. and Irvine, J. W., *Can. J. Chem.*, **46**, 662 (1968).
[289] Zolotov, Yu. A., Golovanov, V. I. and Vanifatova, N. G., *Zh. Analit. Khim.*, **28**, 5 (1973).
[290] Zolotov, Yu. A., Bagreev, V. V. and Rebenko, V. G., *Zh. Analit. Khim.*, **27**, 184 (1972).
[290a] Popandopulo, Yu. I., Bagreev, V. V. and Zolotov, Yu. A., *J. Inorg. Nucl. Chem.*, **39**, 2257 (1977).
[290b] Bargeev, V. V., Fischer, C., Yudushkina, L. M. and Zolotov, Yu. A., *J. Inorg. Nucl. Chem.*, **40**, 553 (1978).
[291] Srinivasan N., Nadkarni, N. M., Rao, M. K., Ramanajan, A., Venkatesen, M. and Gopalkrishnan, V., *India At. Energy Comm., Bhabha At. Res. Cent. Rept.*, **1969**, B.A.R.C.-432; *Chem. Abstr.*, **73**, 81296h (1970).
[292] Komarov, E. V., *Radiokhimiya*, **12**, 306, 312 (1970).
[293] Likhtenshein, I. I., *Zh. Fiz. Khim.*, **44**, 1988 (1970).
[294] Garwin, L. and Hixon, A. N., *Ind. Eng. Chem.*, **41**, 2298 (1949).
[295] Irving, H. and Williams, R. J. P., in *Treatise on Analytical Chemistry*, 1st Ed., Pt. 1, Vol. 3, Kolthoff, I. M. and Elving, P. J., (Eds.), Interscience, New York, 1959, p. 1309.
[296] Specker, H., Jackwerth, E. and Hovermann, G., *Z. Anal. Chem.*, **177**, 10 (1960).
[296a] Sergievskii, V. V., *Radiokhimiya*, **20**, 396, 400, 506 (1978).
[297] Connich, R. E. and Hugus, Z. Z., Jr., *J. Am. Chem. Soc.*, **74**, 6012 (1952).
[298] Glueckauf, E., McKay, H. A. C. and Mathieson, A. R., *Trans. Faraday Soc.*, **47**, 437 (1951).
[299] Gardner, A. W., McKay, H. A. C. and Warren, D. T., *Trans. Faraday Soc.*, **48**, 997 (1952).
[300] Katzin, L. I., *Nature* (*London*), **166**, 605 (1950).
[301] Katzin, L. I. and Sullivan, J. C., *J. Phys. Chem.*, **55**, 346 (1951).
[302] McKay, H. A. C., *Trans. Faraday Soc.*, **48**, 1103 (1952).
[303] McKay, H. A. C., *J. Inorg. Nucl. Chem.*, **4**, 375 (1957).
[304] Moore, T. E., Laran, R. J. and Yates, P. C., *J. Phys. Chem.*, **59**, 90 (1955).
[305] Widmar, M., *J. Phys. Chem.*, **74**, 3618 (1970).
[306] Chuchalin, L. K., Grankina, Z. A. and Peshchevski, B. I., *Zh. Fiz. Khim.*, **44**, 1455 (1970).
[307] Goto, H., Kakita, Y. and Ichinose N. *Nippon Kagaku Zasshi*, **88**, 640 (1967).
[308] Ichinose, N., *Talanta*, **18**, 21, 105 (1971).
[309] Katzin, L. J. and Ferraro, J. R., *J. Am. Chem. Soc.*, **72**, 5451 (1950); **75**, 3825 (1953).
[310] Ferraro, J. R., Katzin, L. I. and Gibson, G., *J. Inorg. Nucl. Chem.*, **2**, 118 (1956).

[311] Katzin, L. I. and Ferraro, J. R., *J. Am. Chem. Soc.*, **75**, 3821 (1953).
[312] Katzin, L. I., Simon, D. M. and Ferraro, J. R., *J. Am. Chem. Soc.*, **74**, 1191 (1952).
[313] Burger, L. L., *J. Phys. Chem.*, **62**, 590 (1958).
[314] Tuck, D. G., *J. Chem. Soc.*, **1959**, 218.
[315] Pagel, H. A. and McLafferty, F. W., *Anal. Chem.*, **20**, 278 (1948).
[316] Tuck, D. G., *J. Chem. Soc.*, **1958**, 2783.
[317] Moore, F. L., *U.S. At. Energy Comm. Rept.*, AECD-3196 (1959).
[318] Hesford, E. and McKay, H. A. C., *Trans. Faraday Soc.*, **54**, 573 (1958).
[319] Nemodruk, A. A. and Glukhova, L. P., *Zh. Neorgan. Khim.*, **8**, 2618 (1963).
[320] Yoshida, H., *J. Inorg. Nucl. Chem.*, **24**, 1257 (1962).
[321] Shevchenko, V. B., Radionov, A. V., Solovkin, A. S., Shilin, I. V., Kirilov, L. M. and Balandina, V. V., *Radiokhimiya*, **1**, 257 (1959).
[322] Shevchenko, V. B., Solovkin, A. S., Shilin, I. V. Kirilov, L. M., Radionov, A. V. and Balandina, V. V., *Radiokhimiya*, **2**, 281 (1960).
[323] Shevchenko, V. B., Solovkin, A. S., Kirilov, L. M. and Ivanchev, A. I., *Radiokhimiya*, **3**, 503 (1961).
[324] Kertes, A. A., *Solvent Extraction Chemistry of Metals*, Macmillan, London 1966, p. 377.
[325] Ulianov, V. C. and Sviridova, R. A., *Radiokhimiya*, **12**, 47 (1970).
[326] Keikichi, N., *J. Inorg. Nucl. Chem.*, **32**, 2265 (1970).
[327] Taube, M., *J. Inorg. Nucl. Chem.*, **12**, 174 (1959); **15**, 171 (1960).
[328] Scargill, D., Alcock, K., Fletcher, J. M., Hesford, E. and McKay, H. A. C. *J. Inorg. Nucl. Chem.*, **4**, 304 (1957).
[329] Best, G. F., McKay, H. A. C. and Woodgate, P. R., *J. Inorg. Nucl. Chem.*, **4** 315 (1957).
[330] Hesford, E., McKay, H. A. C. and Scargill, D., *J. Inorg. Nucl. Chem.*, **4**, 321 (1957).
[331] Solovkin, A. S., *Zh. Neorgan. Khim.*, **5**, 2119 (1960).
[332] Kuznetsov, V. N., *Usp. Khim.*, **23**, 654 (1954).
[333] Vdovenko, V. M. and Kovaleva, G. V., *Radiokhimiya*, **2**, 1682 (1957).
[334] Golovatenko, R. T., *Radiokhimiya*, **8**, 2395 (1963).
[335] Sato, T., *J. Inorg. Nucl. Chem.*, **16**, 157 (1960).
[336] Ionov, V. P. and Tikhomirov, V. I., *Radiokhimiya*, **5**, 559 (1963).
[337] Adamskii, N. M., Karpasheva, S. M., Melnikov, I. N. and Rozen, A. M., *Radiokhimiya*, **2**, 400 (1960).
[338] Zingaro, R. A. and White, J. C., *J. Inorg. Nucl. Chem.*, **12**, 315 (1960).
[339] Siddall, T. H., *Proc. 2nd Intern. Conf. Peaceful Uses At. Energy, Geneva*, **17**, 339 (1958).
[340] Golovatenko, R. T. and Samoilov, O. Ya., *Radiokhimiya*, **4**, 25 (1962).
[341] Fletcher, J. M. and Hardy, C. J., *Nucl. Sci. Eng.*, **16**, 421 (1963).
[342] Peppard, D. F., Mason, G. W. and Gergel, M. V., *J. Inorg. Nucl. Chem.*, **3**, 370 (1957).
[343] Eberle, A. R. and Lerner, M. W., *Anal. Chem.*, **29**, 1134 (1957).
[344] Larsen, R. P. and Seils, C. A., *Anal. Chem.*, **32**, 1863 (1960).
[345] Kertes, A. S. and Halpern, M., *J. Inorg. Nucl. Chem.*, **16**, 308 (1961).
[346] François, C. A., *Anal. Chem.*, **30**, 50 (1958).
[347] Horton, C. A. and White, J. C., *Anal. Chem.*, **30**, 1779 (1958).
[348] Habashi, F., *Talanta*, **2**, 380 (1959).

[349] Yoe, J. H., Will, F., III and Black, R. A., *Anal. Chem.*, **25**, 1200 (1953).
[350] Blanquet, P., *Anal. Chim. Acta*, **16**, 44 (1957).
[351] Clinch, J. and Guy, M. J., *Analyst (London)*, **82**, 800 (1957).
[352] Rodden, C. J., *Analysis of Essential Nuclear Reactor Materials*, Division of Techn. Inform. U.S. At. Energy Comm., Washington, 1964.
[353] Siekierski, S. and Gwóźdź, R., *Nukleonika*, **5**, 205 (1960).
[354] Gwóźdź, R., *Doctoral Thesis*, IBJ, Warsaw, 1963.
[355] Olenkovich, N. L., Mazurenko, E. A., Ermilova, V. I. and Rogachko, M. M. *Zavodsk. Lab.*, **30**, 389 (1964).
[356] Moore, F. L., *Anal. Chem.*, **37**, 1235 (1965).
[357] Moore, F. L., *Anal. Chem.*, **29**, 1660 (1957).
[358] Kuiper, D., *Proc. 5th Intern. Conf. Solvent Extr. Chem.*, **1969**, 227 (see [146]).
[359] Frolov, Yu. G., Ochkin, A. V. and Sergevskii, V. V., *Zh. Analit. Khim.*, **25**, 1433 (1970).
[360] Seeley, F. G. and Crouse, D. J., *J. Chem. Eng. Data*, **11**, 424 (1966).
[361] Awwal, M. A., *Nucleus*, **8**, 19 (1971).
[362] Coleman, C. F., Brown, K. B., Moore, J. G., Allen, K. A., *Proc. 2nd Intern. Conf. Peaceful Uses At. Energy*, Geneva, **15**, (P) 510 (1958).
[363] Rozen, A. M. and Nagibeda, Z. I., *Radiokhimiya*, **13**, 284 (1971).
[363a] Shade, W., *Z. Chem.*, **17**, 302 (1977).
[363b] Schmidt, V. S., Shesterikov, V. N. and Rubisov, V. N., *Radiokhimiya*, **20**, 35, 47 (1978).
[364] Good, M. L. and Bryan, S. E., *J. Am. Chem. Soc.*, **82**, 5636 (1960).
[365] Moore, F. L., *Liquid–Liquid Extraction with High Molecular Weight Amines*, Washington, 1960.
[366] Crouse, D. J. and Brown, K. B., *Ind. Eng. Chem.*, **51**, 1461 (1959).
[367] Coleman, C. F., *Nucl. Sci. Eng.*, **17**, 274 (1963).
[368] Grinstead, R. R., *Proc. Intern. Conf.*, Gothenburg, **1966**, 426 (see [70]).
[369] Makarov, V. M., *Radiokhimiya*, **12**, 584 (1970).
[370] Shmidt, V. S. and Mezhov, E. A., *Radiokhimiya*, **12**, 38 (1970).
[371] Shmidt, V. S., Mezhov, E. A. and Shesterikov, V. N., *Radiokhimiya*, **12**, 590 (1970).
[372] Bucher, J. J. and Diamond R. M. *J. Phys. Chem.*, **69**, 1565 (1963).
[373] Pushlenkov, M., Komarov, E. V. and Shuvalov, O. N., *Radiokhimiya*, **4**, 543 (1963).
[374] Dyrssen, D. and Petkovic, Dj., *J. Inorg. Nucl. Chem.*, **27**, 1381 (1963).
[375] Diamond, R. M., *Proc. Intern. Conf.*, Gothenburg, **1966**, 349 (see [70]).
[376] Smułek, W., *Nukleonika*, **7**, 547 (1962).
[377] Smułek, W. and Siekierski, S., *J. Inorg. Nucl. Chem.*, **24**, 1651 (1962).
[378] Smułek, W., *Doctoral Thesis*, IBJ, Warsaw, 1962.
[379] Brown, K. B., Coleman, C. F., Crouse, D. J. and Ryon, A. D., *U.S. At. Energy Comm. Rept.*, ORNL-2399, 30 (1957).
[380] McDowell, W. J. and Baes, C. F., *J. Phys. Chem.*, **62**, 777 (1958).
[381] Webster, R. K., Dance, D. F. and Morgan, J. W., *Anal. Chim. Acta*, **23**, 101 (1960).
[382] Pyatnitskii, N.V. and Kharienko, R. S., *Ukr. Khim. Zh.*, **28**, 1115 (1962); **29**, 967 (1963); **30**, 635 (1964); **31**, 714 (1965); **32**, 503 (1966); **33**, 734 (1967); **34**, 178 (1968).
[383] Pyatnitskii, I. V. and Tabenskaya, T. V., *Zh. Analit. Khim.*, **25**, 953, 2390 (1970).

[384] Irving, H. and Damodaran, A. D., *Analyst (London)*, **90**, 443 (1965).
[385] Irving, H. and Damodaran, A. D., *Anal. Chim. Acta* **53**, 267 (1971).
[386] Coleman, C. F., Blake, C. A. and Brown, K. B., *Talanta*, **9**, 297 (1962).
[387] Kuzima, M. G. and Lipovskii, A. A., *Radiokhimiya*, **12**, 393 (1970).
[388] Semenov, V. A., Laskorin, V. N. and Skorovarov, D. I., *Zh. Prikl. Khim.*, **43**, 1740 (1970).
[389] Bac, R., *J. Inorg. Nucl. Chem.*, **32**, 3655 (1970).
[390] Gagliardi, E. and Wieland, H., *Mikrochim. Acta*, **1969**, 960.
[391] Sandell, E. B., *Colorimetric Determination of Traces of Metals*, 3rd Ed., Interscience, New York, 1959.
[391a] Sandell, E. B. and Onishi, H., *Photometric Determination of Traces of Metals*, 4th Ed., Part I, *General Aspects*, Wiley, New York, 1978.
[392] Marczenko, Z., *Spectrophotometric Determination of Elements*, Horwood, Chichester, 1976.
[392a] Cresser, M. S., *Solvent Extraction in Flame Spectroscopic Analysis*, Butterworths, London, 1978.
[392b] Fritz, J. S., *Pure Appl. Chem.*, **49**, 1547 (1977).
[393] Koshimura, H. and Okubo, T., *Anal. Chim. Acta.*, 49, 67 (1970).
[394] Mitchell, J. W. and Banks, C. V., *J. Inorg. Nucl. Chem.*, **31**, 2105 (1969).
[395] Brown, W. B., Steinbeck, J. F. and Wagner, W. F., *J. Inorg. Nucl. Chem.*, **13**, 119 (1968).
[396] Krishen, A. and Freiser, H., *Anal. Chem.*, **29**, 288 (1957).
[397] Krishen, A. and Freiser, H., *Anal. Chem.*, **31**, 923 (1959).
[398] Steinbach, J. F. and Freiser, H., *Anal. Chem.*, **25**, 881 (1953).
[399] Steinbach, J. F. and Freiser, H., *Anal. Chem.*, **26**, 375 (1954).
[400] Shigematsu, T. and Tabushi, M., *Bull. Inst. Chem. Res., Kyoto Univ.*, **39**, 35 (1961).
[401] Chalmers, R. A. and Dick, D. M., *Anal. Chim. Acta*, **31**, 520 (1964).
[402] Khady, A., Rudenko, N. P., Kuznetsov, V. I. and Gudym, L. S., *Talanta*, **18**, 279 (1971).
[403] Buchman, J. D., *J. Inorg. Nucl. Chem.*, **7**, 140 (1958).
[404] Alimarin, I. P. and Zolotov, Yu. A., *Zh. Analit. Khim.*, **11**, 389 (1956).
[405] Green, R. W. and Alexander, P. W., *J. Phys. Chem.*, **67**, 905 (1963).
[406] Suzuki, N. and Oki, S., *Bull. Chem. Soc. Japan*, **35**, 233 (1962).
[407] Zolotov, Yu. A., *Acta Chim. Acad. Sci. Hung.*, **32**, 327 (1962).
[408] Zharovskii, F. G., *Ukr. Khim. Zh.*, **25**, 245 (1959).
[409] Van Erkelens, P. C., *Anal. Chim. Acta*, **25**, 129 (1961).
[410] Starý, J. and Růžička, J., *Talanta*, **8**, 775 (1961).
[411] Abrahamczik, E., *Mikrochim. Acta*, **33**, 209 (1947).
[412] Abrahamczik, E., *Angew. Chem.*, **61**, 96 (1949).
[413] Tabushi, M., *Bull. Inst. Chem. Res., Kyoto Univ.*, **37**, 232, 237, 245 (1959).
[414] Kenny, A. W., Maton, W. R. E. and Spragg, W. T., *Nature (London)*, **165**, 483 (1950).
[415] Wold, A., Baird, J. H. and Hough, Ch. R., *Anal. Chem.*, **26**, 546 (1954).
[416] Bulgakova, A. M. and Volkova, A. M., *Zh. Analit. Khim.*, **16**, 715 (1961).
[417] Peshkova, V. M. and Péi Ai, *Zh. Analit. Khim.*, **16**, 2082 (1961).
[418] Rydberg, J., *Ark. Kemi*, **9**, 95 (1956).
[419] Rudenko, N. P., and Starý, J., *Tr. Kom. Anal. Khim. Akad. Nauk SSSR*, **9**, 28 (1958).

[420] Rudenko, N. P. and Starý, J., *Radiokhimiya*, **1**, 52, 700 (1959).
[421] Shigematsu, T. and Tabushi, M., *Nippon Kagaku Zasshi*, **83**, 814 (1962).
[422] Grubitsch, H. and Heggebö, T., *Monatsh. Chem.*, **93**, 274 (1962).
[423] Suzuki, N. and Omori, T., *Bull. Chem. Soc. Japan*, **35**, 595 (1962).
[424] Maddock, A. G. and Miles, G. L., *J. Chem. Soc.*, **1949**, S 248.
[425] Rydberg, J., *Ark. Kemi*, **9**, 109 (1956).
[426] Wilkinson, G., *J. Am. Chem. Soc.*, **74**, 6146 (1952).
[427] Spitsin, V. I., Kuzina, A. F., Zamoshnikova, N. N. and Galil', T. S., *Dokl. Akad. Nauk SSSR*, **144**, 1066 (1962); **5**, 517 (1953).
[428] Rydberg, J., *Ark. Kemi*, **5**, 413 (1953).
[429] McDowell, B. L., Meyer, A. S., Feathers, R. E. and White, J. C., *Anal. Chem.*, **31**, 931 (1959).
[430] West, P. W. and Mukherji, A. K., *Anal. Chem.*, **31**, 947 (1959).
[431] Rydberg, J. and Rydberg, B., *Ark. Kemi*, **9**, 81 (1956).
[432] McKaveney, J. P. and Fresiser, H., *Anal. Chem.*, **30**, 526 (1958).
[433] Satyanarayana, D., Kurmaiah, N. and Rao, V. P. R., *Chemist-Analyst*, **53**, 78 (1964).
[434] Rao, V.P.R. and Satyanarayana, D., *Indian J. Chem.*, **3**, 40 (1965).
[435] Schweitzer, G. K. and Scott, H. E., *J. Am. Chem. Soc.*, **77**, 2753 (1955).
[436] Schweitzer, G. K. and Bishop, W. N., *J. Am. Chem. Soc.*, **76**, 4321 (1954).
[437] Sheperd, E. and Meinke, W. W., *U.S. At. Energy Comm. Rept.*, AECU-3879 (1958).
[438] De, A. K. and Khopkar, S. M., *J. Sci. Ind. Res. India*, **21A**, 131 (1962).
[438a] Ramanujam, A., Nadkarni, M. N., Ramakrishna, V. V. and Patil, S. K., *J. Radioanal. Chem.*, **42**, 349 (1978).
[439] Hagemann, F., *J. Am. Chem. Soc.*, **72**, 768 (1950).
[440] Eshelman, H. C., Dean, J. A., Menis, D. and Rains, T. C., *Anal. Chem.*, **31**, 183 (1959).
[441] Magnusson, L. B. and Anderson, M. L., *J. Am. Chem. Soc.*, **76**, 6207 (1954).
[442] Rangnekar, A. V. and Khopkar, S. M., *Z. Anal. Chem.*, **230**, 425 (1967).
[443] Levy, H. A. and Broido, A., *U.S. At. Energy Comm. Rept.*, CNL-37 (1948).
[444] Solanke, K. R. and Khopkar, S. M., *Anal. Lett.*, **6**, 31 (1973).
[445] Moore, F. L., *Anal. Chem.*, **38**, 1872 (1966).
[446] Kiba, T. and Mizukami, S., *Bull. Chem. Soc. Japan*, **31**, 1007 (1958).
[447] Akaze, L., *Bull. Chem. Soc. Japan*, **39**, 971, 980 (1966).
[448] Sekine, T. and Dyrssen, D., *Anal. Chim. Acta*, **37**, 217 (1967).
[449] Schweitzer, G. K. and Randolph, D. R., *Anal. Chim. Acta*, **26**, 567 (1962).
[450] Khopkar, S. M. and De, A. K., *Anal. Chem.*, **32**, 478 (1960).
[451] Onishi, H. and Banks, C. V., *Anal. Chem.*, **35**, 1887 (1963).
[452] Suttle, J. F., *U.S. At. Energy Comm. Rept.*, AECD-741 (1950).
[453] Onishi, H. and Toita, Y., *Bunseki Kagaku*, **21**, 756 (1972).
[454] De, A. K. and Sahu, Ch. R., *Zh. Analit. Khim.*, **25**, 1759 (1970).
[455] Majumdar, S. K. and De, A. K., *Anal. Chem.*, **32**, 1337 (1960).
[456] De, A. K. and Sahu, Ch., *J. Inorg. Nucl. Chem.*, **31**, 2257 (1969).
[457] Crowther, P. and Moore, F. L., *Anal. Chem.*, **35**, 2081 (1963).
[458] Bronaugh, H. J. and Suttle, J. F., *U.S. At. Energy Comm. Rept.*, LA-1561 (1953).
[459] Rains, T. C., House, H. P. and Menis, O., *Anal. Chim. Acta*, **22**, 315 (1960).
[460] Akiba, K., *Bunseki Kagaku*, **21**, 1630 (1972).

References

[461] Berg, E. W. and McIntyre, R. T., *Anal. Chem.*, **27**, 195 (1955).
[462] Khopkar, S. M. and De, A. K., *Anal. Chim. Acta*, **22**, 223 (1960).
[463] Testa, C., *Anal. Chim. Acta*, **25**, 525 (1961).
[464] Moore, F. L., Fairman, W. D., Ganchoff, J. G. and Surak, J. G., *Anal. Chem.*, **31**, 1148 (1959).
[465] Berg, E. W. and McIntyre, R. T., *Anal. Chem.*, **26**, 813 (1954).
[466] Cefola, M., Andrus, W. S., Miccoli, B. R. and Janowski, L. K., *Mikrochim. Acta*, **35**, 439 (1950).
[467] Maeck, W. J., Booman, G. L., Kussy, M. E. and Rein, J. E., *Anal. Chem.*, **32**, 1874 (1960); **33**, 1775 (1961).
[468] McBride, J. P., *U.S. At. Energy Comm. Rept.*, AECU-339 (1949).
[469] McCarty, D. C., Dearing, B. E. and Flagg, J. F., *U.S. At. Energy Comm. Rept.*, KAPL-180 (1949).
[470] Hála, J., *J. Inorg. Nucl. Chem.*, **29**, 1317 (1967).
[471] Rossotti, F. J. C. and Rossotti, H., *Acta Chem. Scand.*, **10**, 779 (1956).
[472] Sunderman, D. N., Ackermann, I. B. and Meinke, W. W., *Anal. Chem.*, **31**, 40 (1959).
[473] Johnson, D. A. and Lott, P. E., *Anal. Chem.*, **35**, 1705 (1963).
[474] Onishi, H. and Toita, Y., *Talanta*, **11**, 1357 (1964).
[475] De, A. K. and Rahaman, M. S., *Anal. Chem.*, **36**, 685 (1964).
[476] De, A. K. and Rahaman, M. S., *Anal. Chim. Acta*, **27**, 591 (1962).
[477] Mulye, R. R. and Khopkar, S. M., *Sep. Sci.*, **7**, 605 (1972).
[478] Foti, S. C. and Freiling, E. C., *Talanta*, **11**, 385 (1964).
[479] Moore, F. L., *Anal. Chem.*, **29**, 941 (1957); **30**, 1368 (1958).
[480] Boussières, G. and Vernois, J., *Compt. Rend.*, **244**, 2508 (1957).
[481] De, A. K., Singh, T. and Mandal, S., *Proc. 5th Intern. Conf. Solvent Extr. Chem.*, **1969**, 13 (see [146]).
[482] Keenan, T. K. and Suttle, J. F., *J. Am. Chem. Soc.*, **76**, 2184 (1954).
[483] Moore, F. L. and Hudgens, J. E., *Anal. Chem.*, **29**, 1767 (1957).
[484] Cuninghame, J. G. and Miles, G. L., *J. Inorg. Nucl. Chem.*, **3**, 54 (1956).
[485] Yamamoto, T., Muto, H., Kihara, S. and Motojima, K., *Anal. Chim. Acta*, **56**, 191 (1971).
[486] De, A. K. and Rahaman, M. S., *Analyst (London)*, **89**, 795 (1964).
[487] Ashbrook, A. W., *Analyst (London)*, **88**, 113 (1963).
[488] Akki, S. B. and Khopkar, S. M., *Sep. Sci.*, **6**, 455 (1971).
[489] Stokely, J. R. and Moore, F. L., *Anal. Chem.*, **36**, 1203 (1964).
[490] Kolařík, T. and Pánková, H., *Collect. Czech. Chem. Commun.*, **27**, 166 (1962).
[491] Kiba, T. and Kanetani, M., *Bull. Chem. Soc. Japan*, **31**, 1013 (1958).
[492] Perkins, M. and Kahlwarf, P. R., *Anal. Chem.*, **28**, 1989 (1956).
[493] Rozen, A. M. and Nagibeda, Z. I., *Radiokhimiya*, **13**, 284 (1971).
[494] Day, R. A. and Stoughton, R. W., *J. Am. Chem. Soc.*, **72**, 5662 (1950).
[495] Liberti, A., Chiantella, V. and Corigliano, F., *J. Inorg. Nucl. Chem.*, **25**, 415 (1963).
[496] Khopkar, S. M. and De, A. K., *Chem. Ind. (London)*, **1959**, 291.
[497] Hök, B., *Sven. Kem. Tidskr.*, **65**, 106 (1953).
[498] Day, R. A. and Powers, R. M., *J. Am. Chem. Soc.*, **76**, 3895 (1954).
[499] Milich, N., Petrukhin, O. M. and Zolotov, Yu. A., *Zh. Neorgan. Khim.*, **9**, 2664 (1964).

[500] Fukai, R. and Meinke, W. W., *Progress Report*, No. 7, Univ. Michigan, **1958**, 63.
[501] De, A. K. and Rahaman, M. S., *Anal. Chem.*, **35**, 1095 (1963).
[502] Siekierski, S. and Sochacka, R. J., *J. Chromatogr.*, **16**, 376 (1964).
[503] Carnes, W. J. and Dean, J. A., *Anal. Chem.*, **33**, 1961 (1961).
[504] Sekine, T. and Dyrssen, D., *J. Inorg. Nucl. Chem.*, **26**, 1727, 2013 (1964).
[505] Hardy, C. J. and Scargill, D., *J. Inorg. Nucl. Chem.*, **9**, 322 (1959).
[506] Hercules, D. M., *Talanta*, **8**, 485 (1961).
[507] Marsh, S. F., Maeck, W. J., Booman, G. L. and Rein, J. E., *Anal. Chem.*, **33**, 870 (1961).
[508] Moore, F. L., *Anal. Chem.*, **28**, 997 (1956).
[509] Maeck, W. J., Marsh, S. F. and Rein, S. F., *Anal. Chem.*, **35**, 292 (1963).
[510] Přibil, R. and Jelínek, M., *Chem. Listy*, **47**, 1326 (1953).
[511] Shigematsu, T. and Tabushi, M., *Nippon Kagaku Zasshi*, **81**, 265 (1960).
[512] Jensen, B. S., *Acta Chem. Scand.*, **13**, 1890 (1959).
[513] Navrátil, O. and Jensen, B. S., *J. Radioanal. Chem.*, **5**, 313 (1970).
[514] Navrátil, O., *Collect. Czech. Chem. Commun.*, **38**, 1333 (1973).
[515] Navrátil, O. and Mikulec, Z., *Collect. Czech. Chem. Commun.*, **38**, 2430 (1973).
[516] Navrátil, O., *Chem. Listy*, **68**, 470 (1974).
[517] Navrátil, O., and Smola, J., *Collect. Czech. Chem. Commun.*, **36**, 3549 (1971).
[517a] Navrátil, O., *Radiokhimiya*, **19**, 626 (1977).
[518] Jensen, B. S., *Acta Chem. Scand.*, **13**, 1668 (1959).
[519] Zolotov, Yu. A., Lambrev, V. G., Chmutova, M. K. and Sizonenko, N. T., *Dokl. Akad. Nauk SSSR*, **165**, 117 (1963).
[520] Zolotov, Yu. A., Sizonenko, N. T., Zolotovitskaya, E. S. and Yakovenko, E. I., *Zh. Analit. Khim.*, **24**, 20 (1969).
[521] Sizonenko, N. T. and Zolotov, Yu. A., *Zh. Analit. Khim.*, **24**, 1305 (1969).
[522] Skorko-Trybuła, Z., Różycki, C. and Kosiarska, E., *Chem. Anal. (Warsaw)* **22**, 311 (1977).
[523] Karalova, Z. K., Pyzhova, Z. I. and Rodionova, L. M., *Zh. Analit. Khim.*, **25**, 909 (1970).
[524] Zolotov, Yu. A. and Lambrev, V. G., *Zh. Analit. Khim.*, **20**, 659 (1965).
[525] Zolotov, Yu. A. and Lambrev, V. G., *Radiokhimiya*, **8**, 627 (1966).
[526] Arora, H. C. and Rao, N. G., *Indian J. Chem.*, **11**, 488 (1973).
[527] Efimov, I. P., Tomilova, L. G., Voronets, L. S. and Peshkova, V. M., *Zh. Analit. Khim.*, **28**, 267 (1973).
[528] Freger, S. V., Pozovik, A. S. and Obrutskii, M. I., *Zh. Analit. Khim.*, **26**, 2380 (1971).
[529] Al-Niaimi, N. S., Al-Karaghouli, A. R. and Aliwi, S. M., *J. Inorg. Nucl. Chem.*, **35**, 577 (1973).
[530] Lambrev, V. G. and Vlasov, V. S., *Zh. Analit. Khim.*, **25**, 1638 (1970).
[531] Zolotov, Yu. A. and Sizonenko, N. T., *Zh. Analit. Khim.*, **25**, 54 (1970).
[532] Chmutova, M. K. and Kochetkova, N. E., *Zh. Analit. Khim.*, **24**, 216 (1969).
[533] Chmutova, M. K., Kochetkova, N. E. and Zolotov, Yu. A., *Zh. Analit. Khim.*, **24**, 711 (1969).
[534] Revenko, V. G., Bagreev, V. V., Zolotov, Yu. A. and Kopantskaya, L. S., *Zh. Analit. Khim.*, **27**, 1571 (1972).
[535] Sizonenko, N. T. and Zolotov, Yu. A., *Zh. Analit. Khim.*, **24**, 1341 (1969).

References

[536] Myasoedov, B. F., Molochnikova, N. P. and Palei, P. N., *Radiokhimiya*, **12**, 829 (1970).
[537] Chmutova, M. K., Palei, P. N. and Zolotov, Yu. A., *Zh. Analit. Khim.*, **23**, 1476 (1968).
[538] Myasoedov, B. F. and Molochnikova, N. P., *Zh. Analit. Khim.*, **24**, 702 (1969).
[539] Zolotov, Yu. A., Chmutova, M. K. and Palei, P. N., *Zh. Analit. Khim.*, **21**, 1217 (1966).
[540] Sovostina, V. M., Shpigun, O. A. and Peshkova, V. M., *Zh. Analit. Khim.*, **26**, 2044 (1971).
[541] Karalova, Z. K. and Pyzhova, Z. I., *Zh. Analit. Khim.*, **23**, 1564 (1968).
[541a] Roy, A. and Nag, K., *Indian J. Chem.*, **A15**, 474 (1977).
[541b] Roy, A. and Nag, K., *J. Inorg. Nucl. Chem.*, **40**, 331 (1978).
[541c] Nurtaeva, A. K., Efimov, J. P. and Peshkova, V. M., *Zh. Analit. Khim.*, **32**, 1735 (1977).
[542] Kuzmin, N. M., Khorkina, L. S., Lebedev, A. I. and Zolotov, Yu. A., *Zh. Analit. Khim.*, **25**, 1257 (1970).
[543] Motojima, K., Yoshida, H. and Izawa, K., *Anal. Chem.*, **32**, 1083 (1960).
[544] Umland, F., *Z. Anal. Chem.*, **190**, 186 (1962).
[545] Freiser, H., *Chemist-Analyst*, **50**, 94 (1961).
[546] Rudenko, N. P. and Sebastyanov, A. I., *Zh. Analit. Khim.*, **26**, 1994 (1971).
[547] Rakovskii, E. E., Serebryany, B. L. and Klyueva, N. D., *Zh. Analit. Khim.*, **29**, 1086, 1710 (1974).
[548] Rakovskii, E. E., Serebryany, B. L. and Klyueva, N. D., *Zh. Analit. Khim.*, **30**, 9 (1975).
[549] Oki, S. and Terada, I., *Anal. Chim. Acta*, **61**, 491 (1972).
[550] Lacroix, S., *Anal. Chim. Acta*, **1**, 260 (1947).
[551] Lacroix, S., *Anal. Chim. Acta.*, **2**, 167 (1948).
[552] Riley, J. P., *Anal. Chim. Acta*, **19**, 413 (1958); **21**, 317 (1959).
[553] Mervel, R. V., *Zh. Analit. Khim.*, **2**, 103 (1947).
[554] Reinis, M. M. and Larionova, Yu. A., *Zavodsk. Lab.*, **14**, 1000 (1948).
[555] Smith, B. H., *Lab. Pract.*, **22**, 100 (1973).
[556] Umland, F., and Meckenstock, K. U., *Z. Anal. Chem.*, **177**, 244 (1960).
[557] Umland, F., Hoffmann, W. and Meckenstock, K.-U., *Z. Anal. Chem.*, **173**, 211 (1960).
[558] Umland, F. and Meckenstock, K. U., *Z. Anal. Chem.*, **165**, 161 (1959).
[559] Goto, H. and Sudo, E., *Bunseki Kagaku*, **10**, 171 (1961).
[560] Hellwege, H. E. and Schweitzer, G. K., *Anal. Chim. Acta*, **28**, 236 (1963); **29**, 47 (1963).
[561] Hellwege, H. E. and Schweitzer, G. K., *J. Inorg. Nucl. Chem.*, **27**, 99 (1965).
[562] Umland, F. and Hoffmann W., *Z. Anal. Chem.*, **168**, 268 (1959).
[563] Westwood, W. and Mayer, A., *Analyst (London)*, **73**, 275 (1948).
[564] Duffield, W. D., *Analyst (London)*, **84**, 455 (1959).
[565] Kuzmin, N. M., *Dissertation*, GEOKhI Akad. Nauk SSSR, Moscow, 1965.
[566] Blair, A. J. and Pantony, D. A., *Anal. Chim. Acta*, **14**, 545 (1956).
[567] Tandon, J. P. and Mehrotra, R. C., *Z. Anal. Chem.*, **176**, 87 (1960).
[568] Sudo, E., *Nippon Kagaku Zasshi*, **72**, 718 (1951).
[569] Rakovskii, E. E., Petrukhin, O. H. and Severin, V. J., *J. Radioanal. Chem.*, **4**, 207 (1970).
[570] Chernitskaya, R. E., *Zh. Obshch. Khim.*, **22**, 406 (1952).

[571] Kuznetsov, V. I., *Zh. Analit. Khim.*, **7**, 226 (1952).
[572] Rudenko, N. P., *Tr. Kom. Anal. Khim. Akad. Nauk SSSR*, **14**, 209 (1963).
[573] Dyrssen, D. and Dahlberg, V., *Acta Chem. Scand.*, **7**, 1186 (1953).
[574] Lawson, K. L. and Kahn, M., *J. Inorg. Nucl. Chem.*, **5**, 87 (1957).
[575] Jankowski, S. J. and Freiser, H., *Anal. Chem.*, **33**, 776 (1961).
[576] Alimarin, I. P. and Gibalo, I. M., *Vestn. Mosk. Gos. Univ.*, (5), 55 (1956).
[577] Alimarin, I. P., Golovina, A. P. and Puzdrenko, I. V., *Vestn. Mosk. Gos. Univ.*, (2), 185 (1959).
[578] Alimarin, I. P., Bilimovich, G. N. and Tszui Syan-Khan, *Zh. Neorgan. Khim.*, **7**, 2725 (1962).
[579] Alimarin, I. P., Bilimovich, G. N. and Yuicen', Ya. N., *Radiokhimiya*, **4**, 510 (1962).
[580] Oki, S., *Anal. Chim. Acta*, **49**, 455 (1970).
[581] Harvey, B. G., Heal, H. G., Maddock, A. G. and Rowley, E. L., *J. Chem. Soc.*, **1947**, 1010.
[582] Alimarin, I. P., Krasikova, V. M. and Puzdrenkova, I. V., *Vestn. Mosk. Gos. Univ.*, (3), 50 (1964).
[583] Umland, F. and Puchelt, H., *Anal. Chim. Acta*, **16**, 334 (1957).
[584] Eberle, A. R. and Lerner, M. W., *Anal. Chem.*, **34**, 627 (1962).
[585] Dyrssen, D., *Sven. Kem. Tidskr.*, **65**, 43 (1953); **67**, 311 (1955).
[586] Chernitskaya, R. E., *Zh. Obshch. Khim.*, **22**, 408 (1952).
[587] Bagreev, V. V. and Zolotov, Yu. A., *Zh. Analit. Khim.*, **17**, 852 (1962).
[588] Moeller, T. and Cohen, A. J., *Anal. Chem.*, **22**, 686 (1950).
[589] Bullwinkel, E. P. and Noble, P., *J. Am. Chem. Soc.*, **80**, 2955 (1958).
[590] Clayton, R. F., Hardwick, W. H., Moreton-Smith, M. and Todd, R., *Analyst (London)*, **83**, 13 (1958).
[591] Corsini, A., Abraham, J. and Thompson, M., *Talanta*, **18**, 481 (1971).
[592] Clifford, W. E., Bullwinkel, E. P., McClaine, L. A. and Noble, P., *J. Am. Chem. Soc.*, **80**, 2959 (1958).
[593] Talvitie, N. A., *Anal. Chem.*, **25**, 604 (1953).
[594] Panova, M. G., Levin, V. I. and Brezhneva, N. E., *Radiokhimiya*, **2**, 197, 208 (1960).
[595] Rudenko, N. P., *Zh. Neorgan. Khim.*, **1**, 1091 (1956).
[596] Chon-Fachun, Q. and Freiser, H., *Anal. Chem.*, **37**, 361 (1965).
[597] Schneider, H. O. and Roselli, M. E., *Analyst (London)*, **96**, 330 (1971).
[598] Borrel, M. and Pâris, R. A., *Anal. Chim. Acta*, **6**, 389 (1952).
[598a] Dagnall, R. M., West, T. S. and Young, P., *Analyst (London)*, **90**, 13 (1965).
[599] Motojima, K. and Hashitani, H., *Bunseki Kagaku*, **9**, 151 (1960).
[600] Rulfs, C. L., De, A. K., Larkitz, J. N. and Elving, P. J., *Anal. Chem.*, **27**, 1802 (1955).
[601] Sekine, T. and Dyrssen, D., *Talanta*, **11**, 864 (1964).
[602] Dyrssen, D., Heffez, M. and Sekine, T., *J. Inorg. Nucl. Chem.*, **16**, 367 (1961).
[603] Moeller, T., Pundsack, F. L. and Cohen, A. J., *J. Am. Chem. Soc.*, **76**, 2615 (1954).
[604] Johnson, J. E., Lavine, M. C. and Rosenberg, A. J., *Anal. Chem.*, **30**, 2055 (1958).
[605] Ruf, E., *Z. Anal. Chem.*, **162**, 9 (1958).
[606] Kuznetsov, V. I. and Fan-Min-E., *Zh. Neorgan. Khim.*, **7**, 422 (1962).
[607] Kuznetsov, V. I., Bankovskii, Yu. A. and Nevinskii, A. F., *Zh. Analit. Khim.*, **13**, 267 (1958).

[608] Bankovskii, Yu. A. and Ievin'sh, A. F., *Zh. Analit. Khim.*, **13**, 507, 643 (1958).
[609] Bankovskii, Yu. A., Ievin'sh, A. F., Luksha, E. A. and Bochkans, P. Ya., *Zh., Analit. Khim.*, **16**, 150 (1961).
[610] Bankovskii, Yu. A., Tsirule, Ya. A. and Ievin'sh, A. F., *Zh. Analit. Khim.*, **16**, 562 (1961).
[611] Bankovskii, Yu. A., Mezharaups, G. P. and Ievin'sh, A. F., *Zh. Analit. Khim.*, **17**, 721 (1969).
[612] Bankovskii, Yu. A., Ievin'sh, A. F. and Luksha, E. A., *Zh. Analit. Khim.*, **14**, 222, 714 (1959).
[613] Magee, R. J. and Witwitt, A. S., *Anal. Chim. Acta*, **29**, 27 (1963).
[614] Agrinskaya, N. A. and Petrashen', V. I., *Zh. Analit. Khim.*, **16**, 701 (1961).
[615] Bankovskii, Yu. A., Shvarts, E. M. and Ievin'sh, A. F., *Zh. Analit. Khim.*, **14**, 313 (1959).
[616] Motojima, K., *Bull. Chem. Soc. Japan*, **29**, 71 (1956).
[617] Keil, R., *Mikrochim. Acta*, **1973**, 919.
[618] Riley, J. P. and Williams, H. P., *Mikrochim. Acta*, **1959**, 825.
[619] Misumi, S. and Nagano, N., *Anal. Chem.*, **34**, 1723 (1962).
[620] Motojima, K. and Hashitani, H., *Anal. Chem.*, **33**, 239 (1961).
[621] Nishikawa, Y., *Nippon Kagaku Zasshi*, **79**, 236, 351 (1958).
[622] Shigematsu, T., *Bunseki Kagaku*, **7**, 787 (1958).
[623] Motojima, K., Hashitani, H. and Imanashi, T., *Anal. Chem.*, **34**, 571 (1962).
[624] Ishiwatari, N. and Motojima, K., *Bunseki Kagaku*, **19**, 1180 (1970).
[625] Motojima, K., *Bull. Chem. Soc. Japan*, **29**, 455 (1956).
[626] Motojima, K., and Hashitani, H., *Bull. Chem. Soc. Japan.*, **29**, 458 (1956).
[627] Dyrssen, D., *Sven. Kem. Tidskr.*, **64**, 213 (1952).
[628] Koch, O. G. and Koch-Dedic, G. A., *Handbuch der Spurenanalyse*, Springer-Verlag, Berlin, 1964.
[629] Starý, J., Růžička, J. and Salamon, M., *Talanta*, **10**, 375 (1963).
[630] Lutwick, G. D. and Ryan, D. E., *Can. J. Chem.*, **32**, 949 (1954).
[631] Pilipenko, A. T., Shpak, A. B. and Shpak, É. A., *Ukr. Khim. Zh.*, **41**, 78 (1975).
[632] Chakrabarti, C. L., Lyles, G. R. and Dowling, F. B., *Anal. Chim. Acta*, **29**, 489 (1963).
[633] Cyranowska, M. and Downarowicz, J., *Chem. Anal. (Warsaw)*, **10**, 67 (1965).
[634] Kiba, T., Ohashi, S. and Maeda, T., *Bull. Chem. Soc. Japan*, **33**, 818 (1960).
[635] Sandell, E. B. and Cummings, P. F., *Anal. Chem.*, **21**, 1356 (1949).
[636] Wakamatsu, S., *Bunseki Kagaku*, **8**, 298 (1959).
[637] Alimarin, I. P. and Sherif Abdel Hamid, *Zh. Analit. Khim.*, **19**, 195 (1964).
[638] Allen, S. H. and Hamilton, M. B., *Anal. Chim. Acta*, **7**, 483 (1952).
[639] Alimarin, I. P. and Gibalo, I. M., *Dokl. Akad. Nauk SSSR*, **109**, 1137 (1956).
[640] Fritz, J. S., Richard, M. J. and Bystroff, A. S., *Anal. Chem.*, **29**, 577 (1957).
[641] Maddock, A. G. and Miles, G. L., *J. Chem. Soc.*, **1949**, S 253.
[642] Moiseev, I. V., Borodina, N. N., Tsvetkova, V. T., *Zh. Neorgan. Khim.*, **6**, 543 (1961).
[643] Das, G. and Banerjee, S., *Indian J. Chem.*, **8**, 284 (1970).
[644] Furman, N. H., Mason, W. B. and Pekola, J. S., *Anal. Chem.*, **21**, 1325 (1949).
[645] Nadezhda, A. A., Ivanova, K. P., Gorbenko, F. P. and Maslieva, F. I., *Ukr. Khim. Zh.*, **42**, 760 (1976).
[646] Munshi, K. N. and Dey, A. K., *Anal. Chim. Acta*, **27**, 89 (1962).

[647] Klylin, A. E. and Kolyada, I. S., *Zh. Neorgan. Khim.*, **6**, 216 (1961).
[648] Stander, C. M., *Anal. Chem.*, **32**, 1296 (1960).
[649] Willard, H. H., Martin, E. L. and Feltham, R., *Anal. Chem.*, **25**, 1863 (1953).
[650] Elinson, S. V., Pobedina, L. I. and Mirzoyan, I. A., *Zh. Analit. Khim.*, **15**, 334 (1960).
[651] Zolotov, Yu. A., Spivakov, B. Ya. and Gavrilina, G. N., *Zh. Analit. Khim.*, **24**, 1168 (1969).
[652] Gagliardi, E. and Raber, H., *Monatsh. Chem.*, **93**, 360 (1962).
[653] Gupta, H. K. and Sogani, N. C., *J. Indian Chem. Soc.*, **40**, 15 (1963).
[654] Rudenko, N. P. and Samari, A. M., *Zh. Neorgan. Khim.*, **15**, 1343 (1970).
[655] Majumdar, A. K. and Das, G., *Anal. Chim. Acta*, **36**, 454 (1966).
[656] Minczewski, J. and Skorko-Trybuła, Z., *Talanta*, **10**, 1063 (1963).
[657] Minczewski, J. and Skorko-Trybuła, Z., *Chem. Anal. (Warsaw)*, **6**, 377 (1961).
[658] Skorko-Trybuła, Z. and Minczewski, J., *Chem. Anal. (Warsaw)*, **6**, 523 (1961).
[659] Tandon, S. G. and Bhattacharya, *J. Indian Chem. Soc.*, **47**, 583 (1970).
[660] Majumdar, A. K. and Das, G., *Anal. Chim. Acta*, **36**, 454 (1966).
[660a] Vernon, F., and Khorassani, J. H., *Talanta*, **25**, 410 (1978).
[661] Skorko-Trybuła, Z., *Chem. Anal. (Warsaw)*, **9**, 397 (1964); **10**, 831 (1965); **12**, 815 (1967).
[662] Skorko-Trybuła, Z. and Dębska, B., *Chem. Anal. (Warsaw)*, **13**, 557 (1968).
[663] Skorko-Trybuła, Z. and Boguszewska, Z., *Chem. Anal. (Warsaw)*, **14**, 549 (1969); **15**, 345 (1970).
[664] Skorko-Trybuła, Z. and Polanowska, J., *Chem. Anal. (Warsaw)*, **15**, 635 (1970).
[665] Skorko-Trybuła, Z. and Krzyżanowska, M. *Chem. Anal. (Warsaw)*, **16**, 99 (1971).
[666] Fouché, K. F., *J. Inorg. Nucl. Chem.*, **30**, 3057 (1968).
[667] Fouché, K. F., le Roux, H. J. and Philips, F., *J. Inorg. Nucl. Chem.*, **32**, 1949 (1970).
[668] Smolina, E. V., *Dissertation*, MGU, Moscow, 1971.
[669] Förster, H. and Schwabe, K., *Anal. Chim. Acta*, **45**, 511 (1969).
[670] Komar', N. P. and Per'kov, I. G., *Zh. Analit. Khim.*, **19**, 145 (1964).
[671] Dyrssen, D., *Acta Chem. Scand.*, **10**, 353 (1956).
[672] Zharovskii, F. G., Shpak, E. A. and Piskunova, E. V., *Ukr., Khim. Zh.*, **29**, 102 (1963).
[673] Alimarin, I. P., *Usp. Khim.*, **31**, 989 (1962).
[674] Chwastowska, J., *Chem. Anal. (Warsaw)*, **17**, 469 (1967).
[675] Riedel, A., *J. Radioanal. Chem.*, **13**, 125 (1973).
[676] Rudenko, N. P. and Latif Al Samurai, *Vestn. Mosk. Gos. Univ. Ser. Khim.*, **1**, 77 (1969).
[677] Fouché, K. F., *Talanta*, **15**, 1295 (1968).
[678] Pilipenko, A. T., Shpak, E. A. and Shevchenko, L. L., *Zh. Neorgan. Khim.*, **12**, 463 (1967).
[678a] Chwastowska, J., Kosiarska, E. and Maciejko, G., *Chem. Anal. (Warsaw)*, **22**, 927 (1977).
[679] Chwastowska, J. and Różańska, B., *Chem. Anal. (Warsaw)*, **21**, 85 (1976).
[680] Lyle, S. J. and Shendrikar, A. D., *Anal. Chim. Acta*, **32**, 575 (1965).
[681] Minczewski, J., Chwastowska, J. and Różańska, B., *Chem. Anal. (Warsaw)*, **19**, 497 (1974).

[682] Riedel, A., *J. Radioanal. Chem.*, **6**, 75 (1970).
[682a] Chwastowska, J. and Różańska, B., *Chem. Anal. (Warsaw)*, **23**, 745 (1978).
[683] Murugaiyan, P. and Das M. S., *Anal. Chim. Acta*, **48**, 155 (1969).
[684] Alimarin, I. P., Sherif Abdel Hamid and Puzdrenkova, I. V., *Zh. Neorgan. Khim.*, **9**, 2475 (1964).
[685] Alimarin, I. P., Sokolova, I. V. and Smolina, E. V., *Vestn. Mosk. Gos. Univ.*, 67 (1968).
[686] Alimarin, I. P., Smolina, E. V., Sokolova, I. V. and Firsova, G. V., *Zh. Analit. Khim.*, **25**, 2287 (1970).
[687] Hála, J., *J. Inorg. Nucl. Chem.*, **29**, 187 (1967).
[688] Lamarg, G., *Rapp. CEA*, **1971**, No. 4215, 63.
[689] Prihoda, J. and Hála, J., *J. Radioanal. Chem.*, **30**, 343 (1976).
[690] Schweitzer, G. K. and Anderson, M. M., *J. Inorg. Nucl. Chem.*, **30**, 1051 (1968).
[691] D'yachkova, R. A. and Spitsyn, V. I., *Radiokhimiya*, **6**, 102 (1964).
[692] Alimarin, I. P., Petrukhin, O. M. and Tsê, Yün-Hsiang, *Dokl. Akad. Nauk SSSR*, **136**, 1073 (1961).
[693] Alimarin, I. P. and Petrukhin, O. M., *Zh. Neorgan. Khim.*, **7**, 1191 (1962).
[694] Erskine, J. S., Sink, M. L. and Varga, L. P., *Anal. Chem.*, **41**, 70 (1969).
[695] Lyle, S. J. and Shendrikar, A. D., *Talanta*, **12**, 573 (1965).
[696] Pal'shin, E. S., Myasoedov, B. F. and Novikov, Yu. P., *Zh. Analit. Khim.*, **18**, 657 (1963).
[697] Myasoedov, B. F., Pal'shin, E. S. and Palei, P. N., *Zh. Analit. Khim.*, **19**, 105 (1964).
[698] Chmutova, M. K. and Zolotov, Yu. A., *Radiokhimiya*, **6**, 640 (1964).
[699] Fouché, K. F., *J. Inorg. Nucl. Chem.*, **33**, 857 (1971).
[700] Rakovskii, E. E. and Petrukhin, O. M., *Zh. Analit. Khim.*, **18**, 539 (1963).
[701] Alimarin, I. P. and Tsê, Yün-Hsiang, *Zavodsk. Lab.*, **25**, 1435 (1959).
[702] Jordanov, N., Mareva, St. and Koeva, M., *Anal. Chim. Acta*, **59**, 75 (1972).
[703] Alimarin, I. P. and Tsê, Yün-Hsiang, *Vestn. Mosk. Gos. Univ. Ser. Khim.*, 53 (1960).
[704] Zharovskii, F. G., Shpak, E. A. and Piskunova, E. V., *Ukr. Khim. Kh.*, **28**, 1104 (1962).
[705] Afghan, B. K., Marryatt, R. G. and Ryan, D. E., *Anal. Chim. Acta*, **41**, 131 (1968).
[706] Hernandez de Pool, D., Díaz Cadavieco, R., *Acta Cient Venez.*, **13**, 157 (1962); *Anal. Abstr.*, **10**, 3185 (1963).
[707] Chideo, T., *Bunseki Kagaku*, **12**, 271 (1963).
[708] Ryan, D. E., *Analyst (London)*, **85**, 569 (1960).
[709] Chakrabarti, C. L., Magee, R. J. and Wilson, C. L., *Talanta*, **10**, 1201 (1963).
[710] Villarreal, R., Young, J. O. and Krsul, J. R., *Anal. Chem.*, **42**, 1419 (1970).
[711] Gańko, T., *Badanie budowy związków kompleksowych metali z ditizonem (Investigations of the Structure of Metal-Dithizone Complexes)*, Doctoral Thesis, Warsaw University, Warsaw, 1971.
[712] Laing, M. and Alsop, P. A., *Talanta*, **17**, 243 (1970).
[713] Freiser, B. S. and Freiser, H., *Anal. Chem.*, **42**, 305 (1970).
[714] Math, K. S. and Freiser, H., *Talanta*, **18**, 435 (1971).
[715] Kiwan, A. M. and Irving, H. M., *Anal. Chim. Acta*, **54**, 351 (1971).

[716] Irving, H. M. and Nowicka-Jankowska, T., *Anal. Chim. Acta*, **54**, 55 (1971).
[717] Irving, H. M. and Kiwan, A. M., *Anal. Chim. Acta*, **56**, 435 (1971); **57**, 59 (1971).
[718] Nowicka-Jankowska, T. and Irving, H. M., *Anal. Chim. Acta*, **54**, 489 (1971).
[719] Irving, H., Risdon, E. J. and Andrew, G., *J. Chem. Soc.*, **1949**, 537.
[720] Irving H. M. N. H., *Anal. Chem.*, **29**, 857 (1957).
[721] Babko, A. K. and Pilipenko, A. T., *Zh. Analit. Khim.*, **1**, 275 (1946); **2**, 33 (1947).
[722] Pilipenko, A. T., *Zh. Analit. Khim*, **8**, 286 (1953).
[723] Iwantscheff, G., *Das Dithizon und seine Anwendung in der Mikro- und Spurenanalyse*, 2.Aufl., Verlag Chemie, Weinheim, 1972.
[724] Duncan, J. F. and Thomas, F. G., *J. Chem. Soc.*, **1960**, 2814.
[725] Math, K. S., Fernando, Q. and Freiser, H. *Anal. Chem.*, **36**, 1762 (1964).
[726] Minczewski, J., Krasiejko, M. and Marczenko, Z., *Chem. Anal. (Warsaw)*, **15**, 43 (1970).
[727] Beardsley, D. A., Briscoe, G. B., Růžička, J. and Williams, M., *Talanta*, **13**, 328 (1966).
[728] Erdey, L. and Rády, G., *Z. Anal. Chem.*, **135**, 1 (1952).
[729] Irving, H., Kiwan, A. M., Rupainwar, D. C. and Sahota, S. S., *Anal. Chim. Acta*, **56**, 205 (1971).
[729a] Irving, H. M. N. H., *Dithizone*, Chemical Society, London, 1977.
[730] Kawahata, M. Mitidzuki, Ch. and Misaki, T., *Bunseki Kagaku*, **11**, 819 (1962).
[731] Lombardi, O. W., *Anal. Chem.*, **36**, 415 (1964).
[732] Zolotov, Yu. A., Demina, L. A. and Petrukhin, O. M., *Zh. Analit. Khim.*, **25**, 2315 (1970).
[733] Busev, A. I. and Bazhanova, L. A., *Zh. Analit. Khim.*, **6**, 2210 (1961).
[734] DeVoe, J. R. and Meinke, W. W., *Anal. Chem.*, **31**, 1428 (1959).
[735] Babko, A. K. and Pilipenko, A. T., *Zh. Analit. Khim.*, **2**, 33 (1947).
[736] Fischer, H., *Mikrochemie* **30**, 38 (1942).
[737] Dawson, E. C., *Analyst (London)*, **73**, 618 (1948).
[738] Pierce, T. B. and Peck, P. F., *Anal. Chim. Acta*, **27**, 392 (1962).
[739] Yamamura, S. S., *Anal. Chem.*, **32**, 1896 (1960).
[740] Mathre, O. B. and Sandell, E. B., *Talanta*, **11**, 295 (1964).
[741] Biefeld, L. P. and Patrick, T. M., *Ind. Eng. Chem., Anal. Ed.*, **14**, 275 (1942).
[742] Guettel, Ch. L., *Ind. Eng. Chem., Anal. Ed.*, **11**, 639 (1939).
[743] Snyder L. J., *Anal. Chem.*, **19**, 684 (1947).
[744] Kawahata, M., Mochizuki, M. and Misaki, T., *Bunseki Kagaku*, **11**, 448 (1962).
[745] Hallam, K. M., *Anal. Chem.*, **34**, 1339 (1962).
[746] Boussières, G. and Ferradini, C., *Anal. Chim. Acta*, **4**, 610 (1950).
[747] Kawahata, M., Mochizuki, M. and Misaki, T.. *Bunseki Kagaku*, **11**, 1020 (1962).
[748] Ramakrishna, R. S. and Irving, H. M., *Chem. Commun.*, **1969**, 1356.
[749] Starý, J. and Marek, J., *Chem. Commun.*, **1970**, 519.
[750] Mabuchi, H., *Bull. Chem. Soc. Japan*, **29**, 842 (1956).
[751] Irving, H., Bell, C. F. and Williams, R. J., *J. Chem. Soc.*, **1952**, 356.
[752] Hulanicki, A., *Dwutiokarbaminiany jako odczynniki kompleksujące (Dithiocarbamates as Complexing Agents)*, Thesis, Warsaw University, Warsaw, 1967.
[753] Hulanicki, A., *Talanta*, **14**, 1371 (1967).

[754] Podchainova, V. N., *Tr. Kom. Anal. Khim. Akad. Nauk SSSR*, **11**, 146 (1960).
[755] Sastri, V. S., Aspila, K. J. and Chakrabarti, C. L., *Can. J. Chem.*, **47**, 2320 (1969).
[756] Förster, H., *J. Radioanal. Chem.*, **4**, 1 (1970).
[757] Bode, H. and Tusche, K. J., *Z. Anal. Chem.*, **157**, 414 (1957).
[758] Eckert, G., *Z. Anal. Chem.*, **148**, 14 (1955), **155**, 23, (1957).
[759] Bode, H. and Neumann, F., *Z. Anal. Chem.*, **169**, 410 (1959); **172**, 1 (1960).
[760] Bode, H., *Z. Anal. Chem.*, **142**, 182, 414 (1954); **144**, 165 (1955).
[761] Malissa, H. and Gomišček, S., *Z. Anal. Chem.*, **169**, 401 (1959).
[762] Fishkova, N. L., *Anal. Tekhnol. Blagorod. Metal.*, **1971**, 248; *Chem. Abstr.*, **77**, 147279w (1972).
[763] Doering, W. von E. and Knox, L. H., *J. Am. Chem. Soc.*, **73**, 828 (1951).
[764] Lacoste, R. J., Earing, M. H. and Wiberley, S. E., *Anal. Chem.*, **23**, 871 (1951).
[765] Claassen, A. and Bastings, L., *Z. Anal. Chem.*, **153**, 30 (1956).
[766] Jean, M., *Anal. Chim. Acta*, **11**, 79 (1954).
[767] Jewsbury, A., *Analyst (London)*, **78**, 363 (1953).
[768] Šedivec, V., and Vašák, V., *Collect. Czech. Chem. Commun.*, **15**, 260 (1950).
[769] Busev, A. I., Zhanondokovskaya, T. N. and Kuznetsova, Z. M., *Zh. Analit. Khim.*, **15**, 49 (1960).
[770] Chernikhov, Yu. A. and Dobkina, B. M., *Zavodsk. Lab.*, **15**, 1143 (1949).
[771] Šedivec, V. and Vašák, V., *Chem. Listy*, **45**, 435 (1951).
[772] Tortoolen, J. W., Buijze, C. and van Kolmeschate, G. J., *Chemist-Analyst*, **52**, 100 (1963).
[773] Rakovskii, E. E. and Baevskaya, G. M., *Zh. Analit. Khim.*, **26**, 1796 (1971).
[774] Dienstl, G. and Hecht, F., *Mikrochim. Acta*, **1962**, 321.
[775] Van Erkelens, P. C., *Anal. Chim. Acta*, **26**, 32 (1962).
[776] Gage, C., *Analyst (London)*, **80**, 789 (1955); **82**, 435 (1957).
[777] Nazarenko, V. A. and Biryuk, E. A., *Zavodsk. Lab.*, **25**, 28 (1959).
[778] Pyle, J. T. and Jacobs, W. D., *Anal. Chem.*, **36**, 1796 (1964).
[779] Agarwal, B. V., Sangal, S. P. and Dey, A. K., *Z. Anal. Chem.*, **207**, 256 (1965).
[780] Agarwal, B. V., Sangal, S. P. and Dey, A. K., *J. Indian Chem. Soc.*, **41**, 119 (1964).
[781] Fritz, J. S. and Bradford, E. C., *Anal. Chem.*, **30**, 1021 (1958).
[782] Fritz, J. S., Richard, M. J. and Lane, W. J., *Anal. Chem.*, **30**, 1776 (1958).
[783] Hardwick, W. H. and Moreton-Smith, M., *Analyst (London)*, **83**, 9 (1958).
[784] Kotlyar, E. E. and Nazarchuk, T. N., *Zh. Analit. Khim.*, **16**, 688 (1961).
[785] Chernikhov, Yu. A., and Dobkina, B. M., *Zavodsk. Lab.*, **16**, 402 (1950); **15**, 1144 (1949).
[786] Malissa, H., *Anal. Chim. Acta*, **27**, 402 (1962).
[787] Belousov, E. A. and Kirilov, V. M., *Zh. Obshch. Khim.*, **40**, 2134 (1970).
[788] Lanin, V. A., Dyadin, Yu. A., Yakovleva, N. I., Mironova, Z. N. and Yakovlev, I. I., *Izv. Sib. Otd. Akad. Nauk SSSR, Ser. Khim. Nauk*, **1970**, 159; *Chem. Abstr.*, **74**, 57827d (1971).
[789] Kolařík, Z., Šístková, N. and Hejná, J., *Proc. 5th Internat. Conf. Solvent Extr. Chem.*, **1969**, 59 (see [146]).
[790] Peppard, D. F., Mason, G. W., Maier, J. L. and Driscoll, W. J., *J. Inorg. Nucl. Chem.*, **4**, 334 (1957).
[790a] Tarasova, V. A., Levin, I. S. and Rodina, T. F., *Zh. Analit. Khim.*, **32**, 719 (1977).

[791] Ferraro, J. R., Mason, G. W. and Peppard, D. F., *J. Inorg. Nucl. Chem.*, **22**, 285 (1961).
[792] Komarov, E. V., Komarov, V. N. and Pushlenkov, M. F., *Radiokhimiya*, **12**, 455 (1970).
[793] Peppard, D. F., Mason, G. W., Driscoll, W. J. and Sironen, R. J., *J. Inorg. Nucl. Chem.*, **7**, 276 (1958); **12**, 141 (1959).
[794] Peppard, D. F., Mason, G. W. and Andrejasich, C. M., *J. Inorg. Nucl. Chem.*, **25**, 1175 (1963).
[795] Mason, G. W. and Peppard, D. W., *Nucl. Sci. Eng.*, **17**, 247 (1963).
[796] Baybarz, R. D. and Leuze, R. E., *Nucl. Sci. Eng.*, **11**, 90 (1963).
[797] Peppard, D. F., Mason, G. W. and Sironen, R. J., *J. Inorg. Nucl. Chem.*, **10**, 117 (1959).
[798] Levin, I. S., *Zh. Analit. Khim.*, **23**, 673 (1968).
[798a] Navrátil, O., *Collect. Czech. Chem. Commun.*, **42**, 2778 (1977).
[798b] Barnes, J. E., Setchfield, J. H. and Williams, G. O. R., *J. Inorg. Nucl. Chem.*, **38**, 1065 (1976).
[799] Busev, A. I. and Shishkov, A. N., *Zh. Analit. Khim.*, **23**, 181 (1968); **22**, 20 (1967).
[800] Levin, I. S., and Balakireva, N. A., *Talanta*, **17**, 915 (1970).
[801] Dyrssen, D. and Liem, D. H., *Acta Chem. Scand.*, **14**, 1100 (1960).
[802] Gureev, E. S., Kosyakov, V. N. and Yakovlev, G. N., *Radiokhimiya*, **6**, 655 (1964).
[803] Hardy, C. J., Greenfield, B. F. and Scargill, D. J., *J. Chem. Soc.*, **1961**, 174.
[804] Baes, C. F., *J. Inorg. Nucl. Chem.*, **24**, 707 (1962).
[805] Stulzoft, O., *Thèse Doct.*, Fac. Sci., Orsay (Univ. Paris), 1970.
[806] Scadden, E. M. and Ballou, N. E., *Anal. Chem.*, **25**, 1602 (1953).
[807] Sekine, T. and Dyrssen, D., *Talanta*, **11**, 867 (1964).
[808] Hála, J. and Taborská, E., *J. Radioanal. Chem.*, **30**, 329 (1976).
[809] Sheka, Z. A. and Kriss, E. E., *Zh. Neorgan. Khim.*, **7**, 658 (1962).
[810] Kononenko, L. I., Poluéktov, N. S. and Nikonova, M. P., *Zavodsk. Lab.*, **30**, 779 (1964).
[811] Gorlach, V. F., Pyatnitskii, I. V. and Kostyuchenko, L. P., *Ukr. Khim. Zh.*, **36**, 1260 (1970).
[812] Shevchenko, V. B., Mikhailov, V. A. and Zavalskii, Yu. P., *Zh. Analit. Khim.*, **13**, 1955 (1958).
[813] Kertes, A. S. and Beck, A. J., *J. Chem. Soc.*, **1961**, 5046.
[814] Samodelev, A. P., *Radiokhimiya*, **6**, 286 (1964).
[814a] Duyckaerts, G. and Dreze, P., *Bull. Soc. Chim. Belg.*, **71**, 306 (1960).
[815] Pyatnitskii, I. V., Glishenko, L. M. and Gerasina, V. K., *Ukr. Khim. Zh.*, **36**, 830 (1970).
[816] Dyrssen, D. and Krašovec, F., *Acta Chem. Scand.*, **13**, 561 (1959).
[817] Dyrssen, D., *Acta Chem. Scand.*, **11**, 1277 (1957).
[818] Hardy, C. J. and Scargill, D. J., *J. Inorg. Nucl. Chem.*, **17**, 337 (1961).
[819] Mohai, M., Upor, E. and Klesch, K., *Proc. Third Anal. Chem. Conf., Budapest 1970*, Akadémiai Kiadó, Budapest, **1**, 81 (1970).
[820] Peppard, D. F., Mason, G. W. and Moline, S. W., *J. Inorg. Nucl. Chem.*, **5**, 141 (1957).
[821] Cattrall, R. W., *Aust. J. Chem.*, **14**, 163 (1961).

[822] Campbell, M. H., *Anal. Chem.*, **36**, 2065 (1964).
[822a] El-Yamani, I. S. and Abd El-Messieh, E. N., *Radiochem. Radioanal. Lett.*, **33**, 353 (1978).
[822b] El-Yamani, I. S., Farah, M. Y. and Abd El-Messieh, E. N., *J. Radioanal. Chem.*, **45**, 147 (1978).
[823] Kimura, K., *Bull. Chem. Soc. Japan*, **33**, 1038 (1960); **34**, 63 (1961).
[824] Levin, I. S., and Yukhin, Yu. M., *Izv. Sib. Otd. Akad. Nauk SSSR, Ser. Khim. Nauk* **3**, 64 (1970).
[825] Butler, F. E., *Anal. Chem.*, **35**, 2069 (1963).
[826] Peppard, D. F., Mason, G. W., McCarty, S. and Johnson, F. D., *J. Inorg. Nucl. Chem.*, **24**, 321 (1962).
[827] Upor, E. and Klesch, K., *Magy. Kem. Foly.*, **80**, 298 (1974).
[828] Peppard, D. F., Mason, G. W. and Hucker, I., *J. Inorg. Nucl. Chem.*, **18**, 245 (1961).
[829] Weaver, B. and Kappekman, F. A., *U.S. At. Energy Comm. Rept.*, ORNL-3559 (1964).
[830] Baes, C. F. and Baker, H. T., *J. Phys. Chem.*, **64**, 89 (1960).
[831] Goliński, M., *Przem. Chem.*, **48**, 213 (1969).
[832] Bosholm, J. and Grosse-Ruyken, H. J., *J. Prakt. Chem.*, **26**, 79, 83 (1964).
[833] Pierce, B. and Peck, P. F., *Analyst (London)*, **88**, 217 (1963).
[834] Shiokawa I., Matsumoto A., Takatsugi, K., and Hirasima Y., *Kogyo Kagaku Zasshi*, **74**, 14 (1971).
[835] Nakamura, E., *Bull. Chem. Soc. Japan.*, **34**, 402 (1961).
[836] Hardy, C. J., Scargill, D. and Fletcher, J. M., *J. Inorg. Nucl. Chem.*, **7**, 257 (1958).
[837] Neirinckx, R. D., *Anal. Chim. Acta*, **54**, 357 (1971).
[838] Alian, A. and Haggag, A., *J. Radioanal. Chem.*, **20**, 429 (1974).
[838a] Yukhin, Yu. M., Levin, I. S. and Turchinskaya, G. A., *Zh. Neorgan. Khim.*, **23**, 142 (1978).
[839] Peppard, D. F., Mason, G. W. and McCarty, S., *J. Inorg. Nucl. Chem.*, **13**, 138 (1960).
[840] Levin, I. S., Rodina, T. F. and Vorsina, I. A., *Zh. Neorgan. Khim.*, **13**, 1611 (1968); **15**, 496 (1970).
[841] Rodina, T. F., Kolomiichuk, V. S. and Levin, I. S., *Zh. Analit. Khim.*, **28**, 1090 (1973).
[842] Bykhovtsov, V. L. and Zimina, T. Ya., *Radiokhimiya*, **12**, 686 (1970).
[843] Sato, T., *J. Inorg. Nucl. Chem.*, **24**, 699 (1962); **25**, 109 (1963).
[844] Sato, T., *Naturwissenschaften*, **50**, 19 (1963).
[845] Peppard, D. F., Mason, G. W. and Andrejasich, C. M., *J. Inorg. Nucl. Chem.*, **24**, 1387 (1962).
[846] Urbański, T. S. and Minc, S., *Nukleonika*, **7**, 703 (1963).
[847] Sato, T., *J. Inorg. Nucl. Chem.*, **26**, 311 (1964).
[848] Ihle, H., Michael, H. and Murrenhoff, A. P., *J. Inorg. Nucl. Chem.*, **25**, 734 (1963).
[849] Ulyanov, V. S., Sviridova, R. A. and Laskorin, B. N., *Radiokhimiya*, **8**, 416 (1966).
[850] Laskorin, B. N. and Smirnov, V. F., *Zh. Prikl. Khim.*, **38**, 2226 (1965).
[851] Schmid, E. R. and Pfannhauser, W., *Mikrochim. Acta*, **1971**, 434.

[852] Rigg, T. and Garner, J. O., *J. Inorg. Nucl. Chem.*, **29**, 2019 (1967).
[853] Vdovenko, V. M. and Varilov, N. G., *Radiokhimiya*, **11**, 384 (1969).
[854] Sato, T. and Takeda, T., *J. Inorg. Nucl. Chem.*, **32**, 3387 (1970).
[855] Sato, T., *Anal. Chim. Acta*, **52**, 183 (1970).
[856] Yagodin, G. A. and Tarasov, V. V., *Radiokhimiya*, **12**, 644 (1970).
[857] Brunisholz, G., Hirsbrunner, W. and Roulet, R., *Helv. Chim. Acta*, **55**, 2947 (1972).
[858] Harada, T. and Smutz, M., *J. Inorg. Nucl. Chem.*, **32**, 649 (1970).
[859] Gusmini, S. and Nonnemacher, R., *Rapp. CEA*, **1970**, No. R 4004, 31.
[860] Tomažič, B. B., *J. Less-Common Met.*, **22**, 495 (1970).
[861] Roelandts, J. and Duyckaerts, G., *Anal. Chim. Acta*, **68**, 131 (1974).
[862] Aly, H. F. and Latimer, R. M., *Radiochim. Acta*, **14**, 27 (1970).
[862a] Mitsugashira, T., Yamana, H. and Suzuki, S., *Bull. Chem. Soc. Japan*, **50**, 2913 (1977).
[862b] Cecille, L., Le Stang, M. and Manone, F., *Radiochem. Radioanal. Lett.*, **31**, 29 (1977).
[863] Nikolaev, A. V., Durasov, V. B. and Mirnova, Z. A., *Dokl. Akad. Nauk SSSR*, **186**, 603 (1969).
[864] Tuck, D. G. and Hála, J., *J. Chem. Soc.*, **1970**, 3242.
[865] Haffenden, W. J. and Tuck, D. G., *Trans. Faraday Soc.*, **66**, 2526 (1970).
[866] Vdovenko, A. M., Bulyanisto, L. S. and Savoskina, G. B., *Radiokhimiya*, **12**, 650, 654 (1970).
[867] Zaer, E. E., Chuchalin, L. K., Peshchevitskii, V. I., Kuzin, I. A. and Shel'kovnikova, O. S., *Zh. Neorgan. Khim.*, **15**, 3128 (1970).
[868] Takezhanov, S. T., Getskin, L. S., Pashkov, G. L. and Svetasheva, I. M., *Sb. Nauchn. Tr., Vses. Nauchno-Issled. Gornometall. Inst. Tsvetn. Met.*, **1970**, 57.
[869] Chuchalin, L. K., Grankina, Z. A., Kranenko, V. I., Peshchevitskii, V. I. and Kuzin, I. A., *Zh. Fiz. Khim.*, **44**, 3140 (1970).
[870] Ishimori, T., Akatsu, E., Kataoka, S., Osakabe, T., *J. Nucl. Sci. Technol.*, **1**, 18 (1964).
[871] Specker, H. and Pappert, W. Z., *Z. Anorg. Allg. Chem.*, **341**, 287 (1695).
[872] Kozlova, M. D. and Levin, V. I., *Radiokhimiya*, **7**, 430, 437, 534 (1965).
[873] Yadav, A. A. and Khopkar, S. M., *Mikrochim. Acta*, **1973**, 464.
[873a] Kuznetsov, R. A., *Zh. Analit. Khim.*, **32**, 2343 (1977).
[874] Morris, D. F. C. and Jones, M. W., *J. Inorg. Nucl. Chem.*, **27**, 2454 (1965).
[875] Yadav, A. A. and Khopkar, S. M., *Sep. Sci.*, **5**, 637 (1970).
[876] Aoki, F. and Tomioka, H., *Bull. Chem. Soc. Japan*, **38**, 1557 (1965).
[877] Gavrilova, I. M., Klyuchnikov, V. M., Korovin, S. S., Voronskaya, G. N. and Apraskin, I. A., *Zh. Neorgan. Khim.*, **20**, 468 (1975).
[878] Jackwerth, E. and Specker, H., *Z. Anal. Chem.*, **177**, 327 (1960).
[878a] Belousov, E. A. and Alovyainkov, A. A., *Zh. Neorgan. Khim.*, **21**, 2175 (1976).
[879] Gorbenko, F. P. and Sachko, V. V., *Zh. Analit. Khim.*, **20**, 309 (1965); **18**, 1497 (1963).
[880] Gorbenko, F. P. and Sachko, V. V., *Zavodsk. Lab.*, **30**, 943 (1964).
[881] Irving, H. and Edgington, D. N., *J. Inorg. Nucl. Chem.*, **10**, 306 (1959).
[882] Morris, D. F. C., Short, E. L. and Slater, D. N., *J. Inorg. Nucl. Chem.*, **26**, 627 (1964).
[883] Chatelet, M. and Nicaud, C., *Compt. Rend.*, **242**, 1891 (1956).

[884] Morris, D. F. C. and Bell, C. F., *J. Inorg. Nucl. Chem.*, **10**, 337 (1959).
[885] Athavale, W. T., Gulanave, S. W. and Tillu, M. M., *Anal. Chim. Acta*, **23**, 487 (1960).
[886] Musil, A. and Weidmann, G., *Mikrochim. Acta*, **1959**, 476.
[887] De, A. K. and Sen, A. K., *Sep. Sci.*, **1**, 641 (1966).
[888] Tuck, D. G., *Anal. Chim. Acta*, **27**, 296 (1962).
[889] Sastri, M. N. and Sundar, D. S., *Z. Anal. Chem.*, **195**, 343 (1963).
[890] Weidmann, G., *Can. J. Chem.*, **38**, 459 (1960).
[891] Specker, H. and Shirodker, R., *Z. Anal. Chem.*, **214**, 401 (1965).
[892] Shevchenko, V. B., Shilin, I. V. and Zhdanov, Yu. F., *Zh. Neorgan. Khim.*, **5**, 1366 (1960).
[893] Specker, H. and Jackwerth, E., *Naturwissenschaften*, **46**, 446 (1959).
[894] Melnick, L. M., Freiser, H. and Beeghly, H. F., *Anal. Chem.*, **25**, 856 (1953).
[895] Aven, M. and Freiser, H., *Anal. Chim. Acta*, **6**, 412 (1952).
[896] Jackwerth, E., *Z. Anal. Chem.*, **206**, 335 (1964).
[897] Specker, H. and Cremer, M., *Z. Anal. Chem.*, **167**, 110 (1959).
[898] Majumdar, S. K. and De, A. K., *Talanta*, **7**, 1 (1960).
[899] Mozońska, D., *Zesz. Nauk. Politech. Slask., Chem.*, **24**, 105 (1964).
[900] De, A. K. and Sen, A. K., *Talanta*, **14**, 629 (1962).
[901] Pfeifer, V., *J. Radioanal. Chem.*, **6**, 47 (1970).
[901a] Kalyanaraman, S. and Khopkar, S. M., *Indian J. Chem.*, **A15**, 1031 (1977).
[902] Lebedeva, E. N., Korovin, S. S. and Rozen, A. M., *Zh. Neorgan. Khim.*, **9**, 1744 (1964).
[903] Apraskin, I. A., Glyubokov, Yu. M., Korovin, S. S. and Reznik, A. M., *Zh. Neorgan. Khim.*, **9**, 2023 (1964).
[904] Ichinose, N., *Talanta*, **19**, 1644 (1972).
[905] Zharovskii, F. G. and Kostova, R. V., *Ukr. Khim. Zh.*, **37**, 939 (1971).
[906] Jackwerth, E. and Specker, H., *Z. Anal. Chem.*, **176**, 81 (1960).
[907] Goliński, M., *Nukleonika*, **17**, 439 (1972).
[908] Belousov, E. A. and Volkin, N. N., *Izv. Vyssh. Uchebn. Zaved. Tsvetn. Metall.*, **13**, 66 (1970); *Chem. Abstr.*, **73**, 102551e (1970).
[909] Golyub, A. M., Olevinskii, M. I. and Lutsenko, E. F., *Ukr. Khim. Zh.*, **30**, 1274 (1964).
[910] Awwal, M. A., *Anal. Chem.*, **35**, 2048 (1963).
[911] Edge, R. A., *Anal. Chim. Acta*, **27**, 369 (1962).
[912] Korpusov, G. V., Patrusheva, E. N., Petrokova, N. P. and Sokolova, V. F., *Redkozem. Elementy, Akad. Nauk SSSR, Inst. Geokhim. Analit. Khim.*, **1963**, 224.
[913] Yoshida, H., *J. Inorg. Nucl. Chem.*, **24**, 4257 (1962).
[914] Surovskaya, N. A. and Kuznetsova, Yu. S., *Nauchn. Soobshch., Inst. Gorn. Dela Moscow*, **47**, 48 (1969); *Chem. Abstr.*, **73**, 113524z (1970).
[915] Yadav, A. A. and Khopkar, S. M., *Sep. Sci.*, **4**, 349 (1969).
[916] Edge, R. A., Dunn, J. D. and Ahrens, L. H., *Anal. Chim. Acta*, **27**, 551 (1962).
[917] Busev, A. I. and Frolkina, V. A., *Zh. Neorgan. Khim.*, **9**, 2481 (1964).
[918] Caiozzi, M., Zunino, H. and Sepulveda, L., *Talanta*, **16**, 1590 (1969).
[919] Aratono, Y., Ueno, K. and Ishimori, T., *J. Nucl. Sci. Technol.*, **8**, 241 (1971).
[920] Pierret, J. A. and Wilhelm, H. A., *U.S. At. Energy Comm. Rept.*, ISC-796, 26 (1956).

[921] Ryabnikov, D. I. and Volinets, M. R., *Zh. Analit. Khim.*, **14**, 700 (1959).
[922] Hardy, C. J. and Scargill, D., *J. Inorg. Nucl. Chem.*, **13**, 174 (1960).
[923] Alcock, K., Best, G. F., Hosford, E. and McKay, H. A., *J. Inorg. Nucl. Chem.*, **6**, 328 (1959).
[924] Keder, W. E., *J. Inorg. Nucl. Chem.*, **16**, 138 (1960).
[925] Khudinov, E. G. and Yakovlev, G. N., *Radiokhimiya*, **4**, 375 (1962).
[926] Usashev, V. N. and Chaikovskii, A. A., *Radiokhimiya*, **8**, 48 (1966).
[927] Bulanova, I. D. and Vorob'ev, A. M., *Radiokhimiya*, **6**, 621 (1964).
[928] Shankar, J., Venkateswarlu, K. S. and Gopinathan, C., *J. Inorg. Nucl. Chem.*, **25**, 57 (1963).
[929] Yadav, A. A. and Khopkar, S. M., *Talanta*, **18**, 833 (1971).
[930] Faye, G. H. and Inman, W. R., *Anal. Chem.*, **35**, 985 (1963).
[931] Berg, E. W. and Lau, E. Y., *Anal. Chim. Acta*, **27**, 248 (1962).
[932] Belousov, E. A. and Volkin, N. N., *Zh. Neorgan. Khim.*, **15**, 2786 (1970).
[933] Morris, D. F. C. and Scargill, D., *Anal. Chim. Acta*, **14**, 57 (1956).
[934] Petrov, K. A., Shevchenko, V. B., Timoshev, V. G., Maklyaev, F. A., Fokin, A. V., Rodionov, A. V., Balandina, V. V., El'kina, A. V., Nagnikeda, Z. I. and Volkova, A. A., *Zh. Neorgan. Khim.*, **5**, 498 (1960).
[935] Goldschmidt, B., Regnaut, P. and Prevot, I., *Proc. 1st Intern. Conf. Peaceful Uses At. Energy, Geneva*, **9**, 349, 492 (1956).
[936] Flanary, J. R., *Proc. 1st Intern. Conf. Peaceful Uses At. Energy, Geneva*, **9**, 528, 539 (1956).
[937] Healy, T. V. and Gardner, A. W., *J. Inorg. Nucl. Chem.*, **7**, 242 (1958).
[938] Nairn, J. S., Collins, D. N. and McKay, H. A., *Proc. 2nd Intern. Conf. Peaceful Uses At. Energy, Geneva*, **17**, 216, 719 (1958).
[939] Beran, M., *Proc. 5th Intern. Conf. Solvent Extr. Chem.*, **1969**, 75 (see [146]).
[940] Rozen, A. M. and Moiseenko, E. I., *Zh. Neorgan. Khim.*, **4**, 1209 (1959).
[941] Yadav, A. A. and Khopkar, S. M., *Bull. Chem. Soc. Japan*, **44**, 693 (1971).
[942] Peppard, D. F., Mason, G. W. and Maier, J. L., *J. Inorg. Nucl. Chem.*, **3**, 215 (1956).
[943] Favorskaya, D. F. and Romanova, A. D., *Zh. Neorgan. Khim.*, **11**, 1227 (1966).
[944] Yoshida, H., *Bunseki Kagaku*, **12**, 169 (1963).
[945] Yadav, A. A. and Khopkar, S. M., *Chem. Anal. (Warsaw)*, **16**, 299 (1971).
[946] Woidich, H. and Pfannhauser, W., *Mikrochim. Acta*, **1973**, 279.
[947] Giganov, G. P. and Ponomarev, V. D., *Izv. Akad. Nauk Kaz. SSSR*, **5**, 119, 125 (1962); *Chem. Abstr.*, **58**, 9671 (1963).
[948] Starshev, V. N., Krilov, E. I. and Kozmin, Yu. A., *Zh. Neorgan. Khim.*, **10**, 2367 (1965).
[949] Zharovskii, F. G. and Vyazovskaya, L. M., *Ukr. Khim. Zh.*, **32**, 747 (1966).
[950] Yadav, A. A. and Khopkar, S. M., *Anal. Chim. Acta*, **45**, 355 (1969).
[951] Kuzin, I. A. and Shutalin, L. K., *Zh. Prikl. Khim.*, **38**, 2422 (1965).
[952] Widmer, H. M. and Dodson, R. W., *J. Phys. Chem.*, **74**, 4289 (1970).
[953] Schmid, E. R., *Monatsh. Chem.*, **101**, 1330 (1970).
[954] Gresky, A. T., *Proc. 1st Intern. Conf. Peaceful Uses At. Energy, Geneva*, **7**, 505, 540 (1956).
[955] Warf, J. C., *J. Am. Chem. Soc.*, **71**, 3257 (1949).
[956] Hahn, H. T., *J. Am. Chem. Soc.*, **79**, 4625 (1957).
[957] Sato, T., *J. Inorg. Nucl. Chem.*, **6**, 147, 334 (1958); **16**, 156 (1960).

[958] Gill, H. H., Rolf, R. F. and Armstrong, G. W., *Anal. Chem.*, **30**, 1788 (1958).
[959] Vdovenko, V. M., Lipovskii, A. A., Nikitina, S. A. and Yakoleva, N. E., *Radiokhimiya*, **7**, 509 (1965).
[960] Nemodruk, A. A., Novikov, Yu. P., Lukin, A. M. and Kalinina, I. D., *Zh. Analit. Khim.*, **16**, 292 (1961).
[961] Paige, B. E., Elliott, M. C. and Rein, J. E., *Anal. Chem.*, **29**, 1029 (1957).
[962] Koppikar, K. S., Korgaonkar, V. G. and Murthy, T. K. S., *Anal. Chim. Acta*, **20**, 366 (1959).
[963] Pushlenkov, M. F. and Shuvalov, O. N., *Radiokhimiya*, **5**, 551 (1963).
[964] Bakos, I., Szabó, E., András, L. and Iyer, N., *Proc. Symp. Coord. Chem.*, *Tihany, Hung.*, **1964**, 241.
[965] McKay, H. A. C. and Streeton, R. J., *J. Inorg. Nucl. Chem.*, **17**, 879 (1965).
[966] Upor, E., Gorbicz, L. and Nagy, G., *Magy. Kem. Foly.* **75**, 549 (1969); *Chem. Abstr.*, **72**, 62426p (1970).
[967] De, A. K. and Rahaman, M. S., *Talanta*, **11**, 601 (1964).
[968] Nazin, A. G., Levin, V. I. and Golotvina, M. M., *Metody Polucheniya Radioaktivn. Preparatov, Sb. Statei*, **1962**, 118; *Chem. Abstr.*, **58**, 6418b (1963).
[969] Goldin, A. S. and Velten, R. J., *Anal. Chem.*, **33**, 149 (1961).
[970] Velten, R. J. and Goldin, A. S., *Anal. Chem.*, **33**, 128 (1961).
[971] Kirby, H. W., *Anal. Chem.*, **29**, 1599 (1957).
[972] Rolf, R. F., *Anal. Chem.*, **33**, 125, 149 (1961).
[973] Huré, J., Rastoix, M. and Saint-James, R., *Anal. Chim. Acta*, **25**, 1 (1961).
[974] Wallace, R. M. and Pollock, H., *U.S. At. Energy Comm. Rept.*, DP-308, 26 (1958).
[975] Golub, A. M. and Sergunkin, V. N., *Zh. Prikl. Khim.*, **43**, 1203 (1970).
[976] Gorbenko, F. P., Kuchkina, E. D. and Olevinskii, M. I., *Radiokhimiya*, **12**, 661 (1970).
[977] White, J. C. and Ross, W. J., *U.S. At. Energy Comm. Rept.*, NAS-NS 3102 (1961).
[978] White, J. C. and Ross, W. J., *U.S. At. Energy Comm. Rept.*, ORNL-2326 (1957).
[979] Ross, W. J. and White, J. C., *U.S. At. Energy Comm. Rept.*, ORNL-2328 (1957).
[980] Varga, L. P. and Hume, D. N., *Inorg. Chem.*, **3**, 77 (1964).
[981] Alian, A. and Moustapha, Z., *Indian J. Chem.*, **3**, 390 (1965).
[982] Horton, C. A., *U.S. At. Energy Comm. Rept.*, ORNL-2662, 61 (1958).
[983] Weaver, B. and Horner, D. E., *J. Chem. Eng. Data*, **5**, 260 (1960).
[984] Martin, B. and Ockenden, D. W., *U.S. At. Energy Comm. Rept.*, 165 (W) (1960).
[985] Horner, D. F. and Coleman, C. F., *U.S. At. Energy Comm. Rept.*, ORNL-3051 (1961).
[986] Varga, L. P., Wakley, W. D., Nicolson, L. S., Madden, M. L. and Patterson, J., *Anal. Chem.*, **37**, 1003 (1965).
[987] Zaitsev, A. A., Lebedev, I. A., Piroznikov, S. V. and Yakovlev, G. N., *Radiokhimiya*, **6**, 440 (1964).
[988] Sato, T. and Yamatake, M., *J. Inorg. Nucl. Chem.*, **31**, 3633 (1969).
[989] Sato, T., Noshida, T. and Yamatake, M., *J. Appl. Chem. Biotechnol.*, **23**, 909 (1973).
[990] Young, J. P. and White, J. C., *Anal. Chem.*, **31**, 393 (1959).

[991] Hibbits, J. O., Kallman, S., Giustetti, W. and Oberthin, H. K., *Talanta*, **11**, 1462 (1964).
[992] Hibbits, J. O. and Rosenberg, A. F., Williams, R. T. and Kallman, S., *Talanta*, **11**, 1509 (1964).
[993] Rolf, R. F., *Anal. Chem.*, **36**, 1398 (1964).
[994] Shults, V. D. and Dunlap, L. B., *Anal. Chim. Acta*, **29**, 254 (1963).
[995] Korkisch, J. and Koch, W., *Mikrochim. Acta*, **1973**, 157.
[996] White, J. C. and Ross, W. J., *U.S. At. Energy Comm. Rept.*, ORNL-2498 (1958).
[997] Ziegler, M., *Naturwissenschaften*, **46**, 353 (1959).
[998] Bravo, O. and Iwamoto, R. T., *Anal. Chim. Acta*, **47**, 209 (1969).
[999] Adam, J. and Přibil, R., *Talanta*, **18**, 405 (1971).
[1000] Sheppard, J. C. and Warnock, R., *J. Inorg. Nucl. Chem.*, **26**, 1421 (1964).
[1001] Shevchuk, I. A. and Degtyarenko, L. I., *Ukr. Khim. Zh.*, **9**, 1112 (1962).
[1002] De, A. K., Khopkar, S. M. and Chalmers, R. A., *Solvent Extraction of Metals*, Van Nostrand Reinhold, London, 1970.
[1003] Knapp, J. R., van Aman, R. E. and Kanzelmeyer, J. H., *Anal. Chem.*, **34**, 1374 (1962).
[1004] McDonald, C. W. and Moore, F. L., *Anal. Chem.*, **45**, 983 (1973).
[1005] Suzuki, T. and Sotobayashi, T., *Bunseki Kagaku*, **13**, 1103 (1964).
[1006] Nakagawa, G., *Nippon Kagaku Zasshi*, **81**, 444, 747, 1255, 1533 (1960); **82**, 1042 (1961).
[1007] Vosyatninskii, A. I. and Kisel, N. A., *Ukr. Khim. Zh.*, **36**, 712 (1970).
[1008] Tonouchi, S., Suzuki, T., Sotobayashi, T. and Koyama, S., *Bunseki Kagaku*, **20**, 1453 (1971).
[1009] Mahlman, A. A., Leddicotte, G. W. and Moore, F. L., *Anal. Chem.*, **26**, 1939 (1954).
[1010] Moore, F. L., *Anal. Chem.*, **30**, 908 (1958).
[1011] Dono, T., Nakagawa, G. and Wada, H., *Bunseki Kagaku*, **11**, 654 (1962).
[1012] Fujimoto, M. and Nakatsukasa, Y., *Anal. Chim. Acta*, **27**, 373 (1962).
[1013] Selmer-Olsen, A. R., *Anal. Chim. Acta*, **31**, 33 (1964).
[1014] Pyatnitskii, I. V. and Zharchenko, R. S., *Ukr. Khim. Zh.*, **32**, 503 (1966).
[1015] McClellan, B. E., Meredith, M. K., Parmele, R. and Beck, J. P., *Anal. Chem.*, **46**, 306 (1974).
[1016] Moore, F. L., *Anal. Chem.*, **29**, 1660 (1957).
[1017] Żmijewska, W., *J. Radioanal. Chem.*, **10**, 187 (1972).
[1018] Adam, J. and Přibil, R., *Talanta*, **18**, 91 (1971); **21**, 616 (1974).
[1019] Shevchuk, I. A. and Simonova, T. N., *Ukr. Khim. Zh.*, **30**, 983 (1964).
[1020] Sastri, M. M. and Sundar, D. S., *Anal. Chim. Acta*, **33**, 340 (1965).
[1021] Davies, C. W. and Pabel, N. B., *J. Inorg. Nucl. Chem.*, **31**, 2615 (1969).
[1022] Milyukova, M. S., Myasoedov, B. F. and Ryshkova, L. V., *Zh. Analit. Khim.*, **27**, 67 (1972).
[1023] von Baeckmann, A. and Glemser, O., *Z. Anal. Chem.*, **187**, 429 (1962).
[1024] Sharkov, A. I., Perfilev, A. I. and Avlasovich, L. M., *Zh. Prikl. Khim.*, **45**, 2734 (1972).
[1025] Coleman, C. F., Brown, K. B., Moore, J. G. and Crouse, D. J., *Ind. Eng. Chem.*, **50**, 1756 (1958).
[1026] Cerrai, E. and Testa, C., *Energ. Nucl. (Milan)*, **6**, 707, 768 (1959).

[1027] Moore, J. G., Blake, C. A. and Schmitt, J. M., *U.S. At. Energy Comm. Rept.*, ORNL-2346 (1957).
[1028] Bulloch, J. I. and Tuck, D. G., *J. Inorg. Nucl. Chem.*, **28**, 1103 (1966).
[1029] Irving, H. M. and Damodaran, A. D., *Anal. Chim. Acta*, **53**, 277 (1971).
[1030] Irving, H. M. and Damodaran, A. D., *Anal. Chim. Acta*, **50**, 277 (1970).
[1030a] Fedorenko, I. V. and Ivanova, T. I., *Zh. Neorg. Khim.*, **10**, 721 (1965).
[1031] Coute, D. R. and Markham, J. J., *J. Inorg. Nucl. Chem.*, **33**, 2247 (1971).
[1032] Laskorin, V. M., Ulyanov, V. S. and Sviridova, R. A., *Zh. Prikl. Khim.*, **35**, 2409 (1962).
[1033] Panomareva, A. A. and Agrinskaya, N. A., *Zavodsk. Lab.*, **38**, 790 (1972).
[1034] Amand, J. and Vieux, A. S., *Bull. Soc. Chim. Fr.*, **1969**, 3366.
[1035] Kollar, R., Plichon, V. and Saulnier, J., *Bull. Soc., Chim. Fr.*, **1969**, 2193.
[1036] Yatirajam, V. and Ram Jaswant, *Anal. Chim. Acta*, **59**, 381 (1972).
[1037] Ellenburg, J. Y., Leddicotte, G. W. and Moore, F. L., *Anal. Chem.*, **26**, 1045 (1954).
[1038] Sato, T. and Kikuchi, S., *Z. Anorg. Allg. Chem.*, **365**, 330 (1969).
[1039] Alimarin, I. P., Ivanov, I. A. and Gibalo, I. M., *Zh. Analit. Khim.*, **24**, 1521 (1969).
[1040] Moore, F. L., *U.S. At. Energy Comm. Rept.*, ORNL-1635 (1954).
[1041] Marchart, H. and Hecht, F., *Mikrochim. Acta*, **1962**, 1152.
[1042] Nevzorov, A. N. and Buchkov, L. A., *Zh. Analit. Khim.*, **19**, 1336 (1964).
[1043] Pietrov, Ch. T., *U.S. At. Energy Comm. Rept.*, TID-5772 (1960).
[1043a] McDonald, C., Mahayni, M. M. and Kanjo, M., *Sep. Sci. Technol.*, **13**, 429 (1978).
[1044] Gindin, L. M. and Ivanova, S. N., *Izv. Sib. Otd. Akad. Nauk SSSR*, **1964**, 28.
[1045] Khattak, M. A. and Magee, R. J., *Talanta*, **12**, 733 (1965).
[1045a] Agrinskaya, N. A. and Petrashen, V. J., *Zh. Analit. Khim.*, **15**, 155 (1970).
[1046] Yatirajam, V. and Kakkar, L. R., *Anal. Chim. Acta*, **52**, 555 (1970).
[1047] Hallaba, E., Azzam, R. and Ragab, A., *Mikrochem. J.*, **14**, 126 (1969).
[1048] Pal'shin, E. S., Palei, P. N., Davydov, A. V. and Ivanova, L. A., *Zh. Analit. Khim.*, **24**, 797 (1969).
[1049] Nakagawa, G., *Bunseki Kagaku*, **9**, 10, 821 (1960).
[1050] Ko, R., *Anal. Chem.*, **36**, 1290, 2513 (1964).
[1051] Coleman, C. F., Kappelmann, F. A. and Weaver, B., *Nucl. Sci. Eng.*, **8**, 507 (1960).
[1052] Nakagawa, G., *Nippon Kagaku Zasshi*, **81**, 1258 (1960).
[1053] Moore, J. G., Brown, K. B. and Coleman, C. F., *U.S. At. Energy Comm. Rept.*, ORNL-1922 (1955).
[1054] Pyatnitskii, I. V. and Zharchenko, R. S., *Ukr. Khim. Zh.*, **36**, 714 (1965).
[1055] Suzuki, T. and Sotobayashi, T., *Bunseki Kagaku*, **14**, 414 (1965).
[1056] Brown, K. B., Coleman, C. F., Crouse, D. J. and Ryon, A. D., *U.S. At. Energy Comm. Rept.*, ORNL-2268 (1957).
[1056a] Vieux, A. S., Bassolila, L. and Rutagengwa, N., *Hydrometallurgy*, **2**, 351 (1977).
[1057] Scroggie, L. E. and Dean, J. S., *Anal. Chim. Acta*, **21**, 282 (1959).
[1058] Andrew, T. R. and Nichols, P. N., *Analyst (London)*, **90**, 161 (1965).
[1059] Cerrai, E. and Testa, C., *Anal. Chim. Acta*, **26**, 204 (1962).
[1060a] Sato, T. and Watanabe, H., *Anal. Chim. Acta*, **49**, 463 (1970).

[1061] Sato, T. and Watanabe, H., *Anal. Chim. Acta*, **54**, 439 (1971).
[1062] Moore, F. L., *Anal. Chem.*, **33**, 748 (1961).
[1062a] Yonezawa, Ch. and Onishi, H., *Anal. Chim. Acta*, **96**, 211 (1978).
[1063] Moore, F. L., *Anal. Chem.*, **41**, 1658 (1969).
[1064] Keder, W. E., Sheppard, J. C. and Wilson, A. S., *J. Inorg. Nucl. Chem.*, **12**, 327 (1960).
[1065] Brown, K. B., *U.S. At. Energy Rept.*, ORNL-CF-60-1-119 (1960).
[1066] Zakharin, B. S., Zemlyanukhin, V. I. and Shevchenko, V. B., *Radiokhimiya*, **12**, 577 (1970).
[1067] Moore, F. L., *Anal. Chem.*, **27**, 70 (1955).
[1068] Ichikawa, F. and Urono, S., *Bull. Chem. Soc. Japan*, **33**, 569 (1960).
[1069] Moore, F. L., and Reynolds, S. A., *Anal. Chem.*, **29**, 1596 (1957).
[1070] Wilson, A. S., *Proc. 2nd Intern. Conf. Peaceful Uses At. Energy*, Geneva, **17**, 348 (1958).
[1071] Taube, M., *Polish Acad. Sci. Inst. Nucl. Res. Rept.*, 165/V (1960).
[1072] Crouse, D. J. and Brown, K. B., *U.S. At. Energy Comm. Rept.*, ORNL-2720 (1959).
[1073] Reynolds, S. A. and Eldridge, J.S., *U.S. At. Energy Comm. Rept.*, ORNL-CF-57-1-3 (1957).
[1074] Hodora, I. and Balonka, I., *Anal. Chem.*, **43**, 1213 (1971).
[1075] Medvedeva, E. and Gromov, B. V., *Tr. Mosk. Khim.-Tekhnol. Inst.*, **47**, 140 (1964).
[1076] Matsuo, S., *Nippon Kagaku Zasshi*, **82**, 459 (1961).
[1077] Moore, F. L., *Anal. Chem.*, **32**, 1075 (1060).
[1078] Seeley, F. G. and Crouse, D. J., *J. Chem. Eng. Data*, **16**, 393 (1971).
[1078a] Irving, H. M. N. H. and Al-Jarrah, R. H., *Anal. Chim. Acta*, **74**, 321 (1975).
[1079] Zolotov, Yu. A., *Zh. Analit. Khim.*, **26**, 20 (1971).
[1080] Boswell, C. R. and Brooks, R. R., *Mikrochim. Acta*, **1965**, 814.
[1081] Boswell, C. R. and Brooks, R. R., *Anal. Chim. Acta*, **33**, 117 (1965).
[1082] Iofa, B. Z. and Yushchenko, A. S., *Radiokhimiya*, **8**, 621, 707 (1966).
[1083] Dakar, G. M., Iofa, B. Z. and Nesemyanov, A. N., *Radiokhimiya*, **5**, 428 (1963).
[1084] Iofa, B. Z. and Dakar, G. M., *Radiokhimiya*, **5**, 490 (1960); **7**, 25 (1965).
[1085] Tanaka, K. and Takagi, N., *Anal. Chim. Acta*, **48**, 357 (1969).
[1086] Stará, V., *Talanta*, **17**, 341 (1970).
[1087] Jordanov, N. and Futekov, L., *Talanta*, **13**, 163 (1966).
[1088] Busev, A. I. and Frolkina, V. A., *Vestn. Mosk. Univ., Ser. II, Khim.*, **20** (4), 69 (1965); *Chem. Abstr.*, **63**, 17215f (1965).
[1089] Morgunov, A. F. and Fomin, V. V., *Zh. Neorgan. Khim.*, **8**, 508 (1963).
[1090] Chwastowska, J. and Podgórny, W., *Chem. Anal. (Warsaw)*, **20**, 53 (1975).
[1091] Študlar, K., *Collect. Czech. Chem. Commun.*, **31**, 1999 (1966).
[1092] Grimanis, A. P. and Hadzistelios, I., *Anal. Chim. Acta*, **41**, 15 (1968).
[1093] Goto, H. and Kakita, Y., *Nippon Kagaku Zasshi*, **82**, 1212 (1961).
[1094] Goto, H., Kakita, Y. and Ichinose, N., *Nippon Kagaku Zasshi*, **87**, 962 (1966).
[1095] Ichinose, N., *Z. Anal. Chem.*, **255**, 109 (1971).
[1096] Rόżycki, C., *Chem. Anal. (Warsaw)*, **14**, 755 (1969); **15**, 3 (1970).
[1097] Rόżycki, C. and Lachowicz, E., *Chem. Anal. (Warsaw)*, **15**, 255 (1970).
[1098] Vinarov, I V, Orlova, A. I., Grigoreva, L. P, Iltsenko, A. L and Egorova, A. L., *Zh. Analit. Khim.*, **25**, 2132 (1970).

[1099] Mironov, V. E. and Rutkovskii, Yu. I., *Zh. Neorgan. Khim.*, **10**, 1069 (1965).
[1100] Doll, W. and Specker, H., *Z. Anal. Chem.*, **161**, 354 (1958).
[1101] Specker, H. and Doll, W., *Z. Anal. Chem.*, **152**, 178 (1956).
[1102] Green, M. and Kafalas, J. A., *J. Chem. Phys.*, **22**, 760 (1954).
[1103] Baiulescu, G. E. and Craciun, I., *Rev. Chim. (Bucharest)*, **25**, 578 (1974).
[1104] Wells, J. F. and Hunter, D. P., *Analyst (London)*, **73**, 671 (1948).
[1105] Gagliardi, E. and Wöss, H. P., *Z. Anal. Chem.*, **248**, 302 (1969).
[1106] Antipova-Korataeva, I. I., Zolotov, Yu. A. and Seryakova, I. V., *Zh. Neorgan. Khim.*, **9**, 1712 (1964).
[1107] Malyutina, T. M., Orlova, V. A. and Spivakov, B. Ya., *Zh. Analit. Khim.*, **29**, 790 (1974).
[1108] Keil, R., *Z. Anal. Chem.*, **249**, 172 (1970).
[1109] Rafaeloff, R., *Anal. Chem.*, **43**, 272 (1971).
[1110] Fischer, W., Hurre, W., Freese, W. and Hackstein, K. G., *Angew. Chem.*, **66**, 165 (1954).
[1111] Saginashvili, R. M. and Petrashen, V. I., *Zavodsk. Lab.*, **32**, 661 (1966).
[1112] Murach, N. N., Krapukhin, V. V., Kulikov, F. S., Chernyshev, V. I and Nekhamkin, L. G., *Zh. Prikl. Khim*, **34**, 2188 (1961).
[1113] Tandon, S. N. and Gupta, C. B., *Talanta*, **18**, 109 (1971).
[1114] Plaskin, I. N., Anisimova, Z. A. and Gintsova, V. G., *Zh. Neorgan. Khim.*, **11**, 854 (1966).
[1114a] Futekov, L., Stojanov, S. and Specker, H., *Z. Anal. Chem.*, **295**, 7 (1979).
[1115] Yatirajam, V. and Prosad, P., *Indian J. Chem.*, **3**, 544 (1965).
[1116] Gibalo, I. M., Albadri, J. S. and Eremina, G. V., *Zh. Analit. Khim.*, **22**, 816 (1976).
[1117] De, A. K. and Sen, A. K., *Talanta*, **13**, 853 (1966).
[1118] Moore, F. L., *U.S. At. Energy Comm. Rept.*, ORNL-175.
[1119] Golden, J. and Maddock, A. G., *J. Inorg. Nucl. Chem.*, **8**, 46 (1956).
[1120] Yadav, A. A. and Khopkar, S. M., *Talanta*, **18**, 833 (1971).
[1121] White, C. E. and Rose, N. J., *Anal. Chem.*, **25**, 351 (1953).
[1122] Iofa, B. Z., Mitrofanov, K. L., Plotnikova, M. V. and Kopach, S., *Radiokhimiya*, **6**, 419 (1964).
[1123] Havezov, I. and Nürnberg, H. W., *Z. Anal. Chem.*, **262**, 179 (1972).
[1124] Srivastava, T. N. and Rupainwar, D. C., *Bull. Chem. Soc. Japan*, **38**, 1792 (1965).
[1125] Bock, R., Kusche, H. and Bock, E., *Z. Anal. Chem.*, **138**, 167 (1953).
[1126] Irving, H. M. and Rossotti, F. J. C., *Analyst (London)*, **77**, 801 (1952).
[1127] Pohl, F. A. and Bonsels, W., *Z. Anal. Chem.*, **161**, 108 (1958).
[1128] Spoustová, J., *Hutn. Listy*, **25**, 129 (1970).
[1129] Kish, P. P., Zimomrya, I. I., Balog, I. S., Roman, V. V. and Ugrin, V. P., *Zh. Analit. Khim.*, **29**, 1539 (1974).
[1130] Denaro, A. R. and Occleshaw, V. J., *Anal. Chim. Acta*, **13**, 239 (1955).
[1131] Tepuia, T., Kawamura, M. and Nakamori, I., *Bunseki Kiki*, **9**, 228 (1971).
[1132] Tanaka, K., *Bunseki Kagaku*, **18**, 315 (1969); *Chem. Abstr.*, **71**, 16373a (1969).
[1133] Kono, T. and Nemori, A., *Bunseki Kagaku*, **24**, 419 (1975).
[1134] Kitahara, S., *Bull. Inst. Phys. Chem. Research*, (Tokyo), **24**, 454 (1948).
[1135] Jackwerth, E. and Specker, H., *Z. Anal. Chem.*, **177**, 327 (1960).
[1136] Kono, T. and Kobayashi, S., *Bunseki Kagaku*, **19**, 1491 (1970).

[1137] West, P. W. and Carlton, J. K., *Anal. Chim. Acta*, **6**, 406 (1952).
[1138] Stella, R. and Di Casa, M., *J. Radioanal. Chem.*, **16**, 183 (1973).
[1139] Zolotov. Yu. A., Fomina, A. I., Agrinskaya, N. A. and Antipova-Karataeva, I. I., *Zh. Analit. Khim.*, **27**, 2257 (1972).
[1140] Stará, V., *Talanta*, **18**, 228 (1971).
[1141] Paul, A. D. and Gibson, J. A., *Anal. Chem.*, **36**, 2321 (1964).
[1142] Kitahara, S., *Bull. Inst. Phys. Chem. Research (Tokyo)*, **25**, 165 (1959).
[1143] Bock, R. and Herrmann M., *Z. Anorg. Chem.*, **284**, 288 (1956).
[1144] Milner, G. W. C., Barnett, G. A. and Smales, A. A., *Analyst (London)*, **80**, 380 (1955).
[1145] Stevenson, P. C. and Hicks, H. G., *Anal. Chem.*, **25**, 1517 (1953).
[1146] Balchin, L. A. and Williams, D. J., *Analyst (London)*, **85**, 503 (1960).
[1147] Waterbury, G. R. and Bricker, C. E., *Anal. Chem.*, **30**, 5, 1007 (1958).
[1148] Chwastowska, J. and Szymczak, S., *Chem. Anal. (Warsaw)*, **16**, 143 (1971).
[1149] Yoshida, M. and Kitamura, N., *Bunseki Kagaku*, **14**, 323 (1965).
[1150] Bock, R., *Z. Anal. Chem.*, **133**, 110 (1951).
[1151] Kurz, E. and Koberg, G., *Z. Anal. Chem.*, **254**, 127 (1971).
[1152] Sukhanovskaya, A. I., Solovev, E. A., Tikhonov, G. P., Golobev, V. Yu. and Bezhevalnov, E. A., *Zh. Analit. Khim.*, **25**, 1563 (1970).
[1153] Kuchkina, E. D., Gorbenko, F. P. and Litvinova, G. V., *Zh. Analit. Khim.*, **28**, 595 (1973).
[1154] Peng, P. Y. and Sandell, E. B., *Anal. Chim. Acta*, **29**, 325 (1963).
[1155] Alimarin, I. P. and Polyanskii, V. N., *Zh. Analit. Khim.*, **8**, 266 (1953).
[1156] Yatirajam, V. and Ram Jaswant, *Mikrochim. Acta*, **1973**, 77.
[1157] Kim, C. H., Owens, C. M. and Smythe, L. E., *Talanta*, **21**, 445 (1974).
[1158] Minczewski, J. and Różycki, C., *Chem. Anal. (Warsaw)*, **9**, 601 (1964).
[1159] Trotskii, K. V., *Zh. Analit. Khim.*, **12**, 349 (1957).
[1160] Buksh, M. N. and Hume, D. N., *Anal. Chem.*, **27**, 116 (1955).
[1161] Minczewski, J. and Różycki, C., *Chem. Anal. (Warsaw)*, **10**, 463, 701 (1965).
[1162] Różycki, C., *Chem. Anal. (Warsaw)*, **18**, 145 (1973).
[1163] Różycki, C., *Chem. Anal. (Warsaw)*, **14**, 459 (1969).
[1164] Różycki, C., Lachowicz, E. and Jodełka, J., *Chem. Anal. (Warsaw)*, **19**, 639 (1974).
[1165] Minczewski, J. and Różycki, C., *Z. Anal. Chem.*, **239**, 158 (1968).
[1166] Mathieu, G. and Guiot, S., *Anal. Chim. Acta*, **52**, 335 (1970).
[1167] Kuzmin, N. M., Popova, G. D., Kuzovlev, I. A. and Solomatin, V. S., *Zh. Analit. Khim.*, **24**, 899 (1969).
[1168] Kuzmin, N. M., Vlasov, V. S. and Lambrev, V. G., *Zh. Analit. Khim.*, **27**, 1614 (1972).
[1169] Zelle, A. and Fijałkowski, J., *Chem. Anal. (Warsaw)*, **7**, 321 (1962).
[1170] Yamamoto, T. and Odashima, T., *Bunseki Kagaku*, **21**, 1614 (1972).
[1171] Miyamoto, M., *Bunseki Kagaku*, **9**, 748, 753, 869, 925 (1960).
[1172] Marczenko, Z., Kasiura, K. and Krasiejko, M., *Chem. Anal. (Warsaw)*, **14**, 1277 (1969).
[1173] Marczenko, Z. and Krasiejko, M., *Chem. Anal. (Warsaw)*, **15**, 1233 (1970).
[1174] Pohl, F. A. and Bonsels, W., *Mikrochim. Acta*, **1961**, 314.
[1175] Ackermann, G. and Köthe, J., *Z. Anal. Chem.*, **231**, 252 (1967).
[1176] Jackwerth, E., *Z. Anal. Chem.*, **211**, 254 (1965).
[1177] Jackwerth, E., *Z. Anal. Chem.*, **216**, 73 (1966).

[1178] Kuzmin, N. M., Solomatin, V. S., and Bystrova, V. A., *Sb. Metody analiza khimicheskikh reaktivov i preparatov* (Methods of Analysis of Chemical Reagents, Collection of Papers), IREA, **1970**, 44.
[1179] Kuzmin, N. M., Solomatin, V. S., Galaktionova, A. N. and Kuzovlev, I. A., *Zh. Analit. Khim.*, **24**, 725 (1969).
[1180] Antonov, A. V., Drygina, A. I. and Katykov, Yu. A., *Zavodsk. Lab.*, **33**, 967 (1967).
[1181] Yudelevich, I. G., Buyanova, L. M., Protopopova, N. P. and Yudina, N. G., *Zh. Analit. Khim.*, **25**, 1177 (1970).
[1182] Pohl, F. A., *Mikrochim. Acta*, **1954**, 258.
[1183] Nikitina, O. I. and Ivanova, N. K., *Zavodsk. Lab.*, **30**, 46 (1964).
[1184] Répás, P., Sajó, I. and Gegus, E., *Z. Anal. Chem.*, **207**, 263 (1965).
[1185] Lesbats, A. and Tardy, M. R., *J. Radioanal. Chem.*, **17**, 127 (1973).
[1186] Yudelevich, I. T., Buyanova, L. M., Shabirova, V. P. and Starshinova, N. P., *Izv. Sib. Otd. Akad. Nauk SSSR*, **1969**, No. 2, 98.
[1187] Yudelevich, I. T., Buyanova, L. M., Protopopova, N. P. and Dzhyashakueva, B. K., *Zavodsk. Lab.*, **35**, 426 (1969).
[1188] Sinyakova, S. I., Dydareva, A. G., Markova, I. V. and Talalaeva, I. N., *Zh. Analit. Khim.*, **18**, 337 (1967).
[1189] Goryushina, V. G. and Notkina, M. A., *Sb. Metody opredeleniya i analiza redkikh élementov* (Methods for Determination and Analysis of Rare Elements, Collection of Papers), Akad. Nauk SSSR, **1961**, 230.
[1190] Goryushina, V. G. and Notkina, M. A., *Sb. Metody analiza veshchestv vysokoi chistoty* (Methods of Analysis of High Purity Compounds, Collection of Papers), Izd. Nauka, **1965**, 193.
[1191] Yakovlev, Yu. A., Dogatkin, N. N. and Shulepnikov, M. N., *Sb. Metody analiza veshchestv vysokoi chistoty* (Methods of Analysis of High Purity Compounds, Collection of Papers), Izd. Nauka, **1963**, 198.
[1192] Notkina, M. A., *Sb. Metody analiza veshchestv vysokoi chistoty* (Methods of Analysis of High Purity Compounds, Collection of Papers), Izd. Nauka, **1965**, 203.
[1193] Kasiura, K., *Chem. Anal. (Warsaw)*, **11**, 141 (1966).
[1194] Goryushina, V. G. and Biruykova, E. Ya., *Zh. Analit. Khim.*, **24**, 580 (1969).
[1195] Yudelevich, I. G., Buyanova, L. M., Protopopova, N. P. and Yudina, N. G., *Zh. Analit. Khim.*, **24**, 1719 (1969).
[1196] Buyanova, L. M., Yudelevich, I. G. and Arkirova, V. A., *Zh. Analit. Khim.*, **24**, 135 (1969).
[1197] Owens, E. B., *Appl. Spectrosc.*, **13**, 105 (1959).
[1198] Oldfield, J. H. and Bridge, E. B., *Analyst (London)*, **86**, 267 (1961).
[1199] Oldfield, J. H. and Mack, D. L., *Analyst (London)*, **87**, 778 (1962).
[1200] Artyukhin, P. I., Gilbert, E. N. and Pronin V. A., *Radiokhimiya*, **9**, 341 (1967).
[1201] Kuzmin, N. M., Kuzovlev, I. A. and Chugunova, V. V., *Sb. Metody analiza khimicheskikh reaktivov i preparatov* (Methods of Analysis of Chemical Reagents, Collection of Papers), IREA, **17**, 36 (1970).
[1202] Nazarova, M. G., Solobovnik, S. M. and Lapina, E. F., *Zh. Analit. Khim.*, **28**, 571 (1973).
[1203] Tiptsova, V. G., Malkina, E. I. and Anisimova, Z. A., *Zh. Analit. Khim.*, **21**, 459 (1966).
[1204] Jackwerth, E., *Z. Anal. Chem.*, **202**, 81 (1964).

[1205] Karabash, A. G., Samsonova, Z. N., Simonova-Averina, I. N. and Peizulaev, Sh. I., *Tr. Kom. Anal. Khim. Akad. Nauk SSSR*, **12**, 255 (1960).
[1206] Wieteska, E., and Stolarczyk, U., *Chem. Anal. (Warsaw)*, **15**, 183 (1970).
[1207] Notkina, M. A., Petrova, E. I., Cherkashina, T. V. and Chernikhov, Yu. A., *Tr. Kom. Anal. Khim. Akad. Nauk SSSR*, **15**, 80 (1965).
[1208] Tarasevich, N. I., Elomanova, G. G. and Rakhimbekova, Kh. M., *Vestn. Gos. Mosk. Univ., Ser II.*, **22**, 113 (1967).
[1209] Imai, T., *Bunseki Kagaku*, **15**, 109, 321 (1966).
[1210] Dhumwad, R. R., Joshi, M. V., and Patwardhan, A. B., *Anal. Chim. Acta*, **50**, 237 (1970).
[1211] Le Clainche, C., *Methodes, Phys. Anal.*, **7**, 115 (1971).
[1212] Motojima, K., Yamamoto, T., Muto, Ch. and Kato, I., *Bunseki Kagaku*, **19**, 1099 (1970).
[1213] Shabirova, V. P., Yudelevich, I. G., Torgov, V. G. and Kotyarevskii, I. L., *Zh. Analit. Khim.*, **26**, 930 (1971).
[1214] Yudelevich, I. G., Shabirova, V. P., Torgov, V. G. and Shcherbakova, O. I., *Zh. Analit. Khim.*, **28**, 1049 (1973).
[1215] Lysenko, V. I., *Tr. Kom. Anal. Khim. Akad. Nauk SSSR*, **15**, 195 (1965).
[1216] Artyukhin, P. I., Gilbert, E. N. and Pronin, V. A., *Zh. Analit. Khim.*, **21**, 504 (1966).
[1217] Szücs, A. J., *Chem. Anal. (Warsaw)*, **12**, 459 (1967).
[1218] Yudelevich, I. G. Artyukhin, P. I., Chichalino, L. S. Protopopova, N. P., Skrebkova, L. M., Gilbert, E. N. and Pronin, V. A., *Zh. Analit. Khim.*, **21**, 1457 (1966).
[1219] Marczenko, Z. and Mojski, M., *Chem. Anal. (Warsaw)*, **17**, 881 (1972).
[1220] Joshi, B. D., Bangia, T. R. and Dalvi, A. G., *Govt. India Atom. Energy Commis. (Rept)*, 1970, No. BARC-522.
[1221] Goryushina, V. G. and Notkina, M. A., *Sb. Metody analiza veshchestv vysokoi chistoty (Methods of Analysis of High Purity Compounds, Collection of Papers)*, Izd. Nauka, **1965**, 216.
[1222] Artyukhin, P. I., Gilbert, E. N. and Pronin, V. A., *Tr. Kom. Anal. Khim. Akad. Nauk SSSR*, **16**, 169 (1968).
[1223] Yudelevich, I. G., Buyanova, L. M., Protopopova, N. P. and Voevodina, N. V., *Zh. Analit. Khim.*, **25**, 153 (1970).
[1224] Moseeva, Z. P., Pinchuk, G. P., Sokolov, A. B., Karabash, A. G. and Peizulaev, Sh. I., *Zh. Analit. Khim.*, **29**, 1589 (1974).
[1225] Goryanskaya, G. P., Kaplan, B. Ya., Merisov, Yu. I., Nazerova, M. G. and Shripkin, G. S., *Zavodsk. Lab.*, **38**, 1315 (1972).
[1226] Slyusareva, R. L., Kondrateva, L. I. and Peizulaev, Sh., I., *Zavodsk. Lab.*, **39**, 1465 (1973).
[1227] Vakulenko, Yu. I., Mikhailichenko, A. I. and Skoilkin, G. S., *Zavodsk. Lab.*, **39**, 1342 (1973).
[1228] Kuzmin, N. M., *Zh. Analit. Khim.*, **22**, 451 (1967).
[1229] Kuzmin, N. M., Belyaev, V. P., Kalinachenko, V. R. and Yakimenko, L. M., *Zavodsk. Lab.*, **29**, 691 (1963).
[1230] Motojima, K., Hashitani, H. and Katsuyama, K., *Bunseki Kagaku*, **9**, 517 (1960).
[1231] Kuzmin, N. M., Kuzovlev, I. A., Tsykunova, S. V., Krasnikova, G. V. and Galaktionova, A. N., *Zavodsk. Lab.*, **34**, 1058 (1958).

[1232] Motojima, K., Hashitani, H. and Yoshida, H., *Bunseki Kagaku*, **11**, 659 (1962).
[1233] Yagi, I., Katsumata, S., Mukoyama, T. and Hirono, S., *Kogyo Kagaku Zasshi*, **64**, 972 (1961).
[1234] Marshall, G. B. and West, T. S., *Talanta*, **14**, 823 (1967).
[1235] Hashitani, H. and Motojima, K., *Bunseki Kagaku*, **7**, 478 (1958).
[1236] Kuzmin, N. M., Zhuravlev, G. I., Kuzovlev, I. A., Galaktionova, A. N. and Zakharova, T. I., *Zh. Analit. Khim.*, **24**, 429 (1969).
[1237] Minczewski, J., Chwastowska, J., Marczenko, Z., *Chem. Anal. (Warsaw)*, **6**, 509 (1961).
[1238] Vinogradov, A. V., Dronova, M. I. and Korovin, Yu. I., *Zh. Analit. Khim.*, **18**, 29 (1963).
[1239] Kartha, V. B., Krishman, T. S., Patel, N. D. and Gopal, S., *U.S. At. Energy Comm. Rept.*, BARC-284, 9 (1967).
[1240] Blank, A. B., Sizonenko, N. T. and Bulghakova, A. M., *Zh. Analit. Khim.*, **18**, 1046 (1963).
[1241] Belyaev, V. P., Kuzmin, N. M., Kalinachenko, V. P. and Yakimenko, L. M., *Zavodsk. Lab.*, **28**, 685 (1962).
[1242] Negina, V. P. and Zamyatina, V. N., *Zh. Analit. Khim.*, **16**, 209 (1961).
[1243] Goode, G. C., Herrington, J. and Bundy, J. K., *Analyst (London)*, **91**, 719 (1966).
[1244] Dean, J. A. and Cain, C., *Anal. Chem.*, **29**, 530 (1957).
[1245] Brodskaya, B. D., Notkina, M. A., Korneeva, S. A. and Men'shova, N. P., *Zh. Analit. Khim.*, **21**, 1447 (1966).
[1246] Grekova, I. M. and Nazarenko, V. A., *Zavodsk. Lab.*, **35**, 537 (1969).
[1247] Notkina, M. A., Petrova, E. I., Cherkashina, T. V. and Chernikhov, Yu. A., *Tr. Kom. Anal. Khim. Akad. Nauk SSSR*, **18**, 80 (1965).
[1248] Grossmann, O. and Grosse-Ruyken, H., *Z. Anal. Chem.*, **233**, 14 (1968).
[1249] Marchenko, P. V. and Lugina, A. N., *Ukr. Khim. Zh.*, **32**, 1343 (1966).
[1250] Jones, M., Kirkbright, G. F., Ranson, L. and West, T. S., *Anal. Chim. Acta*, **63**, 210 (1973).
[1251] Nix, J. and Goodwin, T., *At. Absorpt. Newsl.*, **9**, 119 (1970).
[1252] Rooney, R. C., *Analyst (London)*, **83**, 83 (1958).
[1253] Zagórski, Z. and Cyranowska, M., *Chem. Anal. (Warsaw)*, **3**, 495 (1958).
[1254] Burrell, D. C., *Anal. Chim. Acta*, **38**, 447 (1967).
[1255] Minczewski, J., Chwastowska, J. and Marczenko, Z., *Chem. Anal. (Warsaw)*, **6**, 509 (1961).
[1256] Schöffmann, E. and Malissa, H., *Mikrochim. Acta*, **1961**, 319.
[1257] Ravnik, V., Dermelj, M. and Kosta, L., *J. Radioanal. Chem.*, **20**, 443 (1974).
[1258] Yamamoto, Y., Kumamaru, T., Hayashi, Y. and Kanke, M., *Bunseki Kagaku*, **20**, 347 (1971).
[1259] Busev, A. I., Byrko, V. M., Lerner, L. A. and Migunova, I. V., *Zh. Analit. Khim.*, **27**, 607 (1972).
[1260] Kreimer, S. E., Mikhailov, L. M. and Lamekov, A. S., *Zh. Analit. Khim.*, **22**, 1105 (1967).
[1261] Sugihara, K. and Saito, T., *Bunseki Kagaku*, **7**, 139 (1955).
[1262] Marczenko, Z., *Mikrochim. Acta*, **1965**, 281.
[1263] Marczenko, Z., Krasiejko, M. and Chołuj, Ł., *Chem. Anal. (Warsaw)*, **8**, 375 (1963).
[1264] Galík, A., *Talanta*, **13**, 109, 589 (1966).

[1265] Jędrzejewska, H. and Maluszecka, M., *Chem. Anal. (Warsaw)*, **12**, 597 (1967).
[1266] Sincz, A., *Chem. Anal. (Warsaw)*, **10**, 1093 (1965).
[1267] Häberli, E., *Z. Anal. Chem.*, **160**, 15 (1958).
[1268] Marchenko, P. V., *Ukr. Khim. Zh.*, **32**, 1216 (1966).
[1269] Hubbard, G. L. and Green, T. E., *Anal. Chem.*, **38**, 428 (1966).
[1270] Ishihoro, I., Kishi, H. and Komyro, H., *Bunseki Kagaku*, **11**, 932 (1962).
[1271] Miller, A. D. and Libina, R. I., *Zh. Analit. Khim.*, **13**, 664 (1958).
[1272] Gurkina, T. V. and Igotin, A. M., *Zh. Analit. Khim.*, **20**, 778 (1965).
[1273] Minczewski, J., Karczewska, B. and Marczenko, Z., *Chem. Anal. (Warsaw)*, **6**, 501 (1961).
[1274] Jędrzejewski, S. and Jędrzejewska, H., *Chem. Anal. (Warsaw)*, **15**, 729 (1970).
[1275] Beloglazova, A. D. and Krupnov, V. K., *Zavodsk. Lab.*, **35**, 451 (1969).
[1276] Bazhanova, L. A. and Shafran, I. G., *Tr. Vses. Nauchn.-Issled. Inst. Khim. Reaktivov Osobo Chist. Khim. Veshchestv.*, **29**, 34 (1966); *Chem. Abstr.*, **68**, 18303q (1968).
[1277] Lee, Meng-Lein and Burrell, D. C., *Anal. Chim. Acta*, **62**, 153 (1972).
[1278] Chao, T. T., Ball, J. W. and Makagawa, H. M., *Anal. Chim. Acta*, **54**, 77 (1971).
[1279] Sizonenko, N. T., Zolotovitskaya, E. S., Yakovenko, E. I. and Balenko, L. E., *Sb. Metody analiza shchelochn. i shchelochnozemelnykh met. vysokoi chistoty (Methods of Analysis of High Purity Alkali and Alkaline Eearth Metals)*, Kharkov, **1971**, 57.
[1280] Sparks, R. W., Vita, O. A. and Walker, C. R., *Anal. Chim. Acta*, **60**, 222 (1972).
[1281] Mizumo, T., *Kogyo Kagaku Zasshi*, **1973**, 1904.
[1282] Fedyashina, A. F., Yudelevich, I. G., Gindin, L. I. and Strokina, T. G., *Izv. Akad. Nauk SSSR, Ser. Khim.*, **1**, 83 (1966).
[1283] Scribner, W. G., Treat, W. J., Weis, J. D. and Moshier, R. W., *Anal. Chem.*, **37**, 1136 (1965).
[1284] Morie, G. P. and Sweet, T. R., *Anal. Chim. Acta*, **34**, 314 (1966).
[1285] Moshier, R. W. and Schwarberg, J. E., *Talanta*, **13**, 445 (1966).
[1286] Kawabuti, K., *Bunseki Kagaku*, **14**, 52 (1965).
[1287] Qureshi, I. H., McClendon, L. T. and La Fleur, P. D., *Natl. Bur. Stand. Spec. Publ.*, **1969**, No. 312/1, 666.
[1288] Artyukhin, P. I., Startseva, E. A. and Nikolaev, A. V., *J. Radioanal. Chem.*, **17**, 173 (1973).
[1289] Jędrzejewska, H., *Chem. Anal. (Warsaw)*, **19**, 117 (1973).
[1290] Szücs, A. J. and Klug, O. N., *Chem. Anal. (Warsaw)*, **12**, 939 (1967).
[1291] Kasiura, K., *Chem. Anal. (Warsaw)*, **13**, 849 (1968).
[1292] Zemskova, M. G., Lebedev, N. A., Melamed, Sh. G., Saunkin, O. F., Sukhov, G. V., Khalkin, V. A., Kherrmann, E. and Shmanenkova, G. I., *Zavodsk. Lab.*, **33**, 667 (1967).
[1293] Zhivopistsev, V. P., Petrov, B. I., Selezneva, E. A., and Sibiryakov, N. F., *Tr. Kom. Anal. Khim. Akad. Nauk SSSR*, **16**, 80 (1968).
[1294] Conradi, G. and Kopanica, M., *Chemist-Analyst*, **53**, 4 (1964).
[1295] Notkina, M. A., *Sb. Metody analiza veshchestv vysokoi chistoty (Methods of Analysis of High-Purity Compounds)*, Izd. Nauka, **1965**, 196.
[1296] Luke, C. L., *Anal. Chim. Acta*, **39**, 447 (1967).
[1297] Headridge, J. B. and Richardson, J., *Analyst (London)*, **95**, 930 (1970).
[1298] Brooks, R. R., *Anal. Chim. Acta*, **24**, 456 (1961).

[1299] Zhivopistsev, V. P., Makhnev, Yu. A. and Petrov, I. B., *Zh. Prikl. Spektrosk.*, **11**, 779 (1969).
[1300] Gorbenko, F. P., Kikhkina, E. D. and Olevinskii, M. I., *Radiokhimiya*, **12**, 661 (1970).
[1301] Danilova, V. N. and Lugina, L. N., *Ukr. Khim. Zh.*, **32**, 290 (1968).
[1302] Jachwerth, E. and Schneider, E. L., *Z. Anal. Chem.*, **207**, 188 (1965).
[1303] Babko, A. K., Kuzmin, N. M., Lisetskaya, G. S., Ovrutskii, M. I. and Freger, S. V., *Ukr. Khim. Zh.*, **33**, 828 (1967).
[1304] Kuz'min, N. M., *Zavodsk. Lab.*, **32**, 1349 (1966).
[1305] Gorczyńska, K., Gluzińska, M. and Ciecierska-Stokłosa, D., *Chem. Anal. (Warsaw)*, **14**, 591 (1969).
[1306] Pepin, D., Gardes, A., Petit, J. and Berger, J. A., *Analusis*, **2**, 549 (1973).
[1307] Schulte, K. E., Henke, G. and Tjan, K. S., *Z. Anal. Chem.*, **252**, 358 (1970).
[1308] Concialini, V., Lanza, P. and Lippolis, M. T., *Anal. Chim. Acta*, **52**, 529 (1970).
[1309] Shkrobot, É. P., Tarayan, M. G. and Blyakhman, A. A., *Zavodsk. Lab.*, **32**, 18 (1966).
[1310] Gorbach, G. and Pohl, F., *Mikrochemie*, **36/37**, 486 (1951).
[1311] Koch, O. G. and Dedic, G. A., *Chemist-Analyst*, **46**, 88 (1957).
[1312] Pavlenko, L. I., Petrukhin, O. M., Zolotov, Yu. A., Karyokin, A. V., Gavrilina, G. N. and Tumanova, I. E., *Zh. Analit. Khim.*, **29**, 933 (1974).
[1313] Pohl, F. A., *Spectrochim. Acta*, **6**, 19 (1953).
[1314] Pohl, F. A., *Z. Anal. Chem.*, **139**, 241, 423 (1953); **141**, 81 (1954).
[1315] Schüller, H., *Mikrochim. Acta*, **1956**, 393.
[1316] Ilcheva, L. and Todorova, G., *Acta Chim. Acad. Sci. Hung.*, **102**, 113 (1979).
[1317] Whewell, R. J., Huges, M. A. and Middle-brook, P. D., *Hydrometallurgy*, **4**, 125 (1979).
[1318] Kuznetsova, E. M., Eliseeva, N. M. and Rashidova, D. Sh., *Zh. Fiz. Khim.*, **52**, 2427 (1978).
[1319] Kandil, A. T. and Farah, K., *Radiochim. Acta*, **26**, 123 (1979).
[1320] Bagawde, S. V., Rao, P.R.V., Ramakrishna, V. V. and Patil, S. K., *J. Inorg. Nucl. Chem.*, **40**, 1913 (1978).
[1321] Kupryunin, G. I. and Renard, E. V., *Radiokhimiya*, **22**, 126 (1980).
[1322] Filippov, E. A., Yakshin, V. V., Serebryakov, I. S., Arkhipova, G. G., Belov, V. A. and Dakalova, T. S., *Radiokhimiya*, **22**, 87 (1980).
[1323] Kandil, A. T., Abdel Gawad, A. S., and Ramadan, A., *Radiochim. Acta*, **27**, 39 (1980).
[1324] Nitsch, W. and Kruis, B., *J. Inorg. Nucl. Chem.*, **40**, 857 (1978).
[1325] Vashman, A. A., Pronin, I. S. and Brylkina, T. V., *Radiokhimiya*, **22**, 115 (1980).
[1326] Fomin, V. V., *Zh. Neorgan. Khim.*, **23**, 2167 (1978).
[1327] Dukov, I. and Genov, L., *Acta Chim. Acad. Sci. Hung.*, **102**, 201 (1979).
[1328] Kandil, A. T. and Ramadan, A., *J. Radioanal. Chem.*, **52**, 15 (1979).
[1329] Akaiwa, H., Kawamoto, H. and Hiyamuta, E., *Bunseki Kagaku*, **28**, 477 (1979).
[1330] Ramanujam, A., Ramakrishna, V. V. and Patil, S. K., *Sep. Sci. Technol.*, **14**, 13 (1979).
[1331] Patil, S. K., Ramakrishna, V. V. and Prakas, B. H., *Sep. Sci. Technol.*, **15**, 133 (1980).
[1332] Patil, S. K., Ramakrishna, V. V. and Sajun, M. S., *Radiochem. Radioanal. Lett.*, **44**, 239 (1980).

[1333] Nishimura, S., Suiyo Kaishi, **19**, 109 (1979), *Chem. Abstr.*, **92**, 150560w (1980).
[1334] Mirza, M. Y., Nowabue, F. J., *Talanta*, **28**, 49 (1981).
[1335] Vaezi-Nasr, F., Duplessis, J. and Guillaumont, R., *Radiochem. Radioanal. Lett.*, **37**, 153 (1979).
[1336] Borus-Boshormenyj, N., Kovacas, J. and Fekete, Z., *Radiochem. Radioanal. Lett.*, **34**, 51 (1978).
[1337] Kosyakov, V. N. and Yerin, E. A., *J. Radioanal. Chem.*, **56**, 93 (1980).
[1338] Kupryunin, G. I. and Renard, E. V., *Radiokhimiya*, **22**, 119 (1980).
[1339] Kosyakov, V. N., Yerin, E. A. and Vitutnev, U. M., *J. Radioanal. Chem.*, **56**, 83 (1980).
[1340] Reznik, A. M., Semenov, S. A., Yurchenko, L. D., *Zh. Neorgan. Khim.*, **24**, 461 (1979).
[1341] Lukachina, V. V., *Zh. Neorgan. Khim.*, **24**, 2748 (1979).
[1342] Janousek, I., *Chem. Listy*, **73**, 136 (1979).
[1343] Byrne, A. R., *Radiochem. Radioanal. Lett.*, **40**, 1 (1979).
[1344] Futekov, L., Stoyanov, S. and Specker, H., *Z. Anal. Chem.*, **295**, 7 (1979).

Chapter 6

ION-EXCHANGE CHROMATOGRAPHY

6.1 INTRODUCTION

In the last twenty-five years the theory and practice of ion-exchange have progressed rapidly. Ion-exchange chromatography has become a valuable and sometimes irreplaceable method of separation in inorganic analysis, and it is becoming increasingly important in organic analysis. The greatest successes have been achieved in trace analysis, however. Let it suffice to mention the first unambiguous identification of element 61, promethium [1], and the discovery of trans-curium elements [2-6] prepared by nuclear reactions, including element 101, mendelevium [7], which was identified when as few as 17 atoms of the newly synthesized element were available.

Today, many monographs on ion-exchange and its applications are available [8-43], but extensive surveys concentrating on applications to trace analysis [44,45-45b] have been a rarity. At the same time, primary research reports are dispersed in such a large number of journals and other scientific publications that very often they are accessible only with difficulty. This chapter is intended to bridge the gap and to review the current status and potential of ion-exchange chromatography in trace analysis.

6.2 GENERAL INFORMATION ON ION-EXCHANGERS

Ion-exchangers are insoluble solid materials which contain exchangeable cations or anions. These ions can be exchanged for a stoichiometrically equivalent amount of other ions initially present in an electrolyte solution when an ion-exchanger is brought into contact with it. Substances containing exchangeable cations are called cation-exchangers and those containing exchangeable anions are called anion-exchangers. Materials capable of both cation- and anion-exchange are called amphoteric ion-exchangers. The term ion-exchanger usually refers to a solid ion-exchanger, but it can also have a broader meaning, and apply to all substances capable of ion-exchange, including liquid ion-exchangers. Certain solid ion-exchangers, and particularly those prepared by modification of naturally occurring substances (sulphonated coals, etc.) possess sorption as well as ion-exchange properties. Therefore, they are more correctly referred to as sorbent ion-exchangers or ion-exchange sorbents [46].

On the basis of the nature of the framework material, two main classes of ion-exchangers can be distinguished, inorganic and organic. Inorganic ion-exchangers form a large group of natural (clays, zeolite-type minerals, etc.) and synthetic (synthetic zeolites and especially molecular sieves, hydrous metal oxides, heteropoly acid salts, etc.) compounds. Many other materials, such as glass and silica, show ion-exchange properties. Naturally occuring organic ion-exchangers such as peat are of little practical importance. Some other substances, e.g. human hair, also show anion-exchange properties [46a]. Semisynthetic ion-exchangers prepared by chemical treatment of naturally occuring materials (coal, cellulose, cotton, etc.) have been used only rarely, except for cellulose ion-exchangers. Those most widely used are the synthetic ion-exchange resins, which have as charge carriers ionogenic groups bonded to a framework (matrix) that is a three-dimensional network of hydrocarbon chains. Owing to their high mechanical and chemical stability, practically complete insolubility in any solvent, and the possibility of having various different ionogenic groups incorporated into the matrix, ion-exchange resins have been commonly used both in analysis and in various commercial processes. Among inorganic ion-exchangers zirconium phosphate, ammonium phosphomolybdate, and hydrous antimony pentoxide have recently gained in importance.

6.2.1 Organic Ion-Exchangers

6.2.1.1 CATION-EXCHANGERS

Semisynthetic ion-exchangers of the type made from sulphonated coal (e. g. Escarbo) have found little use for analytical purposes because of their low ion-exchange capacity, limited mechanical and chemical stability, and tendency to peptize. Cellulose cation-exchangers [32,47,48] have been applied primarily in organic chemistry and biochemistry (separation and isolation of macromolecular ionic compounds such as nucleic acids, enzymes, etc.). In inorganic analysis only synthetic ion-exchange resins have found widespread application.

Synthetic organic cation-exchangers are cross-linked polyelectrolytes which consist of a three-dimensional network of hydrocarbon chains carrying ionic groups such as sulphonate —SO_3^-, carboxylate —COO^-, phenolate —O^-, phosphonate —PO_3^-, phosphinate —HPO_2^-, and others. The negative charges of the functional groups are compensated by the positive charges of a stoichiometrically equivalent number of mobile cations known as the counterions. Within the matrix, counter-ions can be exchanged for other ions taken up from the external solution in contact with the ion-exchanger. The structure of a polymer-type sulphonic acid ion-exchanger is illustrated in Fig. 6.1.

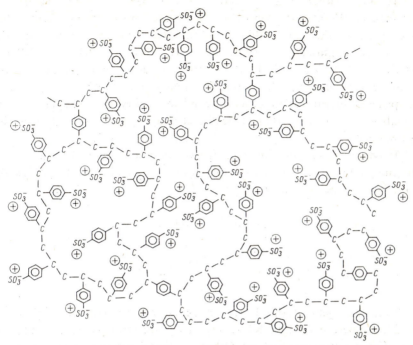

Fig. 6.1 Representation of a swollen sulphonic-acid cation-exchanger with a styrene–divinylbenzene matrix containing univalent ions: ⊕-counterions; solvent not indicated (from [49], by permission of the New York Academy of Sciences)

The properties of an ion-exchanger depend on the nature and number of functional groups, the degree of ionization, the type and extent of cross-linking in the matrix, and the configuration of the functional groups. Ion-exchangers are synthesized by condensation or polymerization. Thus, sulphonation of phenol and subsequent condensation with formaldehyde yields a bifunctional ion-exchanger

which contains both strong-acid sulphonate and weak-acid phenolate groups. This type of structure is found in such ion-exchangers as Amberlite IR-100, Dowex-30, KU-1, etc. The alkaline condensation of phenol, sodium sulphite and formaldehyde yields a similar ion-exchanger except that the sulphonate groups are now attached to the side-chain ($-CH_2SO_3^-$) rather than directly to the benzene ring. Such ion-exchangers (Wofatit P, Amberlite IR-1) are more weakly acidic but have greater thermal stability. With other starting materials, ion-exchangers containing $-COO^-$, $-CH_2COO^-$, $-PO_3^{2-}$, $-OPO_3^{2-}$, $-AsO_3^{2-}$, $-HPO_2^-$, $-SeO_3^{2-}$, and other groups can be formed. More complete information is available in monographs [15,32,39,50] and in the primary literature. In general, condensation-type ion-exchangers are available as irregularly shaped beads, although spherical beads can also be obtained.

Polymerization-type ion-exchangers are based on a styrene-divinylbenzene (S–DVB) copolymer. Sulphonation of the copolymer yields the following product (cf. also Fig. 6.1)

$$\left[\begin{array}{c} -CH-CH_2- \\ | \\ \bigcirc \\ | \\ SO_3^-H^+ \end{array} \right]_n \quad \begin{array}{c} -CH-CH_2- \\ | \\ \bigcirc \\ | \\ CH-CH_2- \\ | \end{array}$$

Ion-exchangers of this type include Dowex 50, Dowex 50W, Zerolit 225, Amberlite IR-120, KU-2, etc. They are strongly acidic and fully ionized over a wide range of pH. Other functional groups can be introduced into the S-DVB copolymer to yield ion-exchangers carrying phosphinate, phosphonate groups, etc. Copolymerization of acrylic or methacrylic acid with divinylbenzene yields resins with carboxylate groups. The structure of such an ion-exchanger is as follows

$$\left[\begin{array}{c} CH_3 \\ | \\ -C-CH_2- \\ | \\ COOH \end{array} \right]_n \quad \begin{array}{c} -CH-CH_2- \\ | \\ \bigcirc \\ | \\ CH-CH_2- \\ | \end{array}$$

Resins of this type are Amberlite IRC-50, Zerolit 226, Duolite CS-11, KB-1, and others. These are weakly acidic ion-exchangers which are completely ionized only in alkaline solutions. Polymerization-type ion-exchangers are produced by pearl polymerization and the resulting beads are spherical in shape. Commercially available ion-exchangers are supplied in a range of particle sizes, e.g. 20–50, 100–200, 200–400, etc. mesh sizes.

General Information

An approximate expression for converting U.S. Standard Sieve sizes into metric dimensions (mm) is $\frac{16}{\text{U.S. mesh number}}$ = sieve hole diameter, Φ mm. Tables listing the nominal hole sizes (mm) for the U.S. and British Standard Sieve series can be found in the literature [51].

The functional groups incorporated into the hydrophobic hydrocarbon matrix impart hydrophilic properties to the ion-exchanger. When brought into contact with an aqueous solution, the ion-exchanger takes up water and swells. The fewer the cross-links interconnecting the hydrocarbon chains, the greater the expansion. With polymer-type ion-exchangers, the degree of cross-linking is directly related to the content (mole percentage) of divinylbenzene (DVB) in the reaction mixture, and is expressed as % DVB or X followed by a number which is the mole per cent of DVB. Most ion-exchangers currently manufactured have gel structures and contain no 'pores' in the true meaning of the word [52–54]. The ions to be exchanged have to diffuse through the gel structure to reach individual exchange sites. In modern ion-exchangers the functional groups incorporated throughout the entire matrix of the resin are virtually all accessible to ion-exchange, provided the ions to be exchanged are not too bulky. The interparticle distances which restrict the size of the ions that can penetrate the matrix are often referred to as the 'apparent pore size'. Usually this size does not exceed 40 Å [54], even in ion-exchangers with a low degree of cross-linking.

Macroporous or macroreticular ion-exchange resins have a sponge-like matrix which contains pores larger than molecules in size. Actually, the pores may be as large as 1400 Å in diameter; in an air-dried resin the pores are filled with air [48,52–54]. The specific surface area which in macroporous resins is usually smaller than $0 \cdot 1$ m^2/g, in macroporous resins ranges from 2 to 120 m^2/g [48,53]. Macroporous resins have less extensive swelling and smaller differences between the amount of swelling in polar and non-polar solvents than gel-type ion-exchangers. Exchange of large ions is faster, and they are more resistant to oxidants, etc. Macroporous resins may be used for removing high molecular-weight ions and for ion-exchange in non-aqueous and mixed media.

The superiority of macroreticular over microreticular cation exchange resin for the chromatographic separation of transition metals in aqueous-acetone solutions of hydrochloric acid has been reported [54a].

A list of the more important commercially available ion-exchangers is given in Table 6.1.

Table 6.1 Principal properties of the most common commercial ion-exchanger
1. Synthetic Organic Ion-Exchangers
a. Cation-Exchangers

Name	Matrix	Functional group(s)	Manufacturer
Dowex 50W-X1	polymerized resin	$-SO_3^-$	Dow
Dowex 50W-X2	polymerized resin	$-SO_3^-$	Dow
Dowex 50W-X4	polymerized resin	$-SO_3^-$	Dow
Dowex 50W-X5	polymerized resin	$-SO_3^-$	Dow
Dowex 50W-X8	polymerized resin	$-SO_3^-$	Dow
Dowex 50W-X10	polymerized resin	$-SO_3^-$	Dow
Dowex 50W-X12	polymerized resin	$-SO_3^-$	Dow
Dowex 50W-X16	polymerized resin	$-SO_3^-$	Dow
Amberlite IR-120	polymerized resin	$-SO_3^-$	R & H
Amberlite IR-122	polymerized resin	$-SO_3^-$	R & H
Amberlite IR-124	polymerized resin	$-SO_3^-$	R & H
Amberlite XE-100	polymerized resin	$-SO_3^-$	R & H
Amberlite 200	polymerized resin	$-SO_3^-$	R & H
Amberlite 200 C	polymerized resin	$-SO_3^-$	R & H
Amberlyst 15	polymerized resin	$-SO_3^-$	R & H
Diaion SK 1A	polymerized resin	$-SO_3^-$	Mitsubishi
Diaion SK 1B	polymerized resin	$-SO_3^-$	Mitsubishi
Diaion SK 102	polymerized resin	$-SO_3^-$	Mitsubishi
Diaion SK 103	polymerized resin	$-SO_3^-$	Mitsubishi
Diaion SK 104	polymerized resin	$-SO_3^-$	Mitsubishi
Diaion SK 106	polymerized resin	$-SO_3^-$	Mitsubishi
Diaion SK 110	polymerized resin	$-SO_3^-$	Mitsubishi
Diaion SK 112	polymerized resin	$-SO_3^-$	Mitsubishi
Diaion SK 116	polymerized resin	$-SO_3^-$	Mitsubishi
Zerolit 225	polymerized resin	$-SO_3^-$	Zerolit
Zerolit 625	polymerized resin	$-SO_3^-$	Zerolit
Zerolit 215	condensation polymer	$-CH_2SO_3^-$ $-OH$	Zerolit
KU-2	polymerized resin	$-SO_3^-$	USSR
KU-1	condensation polymer	$-SO_3^-$ $-OH$	USSR

| Capacity | | Maximum | | |
/g ry resin	meq/ml of resin bed	temperature °C	Physical form	Remarks
5·0	0·4	150	spherical beads	1% DVB
5·2	0·7	150	spherical beads	2% DVB
5·2	1·2	150	spherical beads	4% DVB
5·2	1·3	150	spherical beads	5% DVB
5·1	1·7	150	spherical beads	8% DVB
5·0	1·9	150	spherical beads	10% DVB
5·0	2·3	150	spherical beads	12% DVB
4·9	2·6	150	spherical beads	16% DVB
3–5·0	1·9	120	spherical beads	8% DVB
3–5·0	2·1	120	spherical beads	10% DVB
3–5·0	2·2	120	spherical beads	12% DVB
4·5	1·5	120	spherical beads	5% DVB
4·3	1·75	150	spherical beads	macroporous ca. 20% DVB
4·3	1·75	150	spherical beads	macroporous ca. 20% DVB, special particle size
4·9	1·2	120	spherical beads	macroporous for non-aqueous solvents
—	1·9	120	spherical beads	ca. 8% DVB
—	1·9	120	spherical beads	ca. 8% DVB
—	0·6	120	spherical beads	ca. 2% DVB
—	0·9	120	spherical beads	ca. 3% DVB
—	1·2	120	spherical beads	ca. 4% DVB
—	1·6	120	spherical beads	ca. 6% DVB
—	2·0	120	spherical beads	ca. 10% DVB
—	2·1	120	spherical beads	ca. 12% DVB
—	2·1	120	spherical beads	ca. 16% DVB
5–5·0	2·0–2·1	120 [H$^+$] 140 [Na$^+$]	spherical beads	standard cross-linking 8%; also available 1, 2, 4, 12, and 20% DVB
	1·8	120	spherical beads	macroporous resin with high degree of cross-linking
strong- idic ps)	0·92	40 [H$^+$] 95 [Na$^+$]	irregular	
7–5·1	1·3–1·8	120–130	spherical beads	ca. 8% DVB; other cross-linkings also available
1·9 = 3) 5·6 = 13)	1·3	90	irregular	

Table 6.1 (*continued*)

Name	Matrix	Functional group(s)	Manufacturer
MK-3	condensation polymer	$-SO_3^-$ $-OH$	Poland
Wofatit KPS-200	polymerized resin	$-SO_3^-$	GDR
Lewatit S-100	polymerized resin	$-SO_3^-$	GFR
Kastel C 300	polymerized resin	$-SO_3^-$	Montecat
Kastel C 300 AGR	polymerized resin	$-SO_3^-$	Montecat
Kastel C 100	polymerized resin	$-COO^-$	Montecat
Duolit CS-101	polymerized resin	$-COO^-$	Diamond
Amberlite IRC-50	polymerized resin	$-COO^-$	R & H
Amberlite IRC-84	polymerized resin	$-COO^-$	R & H
Zerolit 236	polymerized resin	$-COO^-$	Zerolit
Kationit KB-4	polymerized resin	$-COO^-$	USSR
Wofatit CP	polymerized resin	$-COO^-$	GDR
Duolit C-63	polymerized resin	$-PO_3^{2-}$	Diamond

1a. Chelating Ion-Exchangers

Name	Matrix	Functional group(s)	Manufacturer
Dowex A1	polymerized resin	$-N\begin{pmatrix} CH_2COO^- \\ CH_2COO^- \end{pmatrix}$	Dow
Srafion NMRR	polymerized resin	$\left[R-C\begin{pmatrix} NH \\ NH_3^+ \end{pmatrix} Cl_n^- \right]_n$	Israel

b. Anion-Exchangers

Name	Matrix	Functional group(s)	Manufacturer
Dowex 1-X1	polymerized resin	$-N(CH_3)_3^+$	Dow
Dowex 1-X2	polymerized resin	$-N(CH_3)_3^+$	Dow
Dowex 1-X4	polymerized resin	$-N(CH_3)_3^+$	Dow
Dowex 1-X8	polymerized resin	$-N(CH_3)_3^+$	Dow
Dowex 1-X10	polymerized resin	$-N(CH_3)_3^+$	Dow
Dowex 21K	polymerized resin	$-N(CH_3)_3^+$	Dow
Amberlite IRA-400	polymerized resin	$-N(CH_3)_3^+$	R & H

General Information

Capacity meq/g of dry resin	Capacity meq/ml of resin bed	Maximum temperature °C	Physical form	Remarks
1·87 (strongly acidic groups) 3·46 (total)		40	irregular	
4·5	2·0	115	spherical beads	
4·8	2·3	100	spherical beads	
4·25	1·4	100 [Na$^+$]	spherical beads	ca. 8% DVB
4·5–4·7	1·5	100 [Na$^+$]	spherical beads	12% DVB
—	3·0	110	spherical beads	10% DVB
10·0	3·5	100	spherical beads	ca. 10% DVB
10–10·2	3·5	120	spherical beads	pK 6·1
10·0	3·5	120	spherical beads	pK 5·3
9–10	3·6	100	spherical beads	2·5 and 4·5% DVB
0·16 (pH 3) 8·55 (pH 13)	4·2	200	spherical beads	10% DVB
10·0	3·5	100	spherical beads	ca. 5% DVB
6·6	3·1	100	spherical beads	
1–1·2	0·33	75	spherical beads	
3·2	0·4	50 [OH$^-$] 150 [Cl$^-$]	spherical beads	1% DVB
3·5	0·8	50 [OH$^-$] 150 [Cl$^-$]	spherical beads	2% DVB
3·5	1·2	50 [OH$^-$] 150 [Cl$^-$]	spherical beads	4% DVB
3·2	1·4	50 [OH$^-$] 150 [Cl$^-$]	spherical beads	8% DVB
3·0	1·5	50 [OH$^-$] 150 [Cl$^-$]	spherical beads	10% DVB
4·5	1·25	50 [OH$^-$] 150 [Cl$^-$]	spherical beads	mech. stronger than Dowex 1
3·7	1·4	60 [OH$^-$] 75 [Cl$^-$]	spherical beads	ca. 8% DVB

Table 6.1 (*continued*)

Name	Matrix	Functional group(s)	Manufacturer
Amberlite IRA-400C	polymerized resin	$-N(CH_3)_3^+$	R & H
Amberlite IRA-401	polymerized resin	$-N(CH_3)_3^+$	R & H
Amberlite IRA-401S	polymerized resin	$-N(CH_3)_3^+$	R & H
Amberlite IRA-900	polymerized resin	$-N(CH_3)_3^+$	R & H
Amberlite IRA-904	polymerized resin	$-N(CH_3)_3^+$	R & H
Amberlyst A-26	polymerized resin	$-N(CH_3)_3^+$	R & H
Zerolit FF-(ip)	polymerized resin	$-N(CH_3)_3^+$	Zerolit
Diaion SA 10A	polymerized resin	$-N(CH_3)_3^+$	Mitsubishi
Diaion SA 101	polymerized resin	$-N(CH_3)_3^+$	Mitsubishi
Diaion SA 11A	polymerized resin	$-N(CH_3)_3^+$	Mitsubishi
Diaion SA 100	polymerized resin	$-N(CH_3)_3^+$	Mitsubishi
Kastel A 500	polymerized resin	$-N(CH_3)_3^+$	Montecatini
Kastel A 500P	polymerized resin	$-N(CH_3)_3^+$	Montecatini
Wofatit SBW	polymerized resin	$-N(CH_3)_3^+$	GDR
AV-17	polymerized resin	$-N(CH_3)_3^+$	USSR
Dowex 2-X4	polymerized resin	$-N\begin{smallmatrix}(CH_3)_2^+\\C_2H_4OH\end{smallmatrix}$	Dow
Dowex 2-X8	polymerized resin	$-N\begin{smallmatrix}(CH_3)_2^+\\C_2H_4OH\end{smallmatrix}$	Dow
Dowex 2-X10	polymerized resin	$-N\begin{smallmatrix}(CH_3)_2^+\\C_2H_4OH\end{smallmatrix}$	Dow
Amberlite IRA-410	polymerized resin	$-N\begin{smallmatrix}(CH_3)_2^+\\C_2H_4OH\end{smallmatrix}$	R & H
Amberlite IRA-910	polymerized resin	$-N\begin{smallmatrix}(CH_3)_2^+\\C_2H_4OH\end{smallmatrix}$	R & H
Amberlite IRA-911	polymerized resin	$-N\begin{smallmatrix}(CH_3)_2^+\\C_2H_4OH\end{smallmatrix}$	R & H

General Information

Capacity		Maximum temperature °C	Physical form	Remarks
meq/g of dry resin	meq/ml of resin bed			
3·7	1·4	60 [OH⁻] 75 [Cl⁻]	spherical beads	ca. 8% DVB, special particle size
3·5	1·0	60 [OH⁻] 75 [Cl⁻]	spherical beads	ca. 4% DVB
3·4	0·8	60 [OH⁻] 75 [Cl⁻]	spherical beads	ca. 4% DVB, strong, decolorant, e.g. for syrups
4·4	1·0	60 [OH⁻] 75 [Cl⁻]	spherical beads	macroporous, av. pore size 250 Å
2·6	0·7	60 [OH⁻] 75 [Cl⁻]	spherical beads	macroporous, av. pore size 645 Å
4·1–4·6	1·1	60 [OH⁻] 75 [Cl⁻]	spherical beads	macroporous for non-aqueous solvents
4·0	1·2	60	spherical beads	isoporous, 2–3, 3–5, or 7–9% DVB
—	1·2	60 [OH⁻]	spherical beads	ca. 8% DVB
—	0·85	60 [OH⁻]	spherical beads	low cross-linking
—	0·85	60 [OH⁻]	spherical beads	ca. 4% DVB
—	1·0–1·3	—	spherical beads	
3·0	0·7	60 [OH⁻]	spherical beads	ca. 7% DVB
3·3	0·6	60 [OH⁻]	spherical beads	ca. 5% DVB
3·5	0·9	60 [OH⁻]	spherical beads	
3·8–4·5		50	spherical beads	6 or 8% DVB
3·2	1·2	30 [OH⁻] 150 [Cl⁻]	spherical beads	2% DVB
3·2	1·4	30 [OH⁻] 150 [Cl⁻]	spherical beads	8% DVB
3·0	1·5	30 [OH⁻] 150 [Cl⁻]	spherical beads	10% DVB
3·3	1·4	40 [OH⁻] 75 [Cl⁻]	spherical beads	ca. 8% DVB
—	1·1	40 [OH⁻] 75 [Cl⁻]	spherical beads	macroporous
2·7	0·9	40 [OH⁻] 75 [Cl⁻]	spherical beads	macroporous, pore size larger than in IRA-910

Table 6.1 (*continued*)

Name	Matrix	Functional group(s)	Manufacturer
Amberlyst A-29	polymerized resin	$-N{\begin{smallmatrix}(CH_3)_2^+\\C_2H_4OH\end{smallmatrix}}$	R & H
Diaion SA 20A	polymerized resin	$-N{\begin{smallmatrix}(CH_3)_2^+\\C_2H_4OH\end{smallmatrix}}$	Mitsubishi
Diaion SA 21A	polymerized resin	$-N{\begin{smallmatrix}(CH_3)_2^+\\C_2H_4OH\end{smallmatrix}}$	Mitsubishi
Kastel A 300	polymerized resin	$-N{\begin{smallmatrix}(CH_3)_2^+\\C_2H_4OH\end{smallmatrix}}$	Montecatini
Zerolit N (ip)	polymerized resin	$-N{\begin{smallmatrix}(CH_3)_2^+\\C_2H_4OH\end{smallmatrix}}$	Zerolit
Amberlite IR-45	polymerized resin	$-N(R)_2$ $-NH(R)$ $-NH_2$	R & H
Amberlite IR-68	polymerized resin	$-N(R)_2$	R & H
Amberlite IR-93	polymerized resin	$-N(R)_2$	R & H
Amberlyst A-21	polymerized resin	$-N(CH_3)_2$	R & H
Dowex 3-X4A	polymerized resin	$-N(R)_2$	Dow
Zerolit M (ip)	polymerized resin	aliphatic polyamines	Zerolit
Dowex 44	condensation polymer	$-N(R)_3$	Dow
EDE-10P	condensation polymer	$-NH(R)$	USSR
Kastel A 100	—	$-N(R)_2$, $-\overset{\mid}{\underset{\mid}{N}}-$ amine and quaternary ammonium groups	Montecatini

c. Amphoteric Ion-Exchangers

Name	Matrix	Functional group(s)	Manufacturer
Retardion 11A8	polymerized resin	$-N(CH_3)_3^+$ $-COO^-$	R & H

Capacity		Maximum temperature °C	Physical form	Remarks
eq/g dry resin	meq/ml of resin bed			
2·6–2·8	1·1	50 [OH^-] 65 [Cl^-]	spherical beads	macroporous for non-aqueous solvents
—	1·3	40 [OH^-]	spherical beads	
—	0·8	40 [OH^-]	spherical beads	
3·2	0·8	40 [OH^-]	spherical beads	ca. 7% DVB
—	1·1	40 [OH^-]	spherical beads	isoporous
5·0	1·9	100	spherical beads	
5·6	1·6	60	spherical beads	aliphatic matrix
4·8	1·4	100	spherical beads	macroporous
·7–5·0	1·7	100	spherical beads	macroporous for non-aqueous solvents
2·8	1·9			
—	1·9	60	spherical beads	isoporous
—	1·4	—	irregular	
9–10	—	60	irregular	
·5–8·0	1·4	40 [OH^-]	spherical beads	
2·95	—	—	spherical beads	Dowex 1X8 with polyacrylate anion

Table 6.1 (*continued*)

Name	Matrix	Functional group(s)	Manufacturer
2. Cellulose and related Ion-Exchangers			
Cellex CM	cellulose	$-OCH_2COO^-$	Bio-Rad
Cellex P	cellulose	$-OPO_3^{2-}$	Bio-Rad
Cellex SE	cellulose	$-OC_2H_4SO_3^-$	Bio-Rad
Cellex T (TEAE)	cellulose	$-OC_2H_4-N^+(C_2H_5)_3$	Bio-Rad
Cellex D (DEAE)	cellulose	$-OC_2H_4-N(C_2H_5)_2$	Bio-Rad
Cellex AE	cellulose	$-OC_2H_4NH_2$	Bio-Rad
Cellex PAB	cellulose	$-OCH_2C_6H_4NH_2$	Bio-Rad
Cellulose phosphate	cellulose	$-O-PO_3^{2-}$	Hungary
Carboxymethyl cellulose	cellulose	$-O-CH_2-COO^-$	Hungary
Ecteola Cellulose	cellulose	$-N(R)_3$	Hungary
DEAE-Cellulose	cellulose	$-OC_2H_4-N(C_2H_5)_2$	Hungary
Cellex E	cellulose	$-N(R)_3$	Bio-Rad
Cellex QAE	cellulose	2-hydroxypropylamino	Bio-Rad
Cellex BD	cellulose	benzoylated DEAE cellulose	Bio-Rad
DE32	cellulose	DEAE	Whatman
CM32	cellulose	carboxymethyl	Whatman
DE23	cellulose	DEAE	Whatman
CM23	cellulose	carboxymethyl	Whatman

Capacity		Maximum temperature °C	Physical form	Remarks
meq/g of dry resin	meq/ml of resin bed			
0·7	—	—	fibres	weakly acidic cation-exchanger
0·85	—	—	fibres	moderately acidic cation-exchanger
0·2	—	—	fibres	strongly acidic cation-exchanger
0·5	—	—	fibres	strongly acidic cation-exchanger
0·4 0·7 0·9	—	—	fibres	moderately basic anion-exchanger
0·8	—	—	fibres	weakly basic anion-exchanger
0·2	—	—	fibres	weakly basic anion-exchanger
0·7	—	—	fibres	weakly acidic cation-exchanger
0·7	—	—	fibres	weakly acidic cation-exchanger
0·35	—	—	fibres	moderately basic anion-exchanger
0·6	—	—	fibres	moderately basic anion-exchanger
0·3			fibres	moderately basic anion-exchanger
0·8			fibres	strongly basic anion-exchanger
0·4			fibres	weakly basic anion-exchanger
1·0			micro-granular	moderately basic anion-exchanger, available pre-swollen as DE 52
1·0			micro-granular	weakly acidic cation-exchanger, pre-swollen—CM 52
1·0			fibrous, fines reduced. Standard—DE22	moderately basic anion-exchanger
0·6			fibrous, fines reduced. Standard—CM22	weakly acidic cation-exchanger

Table 6.1 (*continued*)

Name	Matrix	Functional group(s)	Manufacturer
DEAE Sephadex A-25, A-50	cross-linked dextran gel	DEAE	Pharmacia
QAE-Sephadex A-25, A-50	cross-linked dextran gel	diethyl (2-hydroxypropyl)aminoethyl	Pharmacia
CM-Sephadex CM-25, CM-50	cross-linked dextran gel	carboxymethyl	Pharmacia
SP-Sephadex C-25, C-50	cross-linked dextran gel	sulphopropyl	Pharmacia
DEAE-Sephacel	micro-crystalline cellulose	DEAE	Pharmacia
DEAE Bio-Gel A	cross-linked agarose	DEAE	Bio-Rad
CM Bio-Gel A	cross-linked agarose	carboxymethyl	Bio-Rad

3. Inorganic Ion-Exchangers

Name	Matrix		Manufacturer
Bio-Rad ZP-1	microcrystalline Zr phosphate gel		Bio-Rad
Bio-Rad ZT-1	microcrystalline Zr tungstate gel		Bio-Rad
Bio-Rad ZM-1	microcrystalline Zr molybdate gel		Bio-Rad
Bio-Rad AMP-1	microcrystals of ammonium phosphomolybdate		Bio-Rad
Bio-Rad KCF-1	$K_2[CoFe(CN)_6]$		Bio-Rad
Bio-Rad HZO-1	microcrystalline hydrous Zr oxide gel		Bio-Rad
Bio-Rad HTO-1	microcrystalline hydrous Ti oxide gel		Bio-Rad

4. Ion-Exchangers for HPLC Applications

Name	Matrix	Functional group(s)	Manufacturer
LiChrosorb KAT			Merck
LiChrosorb AN			Merck
Aminex A-4, A-5, A-6, A-7, Q-155, Q-150S	S–DVB co-polymer	—SO_3H	Bio-Rad
Aminex 50W-X2, 50W-X4	S–DVB co-polymer	—SO_3H	Bio-Rad
Aminex A-8, A-9	S–DVB co-polymer	—SO_3H	Bio-Rad
Aminex A-25, A-27, A-28, A-29	S–DVB co-polymer	quaternary ammonium	Bio-Rad

§ 6.2] General Information 299

Capacity		Maximum temperature °C	Physical form	Remarks
meq/g of dry resin	meq/ml of resin bed			
3·5		120	spherical beads	A-25 best for molecules of MW below 30000 or above 200000 and A-50 for 30000–200000
3·0		120	spherical beads	As DEAE-Sephadex
4·5		120		As DEAE-Sephadex
2·3		120	spherical beads	As DEAE-Sephadex
1·4		120	porous spherical beads	Exclusion limit, MW 10^6
	0·015	100	gel beads	Exclusion limit, MW 10^7
	0·02	100	gel beads	Exclusion limit, MW 10^7
1·5	1·5	300	irregular	cation-exchanger, stable in strong acid to pH 13
0·8	1·5	300	irregular	cation-exchanger, stable at pH 1–6
1·0	1·0	300	irregular	cation-exchanger, stable at pH 1–5
1·2	—	—	irregular	cation-exchanger, stable in stron acid to pH 6
0·5	0·4	—	irregular	cation-exchanger, stable in strong acid to pH 12
1·5	1·4	—	irregular	anion- and cation-exchanger stable from pH 1 to $5N$ base
1·0	1·1	—	irregular	anion-exchanger
1·2			irregular porous (10μ)	cation-exchanger, strongly acidic
0·55			irregular porous (10μ)	anion-exchanger, strongly basic
5·0			fine spherical beads	8% cross-linked
5·0			fine spherical beads	2% and 4% cross-linked
5·0			spherical beads	rigid, less susceptible to compression
3·2			spherical beads	8% cross-linked

Table 6.1 (*continued*)

Name	Matrix	Functional group(s)	Manufacturer
Aminex A-14	S–DVB co-polymer	quaternary ammonium	Bio-Rad
Chelex 100	S–DVB co-polymer	iminodiacetate	Bio-Rad
DA-X4	S–DVB co-polymer	—NR$_3$	Durrum
DA-X8	S–DVB co-polymer	—NR$_3$	Durrum
DC-A	S–DVB co-polymer	—SO$_3$H	Durrum
Partisil-10 SAX	silica gel	—NR$_3$	Whatman
Partisil-10 SCX	silica gel	—SO$_3^-$	Whatman
Pellicular Anion		—NR$_3$	Varian
Pellicular Cation		—SO$_3$H	Varian
Bondapak (AX) Porasil	Bonded substituted silicone	—NR$_3$	Waters Ass.
AS Pellionex SAX	S–DVB layer on glass bead	—NR$_3$	Whatman
AE Pellionex SAX	cross-linked polyaromatic resin incorporating polyester type material	—NR$_3$	Whatman
AL Pellionex WAX	aliphatic polymer backbone on glass bead	amino groups	Whatman
HC Pellionex SCX	cross-linked polystyrene on glass	—SO$_3$H	Whatman
Perisorb KAT	bonded substituted silicone	—SO$_3$H	Merck
Perisorb AN	bonded substituted silicone	—NR$_3$	Merck
Vydac Cation Exchange	polymer layer on glass	—SO$_3$H	Applied Science
Vydac Anion Exchange	polymer layer on glass	—R$_3$N	Applied Science
Zipax SAX	Lauryl methacrylate polymer base	—NR$_3$	Du Pont
Zipax SCX	Fluoropolymer base	—SO$_3$H	Du Pont

Addresses of manufacturers:

Zerolit Limited, *now* Dia-prosim Ltd., The Lawn, 100 Lampton Road, Hounslow, Middlesex, TW3 4EB, UK (The resins produced by the Company have been renamed Duolit resins. The reference numbers of these resins are often similar but not always identical to those of Zerolit resins.)

General Information

Capacity meq/g of dry resin	Capacity meq/ml of resin bed	Maximum temperature °C	Physical form	Remarks
3.4			spherical beads	4% cross-linked
			spherical beads	1% cross-linked chelating resin
2.0			20-μm particles	4% cross-linked
4.0			8 and 20-μm particles	8% cross-linked
5.0			8, 12 and 18-μm particles	8% cross-linked
		70	narrow particle-size range, big surface area	HETP 0.09 mm
		70	narrow particle-size range, big surface area	HETP 0.045 mm
0.01			40-μm particles	
0.01			40-μm particles	
0.01			37–50-μm particles	
0.01		85	40-μm glass beads, active layer, 1 μm thick	strongly basic
0.01		90	40-μm glass beads	strongly basic, less polar than S-DVB
		75	40-μm glass beads	weakly basic, slow adsorption kinetics
0.06		90	40-μm glass beads	strongly acidic
0.05			superficially porous, spherical, 1-μm layer	strongly acidic
0.03			superficially porous, spherical, 1-μm layer	strongly basic
0.1		100	solid-core spherical	
0.1		100	solid-core spherical	
0.01			25–37-μm particles	
0.032			25–37-μm particles	

The Dow Chemical Company, Midland, Michigan, USA
Rohm and Haas Company, Philadelphia, Pensylvania 19105, USA
Diamond Alkali Company, Western Division, P.O. Box 829, Redwood City, California, USA
Mitsubishi Chemical Industries Ltd., Japan
Montecatini, Societa Generale per l'Industria e Chimica, Milano, Italy

Table 6.1 (*continued*)

Bio-Rad Laboratories, 2200 Wright Avenue, Richmond, California, 94804, USA and Bio-Rad Laboratories Ltd., Caxton Way, Watford, Herts., UK
E. Merck, 61 Darmstadt, Frankfurterstrasse 250, GFR, and from BDH Chemicals Ltd., Poole, Dorset, BH12 4NN, UK
Durrum Chemical Co., 3950 Fabian Way, Palo Alto, California 94303, USA
Varian Aerograph, 2700 Mitchell Drive, Walnut Creek, California 94598, USA
E. I. Du Pont de Nemours and Co. Inc., Instrument Product Division, 1007 Market Street, Wilmington, Delaware 19898, USA
Pharmacia, Box 175, S-75104, Uppsala, Sweden
Applied Science Inc., P.O. Box 440, State College, Pennsylvania 16801, USA, and from Field Instruments Co., Ltd., Queens House, Holly Road, Twickenham, Mddx., UK
Whatman Labsales Ltd., Springfield Mill, Maidstone, Kent ME14 2LE, U K

6.2.1.2 ANION-EXCHANGERS

Most of the general remarks made in the preceding Section ere also true for anion-exchangers. Condensation-type anion-exchange resins, which are prepared e.g. by condensing aromatic amines with formaldehyde or aliphatic polyamines with aldehydes or by other reactions, are usually polyfunctional. They may contain primary amine —NH_2, secondary amine $>NH$, tertiary amine $>N-$, and sometimes even quaternary ammonium groups —$N^{\pm}-$. Anion-exchange resins containing primary, secondary and tertiary amine groups are weakly basic, do not retain anions of very weak acids (silicates, borates, etc.) and achieve their total ion-exchange capacity only in acidic solutions.

Polymer-type anion-exchange resins are mostly based on S–DVB copolymers. Generally, the synthesis involves chloromethylation and subsequent amination reactions. Treatment of the chloromethylated copolymer with ammonia and aliphatic primary amines leads to weak-base polyfunctional anion-exchange resins (e.g. Amberlite IR-45, Dowex 3, etc.). Amination with tertiary amines yields strongly basic anion-exchange resins containing quaternary ammonium functional groups.

The anion-exchange resins most frequently encountered are the strong-base types carrying the groups —$N^+(CH_3)_3$ (e.g. Amberlite IRA-400,

Dowex 1, Zerolit FF, AV-17, etc.) or $-N^+(CH_3)_2C_2H_2OH$ (Dowex 2, Amberlite IRA-410, etc.). Anion-exchange resins containing quaternary phosphonium $-\overset{|}{\underset{|}{P^\pm}}-$ and tertiary sulphonium groups $\underset{}{>}S^\pm-$ have also been prepared [15,32,50].

Most commercially available ion-exchange resins are 'heteroporous' structures, i.e. their pores are non-uniform in size (actually in apparent pore size, cf. Section 6.2.1.1), the average pore size being related to the degree of cross-linking. Recently anion-exchange resins have been prepared with frameworks cross-linked in a highly regular fashion and thus highly uniform in pore size. They are known as the 'isoporous' type [52], e.g. Zerolit FF-IP, Zerolit N-IP, etc. The structural differences between heteroporous, macroporous, and isoporous ion-exchange resins are illustrated in Fig. 6.2.

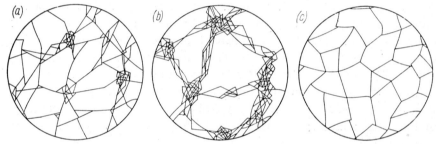

Fig. 6.2 Cross-linking structures (not to scale) of (a) heteroporous, (b) macroporous and (c) isoporous ion-exchangers (from [52], by permission of Dia-prosim Ltd.)

Cellulose anion-exchangers have been applied in trace analysis [55,56, 932]. Most recently, dextran anion-exchangers [39,48] have become commercially available; they are used mainly for separation of high molecular-weight organic compounds. In strongly acidic solutions these exchangers are hydrolysed and so far have little application in inorganic analysis.

6.2.1.3 SELECTIVE ION-EXCHANGE RESINS

Incorporation of functional groups which are capable of selectively or specifically interacting with individual metal ions into a polymer or polycondensate matrix allows selective or chelating ion-exchangers to be developed. So far selective resins have been prepared mostly on a bench scale. The literature [15,29,30,57–68,933 936] provides comprehensive information on these resins. The only commercially available chelating resin, Dowex A-1 (Chelex 100), contains $-N(CH_2COOH)_2$ functional groups [47, 50,57,68a–68f]. A propylenediaminetetra-acetic acid resin prepared by attaching the functional group to a carboxylic acid divinylbenzene resin by means of an esterification reaction was found useful for concentrating trace elements from sea water [68g]. Another chelating resin, Srafion NMRR, which is selective to noble metals, mercury and methylmercury [69–72], has been increasingly used. Sometimes 'ordinary' ion-exchangers can act as selective resins, e.g. a phenol-sulphonic acid resin containing both sulphonate and phenolate groups exhibits a preference for Cs^+ ions through chelate formation [73,74]. Again, a specific chelating affinity for certain ions has been found with phosphonate ion-exchangers [50,75], and certain anion-exchange resins [75] and an amide resin was found to be selective for uranium [75a]. Ion-exchangers in which cyclic polyethers or cryptands [75b,937], or non-cyclic neutral ionophores [75c,d] are attached to a polymeric matrix can be made selective for alkali and/or alkaline-earth metals.

6.2.1.4 Amphoteric Ion-Exchangers

Amphoteric resins, which contain both cation- and anion-exchange groups, can be prepared by condensation or polymerization [15,32]. The resin Retardion 11–A–8, a commercial product of this type, contains quaternary ammonium groups and an equivalent number of carboxylic groups. This resin has been used for separating salts from non-ionic or undissociated compounds by the ion-retardation technique [47,57,76]. Recently resins of this type have been used for several interesting separations of inorganic ions, making use of both the anion- and cation-exchange functions [77].

Chelating resins with functional groups of the iminedialkanecarboxylic acid type (e.g. Chelex 100) are in fact amphoteric ion-exchangers and behave like weakly basic anion-exchangers in acidic solutions [77a].

6.2.1.5 Oleophilic Ion-Exchangers

Ordinary ion-exchangers, except for macroporous ones, cannot be used in non-polar solvents because of the very low exchange reaction rates. Oleophilic resins [78] swell in non aqueous media, and in non-polar solvents their exchange rates are comparable to those observed with conventional resins in aqueous solutions. The resins have been used, for example, for the separation of organic bases [79]. So far no applications of the resins in trace analysis have been reported.

6.2.2 Inorganic Ion-Exchangers

6.2.2.1 Zeolites

Zeolites are crystalline aluminosilicates with ion-exchange properties. They consist of SiO_4 and AlO_4 tetrahedra which have oxygen atoms in common. Filament, loose-layer, and rigid three-dimensional zeolite structures have been described. With some quadrivalent silicon atoms in the crystal lattice replaced by tervalent aluminium atoms, the matrix lattice carries a net negative electric charge, which is balanced by an appropriate number of exchangeable counter-ions. In naturally occurring zeolites these are usually alkali-metal or alkaline-earth ions. Zeolites endowed with a regular three-dimensional structure, whether natural or synthetic (e.g. Linde's Molecular Sieves), have an extremely inflexible rigid framework with a uniform 'pore' or 'channel' size (Fig. 6.3).

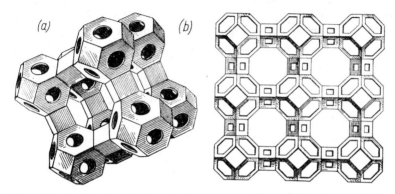

Fig. 6.3 Zeolite (molecular sieve) structures (from [82], by permission of the copyright holders, PWN)

This structure enables the zeolites to act as 'ion sieves' or 'molecular sieves', i.e. to take up small species, ions or molecules, and to exclude completely those which are too bulky to fit into the cavities in the crystal framework. More details on this class of compounds can be found in the relevant literature [23,80–84,938,939].

6.2.2.2 Heteropoly-Acid Salts

Heteropoly-acid salts such as molybdophosphates, molybdoarsenates, molybdosilicates, tungstophosphates, tungstoarsenates and tungstosilicates have been widely used as specific cation-exchangers for certain univalent ions, especially for the heavier alkali metal ions [23,84–93,940–942]. These exchangers are much more resistant to ionizing radiation than organic resins, and they have been used for treatment of highly radioactive waste waters, for recovery of caesium from uranium fission products and sea-water, etc. [81,86,88–90]. Drawbacks to these exchangers include their appreciable solubility in aqueous solutions [90–93] and difficulties encountered in preparing columns through which satisfactory liquid flow-rates can be obtained. Ammonium molybdophosphate suitable for chromatography is commercially available [47].

6.2.2.3 Hydrous Oxides and Insoluble Salts

Insoluble hydrous oxides of multivalent metals such as Al(III), Cr(III), Fe(III), Bi(III), Ce(IV), Ti(IV), Zr(IV), Si(IV), Th(IV), Sn(IV), Nb(V), and Ta(V) exhibit ion-exchange properties [23,94–98b,154,943,944]. They are mostly amphoteric in nature, i.e. in acid solutions they behave like anion-exchangers and in alkaline solutions like cation-exchangers, but some of them display only a single function e.g. silica gel acts only as a cation-exchanger [97,98]. Certain hydrous oxides are highly selective toward multicharged inorganic anions [95].

Recently a number of inorganic ion-exchangers of the salt type have been reported. These include phosphates, tungstates, molybdates, antimonates, vanadates, selenates, arsenates and similar salts of thorium, zirconium, titanium, cerium, tin and other metals [50,94,99–114,945–948], composite ferrocyanides [84,115–118], ferricyanides [949], sulphides [84, 119,120], etc. Zirconium phosphate is the most popular of these; its structure may be represented as follows [121]:

$$\begin{bmatrix} & H_2PO_3 \cdot O & H_2O & & H_2PO_3 \cdot O & H_2O & & HO & H_2O & O \\ & | & \swarrow & & | & \swarrow & & | & \swarrow & \| \\ -& Zr & -O- & & Zr & -O- & & Zr & -O-P-O- \\ & | & & & | & & & | & & | \\ & OH & & & H_2PO_3 \cdot O & & & H_2PO_3 \cdot O & & OH \end{bmatrix}_n$$

The ion-exchange properties of these salts depend on their stoichiometric composition, preparative history, calcination temperature etc. Most are resistant to elevated temperature and ionizing radiation. Generally, they are quite selective toward certain metal ions, but unfortunately are appreciably soluble in aqueous solutions, and their ion-exchange properties are affected by their preparative histories. Some are available commercially [47]. In some of them ion-exchange is not the only process responsible for the uptake of ions from the solution. This is particularly important with materials such as hydrous antimony pentoxide (HAP), hydrous manganese oxide, cupric sulphide, cerous oxalate, and other salts [122] used in radiochemistry for selective absorption of certain nuclides. Of these, HAP has gained particular importance through its use in activation analysis for selective separation of sodium from other elements in concentrated acid solutions [122–125]. These, and other selective sorbents are manufactured by the Carlo Erba Company (Italy).

A comprehensive review of the literature on synthetic inorganic ion-exchangers from 1970 to 1976 is available [125a].

6.2.2.4 Other Inorganic Ion-Exchangers

Other inorganic materials, natural (clays, minerals) or artificial (glass [126], etc.), show ion-exchange properties to a greater or lesser extent. They have been used for the separation of traces of various ions [127,128].

6.2.3 Liquid Ion-Exchangers

Water-insoluble hydrocarbon compounds substituted with ionogenic functional groups, and soluble in water-immiscible organic solvents, have recently been used for liquid-liquid extraction of inorganic ions. The mechanism is effectively that of ion-exchange [50,129–134]. High molecular-weight amines, e.g. tri-n-octylamine, have been used as liquid anion-exchangers and high molecular-weight organic acids, e.g. bis-(2-ethylhexyl)phosphoric acid, as cation-exchangers. Liquid ion-exchangers on filter paper or other appropriate carriers have been used in ion-exchange paper chromatography and in reversed-phase column partition chromatography (extraction chromatography) (cf. Section 6.2.4 and Chapter 7).

6.2.4 Ion-Exchange Paper

Ion-exchange papers can be prepared by chemically modifying cellulose [15,32], impregnating filter paper with colloidal ion-exchangers [135], adding ion-exchangers to the pulp from which the paper is made [136,137], treating the paper with liquid ion-exchangers [131,138], or precipitating inorganic ion-exchangers in the filter-paper phase [139–141]. Ion-exchange paper chromatography and thin-layer chromatography with ion-exchangers used as stationary phases have found widespread use for the separation of trace amounts of substances [61,71,131,134,138–174, 942].

6.3 FUNDAMENTAL CHARACTERISTICS OF AND METHODS FOR EXAMINATION OF ION-EXCHANGERS

6.3.1 General Information

Chromatographic separation depends on the physical properties (particle size, swelling behaviour, etc.) and chemical nature (type of matrix, nature and number of functional groups) of the ion-exchanger used. In general, the properties of the more recent commercial ion-exchangers are very reproducible. Nevertheless, the manufacturer's data should never be relied on as the only basis of reference. The manufacturing process has been refined over the years and it may happen that batches of an ion-exchanger of the same designation and very similar appearance will differ considerably in properties just because they have been produced at different times. By way of illustration, it is well known that, over the period 1952–1967, the ion-exchange capacity of Amberlite IRA-400 rose from 2·3 to 3·7 meq/g [175,176]. Again, manufacturers' data are often incomplete, e.g. they do not specify all the functional groups present in a given ion-exchanger. For example, the ion-exchangers Levatite MII and Wofatit L-160, once broadly advertised as strongly basic, were found on more thorough examination to contain more than 70% of weakly basic groups [177]. Also, monofunctional ion-exchangers may contain small amounts of other functional groups, the presence and proportion of which cannot be foreseen as these depend on the preparative history of the ion-exchanger. It was shown for instance that strongly acidic and strongly basic ion-exchange resins derived from polystyrene often contain small amounts of carboxylic groups [177a].

Therefore, it is extremely important to have methods for precise determination of the properties of the ion-exchangers used in separation processes.

6.3.2 Particle Size

Sieve testing is resorted to for the determination of particle size. A known amount of a representative sample is sieved through a graded set of superimposed standard sieves and the amount retained on each sieve is weighed. In laboratory practice, air-dried samples are usually used. An absolutely dry sample may also be used, or wet screening may be employed [178]. For fine-particle grades (particle size up to 15 μm) obtained by wet classification according to the rate of settling in water, the particle size is examined by microscopic analysis [73,179].

6.3.3 Ion-Exchange Capacity and Polyfunctionality

The ion-exchange capacity is defined as the number of functional groups per unit weight or unit volume of the ion-exchanger. Usually the capacity is expressed in milliequivalents (meq) per gram of absolutely dry ion-exchanger or per cm^3 of a bed of ion-exchanger in suitable form. For cation- and anion-exchangers, the H^+ and the Cl^- forms are the reference forms. An absolutely dry ion-exchanger is assumed to be dried at 105°C, or at 60°C over phosphorus pentoxide or 'Anhydrone', to constant weight. Usually a distinction is made between the exchange capacity for strongly acidic or strongly basic groups, Z_s, and the overall (total) exchange capacity, Z_c. Strongly acidic or strongly basic groups are practically completely ionized, so the corresponding capacity remains constant over a wide pH range. These groups are capable of removing ions from neutral salts such as

NaCl, CaCl$_2$ etc. The capacity of weakly acidic or weakly basic groups is affected by the pH of the solution in contact with the ion-exchanger and complete uptake of cations or anions by these groups can be achieved only in alkaline or acid solutions, respectively. These groups cannot remove ions from neutral salts, and their salt-forms hydrolyse in pure water. Figure 6.4 shows titration curves for various cation-exchangers. Potentiometric titration is a convenient method for determining the capacity, especially for ion-exchangers with highly ionized functional groups. It is also possible to determine Z_s and Z_c simultaneously, particularly when the content of weakly ionized groups is low. Figure 6.5 shows titration curves for a representative strongly basic anion-exchanger, viz., Amberlite IRA-400, and for the polyfunctional Wofatit L-160 which also contains strongly basic groups. A careful potentiometric titration of Amberlite IRA-400 revealed the presence of a small proportion of weakly basic groups ($\sim 4\%$). Wofatit L-160 contains well above 70% of weakly basic groups [177], and the total capacity could not be determined by potentiometric titration. It is, however, of interest to find that the ion-exchanger contains two types of strongly basic groups [177]. Potentiometric titration is most conveniently carried out in the presence of a metal salt (KCl) at low concentration ($\sim 0.01M$) [177,180,181]. It is also essential to select a suitable particle size for the ion-exchanger and to provide efficient agitation.

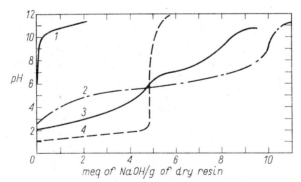

Fig. 6.4 Titration curves of various cation-exchange resins: 1 — phenolic, 2 — carboxylic acid, 3 — phosphinic acid, 4 — sulphonic acid (from [317], by permission of the New York Academy of Sciences)

A static (batch equilibration) method can also be used for determining the overall capacity of cation exchangers by treating a specified amount of an H$^+$-form cation-exchanger with an excess of a standard sodium hydroxide solution and back-titrating the hydroxide in an aliquot of the solution after the exchange equilibrium has been established [73,178].

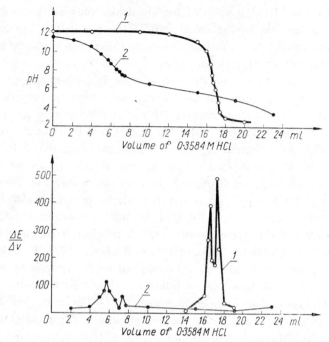

Fig. 6.5 Titration curves of 1 – Amberlite IRA-400 (OH$^-$) (0·1–0·3 mm) and 2 – Wolfatit L-160 (OH$^-$) (0·1–0·3 mm) (from [177], by permission of Z. Błaszkowska and R. Dybczyński)

For strongly acidic groups the ion-exchange capacity is determined by a dynamic method in which 1M sodium chloride is passed in excess through a column packed with a weighed amount of the H$^+$-form of an ion-exchanger and the resulting acid in the eluate is titrated with a standard sodium hydroxide solution [73]. The dynamic method is also useful for determining the content of strongly basic groups. A weighed amount of the anion-exchanger is placed in a column, converted into the OH$^-$–form and then washed with an excess of 1M sodium chloride. The bed is rinsed with water and chloride ions are eluted with an excess of 1M sodium hydroxide, and titrated in the eluate with standard silver nitrate solution [177,178]. A similar procedure may be used for the determination of total capacity except that the ion-exchanger is converted into the Cl$^-$–form with 1M hydrochloric acid and the column is washed with ethanol [178]. The capacity can also be determined by directly titrating a suspension of the Cl$^-$–form of the ion-exchanger with silver nitrate solution using a silver electrode [177]. A method has also been reported for determining primary, secondary and tertiary amine groups in weakly basic anion-exchangers. Three weighed samples of anion-exchanger are treated with known volumes of perchloric acid solution in anhydrous acetic acid and after equilibration the acid is back-titrated potentiometrically with a solution of potassium hydrogen phthalate in acetic acid. Before the reagents are added, the primary and secondary groups of one sample are acetylated with acetic anhydride, and another sample is treated with salicylaldehyde in order to convert the primary groups into a Schiff's base [182].

Other methods used for the determination of ion capacity include pyrolysis-gas chromatography [182a], infrared spectrophotometry [182a] and difference weighing of the exchanger loaded with different counter-ions [182b].

6.3.4 Determination of pK

The negative logarithm of the dissociation constant, $-\log K = pK$, may be determined from pH-titration data, e.g. if a weakly basic cation-exchanger is titrated with sodium hydroxide in the presence of sodium chloride, the pK may be evaluated from equation [15]:

$$pK = pH + \log [Na^+] - \log \frac{[X]}{2} \qquad (6.1)$$

where [X] is the total concentration of ionogenic groups in the ion-exchanger pH is the pH of the external solution at half-conversion, and [Na$^+$] is the concentration of sodium ions in the solution at 50% conversion ($\alpha = 0.5$) of the ion-exchanger.

To calculate the concentration [X] in Eq. (6.1), it is necessary to determine in a separate experiment the water regain in the ion-exchanger at the point of half-conversion into the Na$^+$-form.

A similar method is used for determining pK values of weakly basic anion-exchangers. In particular cases, pK values of individual functional groups in a polyfunctional ion-exchanger can be determined. More information is available in the relevant literature [183].

6.3.5 Mechanical and Chemical Stability

The mechanical stability (resistance to abrasion, etc.) of ion-exchangers is generally satisfactory for analytical purposes and becomes an important factor only in industrial-process applications. Determination of mechanical stability is usually done by comparative sieve testing before and after the ion-exchanger has been used for some specified number of exchange cycles, or before and after it has been ground in a ball-mill, in standard conditions etc. [184,185].

The chemical stability depends chiefly on the chemical nature and structure of the particular ion-exchanger. Certain inorganic ion-exchangers (some zeolites and hydrous oxides) are stable only within a narrow pH range, at about pH 7. Other exchangers, e.g. zirconium phosphate and heteropoly-acid salts, are stable in acid media but hydrolyse in alkaline media. Synthetic organic ion-exchangers are normally insoluble in aqueous solutions and in organic solvents but are subject to thermal degradation at elevated temperatures and to chemical degradation by oxidants [186–188], concentrated alkalis [189,190], etc.

Polymer-matrix ion-exchangers are more resistant to chemical agents than are the condensation types, and cation-exchangers are, on the whole, more stable than anion-exchangers. Generally, chemical stability is examined by bringing an ion-exchanger into contact with a given medium for a specified period of time at a specified temperature, after which the liquid phase is separated and analysed for dry residue, etc. Degradation of ion-exchangers is also studied in terms of variation in ion-exchange capacity, swelling behaviour, water regain, etc. Degradation of the matrix may involve loss or chemical modification of fixed functional groups [191–195], production of new groups, e.g. formation of cation-exchange groups by oxidation of an anion-exchanger [188], cleavage [186,194] or generation of extra cross-links [74,193,195], etc. Ionizing radiation can give rise to effects similar to chemical degradation [196–201]. The inadequate resistance of organic ion-exchangers to such radiation makes the use of inorganic ion-exchangers preferable [176,208] for work with highly radioactive materials.

6.3.5.1 Thermal Stability

Elevated temperatures often aid ion-exchange processes. Therefore, ion-exchangers which are stable at high temperatures are particularly valuable.

Polymer-matrix carboxylic cation-exchangers can withstand temperatures of up to 160°C, and of the organic cation-exchangers containing sulphonate groups, those with the styrene-divinylbenzene matrix are particularly stable. At temperatures higher than 100°C some desulphonation occurs [196–203, 950], the Na^+-form being more resistant than the H^+-form to degradation. Phenolsulphonic acid ion-exchangers are much less thermally stable (cf. Table 6.1). On heating, they are desulphonated, but their mechanical stability [204] and density [193] increase, their exchange kinetics deteriorate [74,193], and they become less selective [74], etc. Evidently, there are further condensation reactions in the ion-exchanger phase, and extra cross-links are formed [74,193–195].

Strongly basic OH^--form [192,194,205] and HCO_3^--form anion-exchangers [205] are thermally labile. Experimental data for a strongly

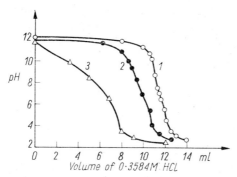

Fig. 6.6 Titration curves of unheated and heated Lewatit MN (OH⁻) (condensation-type matrix): 1 — unheated, 2 — heated for 24 hr at 40°C, 3 — heated for 17 hr at 70°C (from [192], by permission of R. Dybczyński

basic anion-exchanger are shown in Fig. 6.6. Marked signs of degradation appear at a temperature as low as 40°C. The polymer-type anion-exchanger Amberlite IRA-400 proved to be much more stable than the condensation-type anion-exchanger Levatite MN [192]. Strongly basic groups have been shown to decompose by two routes [192], viz.,

Normally, both reactions occur together, and this causes the ion-exchange capacity of strongly basic groups, Z_s, to diminish faster than the total capacity, Z_c. Thus, the polyfunctionality of the ion-exchanger increases, as may be seen from the titration curves in Fig. 6.6. Oxidation with molecular oxygen may also contribute to the degradation of anion-exchangers at high temperatures [206,207]. Anion-exchangers in the Cl⁻-form are much more stable [205,209].

Inorganic ion-exchangers are generally superior to organic ones in thermal stability [99]. However, structural changes cannot be ruled out. For example, the ion-exchange behaviour of zirconium phosphate is known to depend on the temperature at which the exchanger has been dried [210]. This is also true for zirconium oxide [95] and tin(IV) arsenate [211]. The thermal stability of other inorganic ion-exchangers has yet to be studied in more detail.

6.3.6 Swelling

When brought into contact with water, ion-exchangers, especially organic ones, expand or 'swell'. This effect is associated with the tendency of the ions in the ion-exchanger phase to surround themselves with appropriate solvation shells. The elastic forces of the matrix lattice oppose the stretching forces and thus prevent swelling beyond a certain limit. Swelling is directly related to a resin's exchange capacity and to the dissociation of functional groups and inversely related to the number of cross-links. It is also related to the nature of the counter-ion and to the concentration of the external solution. For ion-exchangers with the same chemical composition but different numbers of cross-links (e.g. sulphonic acid ion-exchangers and trimethylbenzylammonium anion-exchangers with an S–DVB matrix), this is a simple relationship between the degree of cross-linkage and the amount of swelling [212–218].

$$\frac{V_{swollen}}{V_{dry}} - 1 = \frac{A}{X} \qquad (6.2)$$

where X is the degree of cross-linkage (mole percent of DVB in the copolymer), $V_{swollen}$ and V_{dry} denote the volumes of the swollen and dry resins, respectively, and A is a constant related only to the nature of the counter-ion [219].

A slightly different treatment of the dependence of the swelling of ion-exchange resins on their cross-linkage has recently been suggested [219a]. The swollen and the dry volumes can be determined pycnometrically with the aid of water and n-heptane [221], or microscopically [222]. Another method is to measure the maximum water content (w/w) in the dry resin, W_{H_2O}, or the specific water regain, W_{H_2O}/Z_s. With sulphonated S–DVB and analogous strong-base ion-exchangers, W_{H_2O} (or W_{H_2O}/Z_s) has been found to be linearly related to the reciprocal of the degree of cross-linkage; appropriate calibration plots have been reported [218,244]. In this method, W_{H_2O} is evaluated either as the difference between the weight of the water-swollen and centrifuged [214,218] or filter-paper-dried [233] resin, and that of the absolutely dry resin (e.g. dried at 105°C to constant weight), or by extrapolating the plot of the swollen resin weight vs. relative humidity up to 100% humidity [223].

In addition to miscellaneous drying techniques, Karl Fischer titration [224,225], proton magnetic resonance [225], and other methods have been used for the determination of water in ion-exchange resins. Low water contents have been determined by calorimetry, viz., by measuring the heat of immersion of a resin in various solvents [226]; tritium-labelled water [227] has also been used for the purpose. Efforts have been made to estimate the degree of cross-linkage by infrared spectroscopy [39,182a], electron microscopy [228], and pyrolysis in combination with gas-liquid chromatography [182a,229,230].

Cross-linkage is an essential parameter which affects the ion-exchange equilibrium [219,220,231–238], kinetics [219,239–241], chromatographic resolution [218,242–246a], dissociation of functional groups [247], etc.

6.3.7 Ion-Exchange Resin Density and Bed Density

The true density of a swollen resin, d_j (g/cm^3), is determined pycnometrically in water [178]. A parameter of importance for laboratory practice is the bed density d_z; this is usually expressed in g of dry resin in a well-defined form per cm^3 of the resin bed in water. Other definitions have also been used, e.g. the apparent density of a hydrated resin bed d_p, expressed in g of swollen resin per cm^3 of resin bed. Bed density is determined by measuring the volume occupied by a known weight of the resin in a column, after the bed has been back-washed [178] or in a measuring cylinder [73]. From these data the free volume (V) of the bed is evaluated as

$$V(\%) = \left(1 - \frac{d_p}{d_j}\right) 100 \qquad (6.3)$$

The true density of a dry ion-exchange resin is determined pycnometrically in n-heptane, n-octane, n-hexane [221,222], or similar solvents, the resin to be examined having been dried at a temperature of 105°–110°C.

6.3.8 Other Properties

The homogeneity of a batch of resin, as regards the degree of cross-linkage and the ion-exchange capacity of individual resin particles, can be examined by the flotation test [248]. A linear gradient of a high-density large-ion salt solution (Na$_2$WO$_4$) is produced along a vertical glass tube equipped with containers at either end. Resin particles collect at definite heights according to density. If more than one ring is formed, the resin is regarded as inhomogeneous.

Macroporous resins are characterized in terms of the BET specific surface, apparent and matrix densities (determined by mercury and helium displacement techniques respectively), and pore-size distribution (established by mercury porometry) [53,249–251].

6.4 ION-EXCHANGE EQUILIBRIUM

In the early attempts to describe quantitatively the distribution of ions between the ion-exchanger and the solution, the process was usually assumed to be sorptive in nature, a supposition justified inasmuch as the early objects of study included clays, soils, and other naturally occurring inorganic ion-exchangers capable of both ion-exchange and physical sorption of ions. The resulting mathematical treatments usually resulted in modified Langmuir or Freundlich isotherms [252–254].

Modern synthetic organic ion-exchangers have structures of the swollen-gel type, produce ion-exchange reactions that are, as a rule, completely reversible, exhibit no hysteresis, and have exchange capacities unaffected by particle size. With organic resins, the ion-exchange reaction is strictly stoichiometric, at least in dilute solutions. Therefore, ion-exchange is commonly treated as a heterogeneous chemical reaction and the law of mass action is applied to it.

Ion-exchange may be treated either in terms of standard thermodynamics or in terms of a special physical model assumed to hold for the resin–solution system. The first approach is universal in nature, affords an exact description of the ion-exchange process, and requires no assumptions to be made about the mechanism of the process. The practical value of the thermodynamic approach is limited, however, because it gives no insight into the physical nature of the ion-exchange process and does not allow prediction of ion-exchange equilibria from other physico-chemical data. The description of ion-exchange equilibria with the aid of models, though thermodynamically less rigorous, does enable the equilibria to be related to the effects of various physical and chemical forces. Although still far from perfect, the models often permit prediction of ion-exchange equilibria, at least qualitatively, from known fundamental data for the solvent, solute, and ion-exchanger.

6.4.1 Fundamental Definitions

The exchange of univalent ions, e.g. sodium for hydrogen ions, can be written as follows

$$RH + Na^+ \rightleftharpoons RNa + H^+ \qquad (6.4)$$

where R denotes a structural unit of the ion-exchanger

The numerical value of the thermodynamic equilibrium constant

$$\mathcal{K}_H^{Na} = \frac{a_{RNa}\, a_H}{a_{RH}\, a_{Na}} \qquad (6.5)$$

(where a denotes the activity of the species indicated by subscript) depends on the choice of standard state. In the formal thermodynamic approach, the ion-exchanger at equilibrium is treated as a homogeneous binary solid solution of the resin species RH and RNa [255–260].

The standard state (activity equal to unity) and the reference state (activity coefficient equal to unity) for RH and RNa are taken to be the respective monoionic forms of the ion-exchanger in equilibrium with water. For the water phase, the normal convention for electrolyte solutions is adopted i.e. the standard state is a hypothetical ca. 1-molal solution (the deviation from $1m$ is given by the activity coefficient) and the reference state is an infinitely diluted solution.

If the activity of the ion-exchanger phase is expressed in terms of the mole fractions N_{RH} and N_{RNa}, the following expression is obtained

$$\mathcal{K}_H^{Na} = \frac{N_{RNa}\, m_H}{N_{RH}\, m_{Na}} \frac{\gamma_{HQ}^2\, f_{RNa}}{\gamma_{NaQ}^2\, f_{RH}} = k_H^{Na} \frac{\gamma_{HQ}^2\, f_{RNa}}{\gamma_{NaQ}^2\, f_{RH}} \qquad (6.6)$$

where m is the molality of the solution, f and γ are the rational activity coefficients of the resin species in the ion-exchanger and the mean molal activity coefficients of the electrolytes HQ and NaQ (Q is a univalent ion) in the solution, respectively, and k_H^{Na} is the rational selectivity coefficient.

For counter-ions of different charge, the ion-exchange reaction is conveniently referred to 1 gram-equivalent, viz.,

$$\frac{1}{q} R_q A + \frac{1}{p} B^p = \frac{1}{p} R_p B + \frac{1}{q} A^q \qquad (6.7)$$

A^q and B^p denoting the counter-ions of charge q and p.

The rational selectivity coefficient is then defined by the equation

$$k_{A^q}^{B^p} = \frac{N_{R_pB}^{|1/p|}\, m_{A^q}^{|1/q|}}{N_{R_qA}^{|1/q|}\, m_{B^p}^{|1/p|}} \qquad (6.8)$$

This coefficient is a measure of the relative affinities of given ions for the ion-exchanger and thus characterizes the selectivity of the ion-exchanger. Selectivity is often expressed with the aid of the 'corrected' rational selectivity coefficient, $k'^{B^p}_{A^q}$, i.e. the selectivity coefficient corrected for the ratio of the mean activity coefficients of the ions in the solution.

$$k'^{B^p}_{A^q} = k^{B^p}_{A^q} \frac{\gamma_{AQ_q}^{\left|\frac{q+1}{q}\right|}}{\gamma_{BQ_p}^{\left|\frac{p+1}{p}\right|}} \qquad (6.9)$$

The selectivity coefficient is not a true constant, since it depends on the composition of the ion-exchanger phase at equilibrium (Fig. 6.7).

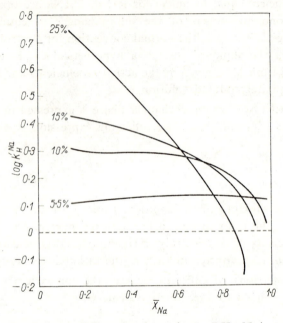

Fig. 6.7 Selectivity coefficients for the exchange $RH + Na^+ = RNa + H^+$ on sulphonic-acid cation-exchange resins of various degrees of cross-linkage as a function of resin phase composition (% DVB marked on curves; each resin is a fully monosulphonated polymer (from [232], by permission of the copyright holders, the Chemical Society)

The corrected rational selectivity coefficient may be shown to be related [261] to the thermodynamic equilibrium constant of the ion-exchange reaction by

$$\ln \mathcal{K}^B_A = \int_0^1 \ln k'^{B^p}_{A^q} d\bar{X}_B \qquad (6.10)$$

where $\bar{X}_B$ is the equivalent ionic fraction of the ion B^p in the ion-exchanger

Although the ion-exchange reaction constant is an unambiguous measure of the relative affinity of ions for the ion-exchanger, its practical value is limited.

Since it is obtained by integration, the constant tells us nothing about the selectivity of the ion-exchanger under the particular experimental conditions of, say, a chromatographic separation. Furthermore, the thermodynamic equilibrium constant can be neither predicted nor determined otherwise than by making a series of measurements of the selectivity coefficient for different compositions of the ion-exchanger.

When traces are being separated, the number of trace ions B^p in the ion-exchanger [Eq. (6.7)] is negligibly small compared with that of the ions A^q originating from the supporting electrolyte. The ion-exchanger is practically completely filled with the ions A^q and its composition may be written as $\overline{X_B} = 0$; $\overline{X_A} = 1$.

Under these conditions, the corrected rational selectivity coefficient, $k'^{B^p}_{A^q}$ [cf. Eq. (6.9)], has a constant value and lends itself well to the characterization of the selectivity of the ion-exchanger.

In addition to rational selectivity coefficients, the ion-exchange literature often reports molal selectivity coefficients. These are derived from a model in which the ion-exchanger phase is treated as a concentrated electrolyte solution separated from the external solution by a semipermeable membrane which is the surface of the ion-exchanger [262]. This is the simple Donnan equilibrium model.

If identical standard states are selected for the two phases, the exchange equilibrium constant described by Eq. (6.7) becomes unity [44,263] and may be written as

$$K \equiv 1 = \frac{\overline{m}_B^{1/p} m_A^{1/q} \overline{\gamma}_B^{1/p} \gamma_A^{1/q}}{\overline{m}_A^{1/q} m_B^{1/p} \overline{\gamma}_A^{1/q} \gamma_B^{1/p}} = \frac{\overline{m}_B^{1/p} m_A^{1/q} \gamma_{AQ_q}^{\left|\frac{q+1}{q}\right|} \overline{\gamma}_B^{1/p}}{\overline{m}_A^{1/q} m_B^{1/p} \gamma_{BQ_p}^{\left|\frac{p+1}{p}\right|} \overline{\gamma}_A^{1/q}} \quad (6.11)$$

where the barred symbols refer to the ion-exchanger phase. The expression

$$K^{B^p}_{A^q} = \frac{\overline{m}_B^{1/p} m_A^{1/q}}{\overline{m}_A^{1/q} m_B^{1/p}} \quad (6.12)$$

defines the molal selectivity coefficient and the corrected molal selectivity coefficient is given by

$$K'^{B^p}_{A^q} = K^{B^p}_{A^q} \frac{\gamma_{AQ_q}^{\left|\frac{q+1}{q}\right|}}{\gamma_{BQ_p}^{\left|\frac{p+1}{p}\right|}} = \frac{\overline{\gamma}_A^{1/q}}{\overline{\gamma}_B^{1/p}} \quad (6.13)$$

It is evident from Eq. (6.13) that the molal selectivity coefficient can be calculated from the ratio of the activity coefficients of the ions in the ion-exchanger and these may be estimated from the reported mean activity coefficients [264,265] in concentrated electrolyte solutions. Such estimates are, however, imprecise and the calculated selectivity coefficients often differ considerably from the observed values. Again, the theory does not provide for the experimentally observed [257,259,261,266] [cf. Fig. (6.7)] variation of K_A^B with the composition of the ion-exchanger phase.

For trace ions B^p, the rational and the molal selectivity coefficients can be shown to be interrelated [267] by

$$k_{A^q}^{B^p} = K_{A^q}^{B^p} \left(\frac{C_r}{W_{H_2O}} \right)^{\left|\frac{p-q}{p \cdot q}\right|} \tag{6.14}$$

where C_r is the concentration of the ion-exchanger phase, (in mmole/g of dry resin $[A^q]$), and W_{H_2O} is the water content, (in g of H_2O/g of dry resin $[A^q]$).

For counter-ions of equal charge ($p = q$), the rational and molal selectivity coefficients are identical.

6.4.2 Selectivity

The selectivity of an ion-exchanger is defined as the capability of the ion-exchanger to select certain counter-ions in preference to others. Utilization of selectivity properties forms the basis for every type of separation. To understand the source of selectivity properties and the nature of the physico-chemical factors involved is of paramount importance for prediction of the course of ion-exchange reactions and their practical use.

6.4.2.1 Ways of Expressing Selectivity

In basic research, the rational [Eqs. (6.8) and (6.9)] and molal selectivity coefficients [Eqs. (6.12) and (6.13)] are usually considered to be a measure of selectivity. If the corrected molal selectivity coefficient is larger than unity ($K_{A^q}^{'B^p} > 1$), the ion-exchanger is selective for B^p ions in preference to A^q ions.

The affinity of an ion for a given ion-exchanger is frequently described by the distribution coefficient, i.e. by the ratio of the concentrations of the ion in the resin and in the solution at equilibrium. The ion-exchange isotherm [Eq. (6.4)] is curved (Fig. 6.8), even with univalent ions, except for the rare cases of non-selectivity ($K_{A^q}^{B^p} = 1$). Therefore, the distribution coefficient remains constant, irrespective of the concentration of a given ion in the solution, only at low concentrations, i.e. when the ion-exchange isotherm may be approximated by a straight line (see dashed lines in Fig. 6.8). Thus, the distribution coefficient is suitable for characterization of the selectivity of the ion-exchanger for ions present in trace amounts. Most often, the coefficient used is the weight distribution coefficient λ_{B^p}, defined as the quotient of the numbers of moles of B^p ions per g of dry resin and per cm^3 of solution at equilibrium.

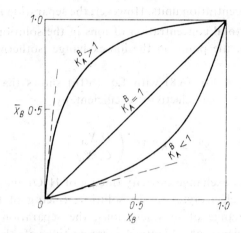

Fig. 6.8 Ion-exchange isotherms for various values of the selectivity coefficient

For traces of B^p, this coefficient [267,268] is related to the rational distribution coefficient [Eq. (6.8)] by

$$\lambda_{B^p} = \frac{(k_{A^q}^{B^p})^{|p|} C_r}{(m_{A^q})^{|p/q|} d} \qquad (6.15)$$

where C_r is the concentration of the ion-exchanger phase, in mmole/g of dry resin [A^q], m_{A^q} is the molality of the major ion (molality of the supporting electrolyte), and d is the density of the solution at equilibrium (practically equal to the density of the supporting electrolyte).

An ion-exchanger may be considered to be highly selective for B^p if $\lambda_{B^p} > 100$ under experimental conditions. If $\lambda_{B^p} < 10$, the selectivity of the ion-exchanger towards B^p ions is considered to be small.

Under given experimental conditions the actual affinity of ions for the ion-exchanger is more adequately described by the separation factor, which is defined as the quotient of the concentration ratios of the two counter-ions in the ion-exchanger and in the solution

$$\alpha_{A^q}^{B^p} = \frac{\bar{m}_{B^p} m_{A^q}}{\bar{m}_{A^q} m_{B^p}} = \frac{\bar{X}_{B^p} X_{A^q}}{\bar{X}_{A^q} X_{B^p}} = \frac{\lambda_{B^p}}{\lambda_{A^q}} \qquad (6.16)$$

A value of $\alpha_{A^q}^{B^p} > 1$ indicates that B^p is selectively taken up by the ion-exchanger in preference to A^q ions. Unlike that of the selectivity coefficient, the numerical value of the separation factor is not affected by the choice of the concentration units. However, the separation factor varies with the temperature, total concentration of ions in the solution, and with $\bar{X}_{B^p}$, i.e. it depends on the point on the ion-exchange isotherm for which it is calculated.

Comparison of Eq. (6.8) with Eq. (6.16) shows that the separation factor is related to the selectivity coefficient by

$$\alpha_{A^q}^{B^p} = (K_{A^q}^{B^p})^q \left(\frac{Q \bar{X}_B}{C_s X_B} \right)^{\left| \frac{p-q}{q} \right|} \qquad (6.17)$$

where Q is the ion-exchange capacity (meq/g of H_2O) and C_s is the overall concentration of electrolytes in the solution, (meq/g of H_2O).

For the exchange of univalent ions, the separation factor and the selectivity coefficients are identical, irrespective of the concentration scale, viz.,

$$\alpha_A^B = K_A^B = k_A^B \qquad (p = q = 1) \qquad (6.18)$$

For trace ions B^p ($m_{A^q} \gg m_{B^p}$), the separation factor $\alpha_{A^q}^{B^p}$ has a constant value at a given temperature and concentration of supporting electrolyte and it characterizes the ability of the ion-exchanger to absorb trace ions in preference to the major ions A^q.

The ion-exchange separation of traces is often described in terms of the separation factor of the ion B^p with respect to other trace ions D^z rather than with respect to the major ions A^q.

$$\alpha_{A^q}^{B^p} = \frac{\lambda_{B^p}}{\lambda_{D^z}} \quad (m_{A^q} = \text{const} \gg m_{B^p} \text{ and } m_{D^z}) \qquad (6.19)$$

The separation factor thus defined (ratio of the distribution coefficients), characterizes the feasibility of separation of two trace elements under given experimental conditions. The actual separation of the two elements is also affected by the dynamics of the column process (cf. Section 6.6.4): if this is unfavourable, poor separation will be obtained even in systems with high separation factors (cf. Section 6.6.4.1).

6.4.2.2 Sources of Selectivity

In its present state, ion-exchange theory does not allow quantitative prediction of the affinity of ions for the ion-exchanger from independent physico-chemical data. Nevertheless, the various factors that determine the selectivity are relatively well known, and numerous approaches involving various models have been developed, which, together with semi-empirical rules and experimental observations permit logical interpretation and sometimes even prediction of the relative affinity of ions for a given ion-exchanger. The rules are not general ones: most often they are valid only for a limited number of ion-exchanger–ion systems. Since the semi-empirical rules and generalizations usually rely on isolated phenomena or properties of ions or ion-exchangers, the observed selectivity may still deviate qualitatively from the predicted selectivity, because of competition between two or more factors, only one of which was taken into consideration. Therefore, exact recognition of all the possible factors determining the selectivity is of paramount practical importance.

Charge on the Counter-Ion

As indicated in Section 6.4.2.1, the separation factor is the best measure of the actual affinity of two counter-ions for a given ion-exchanger under experimental conditions. Even if an ion-exchanger is non-selective for a given ion ($K_{A^q}^{B^p} = 1$), it is evident from Eq. (6.17) that, if $p > q$, $\alpha_{A^q}^{B^p} > 1$, because the concentration of the ion-exchanger phase is usually higher than that of the external solution ($Q > C_s$). The ion of higher charge will thus be taken up selectively, and the more strongly the higher the exchange capacity of the ion-exchanger and the more dilute the external solution. This effect is called 'electroselectivity' [15] in constrast to the 'true' selectivity which requires that the selectivity coefficient K_A^B differs from unity.

For the trace ion B^p the expression

$$\frac{d \log \lambda_{B^p}}{d \log m_{A^q}} = -\frac{p}{q} \qquad (6.20)$$

can be obtained by differentiation of Eq. (6.15).

The graphs of the logarithm of the distribution coefficient for a given ion *vs.* the logarithm of the concentration of the major ion in the supporting electrolyte have differing slopes, depending on the ratio of the charges of the two ions concerned (Fig. 6.9). At suitably high supporting electrolyte concentrations, reversal of selectivity may occur, i.e. the ion of higher charge will have less affinity for the ion-exchanger than the ion of lower charge. This phenomenon is often used to enable elements to be eluted from an ion-exchanger column in a particular sequence.

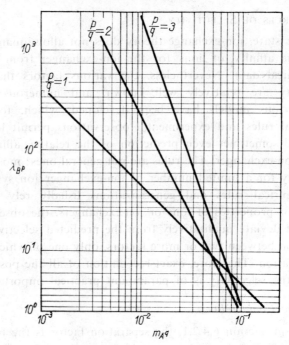

Fig. 6.9 The distribution coefficient of a trace ion as a function of the concentration of supporting electrolyte at various ratios of the ions in question

In dilute solutions, the affinity series is as expected, viz., $La^{3+} > Ca^{2+} > Na^+$; $SO_4^{2-} > Cl^-$, etc. [17].

Deviations from the relationship described by Eq. (6.20) have been found to occur with ion-exchangers which contain only a few functional groups relatively distant from one another [269].

Solvation of Ions

Although the various ion-exchange theories and models provide different interpretations of the effect of solvation on selectivity, they all consistently agree that the greater the solvation of the ion, the lower its affinity for the ion-exchanger. In Gregor's theory [270,271] the ion-exchanger phase is considered to be a system composed of an elastic stretchable polymer network carrying a number of fixed functional groups, the electrical charge on which is compensated by that of the mobile counter-ions. When brought into contact with water, the ion-exchanger swells, the polymer network expands and the pressure exerted upon the internal solution (ion-exchanger phase) increases. At equilibrium, the osmotic pressure inside the ion-exchanger phase is equal to the pressure exerted on the internal solution by the stretched hydrocarbon chains. This model allows the following equation for the molal selectivity coefficient to be derived:

$$K_{A^q}^{B^p} = e^{\frac{\pi}{RT}\left(\frac{V_A}{|q|} - \frac{V_B}{|p|}\right)} \tag{6.21}$$

where π is the swelling pressure and V_A and V_B are the partial molal volumes of the hydrated A^q and B^p ions.

Gregor's theory can explain numerous experimental facts in simple situations, e.g. the affinity series (selectivity sequence) of the alkali metals for sulphonated cation-exchangers, Cs > Rb > K > Na > Li [232,259, 271–273], an analogous series for the alkaline-earth metals [49], Ba^{2+} > Sr^{2+} > Ca^{2+} > Mg^{2+}, etc. It fails, however, to explain the reversal of selectivity sometimes found to occur at higher $\overline{X}_B$ values, the cross-over of the log K_A^B vs. $\overline{X}_B$ curves for ion-exchangers of identical structures but different degrees of cross-linkage, and also selectivities deviating markedly from the rule, especially in anion-exchange [273].

According to Eisenman's theory [274,275] the causes of selectivity are to be looked for in the electrostatic interactions of ions with the functional groups of the ion-exchanger and in the energy associated with the removal (or rearrangement) of water molecules from the coordination shell of a counter-ion as the result of its having entered the ion-exchanger phase. For the exchange of univalent ions, the relationship between the molal selectivity coefficient and these factors is

$$\ln K_A^B = \frac{1}{RT}\left(\frac{e^2}{r_R + r_B} - \frac{e^2}{r_B + r_A}\right) + \frac{1}{RT}(\Delta G_A - \Delta G_B) \tag{6.22}$$

where e is the charge on the electron; r_R is the radius of a functional group; r_A and r_B are the crystallographic radii of the counter-ions: and ΔG_A and ΔG_B are the free energies of the dehydration that is assumed to occur when the exchanged ions pass from the solution into the ion-exchanger; these energies can be estimated from the known hydration enthalpies of the ions.

In contrast to the Gregor theory, which seeks to explain selectivity in terms of differences between the hydrate-ion volumes and assumes that the ion entering the ion-exchanger phase retains its hydration shell virtually unaffected, the theory advanced by Eisenman and refined by Reichenberg [273] emphasizes the energy changes that accompany the solvation process. In this approach, the ion passing from the solution into the ion-exchanger phase and attracted by the functional groups, loses a number of water molecules from its hydration shell.

If the first term in Eq. (6.22) is small (which is the case when the radius r_R of a functional group is large, e.g. for $-SO_3^-$), the selectivity coefficient is determined by the difference between the free energies of dehydration of the ions exchanged. The smaller the free energy of dehydration, the larger the ion-exchange affinity of a given ion. The theory thus explains the observed affinity series of the alkali metals for sulphonic acid cation-exchangers ($Cs^+ > K^+ > Na^+ > Li^+$) and it also explains numerous experimental observations for anion-exchange, e.g. the selectivity sequence $ClO_4^- > I^- > Br^- > Cl^-$, etc.

It should be borne in mind, however, that when r_R is small, the electrostatic-interaction energy [the first term in Eq. (6.22)] may contribute considerably or even decisively to the selectivity coefficient. The 'reversed' selectivity sequence would then be expected to occur, e.g. $Li^+ > Na^+ > K^+ > Cs^+$, which is actually the case with carboxylic-acid ion-exchangers.

These, and some other theories, assume that the causes of selectivity arise in the ion-exchanger phase exclusively. In contrast with this, the theory advanced by Diamond, Chu and Whitney [276–280] treats selectivity as the result of competition between the co-ions (ionic species with the same charge sign as the exchanger's functional groups) in the water phase, the functional groups in the ion-exchanger, and the water molecules in both of the phases, for the solvation shells of the exchanged ions (counter-ions). Ion-exchange follows the course determined primarily by the ion with the greatest 'solvation needs'. If, say, an ion A tends to locate itself mainly in the dilute aqueous solution (where it can acquire a complete hydration shell), then the ion B may be 'displaced' into the ion-exchanger phase. Furthermore, if the ion A interacts more strongly with the co-ion in the external solution than does ion B, the selectivity of the ion-exchanger for B will be still greater, and this increase will arise from the stronger interaction of ion A with the solution components, and not from any specific interaction of the ion-exchanger with ion B. If, however, ion B interacts with the co-ion in the solution more strongly than does ion A, the two effects will oppose each other and the observed selectivity will be the net effect.

In this approach, the structure of water is of major importance. At room temperature water molecules form a hydrogen-bonded network which tends to repel any species except those carrying hydrophilic groups or a charge large enough to interact with water molecules more strongly than do the water molecules with themselves. In the ion-exchanger phase, this structure is considerably distorted by the hydrophobic hydrocarbon chains of the matrix. Large ions of low charge, which are most effective in distorting the structure of water, will thus prefer the ion-exchanger phase and locate themselves as near as possible to the functional groups. Such formation of ion-pairs in the ion-exchanger phase, 'forced' by the structure of water, explains the extremely high affinity of ions such as ClO_4^-, ReO_4^-, $AuCl_4^-$, etc. for strong-base anion-exchangers, and also many other observed affinity series. In the approach of Diamond et. al. the causes of selectivity are considered to arise primarily in water–water, water–ion, and ion–ion interactions occurring in the aqueous phase.

As the concentration of the external solution is increased, the proportion of 'free water' diminishes and it becomes more difficult to accommodate the ions which have the largest solvation needs. In this situation, the functional group in the ion-exchanger competes more effectively with water molecules in meeting the ion's solvation needs. This tendency may result in the reversal of selectivity at suitably high concentrations of the solution. Studies on the distribution coefficients of alkali-metal and alkaline-earth metal ions (trace amounts) over a wide range of supporting electrolyte concentrations have shown that such reversals are indeed observed in practice [276,280,281].

Types of Functional Groups and Specific Interactions in the Ion-Exchange

The functional groups of the commonest strong-acid cation-exchangers $-SO_3^-$, and strong-base anion-exchangers, $-CH_2-N\,(R_1\,R_2\,R_3)^+$, show slight complex-forming tendencies; therefore, these exchangers yield 'normal' affinity series in which the charge and degree of hydration of a given ion determine its position in the series. For ion-exchangers containing other functional groups, the situation is usually much more complicated, because the interaction of counter-ions with the functional groups leads to formation of ion-pairs or even covalent bonds. An example of this is the very strong affinity of anion-exchangers containing primary, secondary and tertiary amine groups for OH^--ions, or of carboxylic cation-exchangers for H^+-ions, in contrast to strong-base (or strong-acid) ion-exchangers which place these ions nearly at the end of the respective affinity series [12,15] (cf. also Section 6.4.3).

With carboxylic-acid resins the affinity series of the alkali metals is $Li^+ > Na^+ > K^+$ [234,282], i.e. reversed, compared with the series observed with sulphonic-acid cation-exchangers. Ion-exchangers with carboxylic acid groups also have high affinity for the alkaline-earth metals and certain transition metals [17], and this appears to be correlated with the tendency of alkanoic acids to form insoluble salts with these metal ions.

Phosphonic-acid resins have extremely high affinity for Th^{4+}, U^{4+}, UO_2^{2+}, Fe^{3+}, Pb^{2+}, Cu^{2+}, and other ions [50,283–285].

Iminodiacetic-acid resins are highly selective for transition metal ions and some others such as Ag^+, Hg^{2+}, UO_2^{2+}, VO_2^+. Pb^{2+}, Th^{4+}, Cu^{2+}, etc. [29,50,286–289].

Numerous specific ion-exchangers have been synthesized by incorporating into the matrix functional groups which can form chelates with the ions [50,63,66,68c–68f,69,290,291,933–937].

Generally, the stronger the ion-pair or complex formed by a given ion with the resin's functional groups, the higher the affinity of this ion for the ion-exchanger. Certain phenomena are explicable in terms of the simple electrostatic model. As already mentioned in Section 6.4.2.2, the 'reversed' selectivity sequence of the alkali-metal ions observed with carboxylic-acid ion-exchangers may be explained by assuming that, in the presence of high field-strength (small r_R) functional groups, the first term in Eq. (6.22) overshadows the second term which reflects the differences in hydration.

The abnormally high affinity of Tl^+ and Ag^+ for sulphonic-acid ion-exchangers is thought to arise from the high polarizability of the ions. This should make it easier for them to attain a shorter distance of nearest approach than would be expected from their ionic radii [15].

Chemical interaction of ions with the ion-exchanger may sometimes result in striking effects like, for instance, the absorption of copper, cobalt and nickel ions by weak-base anion-exchangers from neutral salt solutions [75,292] or partially irreversible sorption of Hg(II) ions by sulphonic-acid cation-exchangers, presumably associated with the formation of organomercury compounds [293].

Specific interactions in the ion-exchanger do not necessarily involve only fixed functional groups. Studies on the ion-exchange of substituted quaternary ammonium ions with a phenolsulphonic-acid cation-exchanger have revealed the affinity series [294] $NH_4^+ < (CH_3)_4N^+ < (C_2H_5)_4N^+ < (CH_3)_3AmN^+ < (CH_3)_2(C_2H_5)PhN^+ < (CH_3)_2PhBzN^+$, which is just the opposite to that predicted from consideration of electrostatic interaction forces or possible steric hindrance. In this case, there are probably van der Waals interactions between the organic cations and the resin matrix, and these increase with increasing molecular size of the organic groups in the counter-ion and increasing number of aromatic substituents. It must also be mentioned that entropy effects [295] are also of consequence. Organic ions promote the formation of the 'iceberg'-type structure of water, in which one water molecule is hydrogen-bonded to four adjacent water molecules [296]. The passage of an organic ion from the water phase into the ion-exchanger phase is accompanied by an increase in the entropy of the system [295]. There are also data indicating that the take-up of halide complexes of certain metals (Fe, Ga, Tl, Au, Sb) by sulphonic-acid cation-exchangers may be due to their chemical interaction with the matrix [297–300c].

Degree of Cross-Linkage

As the degree of cross-linkage is increased, the degree of swelling of the ion-exchanger diminishes and so does the water content. As a result, the concentration of the ion-exchanger phase increases, and this is favourable to specific interactions in the ion-exchanger. With increasing degree of cross-linkage, the swelling pressure rises, and according to Gregor's elastic matrix model (see Section 6.4.2.2), this should affect the ion-exchanger's ability to prefer certain ions to others. As the degree of cross-linkage is increased, the two effects should thus enhance the selectivity of the ion-exchanger. Experimentally, with common inorganic ions, the selectivity coefficient is indeed seen to rise as the degree of cross-linkage is raised, at least when the equivalent fraction of the preferred ion in the ion-exchanger is quite small [212,231–238,242,259,267,272,273,301–309].

If very large ions are to be exchanged, the 'pores' in the ion-exchanger phase may prove to be too small to permit the ions to diffuse freely into the ion-exchanger. As a result, the ions will be partially or completely excluded from the ion-exchanger phase, the action being more pronounced for higher degrees of cross-linkage [218,267,310–316]. Such exclusion, referred to as 'sieve action', is most dramatically displayed by inorganic ion-exchangers which have a regular and rather rigid lattice and a strictly defined pore size [23]. In organic ion-exchangers, the pore size may vary within certain limits owing to the flexibility of the matrix. Again, the inhomogeneity of the ion-exchanger phase gives rise to a certain distribution in pore size rather than a single pore size.

Sieve action may result in reversal of selectivity; for example, tetra-alkylammonium ions show a normal affinity series with slightly cross-linked ion-exchangers ($NMe_4 < NEt_4 < NBu_4$) but with highly cross-linked resins the sequence is reversed [317].

For trace ions, both the selectivity coefficients with respect to the major ion and the separation factors may often increase as the degree of cross-linkage is increased [242–244,267,307,318–321].

The increase in separation factors for trace elements is generally the greater the more different the degrees of hydration [244,267] of the ions to be separated.

In the separation of large ions, e.g. complexes of rare-earth elements with ethylenediaminetetra-acetic acid (EDTA) or *trans*-1,2-diaminecyclohexanetetra-acetic acid (DCTA), a maximum has been found to occur in the curves of the selectivity coefficient and sometimes also of the separation factor plotted *vs.* degree of cross-linkage [218,245,246,267]. The position of this maximum varies with temperature (Fig. 6.10). The decrease in the selectivity coefficient of large ions in highly cross-linked resins is associated with the exclusion of these ions from the ion-exchanger phase by sieve action [218,246].

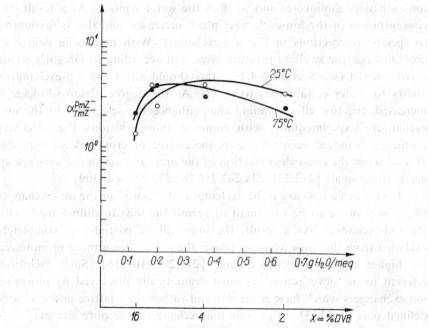

Fig. 6.10 Separation factors for promethium and thulium complexes with DCTA on Dowex 1 as a function of the degree of cross-linkage (from [267], by permission of Inst. Nucl. Res. (Warsaw))

Temperature

The selectivity coefficient may vary with temperature, depending on the magnitude and sign of the reaction enthalpy

$$\Delta H^* = -R \frac{d \ln k_{A^q}^{B^p} \left(\gamma_{AQ_q}^{\left|\frac{q+1}{q}\right|} \middle/ \gamma_{BQ_p}^{\left|\frac{p+1}{p}\right|} \right)}{d(1/T)} \quad (6.23)$$

where R is the gas constant and T is the absolute temperature.

Strickly speaking, ΔH^* is the differential enthalpy change in the ion-exchange reaction described by Eq. (6.7), i.e. the enthalpy change in the process in which one equivalent of the ion A^q is exchanged for one equivalent of the ion B^p, the amounts of ion-exchanger and solution being large enough to prevent any significant change in the composition of either of the phases and any transfer of water from one phase to the other [322, 323].

With organic ion-exchange resins the ion-exchange enthalpy changes range from zero to a few kcal/equivalent [73, 246a, 322–339]. Higher changes have been recorded for some inorganic ion-exchangers [340].

The ion-exchange enthalpy change may be positive or negative, and is usually a function of temperature.

Therefore, selectivity may sometimes increase and sometimes decrease as the temperature is raised (Figs. 6.11, 6.12).

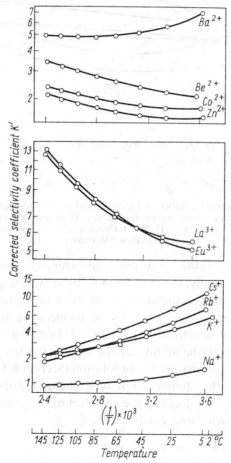

Fig. 6.11 Corrected selectivity coefficients for the exchange of traces of various cations on Dowex 50-X12 (H^+), plotted against reciprocal of the temperature (from [191], by permission of the copyright holders, the American Chemical Society)

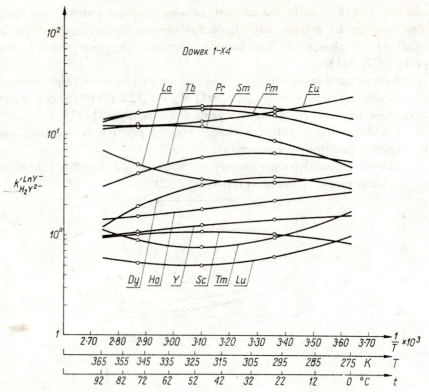

Fig. 6.12 Selectivity coefficient *vs.* reciprocal of the temperature for the exchange of trace amounts of lanthanide (Ln) complexes with EDTA (LnY$^-$) on Dowex 1-X4 (H$_2$Y^{2-}) (from [267], by permission of Inst. Nucl. Res. (Warsaw))

Since the ΔH^*-values may differ considerably even if the ions are closely related chemically, the separation factor may vary quite considerably with temperature and temperature is an essential factor controlling the selectivity. In several cases, a change in temperature has been reported to result in reversal of selectivity [218,267,341] (cf. Figs. 6.11, 6.12).

In systems where the affinity of the ion for the ion-exchanger is also affected by complex formation in the solution (Section 6.4.2.2), temperature may affect not only the equilibrium of ion-exchange but also the equilibrium of complex-formation [318,342], the net effect of temperature on selectivity thus becoming much more complicated.

Formation of Complexes in the Solution

In the preceding sections, selectivity was approached from the viewpoint of the properties of the ions to be exchanged and the properties and structure of the ion-exchanger, with due consideration for parameters such as temperature, etc., which can affect these properties and the strength of the ion–functional group bond. However, ion-exchange equilibria are also affected by interactions of counter-ions with other ions or molecules in the external solution. If, for example, a cation B^p forms complexes with an anion V^- in the solution (including negative complexes which are excluded from the cation-exchanger phase), the ion-exchange equilibrium [Eq. (6.7)] will be displaced to the left to an extent depending on the concentration of the complex-forming anion and the stability of the resulting complexes. As a result, both the selectivity coefficient [Eq. (6.8)] and the separation factor [Eq. (6.15)] will diminish. The selectivity of the ion-exchanger for the trace ion B^p decreases compared with that for the major ion A^q which is not complexed by the ion V^- to a significant extent. If, however, another trace ion D^z is, under the same experimental conditions, complexed by the ion V^- more strongly than is the B^p ion, the separation factor for these two ions will increase. Thus, with respect to the separation of the two trace ions B^p and D^z, the selectivity of the system increases on introduction of the complex-forming agent V^-, although the absolute values of the distribution coefficients of the two ions decrease. Complex formation may thus improve separation and reduce distribution coefficients, which may prove essential for an ion to be removable rapidly from a given ion-exchanger.

Generally, the ion-exchanger prefers the counter-ion which has the lower tendency to form ion-pairs or which associates less strongly with the co-ion in the solution. If cations are complexed by anionic ligands, the cation-exchanger prefers the cation which forms, with the anion, the complex with the smaller average ligand number. In a series of cations which form analogous complexes, the cation producing the weakest complex will have the highest affinity for the ion-exchanger.

Formation of anionic complexes from cations offers the possibility of separation on anion-exchangers. The anion-exchanger prefers the cation which forms, with the anionic ligand, the stronger negative complex or the complex with the greater average ligand number.

Of course, these rules hold only if the effect of complex formation is not outweighed by other factors.

Since complex-formation reactions are often highly selective, the combination of complex formation with ion-exchange allows cations with very similar properties to be separated.

Cation-Exchange in the Presence of Complexing Agents

If the cation B^{p+} is brought into contact with the complex-forming anion V^- the following equilibria will be set up in the solution

$$B^{p+} + V^- \rightleftharpoons BV^{(p-1)+} \qquad K_1 = \frac{[BV^{(p-1)+}]}{[B^{p+}][V^-]}$$

$$BV^+_{(p-1)} + V^- \rightleftharpoons BV_p \qquad K_p = \frac{[BV_p]}{[BV^+_{(p-1)}][V^-]}$$

$$BV_p + V^- \rightleftharpoons BV^-_{(p+1)} \qquad K_{(p+1)} = \frac{[BV^-_{(p+1)}]}{[BV_p][V^-]} \qquad (6.24)$$

In the absence of complexing agents, the distribution coefficient of the B^p ion between the ion-exchanger and the solution may be written as

$$\lambda^0_{B^p} = \frac{[\overline{B^{p+}}]}{[B^{p+}]} \qquad (6.25)$$

where the barred symbol refers to the ion-exchanger phase.

In the presence of V^- ions, the metal B will exist in the solution both as the cation B^p and as other complex ions defined by the equilibria (6.24). If the ion-exchanger is assumed to retain only the B^p ion (an assumption generally plausible and reasonable because anionic complexes are excluded from the cation-exchanger phase, neutral complexes are only slightly sorbed, and cationic complexes are sorbed only negligibly compared with the multivalent metal cation), the distribution coefficient is given by the equation

$$\lambda_{B^p} = \frac{[\overline{B^{p+}}]}{[B^{p+}] + [BV^{(p-1)+}]...[BV_p] + ...} \qquad (6.26)$$

Combination of Eqs. (6.24)–(6.26) gives

$$\lambda_{B^p} = \frac{\lambda^0_{B^p}}{1 + K_1[V^-] + K_1 K_2 [V^-]^2 + ... K_1 K_2 ... K_p [V^-]^p + ...} \qquad (6.27)$$

As evident from Eq. (6.27), complex formation causes a decrease in the distribution coefficient of the metal ion, and the decrease is greater, the more stable the complexes formed and the higher the concentration of ligand in the solution. Even simple inorganic ligands such as Cl^-, Br^-, NO_3^-, etc. can tie up metal cations as complexes of varying stability. This enables the affinity of cations for an ion-exchanger to be modified by suitable choice of supporting electrolyte. The data listed in Table 6.2 illustrate this effect.

Table 6.2 Effect of type of anion on the selectivity of Dowex 50W-X8(H$^+$)* toward certain metal ions

(from [343,344], by permission of the copyright holders, the American Chemical Society)

Metal	λ_M in 1M HCl	λ_M in 1M HNO$_3$	Metal	λ_M in 1M HCl	λ_M in 1M HNO$_3$
Hg (II)	0·28	16·9	Ba (II)	126·9	68·4
Cd (II)	1·54	13·1	Na (I)	5·6	6·3
Fe (III)	35·5	83·3	Co (II)	21·3	28·8
Mg (II)	21·0	24·0	Ga (III)	42·6	94·0
Ca (II)	42·3	39·6	La (III)	265·1	267·0
Sr (II)	60·2	50·3	Th (IV)	2049·0	1180·0

* At 40% loading.

The distribution coefficients for ions which are not complexed to a significant extent by chloride or nitrate (Na$^+$, Ca^{2+}, La^{3+}) are seen to remain constant irrespective of the type of anion. Mercury, which forms strong chloride complexes, has an extremely small distribution coefficient in 1M hydrochloric acid. Therefore, for example, the selectivity of the Hg(II)–Cd(II) pair is the reverse of that in nitric acid solution. The alkaline-earth metals, especially Ba^{2+}, have a stronger tendency to form ion-pairs with nitrates. Therefore the separation factor α_{Sr}^{Ba}, which is 2·1 in 1M hydrochloric acid, drops to 1·36 in 1M nitric acid. Again, complex formation with nitrates has a pronounced effect on Th(IV), and formation of chloride complexes affects the distribution coefficients of Ga(III), Fe(III), and Cd(II).

The strong complexing agents, such as aminopolycarboxylic acids, which form stable complexes with numerous cations, are usually weak acids, so complex formation is influenced by the pH of the solution. A typical multidentate ligand is EDTA, which is a tetrabasic acid, and may be represented by the general formula H$_4$Y. It yields stable complexes, usually of the 1:1 type, with numerous multivalent cations [345–348]. A good illustration of the effect of EDTA and pH on the affinity of cations for a cation-exchanger is the retention of trace amounts of tervalent actinides by Dowex 50-X12 cation-exchanger [349,350].

Americium, californium, and curium form stable complexes with EDTA

$$M^{3+} + Y^{4-} = MY^-, \qquad K_k = \frac{[MY^-]}{[M^{3+}][Y^{4-}]} \qquad (6.28)$$

The order of magnitude of the K_k values is 10^{18}–10^{19} [349,350].

According to Eqs. (6.27) and (6.28)

$$\lambda_M = \frac{\lambda_M^0}{1+K_k[Y^{4-}]} = \frac{\lambda_M^0}{1+\dfrac{K_k[Y']}{\alpha_{Y(H)}}} \qquad (6.29)$$

where

$$[Y'] = [H_4Y]+[H_3Y^-]+[H_2Y^{2-}]+[HY^{3-}]+[Y^{4-}] \qquad (6.30)$$

is the total concentration of all forms of uncomplexed EDTA in the solution and

$$\alpha_{Y(H)} = 1 + \frac{[H^+]}{k_1} + \frac{[H^+]^2}{k_1 k_2} + \frac{[H^+]^3}{k_1 k_2 k_3} + \frac{[H^+]^4}{k_1 k_2 k_3 k_4} \qquad (6.31)$$

is the side-reaction coefficient of EDTA, k_1, k_2, k_3, and k_4 being the successive dissociation constants of H_4Y ($pk_1 = 2 \cdot 0$, $pk_2 = 2 \cdot 76$, $pk_3 = 6 \cdot 16$, $pk_4 = 10 \cdot 26$) [346].

It may be shown for the elements described that, at pH $\geqslant 2$ and EDTA concentration $\geqslant 10^{-3} M$, $K_k[Y']/\alpha_{Y(H)} \gg 1$, and Eq. (6.29) reduces to

$$\lambda_M = \frac{\lambda_M^0 \alpha_{Y(H)}}{K_k[Y']} \qquad (6.32)$$

Therefore, in this system the separation factor of two elements is expressed by the equation

$$\alpha_{M(1)}^{M(2)} = \frac{\lambda_{M(2)}}{\lambda_{M(1)}} = \frac{\lambda_{M(2)}^0 K_{k(1)}}{\lambda_{M(1)}^0 K_{k(2)}} \qquad (6.33)$$

The separation factor is thus seen to be a function of the ratio of the distribution coefficients of the two uncomplexed cations and of the ratio of the stability constants of suitable complexes.

The experimental distribution coefficients for Am, Cm and Cf in the presence and absence of EDTA are plotted in Fig. 6.13 as a function of pH. It is noteworthy that on addition of EDTA and readjustment of the pH to ~ 3, the distribution coefficient for americium diminishes from $\sim 3 \times 10^5$ to ~ 10, i.e. by more than four orders of magnitude.

It is obvious, therefore, that the use of complexing agents increases the potential of ion-exchange chromatography. Complex formation is much more selective than ion-exchange, so the use of complexing agents as eluents increases the selectivity of separations by ion-exchangers. With ions of very similar chemical properties, such as tervalent lanthanide and actinide ions, the ratios of the distribution coefficients for cations adjacent in the periodic table are very close to unity [351–353] and attempts to separate such cations were unsuccessful until organic hydroxy-acids [1,354–359,362] and later aminopolyacetic acids [357–362] had been applied as eluents.

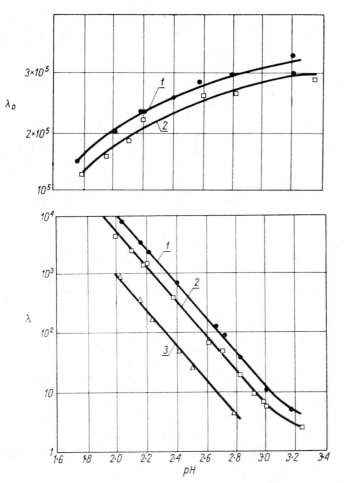

Fig. 6.13 Distribution coefficient *vs.* pH for trans-uranium elements on Dowex 50-X12 (NH_4^+) in the absence (λ_0) and presence (λ) of $0.001M$ EDTA: 1—Am, 2—Cm, 3—Cf (from [349] and [350], by permission of the copyright holders, Pergamon Press Ltd.)

Knowledge of the stability constants of complexes [363] allows potential separations of elements to be predicted on the basis of Eq. (6.33). Obviously, such estimates must always be verified experimentally because unfavourable complex-formation kinetics or a slow ion-exchange step may outweigh the advantages from the large separation factors [359, 362].

When ions with quite different chemical properties are to be separated, it is often possible to select a complexing agent, an ion-exchanger, and a solution pH such that certain cations will be retained by the cation-exchanger with 100% efficiency, while the other cations will occur exclusively as anionic or neutral complexes (Fig. 6.14).

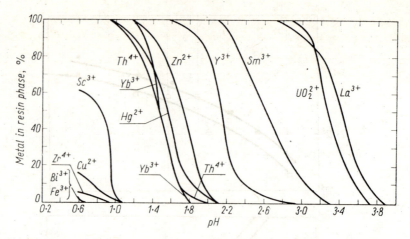

Fig. 6.14 Retention of various metal ions from EDTA solutions in relation to pH. Column: 20 × 40 mm, Dowex 50-X4 (Na$^+$) (0·16–0·32 mm), flow-rate 2·5–3·5 cm/min. Sample: 10 ml of 0·05M metal solution +15 ml of 0·05M EDTA solution +30 ml of buffer; washed with 3 × 30 ml of buffer (from [364], by permission of the copyright holders, the American Chemical Society)

Under these conditions, separation reduces to simply passing the solution through the cation-exchanger, pre-equilibrated with a suitable buffer solution containing a complexing agent of the same pH and composition as the solution of the metals to be separated. Numerous separations based on this principle have been described in the literature [364–367].

Anion-Exchange of Metals in the Presence of Complexing Agents

Metals that normally occur as cations in solution are not retained by anion-exchangers because co-ions are excluded from the ion-exchanger phase by the action of the Donnan potential [11,15]. When the solution contains negative ligands capable of forming anionic complexes with a given metal, the situation changes. To be effectively absorbed by an anion-exchanger, the metal does not need to exist in the solution exclusively in the anionic complex form. The following example will explain this effect.

Suppose that a strong uni-univalent electrolyte solution contains a small amount of the salt of a bivalent cation B^{2+}. Complex-formation equilibria between B^{2+} and V^- ions in the solution are described by Eqs. (6.24). If the maximum coordination number of the metal is 4, the resulting complexes will include the anions BV_3^- and BV_4^{2-}. On contact with a strongly basic anion exchanger in the V-form the following reactions can occur:

$$\overline{BV_3^-} + V^- \rightleftharpoons BV_3^- + \overline{V^-}; \qquad K_V^{BV_3} = \frac{[BV_3^-][\overline{V^-}]}{[\overline{BV_3^-}][V^-]} \qquad (6.34)$$

$$\frac{1}{2}\overline{BV_4^{2-}} + V^- \rightleftharpoons \frac{1}{2}BV_4^{2-} + \overline{V^-}; \qquad K_V^{BV_4} = \frac{[BV_4^{2-}]^{1/2}[\overline{V^-}]}{[\overline{BV_4^{2-}}]^{1/2}[V^-]} \qquad (6.35)$$

$$BV_2 \rightleftharpoons \overline{BV_2}; \qquad K_{BV_2} = \frac{[\overline{BV_2}]}{[BV_2]} \qquad (6.36)$$

where barred symbols refer to the ion-exchanger phase, $K_V^{BV_3}$, $K_V^{BV_4}$ are the selectivity coefficients for suitable ion-exchange recations, and K_{BV_2} is the distribution coefficient of the neutral salt BV_2.

The overall distribution coefficient for the metal is given by the equation

$$\lambda_B = \frac{[\overline{BV_2}] + [\overline{BV_3^-}] + [\overline{BV_4^{2-}}]}{[B^{2+}] + [BV^+] + [BV_2] + [BV_3^-] + [BV_4^{2-}]} \qquad (6.37)$$

Combination with Eqs. (6.24) and (6.24)–(6.36) gives

$$\lambda_B = \frac{[BV_2]\{K_{BV_2} + K_V^{BV_3} K_3[\overline{V^-}] + K_V^{BV_4} K_3 K_4[\overline{V^-}]^2\}}{[B^{2+}] \sum_{i=0}^{4} K_0 K_1 \ldots K_i [V^-]^i} \qquad (6.38)$$

where, by definition, $K_0 = 1$.

Ultimately, the metal distribution coefficient takes the form

$$\lambda_B = F_{BV_2}\{K_{BV_2} + K_V^{BV_3} K_3[\overline{V^-}] + K_V^{BV_4} K_3 K_4[\overline{V^-}]^2\} \qquad (6.39)$$

where F_{BV_2} is the fraction of the total metal in solution as the neutral complex.

Therefore, Eq. (6.39) shows that, although the possibility and extent of metal sorption by an anion-exchanger is controlled by the affinity of the metal–ligand anionic complexes for the ion-exchanger, the distribution coefficient reaches a maximum at a ligand concentration where the neutral complex predominates in the solution.

When most of the metal is converted into anionic complexes, further increase in the concentration of V^- ions in the solution results in the reduction of the distribution coefficient in keeping with the law of mass action [cf. Eqs. (6.34) and (6.35)].

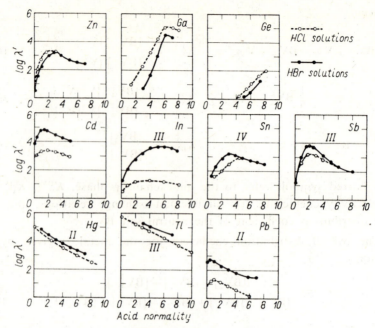

Fig. 6.15 Distribution coefficients of various metals on a Dowex 1-X8 anion-exchanger as a function of hydrochloric or hydrobromic acid concentration (from [368], by permission of the copyright holders, Acta Chem. Scand.)

Examples of this are shown in the plots of the distribution coefficients for Zn, Sn, Sb and Cd on Dowex 1–X8 anion-exchanger *vs.* concentration of hydrobromic acid in the solution [368] (Fig. 6.15).

The expressions given above were derived with numerous oversimplifications, e.g. disregard of activity coefficients and permeation of the ion-exchanger phase with electrolyte, etc. Rigorous description of ion-exchange equilibria in concentrated electrolyte solutions leads to much more complicated expressions [15,263], but the main qualitative conclusions remain essentially unchanged.

As already indicated, complexing agents like EDTA or DCTA react with numerous metals to form stable, usually 1:1 anionic complexes [345–348]. The stability constants [cf. Eq. (6.28)] of many of these complexes are large enough for practically 100% of the metal to occur in the complexed form at not too low a pH value. At pH 4·5–4·7, EDTA is nearly all in the form of the anion H_2Y^{2-} [331,346] so that ion-exchange reactions, e.g. with rare-earth complexes, may be written as:

$$\frac{1}{2}\overline{H_2Y^-} + LnY^- \rightleftharpoons \overline{LnY^-} + \frac{1}{2}H_2Y^{2-} \qquad (6.40)$$

According to Eqs. (6.15) and (6.20), the following relation should hold

$$\frac{d \log \lambda_{\text{LnY}^-}}{d \log [\text{H}_2\text{Y}^{2-}]} = -\frac{1}{2} \quad (6.41)$$

and this has been confirmed by experiment [369] (Fig. 6.16). The complexes of EDTA with Cr(III) and V(IV) gave curves with slopes of $-1/2$ and -1 [370], consistent with the formulae suggested, CrY^- and VOY^{2-}. It is also noteworthy that the distribution coefficients for EDTA complexes of the rare earths on anion-exchangers show much more variation than those of the uncomplexed cations on cation-exchangers [333]. This had made it possible to achieve complete separation of many of the metals in complexed forms [332, 369–371].

It is of interest to study the effect of pH on the selectivity of an anion-exchange separation of two elements which give analogous complexes, e.g. two lanthanides or actinides. At low pH, the cationic form of the metal, Ln^{3+}, and the non-dissociated complex acid form, HLnY, are the most important species in the solution. The distribution coefficient for the metal will take the form

$$\lambda_{\text{Ln}} = \frac{[\overline{\text{LnY}^-}]}{[\text{LnY}^-] + [\text{HLnY}] + [\text{Ln}^{3+}]} \quad (6.42)$$

If the stability constant K_k as defined by Eq. (6.28) and the stability constant of the protonated complex

$$K_{\text{KH}} = \frac{[\text{HLnY}]}{[\text{H}^+][\text{LnY}^-]} \quad (6.43)$$

are introduced, Eq. (6.42) may be rewritten as

$$\lambda_{\text{Ln}} = \frac{[\overline{\text{LnY}^-}]}{[\text{LnY}^-]\left(1 + [\text{H}^+]K_{\text{KH}} + \dfrac{1}{K_k[\text{Y}^{4-}]}\right)}$$
$$= \frac{\lambda_{\text{LnY}^-} K_k[\text{Y}^{4-}]}{K_k[\text{Y}^{4-}](1 + [\text{H}^+]K_{\text{KH}}) + 1} \quad (6.44)$$

where λ_{LnY^-} is the distribution coefficient for the complex ion.

Two extreme cases may be distinguished

(a) $\quad K_k[\text{Y}^{4-}](1 + [\text{H}^+]K_{\text{KH}}) \leqslant 1;$
(b) $\quad K_k[\text{Y}^{4-}](1 + [\text{H}^+]K_{\text{KH}}) \geqslant 1.$

Case (a) occurs in quite strongly acidic solutions where the concentration of Y^{4-} becomes very small [cf. Eqs. (6.29) and (6.31)]. The separation factor for two lanthanides is then defined by the equation

$$\alpha_1^2 = \frac{\lambda_{\text{Ln}(2)}}{\lambda_{\text{Ln}(1)}} = \frac{\lambda_{\text{LnY}^-(2)} K_{k(2)}}{\lambda_{\text{LnY}^-(1)} K_{k(1)}} \quad (6.45)$$

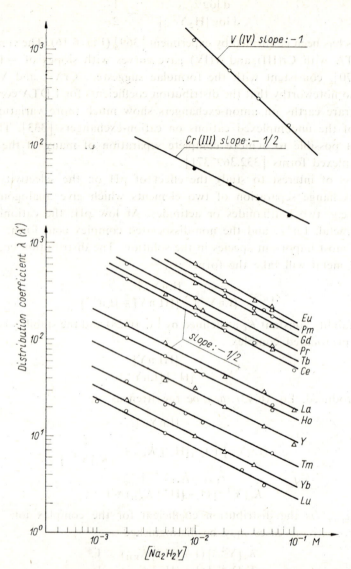

Fig. 6.16 Distribution coefficients for traces of anionic complexes of various metals with EDTA on a strong-base anion-exchanger as a function of Na_2EDTA concentration (from [369] and [370], by permission of the copyright holders, Elsevier Publ. Co. and Pergamon Press Ltd.)

Now, the situation is the opposite of that of two lanthanides separated by cation-exchangers in the presence of EDTA [cf. Eq. (6.33)]. Technically, however, the separation is difficult on account of the low solubility of EDTA in acidic solutions. With other complexing agents such as citric acid, etc., these limitations do not apply.

In case (b), the separation factor is defined by the equation

$$\alpha_1^2 = \frac{\lambda_{Ln(2)}}{\lambda_{Ln(1)}} = \frac{\lambda_{LnY-(2)}(1+[H^+]K_{KH(1)})}{\lambda_{LnY-(1)}(1+[H^+]K_{KH(2)})} \simeq \frac{\lambda_{LnY-(2)}}{\lambda_{LnY-(1)}} \qquad (6.46)$$

This case occurs at pH values high enough for practically 100% of each metal to be converted into the complexed form [cf. Eq. (6.40) and Fig. 6.16]. The separation factor is then equal to the ratio of the distribution coefficients of the complexed ions. It should be borne in mind that for ions complexed by multidentate ligands the distribution coefficients may be affected by the structure of the ions. For example, with the rare-earth elements, the distribution coefficient under standard conditions is a non-monotonic function of the atomic number, a fact interpreted as a result of progressive modification of the denticity of the ligand with changing radius of the central ion [331–333, 369, 371].

The last-mentioned equations for the separation factor [(6.15), (6.46)] continue to hold even at concentrations of EDTA which are very low but still sufficiently high for the EDTA to be able to bind trace elements into complexes in the presence of a simple inorganic ion used as the major ion. Figure 6.17 shows the distribution coefficients for a series of elements

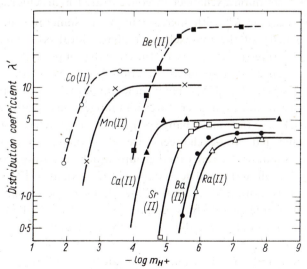

Fig. 6.17 Distribution coefficients for various metals on Dowex 1-X4 (Cl⁻) as a function of acid strength (from [370], by permission of the copyright holders, Pergamon Press Ltd.)

as a function of acidity, determined in a $0.00075M$ EDTA–$0.1M$ Cl⁻ (NH_4Cl–HCl) solution [370]. The plateau in the curve represents the range over which all the metal exists as an anionic complex.

The data in Fig. 6.17 also illustrate the necessity for the reagents used in trace analysis to be sufficiently pure. Minute complex-forming impurities can evidently give rise to quantitative retention of elements which should not normally be taken up by anion-exchangers from chloride solutions. Conversely, in the use of cation exchangers, equally small quantities of complex-forming impurities can result in rapid elution from the ion-exchanger of trace elements which would otherwise exhibit very high distribution coefficients.

Other Factors

The selectivity coefficient usually varies with the composition of the ion-exchanger phase. Most frequently the selectivity coefficient diminishes as the equivalent ionic fraction of the ion with the greater affinity for the ion-exchanger is increased [232, 234, 255, 257, 261, 273, 301, 336].

With trace ions these effects are likely to be absent because the composition of the ion-exchanger phase is always defined by the content of the major ion. Nevertheless, especially when very small columns are used, the effect should not be entirely ignored. The distribution coefficient has been found to vary abruptly with the percentage exchange in the ion-exchanger for the ion investigated, e.g. in the sorption of a Ga(III) chloride complex on Dowex 1-X10 anion exchanger in concentrated hydrochloric acid [44].

The decrease of K_A^B with increase in $\bar{X}_B$ is sometimes attributed to the inhomogeneity of the ion-exchanger phase (local cross-linking fluctuations, varying ring-substitutions of functional groups, etc) which results in non-equivalence of individual functional groups. Recent investigations [222, 273, 336, 372–375] have unequivocally demonstrated that most commercial ion-exchangers are inhomogeneous, both within and between particles, so the ion-exchange properties of a batch of ion-exchanger can vary somewhat. Ion-exchangers are also known to contain small proportions of undesirable functional groups which arise during preparation or storage as a result of thermal degradation [192], oxidation [177a, 188], etc.

Again, it is not unlikely that ion-exchangers occasionally contain small amounts (undetectable by standard analytical methods) of chelating groups which can uncontrollably affect the distribution coefficients of trace metals.

Thermodynamic considerations have indicated, and experiments [376] have borne out, that the selectivity coefficient may be a function of the ionic strength of the external solution. This is the case when swelling of an ion-exchanger varies with the ionic strength of the solution and with the equivalent ionic fraction in the ion-exchanger. Normally, the selectivity coefficient rises as the ionic strength of the electrolyte is increased, although the reverse cannot be ruled out as a possibility. The ionic-strength effect appears to be essentially significant only for very weakly cross-linked ion-exchangers.

6.4.3 Ion-Exchange in Concentrated Electrolyte Solutions

Ion-exchange in concentrated electrolyte solutions is characterized by a number of peculiar features which very often cause numerous elements to exhibit ion-exchange behaviour strikingly different from that observed in dilute solutions. In the first place, as the concentration of the ions in the solution is increased, the activity of the water diminishes, the hydration of the ions becomes less complete and the water molecules occurring in their immediate vicinity are progressively replaced by other ions. At the same time the normal structure of water becomes distorted (cf. Section 6.4.2.2) and the dielectric constant of the water diminishes. The ion-exchanger phase is also modified. As the concentration of the external solution is increased, the Donnan potential [276] diminishes, so penetration of co-ions into the ion-exchanger phase is facilitated. As a result, invasion by the electrolyte takes place. As the activity of the water diminishes, the ion-exchanger shrinks, the average pore size decreases, so the rate of the ion-exchange reaction diminishes and the sieve effect becomes more likely to occur. At the same time the decreased water activity allows the functional groups in the ion-exchanger to compete more effectively with water molecules to meet the solvation needs of the counter-ions. Interaction of counter-ions with co-ions (formation of complexes, ion pairs, etc.) in the solution gains considerably in importance. For all these reasons, the ion-exchange reactions may follow a course considerably deviating from the law of mass action.

In the cation-exchange of alkali-metal ions on a sulphonic-acid cation-exchanger, e.g. in nitric acid solution, the law of mass action expressed in Eq. (6.20) is fairly well satisfied (the slope of the curves of log λ_{H^+} vs. log m_{H^+} is equal to $-1\cdot 0$) over the concentration range 0–$0\cdot 4M$ nitric acid (Fig. 6.18). As the concentration of the acid is raised, individual elements are as likely to demonstrate negative as positive deviations from the law of mass action, and in $10M$ nitric acid, for example, the affinity series is $Na^+ > Rb^+ > Cs^+$, which is the reverse of that usually observed in dilute solutions. The nature and extent of the deviations from the law of mass action depend largely on the nature of the anion in the supporting electrolyte [276,278,280,281]. Variations in the distribution coefficient of the alkaline-earth metals [276,280,281,377] in concentrated mineral acid solutions have been found to produce a similar pattern.

In concentrated acid solutions, certain cations can attain very high distribution coefficients, as is evident from the data for the rare-earth elements in perchloric acid [281] (cf. also Figs. 6.65–6.67).

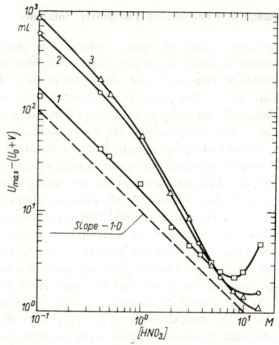

Fig. 6.18 Corrected retention volumes for the elution of traces of alkali metal ions from Dowex 50W [H^+] with HNO_3 solutions: 1 — Na, 2 — Rb, 3 — Cs (from [278], by permission of the copyright holders, the American Chemical Society)

The anion-exchange of simple anions such as halides etc. in concentrated acid or salt solutions is qualitatively similar to the ion-exchange of the alkali metals on cation-exchangers [276,378]. In this case, too, the character of the variation of the distribution coefficient of the trace ion with the concentration of the major electrolyte depends closely on the nature of the electrolyte [276].

The theory of anion-exchange of metal complexes has been briefly reviewed in Section 6.4.2.2. Distribution coefficients have been determined for most elements in hydrochloric [379], hydrofluoric [380,381], nitric [382] and other acid solutions (cf. also Figs. 6.62 and 6.63). The data obtained for the best-examined chloride system allow some generalizations to be made. According to the type of variation of the distribution coefficient with the acid concentration in the solution, the elements can be divided into four classes [263].

1. Elements exhibiting zero or very small distribution coefficients over the entire concentration range. This group consists of metals which do not yield any detectable anionic complexes with a given ligand. For the chloride system, the group includes the alkali metals, alkaline-earth metals, rare-earth elements, etc.

2. Elements having distribution coefficients which decrease over the whole concentration range. This group includes mainly metals which produce stable complexes with the relevant ligand. It also includes elements which normally occur as stable anions, e.g. halogens. For the chloride system, this group includes Au(III), Hg(II), Pd(II), Pt(IV), Rh(III), etc.

3. Elements producing maxima in the curves of distribution coefficient vs. concentration. The metals of this group yield moderately strong complexes with the relevant ligand. Before the maximum is attained, the average charge of the complex ions in the solution is positive, and after the maximum, negative. For the chloride system this group includes Zn(II), Cd(II), Ga(III), In(III), Fe(III), Sn(IV), etc.

4. Elements having distribution coefficients which increase over the whole concentration range. The metals of this group form weak complexes with the relevant ligand so that distribution coefficients that can be estimated are observed only at high ligand concentrations. For the chloride system this group includes Fe(II), Ge(IV), V(V), U(IV), etc.

The curves presented in Fig. 6.15 for Zn, Ga, Sb, Sn and Cd in the bromide system are illustrative of Group 3 elements, for Hg and Tl of Group 2 elements, for Ge of Group 4 elements, etc. The formation of metal complexes with simple inorganic ligands and absorption of these complexes by anion-exchangers can take place not only in the solutions of the corresponding acids but also in concentrated salt solutions. It is of interest that in solutions with identical concentrations of the complexing anion the distribution coefficients of metals are as a rule larger in salt solutions [383]. The distribution coefficient depends also on the type of salt, and it usually increases with the strength of hydration of the major cation in the electrolyte.

The observation that the distribution coefficients of metal complexes in concentrated hydrochloric acid solutions are much less than those in the corresponding chloride salt solutions has been tentatively interpreted by proposing that the anion HCl_2^- is formed, and that, as the anion of an acid stronger than HCl, this would exhibit a stronger affinity than the Cl^- ion for the ion-exchanger [384]. A similar effect has been observed for other acids.

6.4.3.1 ANOMALOUS SORPTION

Elements such as Fe(III), Ga(III), Sb(V), Au(III) and Tl(III) have recently been found to be very strongly sorbed by cation-exchangers from concentrated hydrochloric acid or acidified metal chloride solutions [281, 297–299,300c,385] (cf. Fig. 6.65), These, and also other elements, e.g. In(III) and Po(IV), have large distribution coefficients in concentrated hydrobromic acid and in metal bromide solutions [300a–300c,377,386, 387,570] (cf. Fig. 6.67). Since all these elements are known to yield strong or moderately strong anionic complexes with chlorides and bromides, they would not be expected to be sorbed by cation-exchangers. One tentative explanation of this anomalous behaviour assumes the formation of compounds by the complex anions ($FeCl_4^-$, $SbCl_6^-$, $AuCl_4^-$, etc.) with the resin matrix [297,299,377,385]. Some investigators [298,388] have suggested that, in concentrated acid and salt solutions, the cation-exchanger's sulphonic acid groups are undissociated and, after having added a proton or a metal ion, they acquire a positive charge, thereby becoming capable of holding anions. However, it seems more reasonable [300] that the cation-exchanger sorbs a metal by forming a molecular complex (charge-transfer complex) between the aromatic rings of the matrix and the undissociated complex acid (e.g. $HAuBr_4$) or an ion-pair like ($H^+AuBr_4^-$) or ($M^+AuBr_4^-$). This view is supported by the fact that gold sorbed by a cation-exchanger from hydrobromic acid–bromine solution can rapidly and quantitatively be eluted by an organic solvent (e.g. acetylacetone) [300,389].

6.4.4 Ion-Exchange in Very Dilute Solutions

Some experimental evidence apparently indicates that in very dilute solutions the law of mass action is not obeyed, and that the extent of ion sorption by a given ion-exchanger is related to the ionic strength of the solution [390]. Plots of the molal selectivity coefficient *vs.* ionic strength (μ) for a few ion-exchange reactions (Fig. 6.19) show that the selectivity coefficient diminishes as the ionic strength of the solution is decreased, and levels out toward a constant value only when μ is relatively high.

A consequence of importance for analytical practice is that the 'coefficient of removal' (i.e. the ratio of trace element concentrations before and after equilibrium is attained) reaches a maximum at a particular value

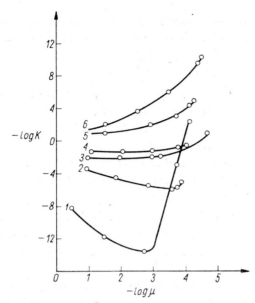

Fig. 6.19 Selectivity coefficient vs. ionic strength curves for several ion-exchange reactions on Dowex 2-X8 anion-exchange resins in dilute and very dilute solutions: $1 - \text{NaI} + \text{PO}_4^{3-}$-form resin, $2 - \text{NaI} + \text{SO}_4^{2-}$-form resin, $3 - \text{NaI} + \text{OH}^-$-form resin, $4 - \text{NaI} + \text{Cl}^-$-form resin, $5 - \text{Na}_2\text{HPO}_4 + \text{Cl}^-$-form resin, $6 - \text{Na}_3\text{PO}_4 + \text{Cl}^-$-form resin (from [390], by permission of the copyright holders, the American Chemical Society)

of the ionic strength and then decreases as μ is further decreased [390] In view of the technical difficulties inherent in the analysis of very dilute solutions, and the possible interferences due to tiny traces of complexing agents in the water and reagents or eluted from ion-exchangers, this phenomenon ought to be re-investigated.

Earlier studies on the retention of ^{140}Ba and ^{140}La in a cation-exchanger column from barium chloride solutions of various concentrations, showed that the amount of 'leakage' of the ions through the column increases as the influent is progressively diluted [391].

6.4.5 Ion-Exchange in Non-aqueous and Mixed Solvents

Ion-exchange reactions proceed not only in aqueous media but also in non-aqueous solvents such as alcohols, ketones, glycols, formamide, liquid ammonia, etc., and also in mixed media, e.g. water–alcohol, water–ketone, water–dimethylsulphoxide (DMSO). In order to proceed, ion-exchange requires a number of conditions to be met, viz.,

1. an ionizing solvent must be used so that the solute and the ion-exchanger's functional groups are at least partially dissociated;

2. the ion-exchanger used must have either an open porous structure or an elastic matrix which is expanded by swelling in the solvent;

3. the ion-exchanger used should undergo neither dissolution nor destruction.

The theory of ion-exchange in non-aqueous solvents and in mixed media has only rarely been discussed. The dependence of selectivity on the nature of solvent cannot be easily or reliably predicted, although some partially successful correlations relate the distribution coefficient of metals in a strong-acid cation-exchanger–$H_2O + DMSO$ system to Walden products [392]. Nevertheless, the ample experimental evidence at hand permits certain generalizations.

The swelling of an ion-exchanger depends mainly on the polarity of the solvent. Most ion-exchangers swell more strongly in highly polar than in less polar solvents. Maximum swelling has often been found to occur in solvents with dielectric constant between 40 and 50. Such solvents, may allow complete ionic dissociation of functional groups, and the lower dielectric constant causes stronger electrostatic repulsion between the fixed charges than that in pure water as medium. In mixed media the composition of the ion-exchanger phase generally differs from that of the liquid phase at equilibrium with the resin. Ion-exchangers take up water with exceptionally high preference, this effect being most strongly pronounced with highly cross-linked resins [393–395]. When put into water-immiscible non-polar organic solvents, water-swollen ion-exchangers lose hardly any water; this has been used for some interesting separations of platinum-metal, rare-earth, and transition elements, the ion-exchanger acting as a solid carrier for the water phase [396–399].

In mixed media, the affinity of cations for a given cation-exchanger generally rises as the proportion of the organic solvent in the medium is increased [400–404]; this rise is sometimes quite appreciable (a few orders of magnitude) [400,401]. It should be emphasised that in mixed media the affinities of individual species, even those closely related in chemical nature, for a given ion-exchanger can change by strikingly different amounts, and this may even result in a reversed selectivity sequence. For example, the selectivity sequence typical of aqueous solutions, viz., $Cs^+ > K^+ > CH_3NH_3^+ > NH_4^+ > Na^+ > Li^+$, on Dowex 50-X8 cation-exchanger reverses in water–methanol 1:4 mole ratio and becomes $K^+ > Cs^+ > Na^+ > CH_3NH_3^+ > NH_4^+ > Li^+$ [403].

Sometimes the affinity of an ion for a given ion-exchanger rises only until the solution attains a definite composition, whereupon it diminishes as the proportion of the organic solvent is further increased [402–406].

Minima have also been observed in the curves of distribution coefficient vs. solution composition [e.g. for Pd(II) on Dowex 50W-X8 in H_2O–DMSO–HCl] [406]. Whenever the organic solvent favours the formation of anionic complexes, the affinity of ionic species for a given ion-exchanger diminishes as the proportion of the organic solvent is increased, as may be seen, e.g., in the cation-exchange of rare-earth elements in water–acetone mixtures containing α-hydroxyisobutyric acid [407].

If an organic solvent alone is able to form complexes with certain ions, it obviously suppresses the distribution coefficient and can sometimes give rise to a reversed selectivity sequence. Such a reversal has been found to occur in the exchange of Cs^+ and other alkali-metal ions on Dowex 50 cation-exchanger, on addition of phenol [408]. According to the chemical nature of the solvent used, the distribution coefficient of a given element may, at constant cation-exchanger and acid concentrations, decrease or increase or pass through a maximum as the proportion of the organic solvent in the mixture is raised. This is the case with the distribution coefficients of Hg(II) on H^+-form Amberlite IR-120 in water–tetrahydrofuran, water–acetone, and water–methanol mixtures that are $0.6M$ in nitric acid [409].

Use of non-aqueous and mixed solvents in the anion-exchange of metal complexes has considerably increased the number of possible separations. In most cases the distribution coefficients of metal complexes in mixed media like water–alcohol, water–acetone, etc. are larger than those in pure water solutions [410–413]. Therefore, such mixed media can be used for separation of elements which do not form complexes strong enough to be retained by anion-exchangers in aqueous solution. The affinity of metal complexes for anion-exchangers generally increases as the proportion of organic solvent in the medium is increased [409–412]. The opposite [409,412,413] has been recorded, however, and occasionally a maximum [412] or minimum [414] in the distribution-coefficient curve has been reported.

Under comparable conditions, the distribution coefficients of metal complexes in mixed media generally rise with increasing hydrocarbon-chain length for solvents from a homologous series (e.g. n-alcohols) [410]. Anion exchange in mixed media has permitted the development of new efficient separations like that of uranium from other elements in water–alcohol mixtures containing hydrochloric acid [410,415,416], selective retention of thorium from water–alcohol solutions of nitric acid [410, 417,418], separation of rare-earth elements in water–alcohol mixtures containing nitrates [419,420], α-hydroxyisobutyrates [420], etc. Ion-exchange in non-aqueous and mixed solvents is often accompanied by a type of extraction [421,422] or sorption of neutral complexes or ion-pairs [423,424] by ion exchangers.

The partition of the complexes of metallic elements [424a] and complexing agents [424b] between a water-immiscible solvent (benzene) and water-swollen ion-exchanger has been studied. The extent of partition was found to depend on the basicity of the complexing agent.

Disadvantages of the use of non-aqueous solvents include possible complications due to limited solubility of electrolytes in such media, and also the generally less favourable kinetics of exchange reactions [425–427, 951].

6.4.6 Prediction of Selectivity from Independent Data

Prediction of selectivity, even qualitatively, may be helpful in the development of new separations with ion-exchangers. Inspection of the activity coefficients of salts chemically related to a given ion-exchanger provides some information. As is evident from Eq. (6.19), the corrected molal selectivity coefficient is the ratio of the activity coefficients of the ions in the ion-exchanger phase, raised to appropriate powers. Since the ion-exchanger phase resembles a concentrated electrolyte solution in nature, it may be expected that the interactions between the counter-ions and the functional groups, reflected in the activity coefficients, will be similar to the analogous interactions in concentrated solutions of salts which are chemically similar to the monomer units of the ion-exchanger or to its functional groups.

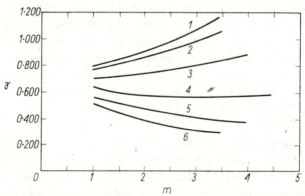

Fig. 6.20 Activity coefficients of alkali metal p-toluenesulphonates (Tol) and acetates as a function of ionic strength of the solution: 1 – AcOK, 2 – AcONa, 3 – AcOLi, 4 – LiTol, 5 – NaTol, 6 – KTol (from [317], by permission of the copyright holders, the New York Academy of Sciences)

Figure 6.20 shows the mean activity coefficients of lithium, sodium and potassium p-toluenesulphonates and acetates. With the p-toluenesulphonates, the activity coefficients follow the sequence $\gamma_{Li} > \gamma_{Na} > \gamma_K$.

According to Eq. (6.13), $K'^K_{Na} = \bar{\gamma}_{Na}/\bar{\gamma}_K$, so the affinity of the alkali metals for sulphonic-acid cation-exchangers (Dowex 50, Amberlite IR-120, etc.) should increase in the series $Li^+ < Na^+ < K^+$, and this is borne out by experimental observations. Conversely, with the acetates, the activity coefficients increase, rather than decrease, in the series: $\gamma_{Li} < \gamma_{Na} < \gamma_K$. Hence, the selectivity of carboxylic-acid ion-exchangers (e.g. Amberlite IRC-50) for the alkali-metal ions should follow a sequence the reverse of that observed with the sulphonic-acid ion-exchangers, viz., $Li^+ > Na^+ > K^+$, which again has been borne out by experiments [317]. However, such considerations are correct only as long as the ion-exchanger is in contact with a dilute electrolyte solution. In concentrated acids (such as perchloric, hydrobromic and hydrochloric) the distribution coefficient of an element undergoing cation-exchange generally increases with the activity coefficient of the acid (and presumably also of the salt) at such high concentrations ($\gamma_{\pm HClO_4} > \gamma_{\pm HBr} > \gamma_{\pm HCl}$) [281,377].

Selectivities may also be predicted from the solubilities of salts chemically related to the functional groups of an ion-exchanger. An example of the correlation of ion-exchange affinity with salt solubility is given in Fig. 6.21, which shows the selectivity coefficients for the ion-exchange of lanthanide (Ln) EDTA complexes (LnY^-) on a strong-base anion-exchanger [369,428] and the solubilities of the NaLnY salts [429]. The functional group of the anion exchanger is —$N(CH_3)_3^+$—a typical quaternary ammonium ion. In view of the well known similarities between the chemical properties of alkali-metal, ammonium, and quaternary ammonium ions, the solubilities of the $RN(CH_3)_3LnY$ and NaLnY salts may be expected to vary similarly in the lanthanide series.

As is evident from Fig. 6.21, the affinity of the LnY^- ion for the anion-exchanger is closely correlated with the solubility of the NaLnY salt: the lower the solubility of the salt, the higher the affinity of the ion for the ion-exchanger. Studies on the affinity of lanthanide complexes of *trans*-1,2-diaminocyclohexanetetra-acetic acid (H_4Z) for Dowex 1-X4 anion-exchanger in relation to the solubility of the NaLnZ salt produced a similar pattern [430]. Similar relations have been found to exist between the affinity of alkaline-earth metals for a sulphonic-acid ion-exchanger and the solubility of the corresponding sulphates [49], the affinity of rare earths for an NO_3^--form anion-exchanger in concentrated lithium nitrate solution [431] and the solubility of lanthanide–magnesium double nitrates, and between the anion-exchange affinity of lanthanide sulphate complexes and the solubility of lanthanide sulphate octahydrates [432].

Analogous relationships exist between the solubilities of alkali-metal phosphates and the selectivity sequence for the alkali metals on phosphonic-acid ion-exchangers [317], between the solubilities of alkali-metal sulphates and the affinities of these ions for sulphonic-acid ion-exchangers, etc. [317]. Thus, the correlation between ion-exchange affinity and salt solubility appears to be quite general in nature.

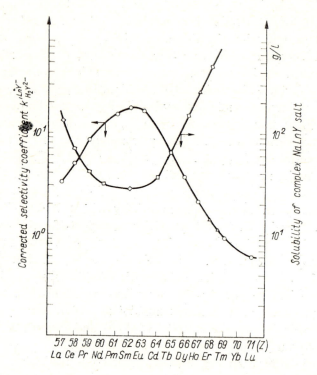

Fig. 6.21 Correlation between the corrected selectivity coefficient for the ion-exchange of rare earth–EDTA complexes on Dowex 1-X4 (H_2Y^{2-}) and the solubility of NaLnY salts (from [428], by permission of R. Dybczyński)

When complexes are formed, selectivity may be approximately predicted from the stability constants of the complexes. The effect of complexation on selectivity has already been discussed in Section 6.4.2.2.

It is also well known that elution of trace quantities of ions from an ion-exchanger can be promoted by the use, as eluents, of acids or salts which normally yield sparingly soluble precipitates with these ions [433]. For example, trace quantities of caesium ion are eluted by a solution of perchloric acid from a cation-exchanger more rapidly than by any other strong acid (nitric, hydrochloric) used at similar concentration.

6.5 ION-EXCHANGE KINETICS

There is not enough space to allow extensive discussion of the problems associated with ion-exchange kinetics. The effects of experimental conditions on the rate of ion-exchange are summarized in Table 6.3.

Table 6.3 Dependence of the ion-exchange rate on experimental conditions

(from [15], by permission of the copyright holder, F. Helfferich)

	Particle diffusion control	*Film diffusion control*
Counter-ion mobility:		
In particle	$\propto D$ *	No effect
In aqueous phase	No effect	$\propto D$
Co-ion mobility	No effect	No effect
Particle size	$\propto 1/r_0^2$	$\propto 1/r_0$
Capacity of ion-exchanger	No effect	$\propto 1/C_r$
Nature of fixed ionic groups	Slow when fixed ionic groups associate with counter-ions †	No effect
Degree of cross-linking	Decreases with increasing cross-linking †	No effect
Selectivity of ion-exchanger	The preferred counter-ion is taken up at higher rate and relased at lower rate, except under boundary conditions	
Concentration of solution	No effect	$\propto C$
Solution volume	Decreases with increasing solution volume	
Temperature	Increases with temperature, ca. 4–8%/°C †	Increases with temperature, ca. 3–5%/°C †
Rate of agitation or flow	No effect	Increases with agitation rate‡

* Note that polyvalent cations and counter-ions which associate with fixed ionic groups are particularly slow.
‡ This effect is reflected in the diffusion coefficients.
† The rate of agitation reaches a limiting hydrodynamic efficiency; beyond a critical agitation rate there is no effect on the rate of ion-exchange.

For an extensive discussion of these problems the reader is referred to the original literature [434–439].

6.6 THE COLUMN PROCESS

Ion-exchange reactions are usually done in columns. Ion-exchanger columns may be used simply for removing one or more ions from a solution and replacing them by another ion present in the ion-exchanger, and also for resolving mixtures of many ions — which is the major objective of ion-exchange chromatography. If a sodium chloride solution of concentration C_0 is fed into a strong-acid H^+-form cation-exchanger column (Fig. 6.22), it will exchange its ions according to Eq. (6.4). At the top of the bed (zone a) the ion-exchanger is completely converted into the Na^+-form. In the ion-exchange zone (marked b) where the actual exchange reaction proceeds, the two forms, Na^+ and H^+, co-exist. An exchange boundary is thus formed. As the feed is continued, the boundary travels farther down the column. In the still unconverted zone (marked c) the ion-exchanger is in the H^+-form and the effluent consists of pure hydrochloric acid of concentration equal to that of the sodium chloride influent

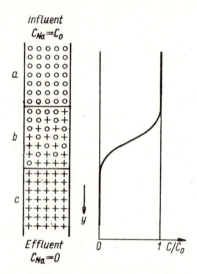

Fig. 6.22 Schematic performance of an ion-exchange column (left) and the axial concentration profile or 'exchange isochrone' (right): $a - Na^+$-form cation exchanger (exhausted zone), b — ion-exchange zone, $c - H^+$-form cation exchanger (still unconverted) (from [17], by permission of the copyright holders, John Wiley Inc.)

If the effluent concentration is recorded as a function of effluent volume, a curve known as the break-through curve is obtained (Fig. 6.23). At point B, the break-through of sodium occurs, i.e. sodium ions first appear in the effluent. Since the break-through point depends on the method of detection, it is usually assumed that it occurs when C/C_0 attains an arbitrarily chosen value (e.g. 0·001, 0·01, or 0·05).

The working exchange capacity, or the break-through capacity, is proportional to the area ABCD, whereas the total capacity is proportional to the area ABED.

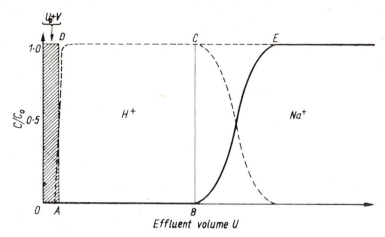

Fig. 6.23 Break-through curve (also called effluent concentration history, 'exchange isoplane') for the exchange reaction $RH + Na^+ = RNa + H^+$

The ratio of break-through to total capacity is a measure of the degree of column utilization in a given process. The steeper the break-through curve, the higher the degree of utilization.

6.6.1 Classification of Chromatographic Development Techniques

Simple exchange of one ion for another is one of the simplest applications of ion-exchange. Ion-exchange chromatography permits resolution of complex mixtures of constituents with closely related properties. As in other types of chromatography [440,441] three fundamental techniques may be distinguished, involving (*a*) frontal analysis, (*b*) elution development, (*c*) displacement development.

Frontal analysis involves feeding continuously a mixture of electrolytes e.g. AQ, BQ, and CQ into a column of ion-exchanger in Z-form. The counter-ion, Z, should be less strongly sorbed by the resin than A, B and C (selectivity sequence: $Z < A < B < C$). The process is shown diagrammatically in Fig. 6.24. The frontal-analysis technique allows only one component (A) to be isolated in pure form and in rather limited amount. In its classical version, the technique is therefore of minor importance as a procedure for resolution. However, when the ions B, C, etc. are trace constituents with high affinity for the ion-exchanger, frontal analysis may be very useful for preconcentration and separation of the traces from the major components which under the experimental conditions are sorbed by the ion-exchanger to a much lesser extent. The theory of

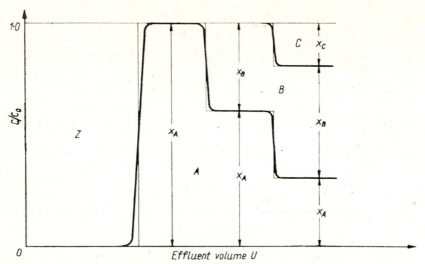

Fig. 6.24 Effluent concentration history in frontal analysis

frontal analysis is discussed at more length in Section 6.6.2. Preconcentration of elements by frontal analysis may be combined with subsequent selective elution of the individual components retained in the column.

Elution development involves introducing a mixture of ions, e.g. A B, and C, at the top of column of ion-exchanger in Z-form, as a narrow band, the counter-ions, Z, having less affinity for the resin than the ions of the mixture to be resolved (selectivity sequence: $Z < A < B < C$). Next, an electrolyte ZQ is fed into the column, i.e. the ions A, B, C are eluted with the counter-ion Z. In this process, the ions travel down the column at different rates (the band-migration rate for a given component is inversely related to the component's distribution coefficient), separating progressively until, in due course, they are completely resolved provided a suitable column length has been used. In the effluent, the ions appear as individual 'peaks' (elution curves) (Fig. 6.25). Since the eluent counter-

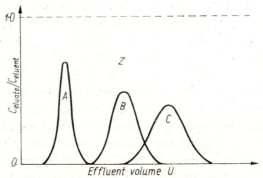

Fig. 6.25 Effluent concentration history in elution development

ions Z have less affinity for the ion-exchanger than the ions being separated, Z ions are continuously overtaking the A, B, C, ion bands on their way downstream during the elution. The boundaries of these bands become progressively more diffuse and the more so the slower the migration rate of the band. As a result, the maximum concentrations of resolved ions in the eluate (peak heights) are usually much lower than the concentration of Z ions in the eluent (Fig. 6.25). The elution development technique is commonly employed for the separation of elements, especially trace elements, and will be described more fully in Section 6.6.3.

Displacement development involves displacing the ions retained in the column by a counter-ion which has more affinity for the ion-exchanger than any ion of the mixture to be resolved. When the ions A, B, and C, retained on a column of ion-exchanger in the Z-form, are displaced by an electrolyte DQ (selectivity sequence: $Z < A < B < C$), they separate into individual bands along the column in the order of increasing affinity for the ion-exchanger. Between these bands arise relatively sharp (self-sharpening) boundaries which do not become very diffuse when moving down the column. Once a certain length of the column has been traversed a steady-state condition is established and further increase in column length does not improve the resolution. Typical effluent concentration profiles are illustrated in Fig. 6.26. In displacement development, the bands

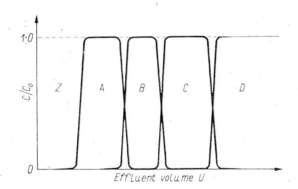

Fig. 6.26 Effluent concentration history in displacement development

migrate down the column at equal rates, determined by the flow-rate of the displacing agent (DQ). The concentrations in g-equivalents of individual elements in the effluent are also identical and equal to that of the displacing ion provided that there is enough of the component to produce a pure band under the experimental conditions. Displacement development is of minor analytical importance, especially with regard to resolution of traces, but it has been widely applied for preparative resolution of rare earths [442], alkali metals [443], etc.

6.6.2 Theory of Break-through Curves

If a mixture of two ions—one a trace constituent of high affinity and the other a major constituent of low affinity for the ion-exchanger—are continuously passed through an ion-exchanger column, then the major constituent—weakly held by the ion-exchanger—will, according to the frontal-analysis mechanism (cf. Fig. 6.24) appear rapidly in the effluent at a concentration equal to that in the influent, whereas the trace constituent—owing to its high affinity and low concentration—will not appear in the effluent until a long period of time has elapsed. This procedure may therefore be used as a convenient method for preconcentrating trace ions from large solution volumes. In practice, the form in which an ion-exchanger is used is often that of the ion which constitutes the major component of the solution (e.g. to remove multiply charged metal ions from dilute acid solutions the H^+-form of a strong-acid cation-exchanger is used).

Formally, this process of retention of trace ions bears many resemblances to a single exchange of one ion for another (cf. Fig. 6.22). However, in the present case the 'exhausted' zone (a in Fig. 6.22) does not correspond, for obvious reasons, to an ion-exchanger completely converted into the trace ion form but rather to one in which the trace-to-major ion ratio reflects the prevailing equilibrium condition (at the experimental composition of the solution). The concentrations of the trace ion, B, in the effluent are shown graphically in Fig. 6.27. The resulting break-through curve is apparently strictly similar to the classical break-through curve (Fig. 6.23) except that the C/C_0 ratios plotted as ordinates refer to concentrations of trace ion rather than to the overall ion concentration in the solution. The curve makes it possible to calculate the weight distribution coefficient of the trace ion (cf. Eq. 6.15), viz.,

$$\lambda_B = \frac{\overline{U} - U_0 - V}{m_j} \qquad (6.47)$$

where $\overline{U}$ is the effluent volume (cm^3) at $C = C_0/2$, U_0 is the dead volume (cm^3) in the column (the volume of liquid between the lower end of the resin bed and the column outlet [369]), V is the void (inter-particle) resin bed volume (cm^3) which amounts to ca. $0 \cdot 4$ of the resin bed volume [44,332,444], and m_j is the dry resin weight (g).

Ion-exchange preconcentration of traces is usually done under conditions chosen so as to prevent the break-through of ions. The lengths of time required for identical ions in solutions of identical compositions to break through in columns of equal length may be different and depend upon column efficiency (cf. curves 1 and 2, Fig. 6.27). The more efficient the column, the steeper the break-through curve. A quantitative measure of column efficiency is the number of theoretical plates, N, which can be calculated from the following equation [445]

$$N = \frac{(\overline{U} - U_0)(U' - U_0)}{(\overline{U} - U')^2} \qquad (6.48)$$

where U' is the effluent volume at $C = 0.159 C_0$.

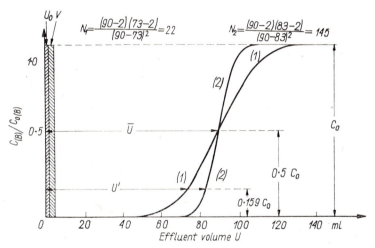

Fig. 6.27 Number of theoretical plates in relation to the shape of the break-through curve

By way of illustration, the numbers of theoretical plates evaluated from curves 1 and 2 in Fig. 6.27 are 22 and 145, respectively, and they indicate that the conditions for curve 2 would result in a more efficient process than those for curve 1.

6.6.2.1 Effect of Ion-Exchange Reaction Equilibrium

It follows from Eq. (6.47) that the higher the affinity for an ion-exchanger of ion B, the larger the volume of the solution that can be passed (under comparable conditions) before break-through. The various parameters affecting the selectivity have been discussed in detail in Section 6.4.2.

Only a few examples of the effects of various parameters break-through curves will be described here.

Figure 6.28 shows the break-through curves for copper ions from H_2PtCl_6 solutions of various hydrochloric acid concentrations, recorded for a strong-acid H^+-form ion-exchange resin column [446]. As the acid strength is increased, the break-through capacity of the column rapidly decreases. This follows not only because the equilibrium of the ion-exchange reaction

$$2RH + Cu^{2+} \rightleftharpoons R_2Cu + 2H^+$$

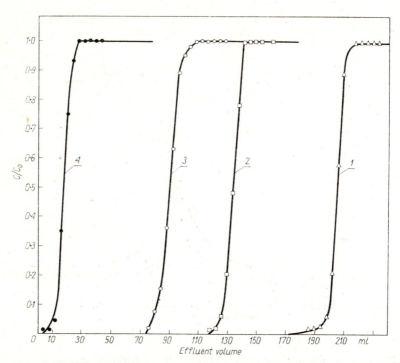

Fig. 6.28 Effect of HCl concentration on the break-through curve of copper(II) ions from chloroplatinate solutions. Column: $6 \cdot 35 \, cm \times 0 \cdot 430 \, cm^2$, Amberlite IR-120 ($H^+$) (0·075–0·15 mm), flow-rate 7 cm/min, $C_{H_2PtCl_6} = 0 \cdot 2N$, $C_{Cu^{2+}} = 0 \cdot 0126N$, C_{HCl}: 1—0, 2—0·15M, 3—0·3M, 4—1·0M (by permission of R. Dybczyński [446])

has been displaced to the left on increase in the hydrogen-ion concentration in the solution, but also because the overall concentration of the solution has been raised and consequently the ion-exchanger's selectivity for multiply charged ions has been suppressed (cf. Section 6.4.2.2).

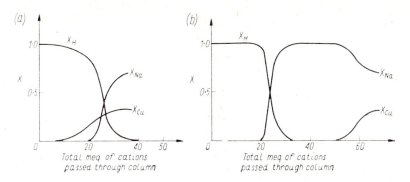

Fig. 6.29 Effect of complex formation on break-through curves. Column: 9.8×140 mm, Amberlite IR-120 (H^+) (0.43 mm), flow-rate 2.7 cm/min: (a) with complexation: $0.012N$ $CuSO_4 + 0.024N$ sodium polymetaphosphate, (b) without complexation: $0.012N$ $CuSO_4 + 0.024N$ Na_2SO_4 (from [9], by permission of the copyright holders, John Wiley Inc.)

Figure 6.29 shows the effect of a complexing agent on the behaviour of copper ions, when it is added to a mixture of copper and sodium ions passed through a column of cation-exchanger in H^+-form [9]. The copper-complexing ions (polymetaphosphates) are seen to reduce the break-through capacity considerably. Thus, the presence of a complexing agent in the solution can cause the prompt appearance of multiply charged ions in the effluent. In work with traces of ions, the accidental addition of even a minute amount of complex-forming agents may result in losses of an element which would otherwise be strongly sorbed by the resin. In the presence of complexants, pH has a vital influence on the break-through capacity for ions capable of forming complexes.

6.6.2.2 Effect of Dynamics of the Column Process

Since ions diffuse in the ion-exchanger and in the solution at finite rates, the chromatographic process proceeds under conditions departing to a greater or lesser extent from a local equilibrium condition. The closer the process approaches a local equilibrium state, the sharper is the boundary of the band of the ion sorbed in the column and the larger the break-through capacity.

Figure 6.30 shows the effect of resin-particle size on the sorption of Cu^{2+} ions from $0.2N$ H_2PtCl_6 [446] by a sulphonic-acid cation-exchanger. As the particle size is increased from 0.075–0.15 mm to 0.30–0.49 mm, the effective utilization of the column's exchange capacity falls from about 44% to about 28%. Increasing the flow-rate has a similar effect (Fig. 6.31) [446].

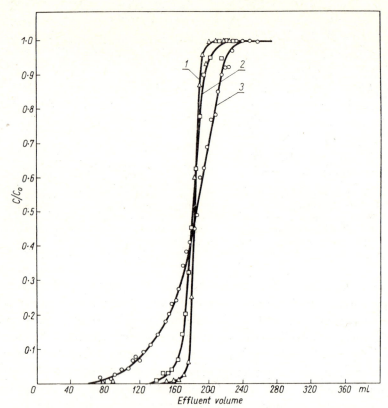

Fig. 6.30 Effect of resin particle size on the break-through curve for copper(II) ions. Column: 6·25 cm × 0·0430 cm^2, Amberlite IR-120 (H$^+$), flow-rate 8 cm/min; $C_{H_2PtCl_6} = 0·2N$, $C_{HCl} \approx 0$, $C_{Cu^{2+}} = 0·126N$. Particle size, 1 — 0·075–0·15 mm, 2 — 0·102–0·30 mm, 3 — 0·30–0·49 mm (by permission of R. Dybczyński [446])

The column can be utilized more effectively if the aspect ratio of the column (i.e. length-to-diameter) is increased [15,447,448]. However, excessive reduction of column diameter (to less than 30 times the particle diameter) may give rise to 'wall effects' leading to ion 'leakage' which distorts the front boundary and results in premature break-through. Highly non-uniform particle size of the ion-exchanger is also favourable to ion leakage. Elevated temperature and reduced cross-linkage often (though not always) result in a sharpened boundary and improved column utilization.

6.6.3 The Elution Development Technique

Elution development is a chromatographic technique which permits quantitative separation of a mixture into its components. Elution development has been widely used in radiochemistry and in analytical chemistry. In-

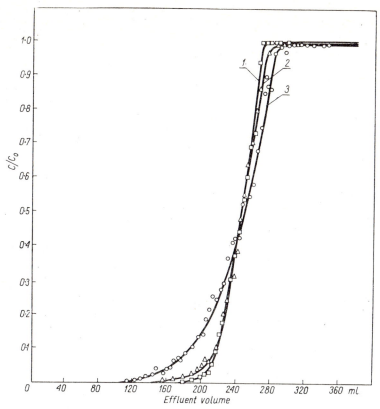

Fig. 6.31 Effect of flow-rate on the break-through curve for copper(II) ions. Column: $6 \cdot 20$ cm $\times 0 \cdot 430$ cm^2, Amberlite IR-120 (H$^+$) (0·30–0·49 mm), $C_{H_2PtCl_6} \approx 0 \cdot 2N$, $C_{HCl} = 0$, $C_{Cu^{2+}} = 0 \cdot 0126N$. Flow rate: $1 - 1 \cdot 2$ cm/min, $2 - 2 \cdot 3$ cm/min, $3 - 7 \cdot 0$ cm/min (by permission of R. Dybczyński [446])

spection of Fig. 6.25 shows that, under suitably chosen experimental conditions, the concentration profiles of individual ions in the effluent are well-resolved peaks. The shape of the peak (or elution curve for an ion) is related to the ion-exchange isotherm (cf. Fig. 6.8). This relationship is illustrated in Fig. 6.32. When the isotherm is convex, the elution curve makes a characteristically steep ascent and a tailing descent. The linear isotherm gives rise to a symmetrical bell-shaped elution curve. When the isotherm is concave, the elution curve makes a gradual ascent and a steep descent. The physical reason for this relationship is the rate of band migration down the column, which in elution development is approximately inversely proportional to the distribution coefficient [cf. Eq. [6.75]]. Only with the linear isotherm does the distribution coefficient have a constant value: the rate of migration of the ions B is then independent of their concentration. Owing to diffusional processes the band becomes diffuse, symmetrically with respect to its centre of gravity. With a non-linear isotherm, the distribution coefficient of an ion is a function of the concentration of the ion in the solution. With a convex isotherm the distribution coefficient is high at low gram-equivalent fractions (X_B) of ions B in the solution and diminishes as X_B is increased. Therefore, ions which happen to occur at the front boundary of the migrating band are slowed, and overtaken by the bulk ions while the band moves down the column, whereas ions which are lagging behind continue to travel more and more slowly, causing the band to tail. When the isotherm is concave, the reverse happens.

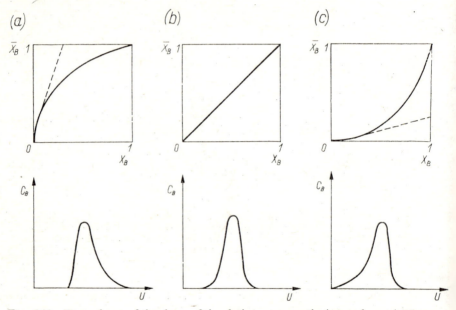

Fig. 6.32 Dependence of the shape of the elution curve on the ion-exchange isotherm

The initial portion of both non-linear isotherms may be approximated by a straight line, so for trace ions, the isotherm is always effectively linear. It is this linearity that has largely stimulated the popularity of the technique. It should be borne in mind, however, that even with trace ions the elution curves need not be symmetrical if the rate of ion sorption is greater than the rate of desorption from the ion-exchanger. (The elution of trace amounts of caesium ions from a phenol–sulphonic-acid cation-exchanger [74] (cf. Fig. 6.51) is a relevant case).

The theories of chromatography may be roughly divided into plate (or equilibrium) theories and rate theories (involving continuous variables).

In either case, relatively simple solutions are obtained only for the linear ion-exchange isotherm, i.e. when the distribution coefficients of the ions resolved are constant.

6.6.3.1 Plate Theory

In the plate theory advanced by Martin and Synge [449] and applied to ion-exchange chromatography by Mayer and Tompkins [450], Glueckauf [445] and other investigators [451,452] the column is treated as a series of stages (theoretical plates). The theoretical plate is defined as a height of packed bed within which the average concentration of the stationary phase is in equilibrium with the solution leaving it [447,449]. Thus local equilibrium is assumed to exist between the mobile phase and the immobile phase at each plate. The faster the ion-exchanger–solution equilibrium is established, the smaller is the height of the theoretical plate and the more efficient is the column (at fixed bed height).

It was shown that, for an ion producing a linear ion-exchange isotherm, the band shape, which is initially asymmetrical (Poisson distribution), may be approximated by the normal (or Gaussian) distribution curve [447,450] after the ion has traversed a certain number of theoretical plates. Accordingly, the theoretical elution curve may be described by the equation [452]:

$$C_B = C_{B(max)} \exp\left\{ -\frac{N(C+1)(U-U_{max})^2}{2C(U_{max}-U_0)^2} \right\} \qquad (6.49)$$

where U is the effluent volume, U_{max} is the retention volume (effluent volume corresponding to the peak concentration of the ion in the effluent), U_0 is the dead volume in the column, N is the number of theoretical plates, C is the distribution ratio (ratio of the quantities of the ion present in the ion-exchanger and in the solution within a single theoretical plate) $\left(C = \frac{U_{max}-U_0-V}{V}\right)$ and V is the free volume of the resin bed.

It may be shown that the theoretical plate height is described by the equation [341,452]

$$H = \frac{L}{N} = \frac{LW^2(C+1)}{8(U_{max}-U_0)^2 C} \approx \frac{LW^2}{8(U_{max}-U_0)^2} = \frac{L\sigma^2}{(U_{max}-U_0)^2} \qquad (6.50)$$

where W is the peak width at a height of $0.368 C_{B(max)}$ (Fig. 6.33), L is the column length and σ is the standard deviation of the chromatographic peak ($\sigma = W/2\sqrt{2}$).

The plate theory thus lends itself to the characterization of the spreading of chromatographic bands and is generally employed for analysis of separations. However, it gives no indication of how the various properties of the mobile and immobile phases, the flow-rate, temperature, and other process parameters affect the column efficiency.

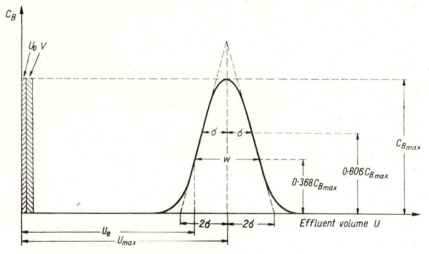

Fig. 6.33 Theoretical elution curve

The functional relationship between the theoretical plate height and the elution process parameters is provided by the rate theory (Section 6.6.3.2).

The plate theory enables distribution coefficients to be evaluated from the elution curves. The weight distribution coefficient [cf. Eq. (6.15)] is evaluated as

$$\lambda = \frac{U_{\max} - (U_0 + V)}{m_j} \qquad (6.51)$$

where m_j is the weight of dry resin in the column (for other symbols see Eq. (6.49)).

Ion-exchange chromatography often utilizes the 'bed distribution coefficient' λ', defined as the ratio of the number of millimoles of the ion in the ion-exchanger per unit bed volume (1 cm^3) to that in solution at equilibrium, per unit solution volume (1 cm^3). The two distribution coefficients are inter-related as follows

$$\lambda' = \lambda d_z \qquad (6.52)$$

where d_z is the bed density (Section 6.3.7).

The bed distribution coefficient may be conveniently evaluated as

$$\lambda' = \frac{U_{max} - U_0}{V_z} - i \qquad (6.53)$$

where V_z is the bed volume, and i ($= V/V_z$) is the fractional void volume of the bed ($i \sim 0.4$) [44,332,454].

The plate theory allows calculation of the total amount of the substance eluted from the column, or the fraction of this amount eluted in a well-defined effluent volume.

Let R be the quantity of ions B eluted from the column after an effluent volume U has been collected (cf. Fig. 6.33) and let r be the fraction of the total quantity (A) of ions B introduced into the column. From Eq. (6.49), we obtain [452]

$$A = C_{B(max)} \frac{W}{2} \sqrt{\pi} = 0.886 \, C_{B(max)} \, W \qquad (6.54)$$

The quantity of ions B eluted in a given volume U is described by the equation

$$R = \frac{C_{B(max)} \sqrt{C} (U_{max} - U_0)}{\sqrt{N(C-1)}} \int_{-\infty}^{t} \exp\left\{-\frac{t^2}{2}\right\} dt = Ar \qquad (6.55)$$

where t is defined by Eq. (6.56) as

$$t = \frac{\sqrt{N(C+1)} \, (U - U_{max})}{\sqrt{C(U_{max} - U_0)}} \qquad (6.56)$$

and r is the normal distribution function $N(0,1)$

$$r = \frac{1}{\sqrt{2\pi}} \int_{-\infty}^{t} \exp\left\{-\frac{t^2}{2}\right\} dt \qquad (6.57)$$

which may be read from the tables [453].

The dead volume of the column U_0, and the fractional void volume of the bed i, which occur in the plate-theory equations can readily be determined in column experiments. For an ion which, under experimental conditions, is not retained by the ion-exchanger ($\lambda = 0$) the retention volume is at every instant equal to $U_0 + V$ [(cf. Eq. (6.51)]; therefore measurements of the retention volume at two bed volumes ($V_{z(1)}$ and $V_{z(2)}$)

will yield U_0 and i as

$$i = \frac{U_{max(2)} - U_{max(1)}}{V_{z(2)} - V_{z(1)}} \quad (6.58)$$

$$U_0 = \frac{U_{max(1)} V_{z(2)} - U_{max(2)} V_{z(1)}}{V_{z(2)} - V_{z(1)}} \quad (6.59)$$

The application of plate theory to the analysis of separations will be discussed in Section 6.6.4.

6.6.3.2 Rate Theory

The rate theory (based on continuous variables) represents a more realistic approach to the chromatographic process than does plate theory. In the rate theory no assumptions concerning local equilibrium are made and column efficiency is described in terms of non-linear differential equations based on finite rates of mass transfer and diffusion and on continuous flow of the liquid through the column. Glueckauf [454], expressed the theoretical plate height by the equation

$$H = 1 \cdot 64 r_0 + \frac{\lambda' \, 0 \cdot 142 r_0^2 u}{(\lambda'+i)^2 \overline{D}} + \frac{(\lambda')^2 \, 0 \cdot 266 r_0^2 u}{(\lambda'+i)^2 (1-i) D (1+70 r_0 u)} \quad (6.60)$$

where r_0 is the particle radius, in cm, $\overline{D}$ is the interdiffusion coefficient in the ion-exchanger, D is the effective diffusion coefficient in the interstitial liquid, and u is the linear flow-rate of the eluent, in cm/sec.

The three terms on the right-hand side represent the contributions due to: (a) eddy diffusion, (b) diffusion within particles, and (c) diffusion within the film adherent to the particle. Equation (6.60) has been shown [455] to be equivalent to the van Deemter equation [456] derived for the plate height in gas–liquid chromatography (ignoring molecular diffusion).

Thus, the rate theories enable the plate height to be calculated from the experimentally available macroscopic process parameters (r_0, u, λ', $\overline{D}, i$, and D), but the plate heights observed in practice are usually much greater than those evaluated from Eq. (6.60) [457]. The factors responsible presumably include irregular column packing, non-uniform particle size, and maldistribution of flow (channelling, etc). Therefore, instead of permitting accurate calculation of H, Eq. (6.60) merely presents the factors which affect H. The first term in Eq. (6.60) has now been recognised to be more correctly expressed as ωr_0 where ω is a constant related to irregularity of packing and to some extent also to column diameter [15,447]. The equation fails to allow for every possible mechanism of distortion of chromatographic bands. At low flow-rates, the plate height can also be affected by longitudinal (molecular) diffusion in the solution [15,74].

§ 6.6] The Column Process

Recent findings have shown that, under certain experimental conditions, longitudinal diffusion in the ion-exchanger phase must not be disregarded [218,244,267].

Ultimately the plate height equation can be written as: [218,267,460]

$$H = \omega r_0 + \frac{\lambda' \, 0 \cdot 142 r_0^2 u}{(\lambda'+i)^2 D} + \frac{(\lambda')^2 \, 0 \cdot 266 r_0^2 u}{(\lambda'+i)^2 (1-i) D (1+70 r_0 u)} + \frac{D i \sqrt{2}}{u} + \frac{2 \gamma_s \, \overline{D} \lambda'}{u i} \quad (6.61)$$

The fourth and fifth terms in Eq. (6.61) represent the contributions due to longitudinal diffusion in the solution and in the ion-exchanger phase, respectively.

6.6.4 Analysis of Chromatographic Separations

One of the most important applications of the theories of chromatography is the analysis of chromatographic separations. The most common lines of approach to this problem will now be presented.

6.6.4.1 Resolution

A measure of the separation of two chromatographic peaks which is known as the resolution, R_n, is defined as follows [341]:

$$R_n = \frac{U_{\max(2)} - U_{\max(1)}}{n(\sigma_1+\sigma_2)} = \frac{\Delta U_{\max}}{n(\sigma_1+\sigma_2)} \quad (6.62)$$

where $U_{\max(1)}$ and $U_{\max(2)}$ are the retention volumes for components 1 and 2, respectively, σ_1 and σ_2 are the standard deviations of the respective peaks and n is an arbitrary number greater than zero.

For not too low distribution coefficients of component 1 ($\lambda_1' > 5$), resolution was found to be a relatively simple function of the distribution coefficient and theoretical plate height [341]

$$R_n = \frac{(\alpha_1^2-1)\sqrt{L}}{n(\alpha_1^2+1)\sqrt{\overline{H}}} = \frac{(\alpha_1^2-1)\sqrt{\overline{N}}}{n(\alpha_1^2+1)} \quad (6.63)$$

where $\overline{H}$ and $\overline{N}$ are respectively the mean plate height and the mean number of theoretical plates for the two peaks involved ($\overline{H} = (H_1+H_2)/2$) and α_1^2 is as defined by Eq. (6.19).

Differentiation of Eq. (6.63) serves to show that (as may also be felt intuitively) the resolution increases as the separation factor is increased and plate height decreased. The fact that, at fixed column length, resolution is a function of only two variables $(\alpha_1^2, \bar{H})$ enables us to classify generally the possible variations in resolution caused by any factors which may influence either the separation factor or plate height or the two together (Table 6.4) [341].

Table 6.4 Possible changes in resolution resulting from simultaneous variation of the separation factor and the theoretical plate height
(from [341], by permission of the copyright holders, Elsevier Publ. Co.)

	$\alpha_1^2 \nearrow$	$\alpha_1^2 \leftrightarrow$	$\alpha_1^2 \searrow$
$\bar{H} \nearrow$	(I) $R_n \nearrow$ (a), $R_n \leftrightarrow$ (b), $R_n \searrow$ (c)	(II) $R_n \searrow$	(III) $R_n \searrow$
$\bar{H} \leftrightarrow$	(IV) $R_n \nearrow$	(V) $R_n \leftrightarrow$	(VI) $R_n \searrow$
$\bar{H} \searrow$	(VII) $R_n \nearrow$	(VIII) $R_n \nearrow$	(IX) $R_n \nearrow$ (a), $R_n \leftrightarrow$ (b), $R_n \searrow$ (c)

$\nearrow$ = increase, $\leftrightarrow$ = no change, $\searrow$ = decrease

The most convenient case, theoretically, is case VII (Table 6.4), where both α_1^2 and $\bar{H}$ change in the desired direction. In cases IV and VIII the resolution also always increases. Under certain circumstances cases I and IX may also result in increased resolution.

Separation of two components is arbitrarily taken to be complete when the resolution exceeds unity, viz.,

$$R_n \geq 1, \quad \text{i.e.} \quad \Delta U_{max} \geq n(\sigma_1 + \sigma_2) \tag{6.64}$$

The value of n is chosen arbitrarily and depends upon the requirements set by the experimenter. If the effluent from the column is divided into two fractions at the point $U = U_{max(1)} + n\sigma_1$, then the quantity of each of the components in the appropriate fraction is determined from the area under the Gaussian curve (normal distribution function)

$$\phi(n) = \frac{1}{\sqrt{2\pi}} \int_{-\infty}^{n} e^{-n^2/2} \, dn \qquad (n \geq 0) \tag{6.65}$$

and contamination with the other component is

$$1 - \phi(n) = 1 - \frac{1}{\sqrt{2\pi}} \int_{-\infty}^{n} e^{-n^2/2} \, dn \tag{6.66}$$

Table 6.5 Values of the normal distribution function for selected values of n

$$\phi(n) = \frac{1}{\sqrt{2\pi}} \int_{-\infty}^{n} e^{-\frac{n^2}{2}} dn; \quad (n \geqslant 0)$$

n	$\phi(n)$	$1-\phi(n)$	n	$\phi(n)$	$1-\phi(n)$
1·0	0·8413	0·1587	5·0	0·99999971	0·00000029
1·5	0·93319	0·06681	6·0	0·999999999	0·000000001
2·0	0·97725	0·02275	7·0		$1·5 \times 10^{-12}$
2·5	0·993790	0·008198	8·0		5×10^{-16}
3·0	0·998650	0·001350	9·0		1×10^{-19}
3·5	0·9997674	0·0002326	10·0		$1·4 \times 10^{-22}$
4·0	0·99996833	0·00003167	15·0		5×10^{-51}
4·5	0·999996602	0·000003398	20·0		$2·5 \times 10^{-89}$
			25·0		3×10^{-138}
			30·0		5×10^{-198}

The function $\phi(n)$ has been tabulated for various values of n (cf. [453]); selected values are given in Table 6.5.

In gas–liquid chromatography the degree of resolution is commonly characterized in terms of $n = 2$ (resolution R_2) [458].

In ion-exchange chromatography, it has been suggested that resolution R_3 should be used ($n = 3$) [332,341]. In that case, a separation is considered to be complete if at least 99·86% of the component has been collected in its appropriate fraction and contamination with the component from the adjacent peak is less than 0·14% (Fig. 6.34). For $R_2 \geqslant 1$, the corresponding critical figures are 97·72% and 2·28% (cf. Table 6.5).

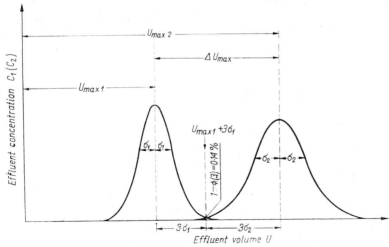

Fig. 6.34 Schematic representation of 'complete' separation of two components ($R_3 = 1$).

If R_3 is known, resolution at another value of n can be calculated from the equation

$$R_n = R_3 \frac{3}{n} \qquad (6.67)$$

which in conjunction with the data in Table 6.5 permits a quantitative estimate of the resolution achieved.

The conditions for complete resolution ($R_n = 1$) can be established by rearranging Eq. (6.63).

If the theoretical plate height is known, the separation factor should be given by [341]

$$\alpha_1^2 = \frac{\sqrt{N} + n}{\sqrt{N} - n} \qquad (6.68)$$

At a given α_1^2, the required column length is

$$L = \frac{\overline{H} n^2 (\alpha_1^2 + 1)^2}{(\alpha_1^2 - 1)^2} \qquad (6.69)$$

Variables Affecting the Separation Factor

As has been indicated in Section 6.4.2, both the affinity of a given trace ion for the ion-exchanger and the differentiation of affinity with the nature of the trace ion are affected by a number of variables; the major factors include the chemical nature, charge and size of ions, types of ion-exchanger (functional groups), ionic composition and concentration of the solution, presence of complexing agents, etc.

In practice, the separation factor is calculated from the distribution coefficients, and these are most readily determined by a static (batch equilibration) method involving measuring the distribution of an element between the ion-exchanger and the solution after sufficient time has elapsed for equilibrium to be reached. The weight distribution coefficient can then be calculated [448] as

$$\lambda_B = \frac{C_{B(0)} - C_{B(equil)}}{C_{B(equil)}} \frac{v}{m_j} \qquad (6.70)$$

where $C_{B(0)}$ and $C_{B(equil)}$ are the initial and equilibrium concentrations of the trace ion in the solution, respectively, v is the volume of the solution, in cm^3, and m_j is the dry resin weight, in g.

Determination of distribution coefficients by batch equilibration is often carried out with the use of radioactive tracers. The factors affecting precision of such determinations have recently been discussed [458a].

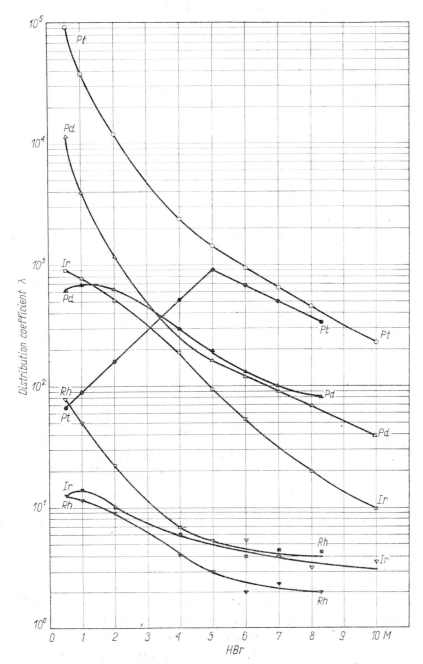

Fig. 6.35 Weight-distribution coefficients of platinum-group elements on Dowex 1-X4 [Br^-] in hydrobromic acid solutions: empty points — $HBr + 0.0035M$ Br_2, full points — $HBr + 2\%$ $N_2H_4 \cdot HCl$ (from [459], by permission of the copyright holders, Akadémiai Kiadó)

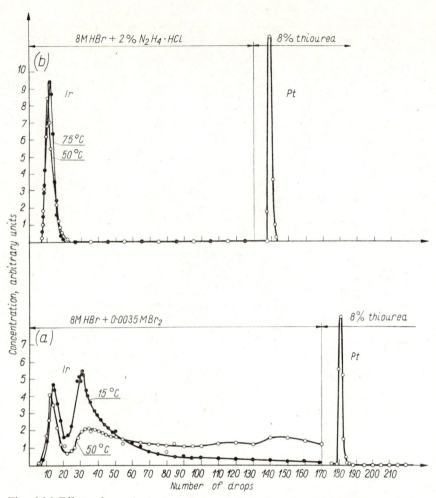

Fig. 6.36 Effect of a reductant on the resolution of traces of iridium from platinum. Column: 2·50 cm × 0·0311 cm², Dowex 1-X8 (Br⁻) (25–62μm), flow-rate 0·48–0·54 cm/min (from [459], by permission of the copyright holders, Akadémiai Kiadó)

The distribution coefficients of many elements in numerous ion-exchanger–solution systems have been collated in the literature [233,276, 281,343,370,377–382] (cf. also Table 6.2, Figs. 6.62–6.69, and Section 6.8.1).

It must be borne in mind that for elements which are present in several co-existing forms in solution, the distribution coefficient determined by the static method is an average value which gives little indication of the behaviour of the element in the column. A good illustration is the separation of platinum and iridium in the Dowex 1-X4 [Br⁻]–HBr system [459]. Figure 6.35 shows the distribution coefficients for Ir and Pt as a function

§ 6.6] **The Column Process** 377

of hydrobromic acid concentration. In a solution of $HBr + Br_2$, differences between the static-method distribution coefficients of the two elements are high enough for a good ion-exchange separation to be expected. However, in column experiments (Fig. 6.36a) iridium was found to be eluted as two peaks corresponding to Ir(III) and Ir(IV), the Ir(IV) peak being highly asymmetric. As the temperature is raised, this asymmetry increases (cf. Fig. 6.36a), probably because of hydrolysis of the Ir(IV) species. In a solution of $HBr + N_2H_4 \cdot HCl$, the hydrazine hydrochloride reduces the iridium quantitatively to Ir(III) which is eluted from the column as a sharp peak and may readily be separated from platinum (Fig. 6.36b). This example indicates clearly that, with elements of variable oxidation state, an oxidant or reductant present in the solution may have a big effect on the feasibility and quality of separation.

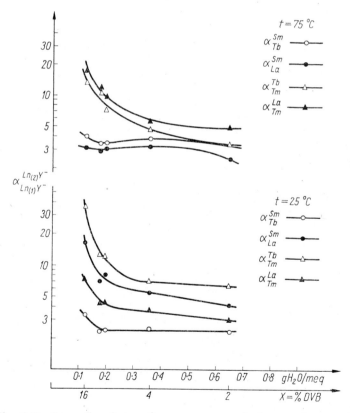

Fig. 6.37 Separation factors for some rare-earth elements (as complexes with EDTA) on Dowex 1 anion-exchanger in relation to the degree of cross-linkage and temperature (from [218], by permission of the copyright holders, Elsevier Publ. Co.)

In a given chromatographic system, the separation factor can be modified by varying the temperature [191,246a,267,341,952] and the degree of cross-linkage [218,245–246a,267,328,463,467]. The effects of these factors on selectivity have already been discussed in Section 6.4.2.2. As a further illustration, Fig 6.37 shows the dependence of the separation factors for complexes of rare earths with EDTA on the degree of cross-linkage of the anion-exchanger and the temperature [218,267].

Factors Affecting Theoretical Plate Height

As may be seen from Eq. (6.63), when the separation factor is kept constant the separation of two elements can be improved by increasing the column efficiency, i.e. the number of theoretical plates. It has been experimentally demonstrated that a considerable increase in column efficiency can be important in achieving difficult separations, even if the separation factor is simultaneously decreasing [218,267,341]. The major factors controlling column efficiency in ion-exchange chromatography are discussed below.

Column Size. As is evident from Eq. (6.63), resolution is directly related to the square root of the column length. Theoretically, when the ions to be separated have different affinities for the ion-exchanger ($\alpha_i^2 > 1$), the mixture can always be resolved provided a long enough column is used. Increasing the column length does not affect the theoretical plate height but increases the total number of theoretical plates. Figure 6.38

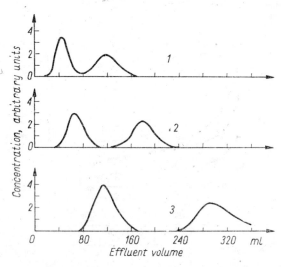

Fig. 6.38 Effect of column-length on the separation of tetrametaphosphate from trimetaphosphate on Dowex 1-X10 (Cl$^-$) (0·074–0·14 mm); eluent 0·5M KCl. Column length: 1–5·8 cm, 2–8·3 cm, 3–13·8 cm (from [451], by permission of the copyright holders, the American Chemical Society)

shows the effect of column on the separation of tetrametaphosphate from trimetaphosphate by elution development with a potassium chloride solution in an anion-exchanger column.

In practice, column length cannot be increased indefinitely. First, pressure must be applied to force the liquid through the column at a suitable flow-rate. Secondly, the time required to carry out a separation increases in direct relation to column length. Thirdly, the maximum concentration at the peak decreases in inverse proportion to the square root of the column length. The last conclusion can be deduced, in agreement with the properties of the normal distribution, from Eq. (6.49).

$$C_{B(max)} = \frac{1}{\sigma \sqrt{2\pi}} \qquad (6.71)$$

Combination of Eqs. (6.50) and (6.53) gives the expression for the standard deviation of the chromatographic peak

$$\sigma = S(\lambda' + i)\sqrt{HL} \qquad (6.72)$$

where S is the cross-sectional area of the column.

Substitution of Eq. (6.72) in (6.71) yields

$$C_{B(max)} = \frac{1}{S(\lambda' + i)\sqrt{HL}\sqrt{2\pi}} \qquad (6.73)$$

Therefore, as the column length is increased, the chromatographic peak spreads and flattens more and more so that ultimately the concentration of the substance in the effluent may be too small to be determined accurately in the fractions collected. At the same time, the effluent volume corresponding to a given peak increases in direct relation to the square root of the column length [cf. Eq. (6.72)].

In ion-exchange chromatography a high aspect ratio of the column is generally considered to be favourable. Columns of very small diameter (less than about 30 particle diameters) tend, however, to suffer from wall effects, i.e. from increased leakage of ions through the column owing to the considerably differing liquid flow-rates at the wall and at the axis [15,447]. This results in a decrease in resolution. A very large column diameter has been found to increase the plate height owing to increased eddy diffusion [first term in Eq. (6.61)] [461].

Type of and Degree of Cross-Linkage of the Ion-Exchanger, and Reactions in the Solution. The literature furnishes no comparative on data theoretical plate height in relation to the type of ion-exchanger functional groups. With similarly cross-linked completely dissociated ion-exchangers, plate heights would be expected to be of the same order of magnitude. If a slow chemical reaction proceeds in the ion-exchanger, the plate height should increase, as is observed, e.g., in the elution of Cs^+ from a phenol–sulphonic-acid cation-exchanger [74]. Relatively large plate heights may well be expected to occur in the separation of metals on chelating ion-exchangers.

Experiments on the elution of the light lanthanides from cation-exchanger columns with buffered citrate and EDTA solutions have shown that with the latter eluent resolution is inferior despite the superior separation factors [359]. The greater plate height in the EDTA elution of lanthanides from a sulphonic-acid cation-exchanger may be explained by a slow internal reaction occuring in the solution (formation or decomposition of a lanthanide–EDTA complex). Reports on complex-formation kinetics [462] and displacement-development resolution of lanthanides with EDTA [495] appear to verify this hypothesis. The effect of a slow internal reaction proceeding in the ion-exchanger or in the solution is not explicit in Eq. (6.61) but only implicit in the ion diffusivities.

It is well known that particle diffusion coefficients decrease with increasing degree of cross-linkage [219,320,447]. If plate heights normalized with respect to the distribution coefficient are compared for ion-exchangers of various degrees of cross-linkage at fixed flow-rate, particle size etc.,

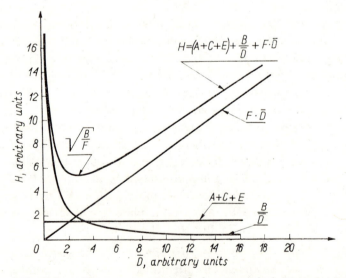

Fig. 6.39 Schematic representation of the theoretical plate height in relation to diffusion coefficient in the resin phase (from [218], by permission of the copyright holders, Elsevier Publ. Co.)

they may be expressed as a function of a single variable, *viz.* particle diffusion coefficient [218]:

$$H = A + \frac{B}{\bar{D}} + C + E + F\bar{D} \quad (6.74)$$

where A, B, C, E, F are constant (under the experimental conditions used).

This predictable variation of the plate height with $\bar{D}$ (Fig. 6.39) shows that there always exists an optimum degree of cross-linkage at which H attains a minimum. Recent investigations have supported this conclusion. Experimental plate heights (normalized to the volume distribution coefficient $\lambda' = 10$ or 20) determined in a few systems [218, 244–246] are plotted in Fig. 6.40 as a function of water content in a swollen ion-

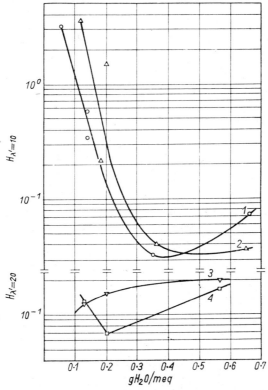

Fig. 6.40 Theoretical plate height (at 25°C) normalized to the bed distribution coefficient $\lambda' = 10$ or $\lambda' = 20$ as a function of specific water content in the ion-exchanger: 1 – rare earths in the system Dowex 1 (H_2Y^{2-})–Na_2H_2Y, 2 – rare earths in the system Dowex 1 (H_2Z^{2-})–Na_2H_2Z, 3 – alkali metals in the system Dowex 50W (H^+)–HCl, 4 – alkaline-earth metals in the system Dowex 50W (H^+)–HCl (from [467], by permission of the copyright holders, Nauk. Inst. Technol., Wrocław)

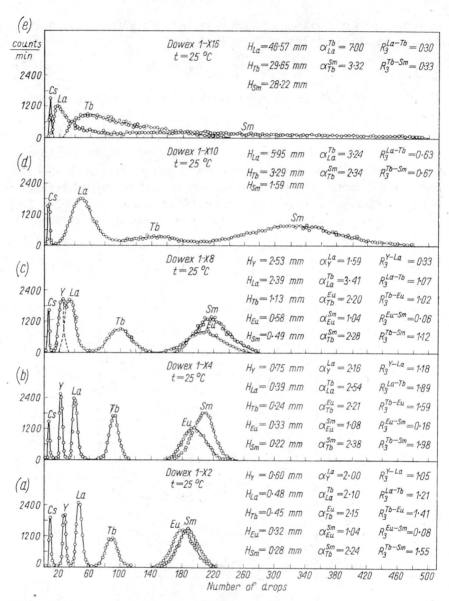

Fig. 6.41 Effect of the degree of cross-linkage on the separation of Y, La, Tb, Eu, and Sm ethylenediaminetetra-acetates at 25°C: (a) Dowex 1-X2 (H_2Y^{2-}) (6–38 μm), column 4·90 cm × 0·0305 cm², eluent 0·011M Na_2H_2Y, flow-rate 0·95 cm/min; (b) Dowex 1-X4 (H_2Y^{2-}) (11–42 μm), column 4·90 cm × 0·0305 cm², eluent 0·036M Na_2H_2Y, flow-rate 1·32 cm/min; (c) Dowex 1-X8 (H_2Y^{2-}) (17–44 μm), column 4·95 cm × 0·0305 cm², eluent 0·078M Na_2H_2Y, flow-rate 1·33 cm/min; (d) Dowex 1-X10 (H_2Y^{2-}) (17–46 μm), column 5·10 cm × 0·0305 cm², eluent 0·063M Na_2H_2Y, flow-rate 1·18 cm/min; (e) Dowex 1-X16 (H_2Y^{2-}) (20–43 μm), column 4·90 cm × 0·0305 cm², eluent 0·048M Na_2H_2Y, flow-rate 1·31 cm/min (from [218], by permission of the copyright holders, Elsevier Publ. Co.)

exchanger, which is regarded as a measure of the degree of cross-linkage [213,214,218,267]. If the particle diffusion coefficients are assumed—to a first approximation—to be proportional to the specific water content in the ion-exchanger, then the curves of H vs. W_{H_2O}/Z_s constructed for the separation of lanthanide–EDTA complexes on strong-base anion-exchangers and of alkaline earth metal ions on sulphonic-acid cation-exchangers (Fig. 6.40) are consistent with the predictions of Eq. (6.74) (cf. Fig. 6.39). The curves of H vs. W_{H_2O}/Z_s obtained in the separation of the alkali metals have revealed no minimum but at still higher degrees of cross-linkage a minimum is likely to occur. The effect of cross-linkage on plate height is particularly noticeable with large complex ions (e.g. complexes of rare earths with EDTA and DCTA [218,246,267]) where the size is presumably comparable with that of the pores in the polymer matrix of more extensively cross-linked ion-exchangers with which sieve effects have been found to occur. A distinct effect of cross-linkage on the plate height was also observed in anion-exchange separation of phosphate complexes of gallium and indium [246a]. With LnY^- ions, an increase in the degree of cross-linkage of an ion-exchanger from X4 to X16 has resulted in the plate height being increased by two orders of magnitude, thus exercising a pronounced effect on the resolving power for the rare earths (Fig 6.41). As a consequence, the resolution of, say, a Tb–Sm mixture, which is more than satisfactory on moderately cross-linked ion-exchangers, with strongly cross-linked ion-exchangers falls much below unity (Fig. 6.42), although the separation factors do rise as the degree of cross-linkage is increased [218,267] (cf. Fig. 6.37) (Table 6.5, case Ic of the classification). The cross-linkage effect has also been observed in the separations of metal ion complexes with halogens [459,463] and other ligands on an anion-exchanger [464]. With small inorganic ions the effect of cross-linkage on the quality of resolution is much less pronounced, though in this case, too, it may be important [244,245,267]. With alkali metals, the quality of separation increases as the degree of cross-linkage is increased (Table 6.5, case VII) (Fig. 6.43). In the separation of rare earths by elution with buffered β-hydroxyethylethylenediaminetriacetic acid solutions on a sulphonic-acid cation-exchanger, the plate height was found to diminish by a factor of more than 7 as the degree of cross-linkage was reduced from X12 to X2 [465]. Interestingly, in the same system the theoretical plate height decreased considerably after a neutral salt (e.g. NaCl) had been added to the eluent. As the salt concentration was increased, the plate height diminished; it was also related to the cation, decreasing in the order Li > Na > K. No theoretical interpretation of this behaviour has been attempted as yet.

Not only the nominal degree of cross-linkage but also its uniformity seems to affect the efficiency of chromatographic separations. When amino acids are separated on sulphonic-acid cation-exchangers, narrower bands and more complete resolution are obtained for a resin cross-linked with pure m-divinylbenzene than for that prepared from styrene and technical divinylbenzene [465a].

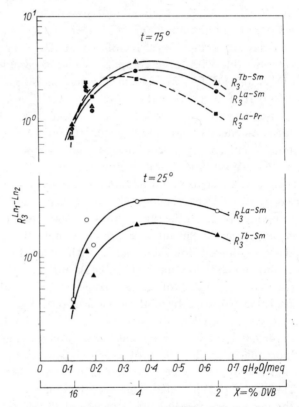

Fig. 6.42 Resolution for some pairs of lanthanide ethylenediaminetetra-acetates in the system Dowex 1 (H_2Y^{2-})–Na_2H_2Y as a function of the degree of cross-linkage (from [218], by permission of the copyright holders, Elsevier Publ. Co.)

Particle size. If theoretical plate height is primarily governed by eddy diffusion or film diffusion, H should be directly related to particle radius and, if particle diffusion control predominates, to the square of the particle radius (cf. Eq. 6.61). Therefore, fine-grained ion-exchangers should be preferable for chromatographic separations. Obviously, the resistance of the bed to flow increases as the particle size is reduced and increasingly

§ 6.6]   The Column Process   385

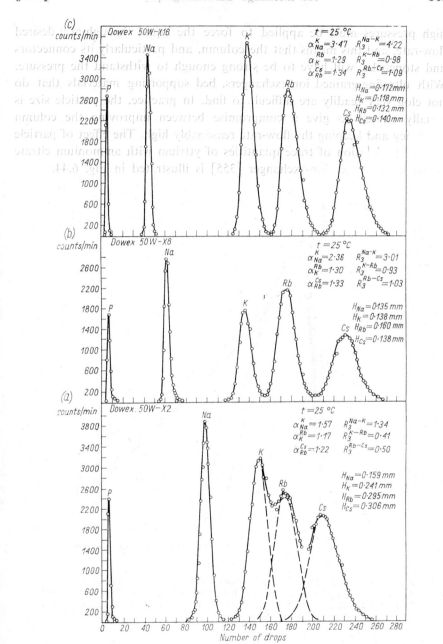

Fig. 6.43 Effect of the degree of cross-linkage on the separation of alkali metals at 25 °C on a sulphonic-acid cation-exchanger: (a) Dowex 50W-X2 (H$^+$) (12–22 μm), column 7·0 cm × 0·0305 cm^2, eluent 0·0679M HCl, flow-rate 0·72 cm/min; (b) Dowex 50W-X8 (H$^+$) (11–23 μm), column 7·0 cm × 0·0305 cm^2, eluent 0·261M HCl, flow-rate 0·72 cm/min; (c) Dowex 50W-X16 (H$^+$) (11–21 μm), column 7·0 cm × 0·0305 cm^2, eluent 0·636M HCl, flow-rate 0·71 cm/min (from [244], by permission of the copyright holders, Elsevier Publ. Co.)

high pressures must be applied to force the liquid through at a desired flow-rate, and this means that the column, and particularly its connectors and stopcocks, will have to be strong enough to withstand the pressure. With ultrafine-grained ion-exchangers, bed supporting materials that do not clog very readily are difficult to find. In practice, the particle size is usually chosen to give a compromise between improving the column efficiency and keeping the flow-rate reasonably high. The effect of particle size on the elution of trace quantities of yttrium with ammonium citrate solution from a cation-exchanger [355] is illustrated in Fig. 6.44.

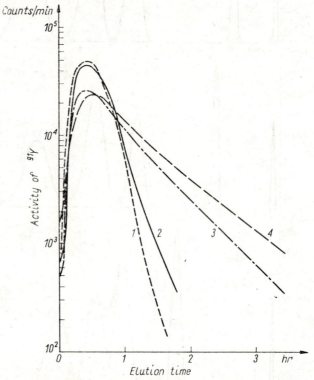

Fig. 6.44 Effect of particle size (mm) on the elution curves of ^{91}Y from Amberlite IR-1 cation-exchanger with citrate solutions: 1 — 0·044–0·55, 2 — 0·074–0·088, 3 — 0·25–0·29, 4 — 0·36–0·50 (from [355], by permission of the copyright holders, the American Chemical Society)

The distribution of particle size within a given fraction is also of importance in ion-exchange chromatography. Uniform particle size ensures regular and reproducible column packing and reduces complications associated with the distortion of the front boundary caused by maldistribution of flow, channelling, etc. A resin of uniform particle size will result

§ 6.6] The Column Process 387

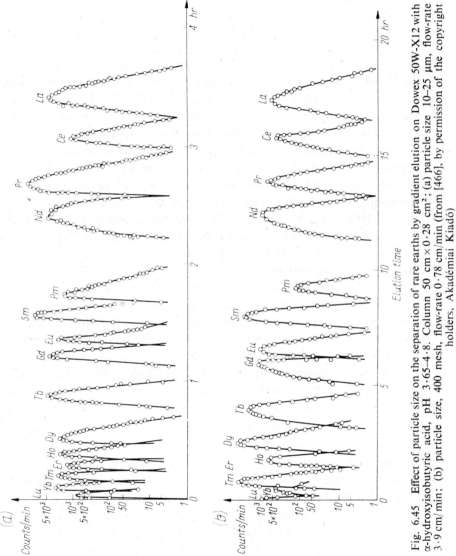

Fig. 6.45 Effect of particle size on the separation of rare earths by gradient elution on Dowex 50W-X12 with α-hydroxyisobutyric acid, pH 3·65-4·8. Column 50 cm × 0·28 cm²; (a) particle size 10–25 μm, flow-rate 3·9 cm/min; (b) particle size, 400 mesh, flow-rate 0·78 cm/min (from [466], by permission of the copyright holders, Akadémiai Kiadó)

25*

in better separations (at constant flow-rate) or separations that are as good as those on an ion-exchanger of non-uniform particle size but are faster because of the higher flow-rate that can be used (Fig. 6.45) [466].

Within recent years, fine-grained spherical ion exchangers have become available commercially [47,468–470b,953].

Particle size as low as 10 μm or even less, say 3–7 μm, is no longer an exception in ion-exchange practice. Progress in design of columns for operation at high pressure (up to 700 atm), with pumps, feeders, detectors, etc. has made such separations possible [469,471–472a]. This has developed into a new branch of chromatography (known as HPLC) which has been discussed in several monographs [472–472d]. LC stands for liquid chromatography, but HP is variously interpreted as meaning high performance or high pressure (or high price). HPLC has mainly been used for the separation of complex organic mixtures. Separation of inorganic ions is more difficult because glass columns can withstand pressures of only 50–70 atm, and the concentrated mineral acids often used in ion-exchange chromatography prevent the use of steel columns. Nevertheless, HPLC has been successfully used for resolving and determining rare earths [473–477b], actinides [477b], alkali metals [478] and some other inorganic ions [75a,479–481]. High-pressure systems require the use either of conventional ion-exchangers of suitably high degrees of cross-linkage [474] or of macroporous ion-exchangers [480]. Special techniques [482] are usually required to pack columns with ultrafine-grain ion-exchangers, e.g. a dynamic slurry technique using a fluid flow greater than the settling velocity of the resin or the dynamic technique whereby a larger prepacked resin bed is extruded into a smaller column.

A liquid chromatograph intended for high speed separations, in which columns, valves and tubings are made only of glass and plastic, was recently constructed [75a]. This permits the use of concentrated acids and other corrosive liquids as eluents.

Flow-Rate. According to theory, the theoretical plate height should be related to the flow-rate as it is in gas–liquid chromatography [440] (this relationship follows a course quite similar to that presented in Fig. 6.39 which describes the variation of H with increasing diffusion coefficient). However, the contributions of the individual types of diffusion to the overall theoretical plate height Eq. (6.61) differ from the corresponding contributions in gas–liquid chromatography, and it is technically difficult to produce high flow-rates, so it is difficult to study the influence on H of a wide range of rates of flow. Therefore, in practice, often only extreme cases are observed. Most frequently the theoretical plate height is controlled by particle diffusion or by film diffusion. Then, as the flow-rate is increased, the plate height increases [cf. Eq. [6.61]] and separations become increasingly poorer [460] (Fig. 6.46).

Whenever longitudinal diffusion has an important influence on plate height, increasing the flow-rate may have a beneficial effect on the quality of separation [460] (Fig. 6.47).

Flow-rate is the fundamental factor determining the time required for separation. The rate of migration, u_B, of the band of substance B down the column is directly related [15,441] to the rate of flow and inversely related to the distribution coefficient, viz.,

$$u_B = \frac{u}{\lambda' + i} \qquad (6.75)$$

Flow-rate can be increased with little deterioration in the quality of separation by reducing the resistance to mass transfer in the stationary phase, that is, the particle diffusion. One way already described (p. 388) involves the use of an ultrafine-grained ion-exchanger in conjunction with HPLC equipment. In recent years a novel type of ion-exchanger, known as the pellicular [469,483,953] or controlled surface porosity type [469,472,484–486] has been developed: in this a thin (a few μm thick) polymer layer containing ionogenic groups is deposited on a spherical impermeable core support (usually glass). Such ion-exchangers, with particle sizes ranging from 20 to 60 μm [469–470b,472,472a] can be used for separations at high pressures (up to a few hundred atmospheres), and plate heights of about 1 mm [484] are obtained at flow-rates of ca. 1 cm/sec. Narrow bore columns (1–2 mm) up to a few meters long [469, 472,472a,486] are used. Separation may often be done in just a few minutes. A drawback is the low exchange capacity, of 3–60 μeq/g [469–470b,484].

Another type of ion-exchange material that ensures rapid mass transfer is prepared from spherical beads of a styrene–DVB copolymer with ionogenic groups incorporated only in a thin surface film [472,487–490]. The exchange capacities range from 0·6 to 650 μeg/q [487,489] according to film thickness. Recently ion-exchange celluloses [491] and exchanger-like porous materials with a silica surface modified by chemically bound ionogenic groups [472,492–494a] have been used in HPLC. So far, the only HPLC applications of these materials have been separations of organic compounds, but there is no doubt they could be used for separation of trace amounts of inorganic ions; and silica gel substituted with amino silyl groups has been used to remove copper and iron from water to allow the hardness to be determined [494b].

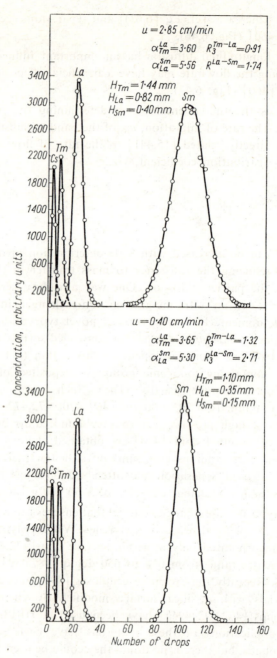

Fig. 6.46 Effect of flow-rate on separation of lanthanide ethylenediaminetetraacetates. Column $2\cdot70$ cm $\times\,0\cdot0341$ cm^2, Dowex 1-X4 (H_2Y^{2-}) (11–42 μm), temperature 25°C, eluent $0\cdot04M$ Na$_2$H$_2$Y (from [460], by permission of the copyright holders, Akadémiai Kiadó)

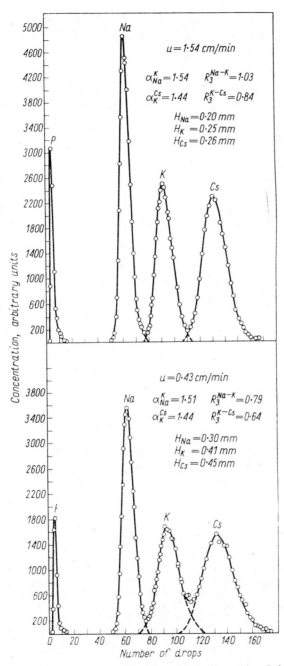

Fig. 6.47 Effect of flow-rate on separation of alkali metals. Column $5 \cdot 0$ cm $\times$ $0 \cdot 0305$ cm^2, Dowex 50W-X2 (H$^+$) (12–22 μm), temperature 25°C, eluent $0 \cdot 063M$ HCl (from [460], by permission of the copyright holders, Akadémiai Kiadó)

Distribution Coefficients of Ions. The plate height is not a constant for a column, not even under fixed experimental conditions, because it is related to the distribution coefficient of the ions to be separated. As evident from Eq. (6.6), under particle diffusion control, the plate height is approximately proportional to $1/\lambda'$. When the contribution due to longitudinal diffusion in the ion-exchanger phase is also important, the

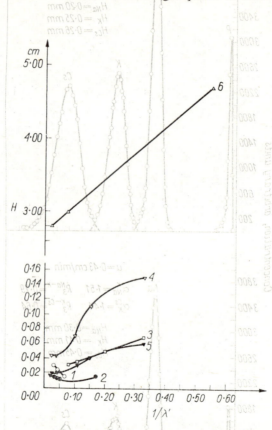

Fig. 6.48 Theoretical plate height as a function of reciprocal of the bed-distribution coefficient in the separation of various ions at a temperature of 25°C: 1—alkali metals in the system Dowex 50W-X2 (H^+)—HCl, (12–22 μm), eluent $0.0679M$ HCl, flow-rate 0.72 cm/min; 2—alkali metals in the system Dowex 50W-X16 (H^+)—HCl (11–21 μm), eluent $0.636M$, HCl, flow-rate 0.71 cm/min; 3—lanthanide 1,2-diaminecyclohexa-netetra-acetates in the system Dowex 1-X4 (H_2Z^{2-})–$Na_2H_2Z_{aq}$, (7–21 μm), eluent $0.0047M$ Na_2H_2Z, pH 4.8, flow-rate 0.69 cm/min; 4—lanthanide ethylenediamine-tetra-acetates in the system Dowex 1-X2 (H_2Y^{2-})–$Na_2H_2Y_{aq}$, (6–38 μm), eluent $0.011M$ Na_2H_2Y, pH 4.7, flow-rate 0.91 cm/min; 5—lanthanide ethylenediaminetetra-acetates in the system Dowex 1-X4 (H_2Y^{2-})–$Na_2H_2Y_{aq}$, (11–42 μm), eluent $0.036M$ Na_2H_2Y, pH 4.6, flow-rate 1.19 cm/min; 6—lanthanide ethylenediaminetetra-acetates in the system Dowex 1-X16 (H_2Y^{2-})–$Na_2H_2Y_{aq}$, (20–43 μm), eluent $0.048M$ Na_2H_2Y, pH 4.6, flow-rate 1.31 cm/min (from [267], by permission of the Inst. Nucl. Research (Warsaw))

plate height may be expected to be proportional to λ' at suitably high distribution coefficients. Longitudinal diffusion in the ion-exchanger phase should be more significant for higher particle diffusion coefficients. Illustrative plots of the plate height vs. the distribution coefficient in various ion-exchanger–solution systems are shown in Fig. 6.48 [218,245,267]. Particle diffusion control (separation of LnY^- complexes on Dowex 1-X16), longitudinal diffusion control (separation of alkali metals on Dowex 50W-X2), and also intermediate cases have been included.

Temperature. Increased temperature results in higher diffusion coefficients in the ion-exchanger and solution phases, so the plate height is bound to be affected. If particle diffusion or film diffusion control prevails [the second and third terms in Eq. (6.61)]—which is commonly true in practice—the plate height diminishes as the temperature is raised (Fig. 6.49a,b).

Figure 6.50 shows the elution curves of lanthanide–EDTA complexes from Dowex 1-X4 (H_2Y^{2-}) at several temperatures. It should be borne in mind that temperature also affects the selectivity coefficient and thus also the separation factor (cf. Section 6.4.2.2). The changes in resolution observed in Fig. 6.50 are not only due to reduced plate height but also to considerable changes in the separation factors. Thus, for example, the separations of Y–La and Eu–Sm mixtures correspond to case VII of the classification given in Table 6.4. With the La–Tb mixture, reversal of the selectivity sequence takes place. Case IXc applies within the temperature range 25–68 °C (cf. Fig. 6.12) whereas at temperatures over 68 °C, case VII of the classification applies.

Improved column efficiency at increased temperatures has been observed in numerous cases, including the anion-exchange separation of Ga and In phosphate complexes [246a], **separation of rare-earth phosphate complexes [952]**, elution of rare earths from cation-exchanger columns with citrate [4,355], various hydroxy-acids [318] or EDTA [359], the elution of Fe, Cu, Ni, Co, Mn, Zn and Cd from a cation-exchanger with sulphosalicylate [496], and also the separation of noble metals on anion-exchangers in hydrobromic acid medium [459].

If longitudinal diffusion (in the ion-exchanger or the solution phase) is the factor controlling the plate height, increasing the temperature increases the plate height (Fig. 6.49d) because in the longitudinal diffusion terms in Eq. (6.61) the diffusion coefficients occur in the numerator. Intermediate cases are also possible (Fig. 6.49c). The effect of temperature on the separation of trace amounts of alkali metals on a phenol–sulphonic-

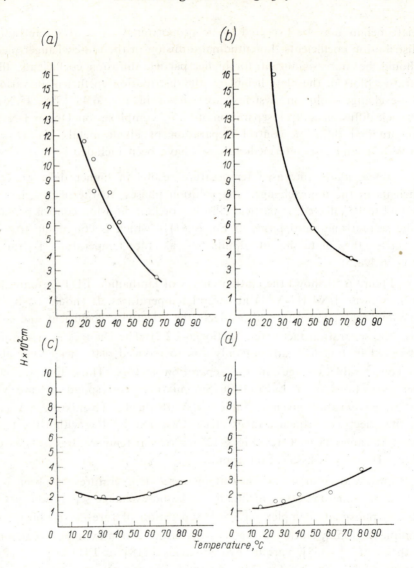

Fig. 6.49 Theoretical plate height normalized with respect to the bed-distribution coefficient, $\lambda' = 10$, as a function of temperature in various systems: (a) lanthanide ethylenediaminetetra-acetates in the system Amberlite IRA-400 (H_2Y^{2-})–$Na_2H_2Y_{aq}$, (10–35 μm), eluent 0.008–$0.009M$ Na_2H_2Y, pH 4.6–4.7, flow-rate 1.08–1.25 cm/min; (b) lanthanide 1,2-diaminecyclohexanetetra-acetates in the system Dowex 1-X8 (H_2Z^{2-})–$Na_2H_2Z_{aq}$, (12–43 μm), eluent $0.0047M$ Na_2H_2Z, pH 4.9, flow-rate 0.69 cm/min; (c) alkali metals in the system cation-exchanger MK-3 (H^+)–HCl, (11–31 μm), eluent $0.0712M$ HCl, flow-rate 0.84–0.91 cm/min; (d) alkali metals in the system Amberlite IR-120 (H^+)–HCl, (17–51 μm), eluent $0.5199M$ HCl, flow-rate 0.47–0.51 cm/min (from [267], by permission of the Inst. Nucl. Res. (Warsaw))

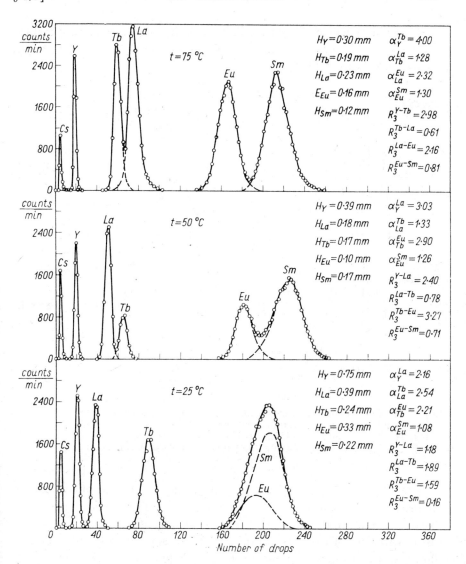

Fig. 6.50 Effect of temperature on the chromatographic separation of lanthanide–EDTA complexes. Column $4.85 \text{ cm} \times 0.0305 \text{ cm}^2$, Dowex 1-X4 ($H_2Y^{2-}$) (11-42 μm), eluent $0.0360M$ Na_2H_2Y, flow-rate 1·23–1·32cm/min (from [267], by permission of the Inst. Nucl. Res. (Warsaw))

acid cation-exchanger is shown in Fig. 6.51. At increased temperature, the resolution is seen to be inferior. This corresponds to cases III and II of the classification (Table 6.4).

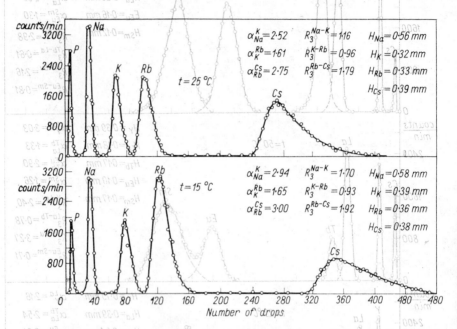

Fig. 6.51 Effect of temperature on the chromatographic separation of alkali metals on a phenol-sulphonic-acid cation-exchanger. Column $4 \cdot 97$ cm $\times 0 \cdot 0330$ cm^2, MK-3 (H$^+$) (11–31 μm), eluent $0 \cdot 1457 M$ HCl, flow-rate $0 \cdot 94$–$0 \cdot 97$ cm/min (from [74], by permission of the copyright holders, Polska Akad. Nauk)

Increasing the temperature may have favourable side-effects, e.g. it is beneficial for elutions carried out at higher flow-rates, which is of particular importance if ion-exchangers of very small particle size are used. Variation of the pressure that has to be applied to obtain a desired flow-rate exactly parallels the variation of solution viscosity with temperature [332].

On the other hand, increasing the temperature may give rise to undesirable effects such as irreversible structure changes in the ion-exchanger [74], evolution of air bubbles in the column, etc. In certain systems, variation of the experimental temperature during the elution yields separations similar to those obtainable by gradient elution [497,497a].

Sample Size (Volume). The plate theory has been derived on the assumption that the number of ions to be separated on the column is very small and can fit into the first theoretical plate [450]. If this assumption is not fulfilled and the sample is much larger, it occupies many theoretical plates at the start of elution development. If the chromatographic processes for the ions occupying different plates are—to a first approximation—assumed to proceed independently, it is easy to see that the net elution curve will be a superposition of the curves 'starting' from successive plates and therefore will obviously be broader than the curve obtained with an infinitely small sample. Therefore, the plate height should increase as the sample size is increased. Quantitatively, the problem has still been very little studied.

Glueckauf [445] has derived a relation defining the limiting volume to be occupied in the column by the sample, in which the equations based on the plate concept are still valid, viz.,

$$V_{pr} < \frac{V_z}{\sqrt{2N}} \qquad (6.76)$$

where V_{pr} is the column volume initially occupied by the sample mixture to be resolved, V_z is the ion-exchanger bed volume, and N is the number of theoretical plates.

If approached in broader terms, the sample-size problem involves not only the volume of solution applied to the column but also the number of gram-equivalents of ions to be separated and its ratio to the overal. exchange capacity of the column. In more difficult chromatographic separations, the quantity of ions resolved may be restricted to a fraction of 1% of the overall exchange capacity [74,332]. Overloading the column not only increases the plate height but may also give rise to altered retention volumes [498,499], increased peak asymmetry [498,499], etcl

6.6.4.2 Other Ways to Express the Quality of Separation

Glueckauf [445] has pointed out that, if the effluent is divided into fractions at a point corresponding to a volume

$$U = \sqrt{U_{max(1)} U_{max(2)}} + \frac{2 U_{max(1)} U_{max(2)} (m_1^2 - m_2^2)}{N (U_{max(2)} - U_{max(1)}) (m_1^2 + m_2^2)} \qquad (6.77)$$

where $U_{max(1)}$ and $U_{max(2)}$ are the retention volumes of components 1 and 2, respectively, m_1 and m_2 are the total numbers of gram-equivalents of components 1 and 2 applied to the column, then the contamination of component 2 by component 1 and vice versa (η) is given by

$$\eta = \frac{\Delta m_1}{m_2} = \frac{\Delta m_2}{m_1} \qquad (6.78)$$

where $\Delta m_1 \ll m_1$ and $\Delta m_2 \ll m_2$.

He also presented a diagram (Fig. 6.52) from which, for a known ratio

$$\frac{U_{max(2)} - U_0}{U_{max(1)} - U_0} = \frac{\lambda'_2 + i}{\lambda'_1 + i} \qquad (6.79)$$

and a known mean number N of theoretical plates, the value of the expression $\eta\,(m_1^2 + m_2^2)/2m_1 m_2$ can be read, and this defines the purity of the two fractions collected and thus also the quality of separation.

To study the optimization of the separation process, Giddings [500] introduced the separation function, defined as

$$F = \frac{(U_{max(2)} - U_{max(1)})^2}{8\,(\sigma_2^2 + \sigma_1^2)} \approx \frac{(\Delta U_{max})^2}{16\,\sigma^2} \qquad (6.80)$$

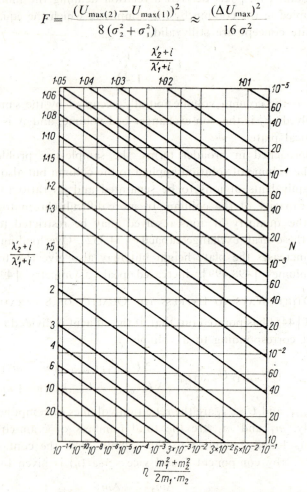

Fig. 6.52 Inter-relation of the number of theoretical plates, product purity, peak-retention-volume ratio, and amounts of the component being separated (from [445], by permission of the copyright holders, the Chemical Society)

which under certain conditions is more convenient than the resolution R_n. As with R_n, separation is considered to be complete when $F \geqslant 1$.

Other quantities have also been used to characterize the separation of two chromatographic peaks, e.g. the valley-to-peak ratio [501] or an analogous parameter [502,503] which can be determined graphically or calculated provided the retention volumes and two peak heights, and the corresponding values for the inter-peak valley are known.

6.7 THE TECHNIQUE OF ION-EXCHANGE CHROMATOGRAPHY

6.7.1 Pretreatment of the Ion-Exchanger

Commercial ion-exchangers are often supplied in batches which are highly non-uniform in particle size, containing a fine-powder fraction which clogs the column, increases flow maldistribution, gives rise to channelling, etc.

As already indicated in Section 6.6.4.1, the quality of chromatographic separation depends not only on the average particle size of the ion-exchanger but also on the size range of the fraction. With medium-fine particle size (> 0.06 mm), a suitable fraction can be produced by sieving, preferably with the aid of a mechanical shaker. If a given batch does not contain enough of the desired fine particles, the (air-dry) ion-exchanger may be comminuted, e.g. in a ball mill. The water-swollen ion-exchanger fraction is stirred well with water, and the supernatant liquid is decanted. This operation is intended to remove the finest particles (fines) and should be repeated until the supernatant is completely clear. Ultrafine-grained fractions are prepared by making use of the different settling rates of particles of different diameters. For this purpose, the ion-exchanger is suspended in water in a tall-form beaker or a cylinder and allowed to settle for a definite time, after which the liquid is decanted [369,503a]. The operation must be repeated several times to yield a relatively uniform fraction. A more accurate technique is to pass a stream of water (or a suitable solution) at a controlled rate upward through the ion-exchanger suspension in a separating funnel or a train of separating funnels or other vessels of increasing diameters [179,466,504] (Fig. 6.53).

The expression relating the volume flow-rate, B (cm^3/min), to particle radius r and maximum vessel radius R (cm) is

$$\frac{B}{\pi R^2} = \frac{120 g (d_1 - d_2) r^2}{9 \eta} \quad \text{(cm/min)} \qquad (6.81)$$

where d_1 is the density of the ion-exchanger particles, d_2 is the liquid density, g is the acceleration due to gravity (981 cm/sec^2), and η is the liquid viscosity.

At a given flow-rate B, the coarsest particles will be retained in the

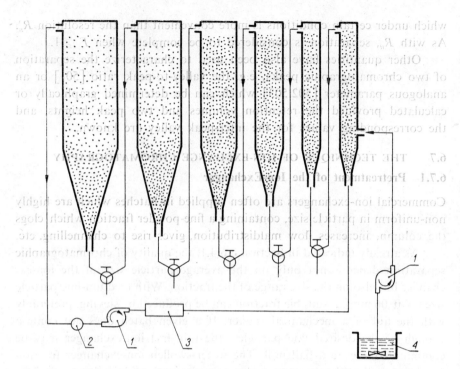

Fig. 6.53 Assembly for separating a batch of ion-exchanger into narrow particle-size fractions: 1 — pump, 2 — flow-meter, 3 — filter, 4 — ion-exchanger suspended in $0.5M$ HCl (from [466], by permission of the copyright holders, Akadémiai Kiadó)

first vessel and less coarse ones will be carried over to the successive vessels in which the linear velocities (cm/min) are lower owing to the larger diameters. After a certain time the liquid flow is discontinued and the ion-exchanger particles are allowed to settle. The fractions thus obtained have superior uniformity and the particle size is usually uniform within ± 5 μm [179, 466].

Generally, commercial ion-exchangers contain some impurities, both ionic and non-ionic in nature, which may be eluted from the resin and may affect the chromatographic process and the determination of the ions in the effluent. Non-ionic impurities include low molecular-weight organic compounds left over from the manufacturing process (unconverted reactants, by-products, etc.) or ion-exchanger degradation products formed on account of elevated temperature, ionizing radiation, oxidants, etc. During the chromatographic process these impurities progressively diffuse outward from the bulk ion-exchanger into the solution. This is one of the reasons that make it undesirable to replace distilled water by demineralized water in the study of the physico-chemical characteristics of surface-active agents, in the preparation of multilayer membrane electrodes [505], etc.

Ionic impurities include iron ions originating from the production reactor walls, and metal ions released from catalysts, etc.

Before being used for analytical purposes, the ion-exchanger must be freed from these impurities by a pretreatment involving alternating washing with $1M$ hydrochloric acid, distilled water, and $1M$ sodium hydroxide, used in excess. For cation-exchangers, especially of the condensation type, the $1M$ sodium hydroxide is often replaced by $1M$ sodium chloride. Ethanol is sometimes also used in the washing. Finally, the ion-exchanger is converted into the desired ionic form. Specially pretreated analytical-purity grade ion-exchangers are available commercially [47]. They are often marked AG (Analytical Grade), e.g. AG 1-X8 and AG 50W-X4 are purified Dowex 1-X8 and Dowex 50W-X4, respectively. AG anion-exchangers contain, for example, $< 5 \times 10^{-4}\%$ of Al, $< 5 \times 10^{-5}\%$ of Fe+Cu, and $< 5 \times 10^{-7}\%$ of Pb. Cation-exchangers are only slightly inferior in quality [47].

6.7.2 Equipment

The equipment used in ion-exchange chromatography is relatively simple. The column may be an ordinary glass burette with a glass-wool or quartz-wool plug at the bottom (Fig. 6.54a) to support the resin bed. Other types

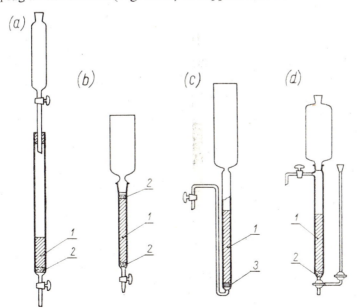

Fig. 6.54 Various types of ion-exchange chromatography columns:
1—resin bed, 2—glass-wool plug, 3—fritted-glass disc

of column are shown in Fig. 6.54b–d. The resin is held in place by a glass-wool plug or a fused-in sintered-glass disc (Fig. 6.54c). In order to pack a column with a water-swollen ion-exchanger, the column is filled with water to a certain level and the ion-exchanger suspended in water is introduced and allowed to settle freely to form the bed. Occlusion of air bubbles must be carefully avoided. The level of liquid in the column must always be kept higher than the bed level. To avoid ingress of air while passing the liquid through the column a compact glass-wool plug (Fig. 6.54b) is sometimes placed at the top of the bed; in routine operations it is convenient to use columns with siphon take-off tubes (Fig. 6.54c). In certain cases, the sorption of ions from an upwards flow, with subsequent elution with a downwards flow proves beneficial; a suitable type of column is shown in Fig. 6.54d. If experimental temperatures other than room temperature are required, thermostatically controlled water-jacketed columns are used (Fig. 6.55).

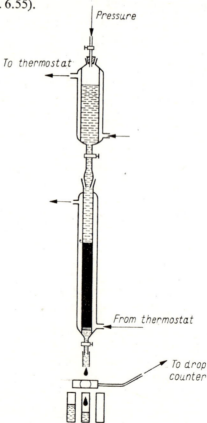

Fig. 6.55 Jacketed column for operation at controlled temperature (from [459], by permission of the copyright holders, Akadémiai Kiadó)

The hydrostatic pressure of the liquid column above the bed is often sufficient to achieve the required flow-rate. If the resistance offered by the bed is too large, a simple arrangement for operation under pressure may be used, or suction may be applied (Figs. 6.56 and 6.57). If microcolumns are used, especially in the separation of radioactive elements, the effluent is conveniently collected drop by drop on a moving paper tape (Fig. 6.57). The drops are dried by infrared radiation. For work with corrosive eluents, paper impregnated with 'Plexiglas' dissolved in toluene or cyclohexanone [300] may be used. Large volumes of effluent are collected with the aid of fraction collectors of various types (Fig. 6.58).

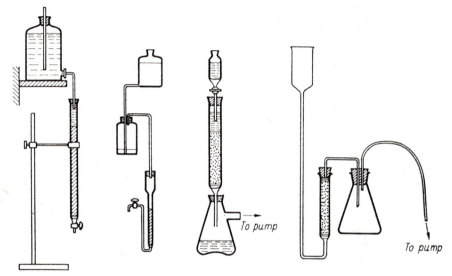

Fig. 6.56 Various types of equipment for increasing or reducing the pressure in the column

In recent years there has been much progress in the automation of the chromatographic process. It is possible to feed the eluent fully or semi-automatically to several columns at a specified rate [506,507]. It is possible to replace one eluent by another, to modify the flow-rate, to control the fraction-collector operation, etc. [507–508a]. Mixtures can thus be resolved automatically into groups of components or into individual components by successive sorption–elution cycles carried out in a single column or in several columns connected in series. The operation of certain sub-units can be programmed electronically [507,508a,509].

A multicolumn assembly has been constructed in which the sample is automatically passed through a series of columns containing various ion-exchangers or extractants supported on a solid carrier, and the comp-

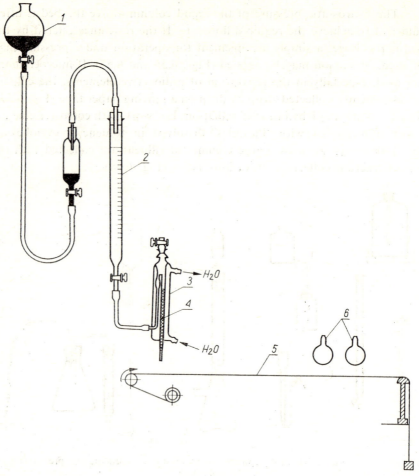

Fig. 6.57 Micro assembly for ion-exchange chromatography: 1 — mercury reservoir, 2 — burette containing eluent, 3 — jacketed column, 4 — resin bed, 5 — paper tape, 6 — infra-red heaters (from [267], by permission of the Inst. Nucl. Res. (Warsaw))

osition and the pH of the individual column feeds are also automatically modified by addition of suitable reagent solutions through inter-column connections. In this manner, a mixture of a large number of ions can be automatically resolved into groups of a few elements selectively retained in individual columns [510,510a]. Such equipment does not require continuous attention, and it can be controlled remotely, a fact of importance for work with radioactive substances. So far this equipment has been used primarily for radiochemical separations.

In contrast to conventional equipment used for routine separations, which should be simple and universally accessible, HPLC requires the use

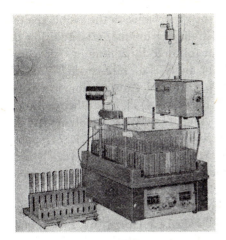 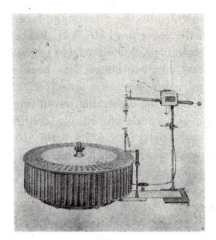

Fig. 6.58 Fraction collectors (LKB-Produkter, Sweden)

of relatively expensive high-precision equipment. Thick-walled glass columns [476–478] or, at very high pressures, stainless-steel columns [473,482,511] are used. The column performance is considerably affected by the smoothness of the internal walls [484]. An interesting development is the steel-embedded glass column which permits operation at pressure exceeding 200 atm [512]. Exacting requirements must also be met by the eluent storage vessels, pumps, feeders, connections, detectors and recorders. For details, the interested reader is referred to the literature [469,471, 472d,486].

6.7.3 Effluent Analysis

So far, ion-exchange chromatography—though a highly valuable tool for preconcentration and separation of traces—has been of much less significance than, for example, gas–liquid chromatography as an instrumental method of analysis.

The effluent is most frequently collected in fractions which are subsequently analysed by suitable methods. It is usual to elute a given element completely in a well-resolved effluent fraction, and then to determine it by any conventional trace analysis method.

Nowadays, continuous effluent analysis is commonly used for analysing highly complicated mixtures of organic compounds by HPLC [468,469]. The detectors usually are based on ultraviolet absorption [512–514] or differential refractometry [492,515]. The earliest fully automated determination of amino-acids by ion-exchange chromatography involved adding ninhydrin to the effluent and measuring the absorbance in the visible region of the spectrum [516,517]. Recently, a similar technique, *viz.*, addition of a colour-producing reagent to the effluent followed by continuous measurement of absorbance, has been used for analysis of rare-earth [477,955] and other inorganic ions [68d,291,518]. Ultraviolet absorption has been used for the determination of inorganic ions [480,481], and other methods have been developed for continuous effluent analysis, involving β- or γ-radiometry [355,473,478,519,520,954], conductometry [476,521,522], polarography [523–525], coulometry [526–528], flame ionization [529], ion-exchange membrane length measurements [530,530a], etc., but few have been widely applied in ion-exchange chromatography.

Recently a novel ion-exchange chromatographic method of analysis known as ion chromatography has been devised [530b,530c,530l–530n]. The method combines the advantages of instrumentation developed for HPLC with a unique combination of two ion-exchange columns, one of which acts as a 'separating column', and the other as 'stripper', which removes the ions of the background electrolyte, suppressing the conductivity of the eluent and leaving only ions of interest as major conducting species in the effluent. This, in turn, permits the use of a flow-through conductivity cell as an universal and sensitive detector for all ionic (both cationic and anionic) species. The conductivity is continuously recorded on a strip chart recorder providing a chromatogram ready for quantification.

Typical combinations of separating and stripper columns and eluents are, e.g.: resin (H^+)–resin (OH^-) (eluent HCl) or resin (Ag^+)–resin (Cl^-) (eluent $AgNO_3$) for analysis of alkali and alkaline-earth cations, and resin (OH^-)–resin (H^+) (eluent NaOH), or resin (phenate$^-$)–resin (H^+) (eluent Na-phenate) for analysis of various anions. This technique has been used with success for rapid, automated determination of Li^+, Na^+, K^+, Rb^+, Cs^+, NH_4^+, Ca^{2+}, Mg^{2+}, F^-, Cl^-, Br^-, I^-, SCN^-, NO_3^-, NO_2^-, IO_3^-, BrO_3^-, SO_4^{2-}, SO_3^{2-}, PO_4^{3-}, CO_3^{2-}, CrO_4^{2-}, organic acids, amines and quaternary ammonium compounds. The samples analysed included surface, potable and rain waters [530b,530f,956], boiler water [530d,530g,530h,957], environmental samples [530e,530f,958,959], biological liquids and fruit juices [530b], brines [530j], nuclear wastes [530k], alloy pickling baths [960] and geological samples [530i]. The method appears to be particularly attractive for the determination of anions, for which there are often no really competitive methods at these concentration levels. Concentrations in the ppm range lend themselves to routine analysis, and with the use of an ion-exchange concentrator, column determinations down to the ppb level can easily be made.

6.7.4 The Elution Development Technique

6.7.4.1 Simple Elution

To achieve a high-quality separation by elution development, it is necessary to add the mixture of elements as a band that is as narrow as possible, at the top of the bed (cf. Section 6.6.3). This not only requires the column to be designed so that the amount of the ions to be separated is small compared with the exchange capacity, but also that the feed is run onto the column in a suitable manner. Three techniques are used to add the sample to the column.

1. The ions to be resolved are sorbed under static conditions onto a small amount of the ion-exchanger which is subsequently transferred to the top of a column packed with a suitable form of the same exchanger. After the ion-exchanger added has settled to form a uniform layer on top of the column, an eluent is added and elution development is carried out in conventional fashion [5,531].

2. The ions to be resolved are obtained in a solution of such a nature that they exhibit high affinity for the ion-exchanger (e.g. retention of multiply charged metal ions on a cation-exchanger from a dilute solution of a simple salt or acid) and this solution is added to the top of the bed. Under these conditions the ions, even if contained in an appreciable volume of solution, form a narrow band at the top of the bed. An eluent is then added (e.g. a buffered solution of a complexing agent) and elution is continued in conventional manner [532].

3. A mixture of the ions to be separated, contained in a small eluent volume, is run onto the top of the bed. The solution is allowed to permeate slowly into the bed, then the column walls are repeatedly rinsed with small portions of eluent, the liquid level each time being run down to the top of the bed. Finally, the column is eluted in conventional manner [74,300,369].

In HPLC, the sample is injected into the stationary-phase bed or into the eluent stream before the inlet to the column by the use of a syringe or a special sample-feeding valve [469,472].

Simple elution involves continuously eluting the sample mixture from the column with the same eluent. Since the chromatographic peak broadens and flattens [cf. Eqs. (6.72) and (6.73)] as the retention volume is increased, elution of elements with very high distribution coefficients takes prohibitively long. Therefore, conventional elution is usually used for mixtures containing at most six components. Examples of simple elution development are shown in Figs. 6.51 and 6.41.

6.7.4.2 STEPWISE ELUTION

Stepwise elution involves eluting a mixture of ions with several eluents arranged in increasing order of eluting power. This technique is particularly useful when the ions to be resolved have very different distribution coefficients. Eluents can be chosen so that each one used will give a sharp peak for one species the others being retained in the column. A classical example of such a separation is the stepwise elution of transition elements from a strong-base anion-exchanger with hydrochloric acid of various concentrations [383] (Fig. 6.59) (cf. Fig. 6.62 for distribution coefficients

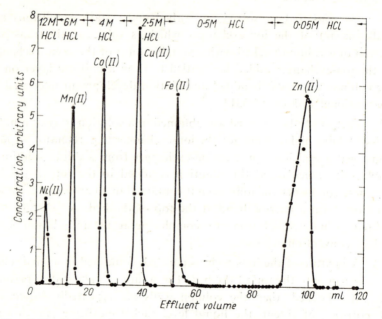

Fig. 6.59 Anion-exchange separation of transition metals by stepwise elution development with HCl solutions. Column 26 cm × 0.29 cm², Dowex 1 (Cl⁻) (200–230 mesh), flow-rate 0.5 cm/min (from [383], by permission of the copyright holders, the American Society for Testing Materials)

in this system). The stepwise elution technique considerably reduces the separation time, and separations that are impracticable by the conventional technique become possible. Generally, the resolution obtained by stepwise elution is not inferior to that obtained by conventional elution, provided the nature and sequence of the eluents have been correctly chosen. An excessively strong eluent used at a particular step may result in rapid migration of several components and deterioration of resolution. In advantageous cases like those in Fig. 6.59, stepwise elution ensures successive

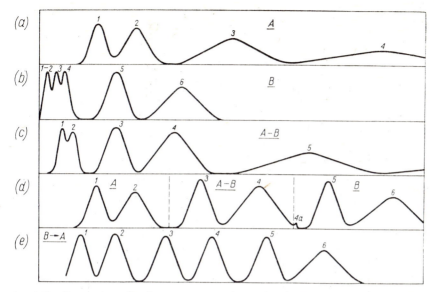

Fig. 6.60 Separation of a six-component mixture by elution development: (a) hypothetical separation by conventional elution with a weak eluent, (b) same with a strong eluent, (c) same with an intermediate-power eluent, (d) stepwise elution, (e) gradient elution (from [534], by permission of the copyright holders, Elsevier Publ. Co.)

elution of each component in a small effluent volume, which is of particular importance if the fractions are to be analysed by certain analytical methods. This version lends itself well to routine determinations. Selective elution can be carried out not only with the same eluent species used at various concentrations but also with eluents of different ionic compositions, and also complexing agents, non-aqueous and mixed solvents, etc. On the other hand, the method is less suitable for the analysis of unknown mixtures, because spurious peaks are likely to occur (cf. Fig. 6.60d, peak 4a).

6.7.4.3 Gradient Elution

The eluent concentration or, generally, composition, may be varied gradually, rather than abruptly, in regular manner by the use of relatively simple equipment (Fig. 6.61).

The earliest and most popular is the equipment presented in Fig. 6.61a,b, producing an exponential (convex) pattern of eluent concentration as a function of effluent volume. If the initial eluent concentrations in vessels R and M are $C_{(R)}$ and $C_{(M)}$, respectively, then the general expression [533] for the eluent concentration gradient is

$$C = C_{(R)} - (C_{(R)} - C_{(M)}) e^{-v/V_M} \tag{6.82}$$

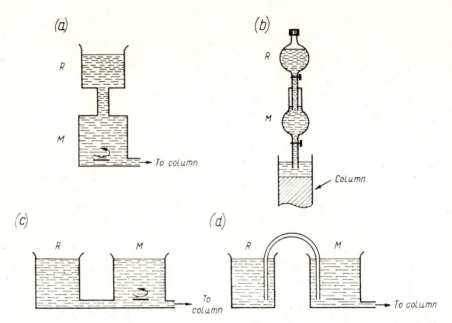

Fig. 6.61 Equipment for producing eluent-composition gradients (from [534], by permission of the copyright holders, Elsevier Publ. Co.)

where U is the effluent volume and V_M is the volume of the solution in the vessel M.

The arrangements shown in Fig. 6.61c,d are gradient vessels of proportional volume: they make it possible to produce convex, linear, or preferably (according to recent opinion [534]) concave concentration gradients. In the simplest case, when the cross-sectional areas of the vessels R and M are identical and the initial concentration in vessel M is $C_{(M)} = 0$, a linear concentration gradient is obtained [533,535]

$$C = \frac{C_{(R)} U}{2 V_M} \tag{6.83}$$

A schematic comparison of the separation of a hypothetical six-component mixture by simple elution with eluents of various powers, by stepwise elution and by gradient elution, is shown in Fig. 6.60. Owing to the reduced effluent volume and sharper peaks (nearly identical peak widths are obtainable for each component), gradient elution allows the elution time to be reduced considerably without sacrificing quality of separation. At the same time spurious peaks, possible in stepwise elution, are unlikely to occur. The retention volumes of the ions separated by

§ 6.7] Techniques 411

gradient elution development can be calculated from the equation [533, 536, 537]

$$\int_0^{(U_{max}-U_0-V)} \frac{dU}{m_j \lambda \{U\}} = 1 \qquad (6.84)$$

where $\lambda\{U\}$ is the distribution coefficient of a component in gradient elution development [for other symbols see Eq. (6.51)].

Calculation of the retention volume from this formula is possible if the dependence of $\lambda\{U\}$ on effluent volume is known. Examples of such calculations can be found in the literature [537–540]. Gradient elution was first used in chromatography for cases when tailing of peaks occurred as a result of non-linearity of the adsorption isotherm or of a low adsorption rate of the given substance [541]. Today the possibility of separation of multicomponent mixtures in a single operation in a relatively short time is regarded as the main advantage of gradient elution.

Giddings [542] proposed a formula for the maximum number of components n of a mixture which can be completely separated by simple elution, with the assumptions that the number of theoretical plates is constant and the peak width $m\sigma$ is related to the retention volume by the equation

$$\frac{U_{max} - U_0}{m\sigma} = \frac{\sqrt{N}}{m} \qquad (6.85)$$

[cf. Eq. (6.50)].

The equation has the form:

$$n = 1 + \frac{\sqrt{N}}{m} \ln \frac{U_{max(k)}}{U_{max(1)}} \qquad (6.86)$$

where $U_{max(1)}$ and $U_{max(k)}$ are the retention volumes of the first and last component, respectively.

An analogous formula [543] for finding the maximum number of components which can be separated by gradient elution, assuming equal width of all peaks, has the form:

$$n' = \frac{\sqrt{N}}{m} \left(\frac{U_{max(k)}}{U_{max(1)}} - 1 \right) \qquad (6.87)$$

Comparison of Eqs. (6.86) and (6.87) shows that application of gradient elution increases the number of components which can be separated, by the factor

$$F = [(U_{max(k)}/U_{max(1)}) - 1]/\ln(U_{max(k)}/U_{max(1)})$$

or reduces the number of theoretical plates necessary for separating the given number of components by a factor of F^2.

In ion-exchange chromatography gradient elution, positive and negative electrolyte concentration gradients and pH gradients have been used for several applications including separating halide ions [537], polyphosphates [544], rare-earth elements [419,476,477,545–548], alkaline earths [549,550], etc. Temperature gradients [550a] can also be used in ion-exchange chromatography. Gradient elution chromatography has been discussed in detail by Liteanu [497a,550b].

6.8 APPLICATIONS AND EXAMPLES

6.8.1 Preconcentration of Trace Amounts of Elements

Most conventional instrumental-analysis techniques such as spectrography, spectrophotometry, polarography, are not sensitive enough to be used for direct determination of most elements at concentrations lower than $10^{-5}\%$. Furthermore, even if a given element is theoretically determinable, the matrix element(s) may often have an adverse effect on the accuracy and precision of the determination. Preconcentration, and separation of the trace elements from the matrix, represents the best and very often the only practicable approach [551–553].

The ion-exchange technique is capable of selectively retaining trace amounts of an element in a small column from solution volumes as large as hundreds of litres. The element retained can then be eluted with a small volume of eluent. The eluate volume can be reduced further by at least one order of magnitude by evaporation. The enrichment factors (for preconcentrations of elements in solution) thus obtained are of the order of 10^3–10^5. If the limiting concentration that can just be determined by a given method is, e.g. $10^{-6}M$, then a preconcentration step involving ion-exchange can shift the limit of determination to concentrations as low as 10^{-9}–$10^{-11}M$.

In ion-exchange separations of trace constituents from matrix constituents, enrichment factors considerably exceeding 10^5 are readily achieved [554–558].

In typical radiochemical separations, the decontamination factors for trace activities with respect to the matrix activity may often exceed a factor of 10^6 [463,559,560].

Ion-exchange preconcentration of trace elements is usually performed as a column or dynamic process, though occasionally static (batch) processes may be used [561–564]. It should be borne in mind, however, that in theory static conditions preclude complete separation of the elements determined (cf. Section 6.4). The static method is justified only when, under the experimental conditions, the trace elements to be separated have a very high affinity for the ion-exchanger. Recently ion-exchange membranes have been used in static [565,566] and in column processes [567] for preconcentration of trace elements.

To recover trace elements successfully from large volumes of solution or to separate micro from macro constituents and also to resolve traces one from another, it is necessary to choose appropriate ion-exchange-solution systems in which either the elements to be separated exhibit very high affinities for the ion-exchanger or the ion-exchange affinities of the ions to be resolved differ quite considerably from one another. The collections of data on the distribution coefficients for nearly all the elements are very helpful for designing separations. Selected data on equilibrium distribution of elements between a strongly basic anion-exchanger and hydrochloric acid [379,568], nitric acid [382,568], hydrofluoric acid [380,381] solutions, and also between a strongly acidic cation-exchanger and hydrochloric [281], perchloric [281,569], hydrobromic [377,570], and hydrofluoric acid [381] solutions, are reproduced in Figs. 6.62-6.68. The analogous table presented in Fig. 6.69 shows changes in ion affinities for an inorganic ion-exchanger (zirconium phosphate) in relation to the pH of the solution [571].

Less complete data covering groups of 10–20 or more than 20 elements have been plotted or tabulated for systems involving a strongly basic anion-exchanger in sodium hydroxide [572], hydrobromic acid [368,572a], acetic acid [573–575], hydrochloric–hydrofluoric acid [576], acetic–hydrochloric acid [577], oxalic—hydrochloric acid [578], nitric–hydrofluoric acid [579], sulphuric–hydrofluoric acid [579], sulphuric acid [568,580,591], phosphoric acid [581,582,582a], magnesium nitrate [961], nitrites [583], ammonium thiocyanate [962], formic acid [584], oxalic acid [585], tartaric acid [586], hydrazoic acid [586a], sodium chloride [586b], sodium carbonate [586b], sodium sulphate [586b], hydrobromic–nitric acid [586c], and EDTA [369,370] solutions and for systems involving a strongly acidic cation exchanger in hydrochloric acid [343], nitric acid [344], sulphuric acid [579], formic acid [584,808], tartaric acid [586], nitrites [583], nitric–hydrofluoric acid [579], sulphuric–hydrofluoric acid [579], hydrochloric–perchloric acid [963] and ammonium sulphate–sulphuric acid [587], solutions; distribution coefficients have been determined for a number of elements in systems involving a weakly basic anion-exchanger in hydrochloric [600], sulphuric [588], and hydrobromic acid solutions [589].

Distribution coefficients have also been reported for a number of elements on a strongly acidic cation-exchanger in equilibrium with mixtures

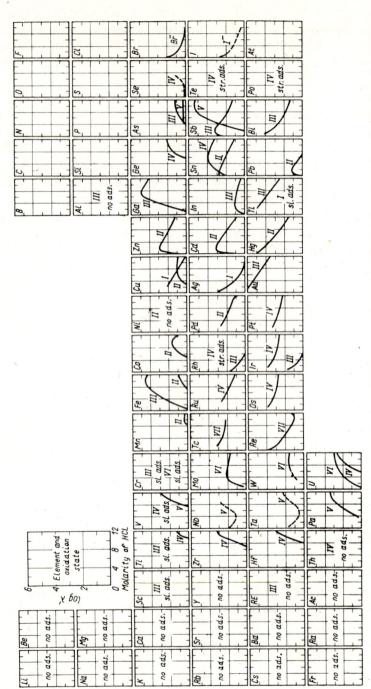

Fig. 6.62 Bed-distribution coefficients of elements on Dowex 1-X10 anion-exchanger in HCl solutions (no ads.—λ' close to zero over [HCl] range 0.1–12M; sl. ads.—slight affinity for the exchanger, $0.3 \le \lambda' \le 1$; str. ads.—high affinity for the exchanger, $\lambda' \gg 1$) (from [379], by permission of the copyright holders, I.A.E.A., Vienna)

§ 6.8] Applications 415

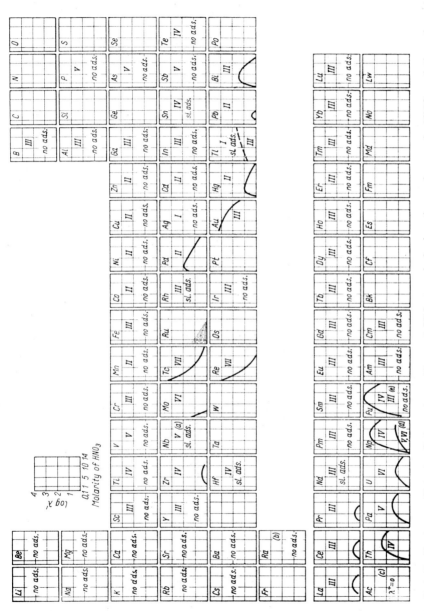

Fig 6.63 Bed-distribution coefficients of elements on Dowex 1-X10 strongly basic anion-exchanger in HNO_3 solutions (no ads.—no adsorption from 0.1-14M HNO_3; sl. ads.—slight affinity for the exchanger) (from [382], by permission of the copyright holders, the American Chemical Society)

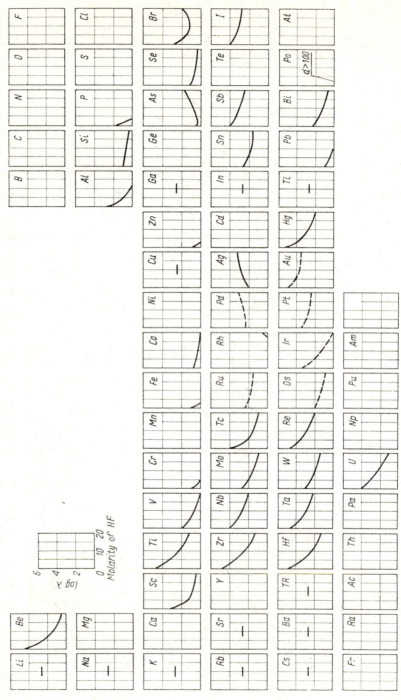

Fig. 6.64 Weight-distribution coefficients of elements on AV-17 X14 anion-exchanger in HF solutions (continuous horizontal line—λ close to zero, dashed line—sorption of chloride complexes from HF solutions) (from [381], by permission of the copyright holders, Akad. Nauk **SSSR**)

§ 6.8] **Applications** 417

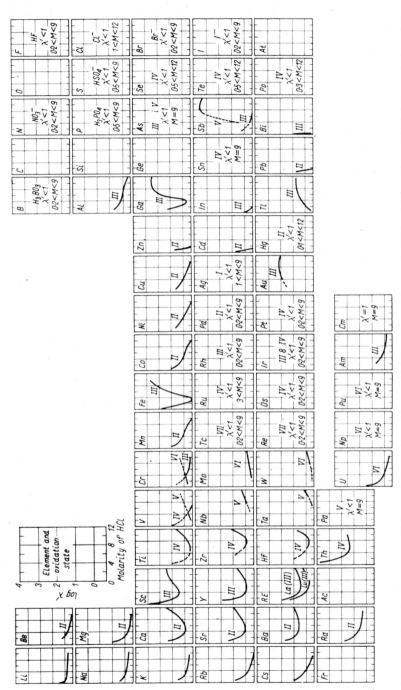

Fig. 6.65 Bed-distribution coefficients of elements on Dowex 50-X4 cation-exchanger in HCl solutions (from [281], by permission of the copyright holders, Elsevier Publ. Co.)

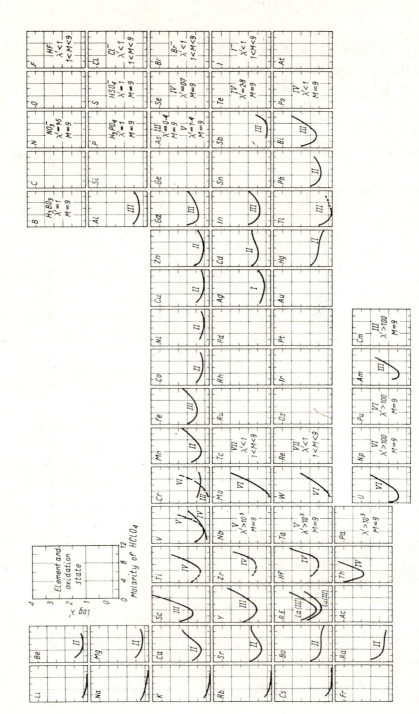

Fig. 6.66 Bed-distribution coefficients of elements on Dowex 50-X4 cation-exchanger in HClO$_4$ solutions (from [281], by permission of the copyright holders, Elsevier Publ. Co.)

6.8] Applications 419

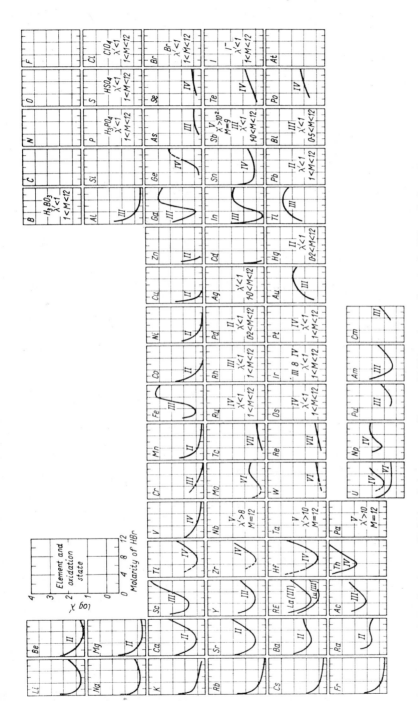

Fig. 6.67. Bed-distribution coefficients of elements on Dowex 50-X4 cation-exchanger in HBr solutions (from [377], by permission of the copyright holders, Elsevier Publ. Co.)

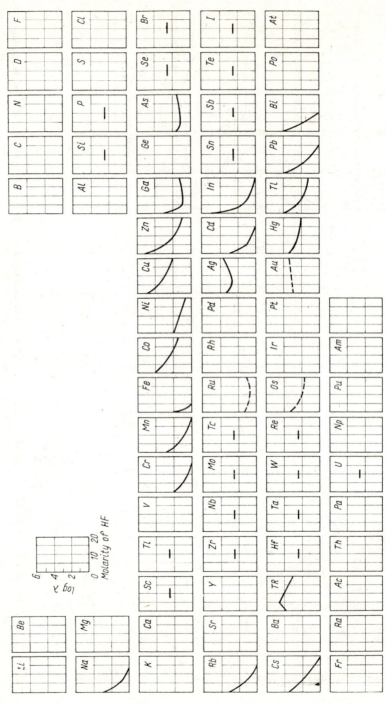

Fig. 6.68 Weight-distribution coefficients of elements on KU-2 X6 cation-exchanger in HF solutions (continuous horizontal line—$\lambda = 0$, dashed line—sorption of chloride complexes from HF solutions) (from [381], by permission of the copyright holders, Akad. Nauk SSSR)

§ 6.8] Applications 421

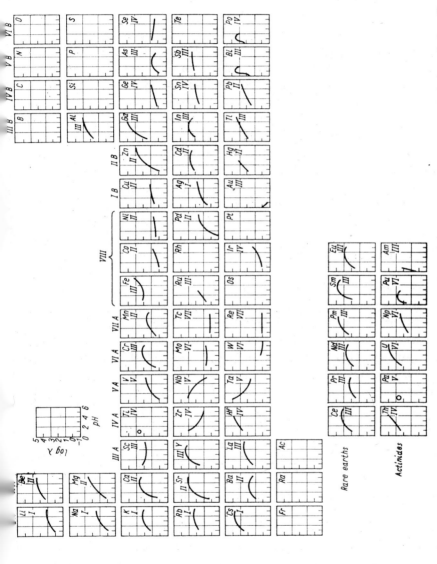

Fig. 6.69 Weight-distribution coefficients of elements on zirconium phosphate as a function of pH of the solution (from [571], by permission of the copyright holders, the American Chemical Society)

of hydrobromic acid with acetone [589a], mixtures of hydrochloric acid and alcohol [412,590], tetrahydrofuran [412], acetone [402,412,592], acetic acid [412], and organic solvent+cupferron [593]); mixtures of hydrofluoric acid with alcohols [594], 1,2-propanediol (methyl glycol) [594], acetone [594], tetrahydrofuran [594], and acetic acid [594]. Distribution coefficients for many elements have also been measured for the systems of strongly basic anion-exchanger in equilibrium with mixtures of hydrofluoric acid and alcohols [410,411,413], glycols [411], formic and acetic acid [411,413], acetone [411,413], tetrahydrofuran [411,413], N,N-dimethylformamide [413], dioxan [413], and pyridine [413]; mixtures of nitric acid with alcohols [410]; mixtures of hydrobromic acid with alcohols [595,596], 1,2-propanediol and 1,2-butanediol [595,596], acetone [595,596], tetrahydrofuran [595,596] and acetic acid [595,596]. Distribution coefficients have been published for various metals on DEAE cellulose in equilibrium with hydrochloric acid–methanol [597], nitric acid–methanol [598], and acetic acid–nitric acid solutions [599].

Interestingly, the distribution coefficients of numerous elements on a weakly basic anion-exchanger in hydrochloric acid solutions [600] differ considerably from those on strongly basic anion-exchangers [379,568]. Similar behaviour has been observed for hydrobromic acid solutions [368,589,601].

The literature also reports retention data for numerous ions on materials conventionally termed inorganic ion-exchangers, although in some of them the mechanism of retention most likely fails to be rigorously ion-exchange in nature. These data cover, e.g. hydrous and anhydrous manganese oxide [122], hydrous antimony pentoxide [122], acidic alumina [122], cerous oxalate [122], cuprous chloride [122], various metal sulphides [122,602], zirconium phosphate [122,571], zinc oxide [122,603], zirconium oxide [571], zirconium molybdate and tungstate [571], cerium tungstate [111], various titanium salts [109] and various tungstates [604].

Most of the distribution coefficients mentioned above have been determined by a static method and for trace amounts of elements. Only a few values have been determined at appreciable mole fractions of ions in the ion-exchangers [343,344]. If the literature data are to be utilized for designing macro–micro separations, it must be borne in mind that, as indicated in Section 6.4.2.2, ion selectivities may vary considerably with the composition of the ion-exchanger phase. It must also be taken into account that distribution coefficients determined by the static method may not always unequivocally describe the feasibility of column separation, often because of the unfavourable dynamics of the column process [218,267]. For elements which may exist in solution in different oxidation states or as various complexes (e.g. various isomers), the distribution coefficient determined by a static method represents an average value which is inconclusive as regards the behaviour of the element during its elution from the column [77,459]. Under the prevailing elution conditions, the element may produce two or more maxima or be partly retained by the ion-exchanger [77]. In spite of these limitations the cited values and references to distribution coefficients of the elements in various systems form an invaluable aid in designing separation schemes for preconcentration or mutual separation of elements.

6.8.1.1 SELECTIVE SORPTION OF TRACES

Preconcentration and separation of trace elements by ion-exchange is most conveniently carried out under conditions enabling trace elements to be retained on a small column, and major elements to pass through the column unsorbed by the ion-exchanger. This is possible when the distribution coefficients of the trace elements are large and those of the major constituents are close to zero. Typical examples are the use of a cation-exchanger for separation of trace elements from macro amounts of other metals converted into neutral or anionic complexes, and the retention by an anion-exchanger of trace metals converted into stable anionic complexes in the presence of macro amounts of other metals remaining in the cationic form. Under these conditions, the column acts as a selective filter which sorbs trace ions only and allows major constituents to pass undisturbed into the eluate. Therefore, small columns may be used and the ion-exchanger loadings may be substantially lower than those corresponding to the total number of milliequivalents of ions present in the sample.

The trace elements retained on a column can be recovered as a group by elution with a small volume of a powerful eluent, or they may be resolved into individual constituents by the use of one or more suitably selected differentiating eluents.

Theoretically, the simplest case occurs when the matrix is a pure solvent (e.g. water). Then the ion-exchanger need not have a particularly high selectivity toward the ions retained; it is enough if it exhibits a high capacity and a high rate of exchange. Therefore, strongly acidic cation- and strongly basic anion-exchangers have often been used to sorb ions from dilute aqueous solutions. In dilute solutions almost every ion has a high distribution coefficient and is quantitatively sorbed on an ion-exchanger column.

Potter and Moresby [605] used this method for determining copper and iron ions in distilled water. After a 100–1000-fold preconcentration on a sulphonic acid cation-exchanger in H^+-form, the ions were determined spectrophotometrically in the eluate. A similar method was used to determine traces of lead and copper [606], silver [607] and vanadium [608] in natural waters. The vanadium was determined by neutron activation analysis.

As little as 0·04 μg/l of uranium were determined in natural waters after preconcentration from 1-l samples on an anion-exchanger (in thiocyanate form) followed by neutron irradiation of the resin and delayed neutron counting [608a]; ng/ml amounts of As, Cd and Zn were determined in river water by preconcentration on anion-exchanger at pH = 10 followed by neutron irradiation and Ge(Li) spectrometry [964].

The preconcentration step may often be combined with simultaneous separation from elements which may interfere with the later determination. Preconcentration of elements as chloride, bromide, nitrate or thiocyanate complexes on strong-base ion-exchangers together with selective elution has been used for the determination of, e.g. cobalt [609], cadmium [610], thorium [611], lead [612], cadmium, copper and lead [612a], zinc [613] and copper [614] in natural waters at the ng/ml level. The elements eluted were determined by spectrophotometry or atomic-absorption spectroscopy. Similarly, chromates have been determined by being first sorbed on an anion-exchanger (with the eluent passed upwards through the column), then eluted with an acidified reductant solution [615].

La, Sm, Eu, Dy, Mn, Cu, and Zn were preconcentrated from river water samples on Chelex 100 resin in NH_4^+-form. The resin was then irradiated with neutrons in a reactor, rare earths and transition metal ions were eluted with $1M$ Na_2CO_3 and $2M$ HNO_3 solutions, respectively, and measured by Ge(Li) spectrometry [615a]. A similar method was used for determining Zn and Cd in stream watrr [965].

As, Co, Mo, Zn, Cd, and Hg were isolated from neutron-irradiated freshwater samples on an anion-exchanger in HCl medium, separated by stepwise elution and measured by NaI(Tl) spectrometry. Determination limits ranging from 0·1ng/ml down to 0·001ng/ml, were obtained [615b]. Sub ng/ml amounts of As, Sb, and W were determined in natural water samples by retaining them on acidic Al_2O_3 columns after neutron irradiation and measuring radionuclides on the column by Ge(Li) spectrometry [615c].

Ion-exchange preconcentration of traces has also been widely utilized in radiochemical analysis. Radiocaesium-137 and radiostrontium-89 and -90 originating from fall-out have been determined in rainfall water, surface water, thawed snow, etc., by being sorbed on Dowex 50W, H^+-form and separated by elution with buffered ammonium glycolate solutions [616]. Dowex 50W and Dowex 1 have also been used for the rapid determination of the radionuclides ^{13}N and ^{18}F which are formed by nuclear reactions occuring in the water of the cooling system of a nuclear reactor [617]. ^{137}Cs in sea water was determined by retaining the radionuclide from a 10-l sample on a small column filled with potassium hexacyanocobalt(II) ferrate(II) and measuring it directly on the column by NaI(Tl) or Ge(Li) spectrometry. The detection limit was 0·1 pCi/l [617a].

Under suitable conditions ion-exchangers can also be used for determining air-borne pollutants. A method developed for determining air-borne fluoride involves passing sampled air through an anion-exchanger, eluting the exchanger-sorbed fluoride with aqueous 0·5% sodium chloride and determining the fluoride in the eluate by photometry [618]. This method can measure air-borne fluoride concentrations of 1–30 ng/ml.

Quantitative sorption of trace elements by an ion-exchanger in the presence of macro amounts of matrix ions is a more complicated task than sample preconcentration of ions from dilute solutions. In this case the ion-exchanger must have high selectivity towards the trace ions and at the same time very slight selectivity towards the matrix ions. Chelating ion-exchangers have been extensively used for this purpose.

Microgram quantities of Zn, Cu, Ni, Co and Cd have been separated, e.g. from sea-water, by the use of Dowex A-1 and, after elution with 2M hydrochloric acid, determined by atomic-absorption spectroscopy [286]. A similar method has been used for determining traces of Cu, Pb, Zn, Cd, Ni and Fe in industrial waters [287]. Dowex A-1 has also been employed for sorbing V and Mo traces from sea-water; these were then eluted with $2M$ ammonia solution, and determined photometrically [619]. Recent studies have revealed, however, that during the preconcentration of e.g. Zn, Cd, Pb and Cu from sea-water by the use of such ion-exchangers, separation is not quantitative, presumably owing to the formation of chelates of the metals, adsorption of ions on the surface of colloidal particles etc. [619a]. After destruction of organic substances, complete separation was readily obtained [620].

Some refinements in the methodology of isolating trace elements on a chelating resin followed by their determination by atomic-absorption spectroscopy [620a] and neutron-activation analysis [620b] have been published recently. Chelating ion-exchangers have been used for isolation of Mn, Fe, Zn, Al, Ti, Cd, Co, Ni, Cu, Ag, Ta, Bi, Ba, Sr, Ca and Mg from aqueous 18–19% sodium chloride solutions [621]. Similarly Cu, Mn, Ca, Al, Mg, Ag and Fe have been determined in concentrated lithium chloride solutions [622].

Trace impurities (Ni, Co, Cu, Fe, and Zn) in Mo and W have been determined by converting the matrix metals into their fluoride complexes and sorbing the impurities on a sulphonic-acid cation-exchanger. The traces were eluted with hydrochloric acid, then separated by stepwise elution from an anion-exchanger and determined by spectrophotometry and anodic-stripping voltammetry. The limit of determination in a 0·5-g sample is 0·5–5 ppm [623].

The high affinity of chloride complexes of certain metals for strong-base anion-exchangers has been utilized for the determination of Au, Tl, Bi and Cd in sea-water. As much as 250 litres of sea-water, slightly acidified with hydrochloric acid, was passed through a Cl^--form Amberlite IRA-400 column (13·2 cm $\times$ 0·5 cm^2). The ion-exchanger was washed and ignited and the residue was analysed spectrochemically. In this procedure the enrichment factor was as high as 2×10^7 [554]. In a similar manner, thallium was determined in sea-water and in sea-bottom sediments [555]. Traces of gold in rinse waters usually discharged as waste have been recovered by sorption on condensation-type Cl^--form weakly basic anion-exchangers [624,625]. Gold was partially reduced and could be recovered only by igniting the ion-exchanger. Again, when strong-base Cl^--form anion-exchangers were used for separation of traces of gold and other noble metals [626] and the determination of traces of gold in copper [562], quantitative elution of gold was impracticable and the ion-exchanger had to be ignited. In the determination of gold in minerals by neutron activation analysis, gold from a 5-g mineral sample was sorbed on a column of a nearly stoichiometric amount of the Cl^--form AV-17 anion-exchanger [627]. Radiochemical purity was thus enhanced without sacrificing the high yield of gold (90–98%). The use of a scintillation spectrometer for activity measurements enabled as little as 10^{-8}% of gold to be determined.

In ore, alloy and similar analyses, silver has been sorbed on an electron ion-exchanger EO-7, in which it was reduced to the metal, and later eluted with $3M$ nitric acid [628]. Preconcentration is feasible from solutions even as dilute as $< 10^{-6}M$. Silver from industrial waste waters can be recovered by sorption on an inorganic ion-exchanger, potassium hexacyanocobaltate(II)–hexacyanoferrate(II) [629]. Trace amounts of Pt(IV) and Pd(II) have been separated from common metals (Fe, Co, Ni, Cu, Zn and Pb) by sorption from thiocyanate solutions on a weekly basic DEAE cellulose anion-exchanger [630].

Microgram quantities of Sn in copper have been determined by atomic absorption after separation on Dowex 1 (Cl^-) anion-exchanger [631]. Thorium has been separated from uranium and long-lived fission products on AV-17 anion-exchanger [632]. Anion-exchangers in the fluoride form have been used for preconcentrating and isolating traces of boron from hydrofluoric acid [633] and also for separating beryllium from aluminium and iron [634].

The strongly acidic cation-exchanger Dowex 50W-X8 (H^+) has been used to separate cations from silicate rocks. The rock was fused with K_2CO_3 and H_3BO_3, the powdered melt was brought into contact with the cation-exchanger in the presence of water, as a result of which it dissolved and the trace elements were separated quantitatively together with some major elements, usually in the form of cations, from macro amounts of Si and B [635]. The cations retained on the resin were eluted with HCl solution and determined by emission-spectrography or atomic-absorption methods. Trace amounts of caesium were separated from macro amounts of sodium and potassium on a phenolsulphonate cation-exchanger in the course of the analysis of mineral salts [556]. The caesium separated was determined by neutron-activation analysis. It was shown that amounts of caesium of the order of 10^{-8} g and less can be separated quantitatively from 0·5 g of KCl or NaCl on the phenolsulphonate cation-exchanger, which shows a chelating activity towards Cs^+ ions. Strongly acidic cation-exchangers have also been used for separating traces of caesium and rubidium [636], and traces of barium and strontium [637] from sea water. Traces of calcium [638] and copper [639] were separated from solutions of alkali metal salts on carboxylate cation-exchangers. A mixture of sulphonate and carboxylate cation-exchangers in hydrogen form was used for separating traces of impurities in arsenic and arsenic trichloride. In weakly acidic solutions (pH = 1·0–1·7) the weakly dissociated arsenic acid is not retained by the cation exchangers, and this facilitates separation of the matrix from the impurities (Cu, Pb, Bi, Co, Cd, Mg, Fe, Cs, Ni, Zn, Al, Mn, Ca, Sr, Ba, Au). The preconcentrated trace elements were determined spectrographically [640].

Trace quantities of plutonium were separated from gram amounts of uranium and long-lived fission products on zirconium phosphate [641]. Micro amounts of light rare earths (La–Eu) in gadolinium oxide and in metallic gadolinium were determined spectrophotometrically after initial enrichment on Amberlite IRA-400 (NO_3^-) in aqueous methanol solutions of ammonium nitrate [419]. Separation on anion-exchangers in similar systems was also used for determining traces of Lu, Yb and Tb in rocks by neutron-activation analysis [642]. Rare earths in rocks and other matrices ng/g levels were determined by mass spectrometry, after concentration and separation from the remaining elements on AG 50W-X8 ion-exchanger [643]. Cation-exchange chromatography was also used in the deterrminination of Zn and Pb by atomic absorption [966] and rare earths in rocks by neutron activation [967]. Noble metals and mercury (including methylmercury) were separated on a chelating ion-exchanger with guanidine groups (Srafion NMMR) [69–72,644]. This method was used in activation analysis of rocks, samples of moon soil, etc.

In photon-activation analysis of biological materials Cr, Mn, Co, Ni, Zn, and Pb were isolated from irradiated samples on anion-exchange resin in the form of tropolone–5–sulphonate complexes [968]. Ag, Hg, Sb, Pb and Cd were concentrated on ion-exchangers with dithiocarbamate groups [645]. Traces of germanium were separated from solutions of zinc sulphate on pyrogallolformaldehyde resin [646]. Beryllium can be separated selectively from a number of elements, including large amounts of aluminium and alkali metals, on Dowex 50W in a dimethylsulphoxide–H_2O–NH_4SCN solution [647].

Many elements such as Be, Ga, Sn, Ti, Mo, W, Th and Cl can be retained selectively on resins obtained by formaldehyde and pyrocatechol or pyrogallol condensation [648]. Trace amounts of Fe, Ca, Zn, Al, Pb, Cd, Ni, Ba, Be, Bi, Mn and Mg were determined in platinum and Pt–Rh alloys by converting the platinum metals into anionic chloride complexes and retaining metals on Dowex 50W-X8 (H^+). After elution with $4M$ HNO_3, the trace elements were determined spectrographically or spectrophotometrically [649].

The high selectivity of some inorganic exchangers towards caesium ions was first utilized for preconcentrating and then determining ^{137}Cs in milk [89,651], sea water [90,651,653,969], radioactive wastes [650] and other materials [651,652].

Sulphonate and phenolsulphonate cation-exchangers were also often used, together with appropriate complexing agents, for separation and determination of ^{90}Sr and other radioactive nuclides in milk, sea-water and other samples from the environment [653–657].

6.8.1.2 SELECTIVE RETENTION OF THE MATRIX

Though this method of separating the matrix from trace elements requires the use of bigger columns and larger volumes of solution, it may be the best procedure for a given set of elements. An additional advantage of this method is that the separated traces are obtained in one operation directly in solution. It is best, of course, when the matrix elements are strongly retained by the ion-exchanger, and the trace elements to be determined have a partition coefficient close to zero. In practice, however, it often happens that initially the ion-exchanger retains both the matrix and the trace elements which must then be eluted individually or in groups with appropriate eluents.

Microgram quantities of gold were separated from macro quantities of iron, copper and nickel by retaining the latter on the cation-exchanger Dowex 50W-X8 from a solution of chlorides of concentration smaller than $0.05M$ and pH ~ 1.5. In these conditions gold passes quantitatively into the effluent where, after evaporation and decomposition of any remaining organic compounds, it can be determined spectrophotometrically [658]. This method was used for determining gold in ores.

Trace quantities of rare-earth elements were separated from milligram quantities of uranium by retaining uranium on the weakly acidic cation-exchanger Amberlite IRC-50 (Na^+) from sodium EDTA solutions of pH ≈ 5.9 [659]. The rare earths in the effluent were determined spectrographically.

Trace quantities of rare earths may also be separated from gram quantities of uranium by sorbing uranium on Dowex 1-X8 from a solution of acetic acid (90%)–HCl (10%) [660]. Rare-earth elements may then be determined in the effluent by activation analysis, emission spectroscopy or spectrophotometry. Manganese at the μg/g level in high purity iron was determined by activation analysis, by retaining iron, before irradiation, on Dowex 1-X8 exchanger, from $6M$ HCl [661]. This made it possible to eliminate interference by the reaction $^{56}Fe\,(n,p)^{56}Mn$ caused by fast neutrons. Light rare-earth impurities in pure heavy rare-earth samples were determined by retaining the matrix on Dowex 1-X4 anion exchanger from aqueous methanol solutions of α-hydroxyisobutyric acid [420]. The light rare earths in the effluent were then determined spectrographically.

Microgram quantities of rhodium in plutonium–uranium alloys were determined by retaining the matrix elements and Pd and Mo from $8M$ HCl on Dowex 1-X10 anion-exchanger. Rhodium passed to the effluent and could be determined spectrophotometrically [662]. Palladium, platinum, rhodium and iridium were separated from macro quantities of common metals by retaining the latter from chloride solutions of pH ~ 1.5 on Dowex 50-X8 cation exchanger [663,664]. Platinum metals were determined spectrophotometrically in the eluate. In this way μg/g levels of platinum metals could be determined in iron and rocky meteorites. An example of such a procedure, in which both the trace elements and the matrix elements are retained on the column and subsequently stripped with appropriate eluents, some of the trace elements being eluted before and some after the matrix, is the separation of trace quantities of Lu, Dy and Ho from macro quantities of Er in the form of EDTA complexes on Amberlite IRA-400 (H_2Y^-) anion exchanger while determining traces of other rare earths in spectrally pure Er_2O_3, by activation analysis [665]. Chromatographic separation of traces of La from macro quantities of Pr in the Dowex 1-X4 (H_2Y^{2-})–Na_2H_2Y system in the determination of lanthanum by activation analysis made it possible to avoid systematic errors which arise in determinations by purely instrumental methods [666].

Trace quantities of Ni, Fe, Cu, Co and Pb were separated from macro quantities of Mn in a similar manner, by retaining the chloride complexes from an HCl–propan-2-ol solution on Dowex 1-X8 anion-exchanger, and subsequent stepwise elution of constituents with hydrochloric acid–propan-2-ol mixtures of various concentrations [667].

In another procedure the matrix and trace elements are retained on the column, then the column is treated with concentrated HCl. Under these conditions, macro quantities of such elements as Na, K, Ba, Sr and Ag precipitate as chlorides which are insoluble in concentrated HCl and remain on the column, while many trace elements pass to the effluent with good yield [668].

In activation analysis, in which the degree of matrix interference during trace determination depends not so much on the absolute matrix element concentration in the sample as on the intensity of its induced radioactivity, one can use 'isotopic ion exchange'. In this method the irradiated sample, after dissolution, is passed through an ion-exchanger saturated with non-radioactive ions of the element constituting the matrix of the analysed sample. The composition and concentration of the solution must be chosen so that the trace elements pass easily to the effluent, while the radioactive ions of the matrix elements are exchanged for nonradioactive ions of the same element. This generally allows relatively easy measurement of the activity of the trace elements by γ-spectrometry. This method was used, e.g., for determining traces of Ag, V, Ca, Rb and Al in magnesium after separating ^{27}Mg from a 75% acetone and $0\cdot 3M$ HCl solution on Dowex 50W-X8 cation-exchanger [402].

6.8.2 Separation of Traces

Separation of elements present in trace quantities is usually easier than macro–micro separations. For small quantities of ions the distribution isotherm is linear and the elution curves are symmetrical (section 6.6.3). Since the quantities of ions separated are small, it is possible to use a small amount of exchanger and to carry out the separation in a short time. Depending on the composition of the initial mixture, and on the methods available for determination of the elements involved, in some cases separation into groups may be sufficient although sometimes complete separation of the individual elements may be required. In the course of the systematic

Applications

resolution of one sample into many components, often known separation schemes are used. Such schemes entail separation of the trace elements from the matrix, separation into groups, and finally separation within the groups with isolation of the individual components.

6.8.2.1 SEPARATION INTO GROUPS

As already mentioned, ion-exchange chromatography is particularly suitable for separating small quantities of elements, owing to its simplicity, high selectivity, and also because ion-exchange of elements does not depend on the presence of other elements in the sample. In the literature, many schemes for separation of traces have been described; often they depend solely on ion-exchange chromatography, or ion-exchange chromatography may play the primary role. The number of elements separated

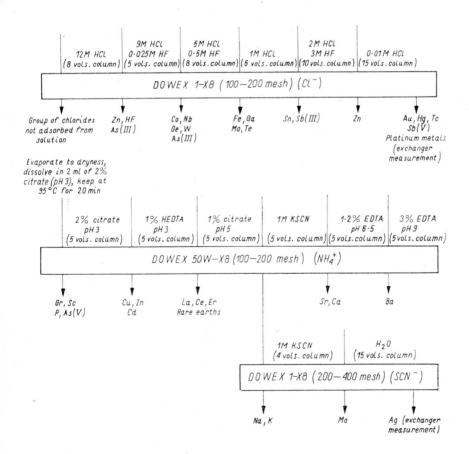

Fig. 6.70 Scheme for separation of several trace elements into groups by ion-exchange (from [669], by permission of the copyright holders, I.A.E.A., Vienna)

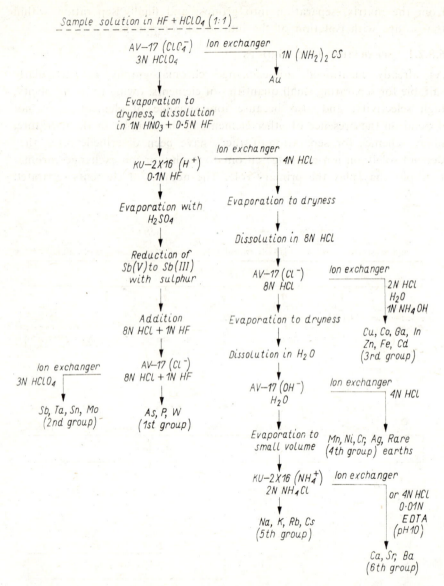

Fig. 6.71 Scheme for separation of a range of trace elements into groups by ion-exchange (from [670], by permission of the copyright holders, Izd. Nauka, Moscow)

ranges from a few to several dozen and the quality of separation is usually good. Separation of a complex mixture into groups normally takes from one to several hours, but in some particular cases the time can be much shorter. In most cases separation involves successive cycles of adsorption and elution, using a number of columns with ion-exchangers in different ionic forms. The solution fed to each successive column is usually obtained by evaporating the effluent from the preceding column to dryness and dissolving the remainder in an appropriate eluent. The group separation diagram has often been extended to obtain separation into individual components. Figures 6.70 and 6.71 illustrate two purely ion-exchange separations of complex trace-element solutions into multicomponent groups. The first one (Fig. 6.70) was used for determining trace quantities of element in biological materials by activation analysis and γ-spectrometry. The second one (Fig. 6.71), also developed for use with activation analysis, was used for trace analysis of very pure silica. Again, the determination of the elements was done by γ-spectrometry. However, this method was not free from problems; to improve its sensitivity, precision and accuracy, ion-exchange methods for separating the groups into individual components [671–674] have been elaborated, and these led the Geiger-Müller counter to be used for the final measurement of activity. Among other purely ion-exchange methods are: systematic separation of Na, Mg, Ca, Al, Ln, Ni, Th, Cr(III), V(IV), Pb, Sc, Mn, Ti, As(V), Fe(II), Zr, Se(IV), Cu, As(III), Cu(II), Fe(III), U(VI), Zn, Cd, Bi, Sn(II), Tl(III) and Hg(II) into four groups on Dowex 1-X8 (Cl$^-$) by elution with HCl solutions of different concentrations [675], and separation of 35 radioactive metallic elements into six groups by elution from a Dowex 50W-X8 (H$^+$) column with solution of complexing agents of appropriate pH and ionic strength [676]. The last two separation schemes should be regarded as exceptions, because stepwise elution from a single column is used. In high temperature alloys, Ni, Cr, Co, Fe, Ti, W, Mo, Nb and Ta were determined, by first separating the elements into groups and then into individual components on anion-exchange columns by using hydrochloric acid and hydrofluoric acid solutions as eluents [677].

Hydrochloric and hydrofluoric acids, solutions of their ammonium salts, and HNO_3 and $HClO_4$ were used as eluents in the trace element separation process of some elements on columns of cation-exchanger Dowex 50W-X12 and anion-exchanger Dowex 1-X10, when the following 6 groups of elements were obtained: (1) W, Mo, Sn, Bn, Hg, Ta, Re, Pt, Au, (2) In, Zn, Cd, (3) Na, K, Rb, Cs, Mn, Ca, Sr, Ba, (4) Cu, Co, Ni, Ga, Fe, (5) S, P, (6) Hf, Zr, Ag, Th, Sc, rare earth elements [678]. This scheme was developed chiefly for activation-analysis purposes.

Strongly acidic cation-exchangers and strongly alkaline anion-exchangers were also applied in different or similar systems for separating Cr, Co, Se, Fe, Zn, Ag, Mo, Hg and Au in activation analysis of biological samples [679], and also for concentrating and separating Mo, S, P, Fe, Zn, Pb, Cu, Co, Mn, Na, K, Ca, Mg in the determination of trace elements in biological materials [680].

Recently several new, sometimes very expanded, separation schemes, used in activation analysis of biological materials, have been developed [681–683]. For instance Velandia and Perkons [683] separated more than 50 elements into 12-15 groups selected to minimize interferences in the determination by γ-spectrometry.

In an interesting method developed by Tjioe *et al.* [507], the sample is decomposed in a closed system, then the volatile elements are distilled off quantitatively from an HBr medium. The distillate and remainder are subjected to further separation on ion exchangers. All the operations may be automated.

In a systematic scheme for spectrographic analysis of silicates the separation into groups is done by ion-exchange and extraction [684]. Radioactive elements Cs, Sr, Zr, Cr, Nb, Np, Co, Fe, Zn and rare earth elements were determined quantitatively in natural samples by anion-exchange and precipitation methods [685].

Ion-exchange chromatography combined with extraction chromatography constitutes the basic of a recent scheme for separating 27 elements: Sb(V), Ga(III), Fe(III), Mo(IV), Sn(IV), Co(II), Cu(II), U(VI), Zn(II), Cd(II), Bi(III), Ti(IV), Sc(III), Th(IV), Zr(IV), Hf(IV), Cr(III), V(IV), Pb(II), Mn(II), Al(III), Ni(II), Mg(II), Ca(II), Ba(II), Sr(II), La(III), [686]. Five columns were used for separating the elements into groups. Most of the elements could be isolated by selective elution. The separated elements were determined by titration, flame photometry, atomic-absorption spectrometry, and other methods. This scheme is suitable for separating macro and micro quantities of the listed elements [686]. In rocks, 45 elements were determined by activation analysis, by using high resolution γ-spectrometry after the elements were separated into groups by distillation, ion-exchange and extraction methods [687]. The same scheme was later used for determining a number of elements in rocks and moon soil brought by Apollo 11 [688], in meteorites [689a], etc. Another group of researchers, studying the composition of moon soil by activation analysis, also used ion-exchange chromatography for separating silver, rare earth elements, titanium, zirconium, hafnium, cobalt, and mercury [689].

It has been suggested [690] that cellulose phosphate could be used for group separation of rare earth elements, alkali metals and alkaline-earth elements. Alkali metals, alkaline-earth and rare-earth elements were separated on columns of strongly basic anion-exchangers in the hydroxide, carbonate and oxalate forms [691]. In this case the column operates as a precipitating agent and as a filter. When determining the composition of meteorites, Al, Be, Ni, Sc, Mn and Ti were separated in a similar manner; these elements form insoluble hydroxides and are quantitatively retained on the anion-exchanger AV-17 (OH^-) from a solution of pH 3–4, while Na, K, Mg, and Ca pass to the effluent [692]. The elements retained on the exchanger can be separated by selective elution. Trace impurities in indium antimonide were determined by activation analysis after separation by ion-exchange, distillation, reduction and extraction [693]. A number of elements were separated on Dowex 50 (H^+) cation-exchanger in tetrahydrofuran–nitric acid solution. Subsequently a solution of dithizone in the same eluent was used for selective separation of elements forming stable dithizonates [694]. Ion-exchange combined with isotopic exchange was used for selective isolation of radioactive gold, copper and antimony ions from mixtures [695]. The anion-exchanger in the molybdate form was used for selective separation of arsenic and phosphorus from mixtures, as a result of formation of the relevant heteropoly acids in the ion-exchanger bed [695]. The method of separating multi-component mixtures of trace elements developed for use in activation analysis of biological materials also deserves mention [510,696–698]. Unlike most of the procedures already described, which consist mainly of a number of adsorption and elution cycles, this method relies on selective adsorption. After removal of organic matter and distillation of elements forming volatile oxides and halides, the irradiated sample is passed through a number of columns connected in series and packed with ion-exchangers in the appropriate ionic form and also with extraction solvents deposited on solid supports. The sample is passed through the columns automatically at a predetermined rate. At the same time, solutions of appropriate reagents are added automatically into the spaces between columns, so as to maintain a constant composition of the solutions fed to the particular columns. As a result, only a definite group of elements is retained on a given column. After separation is concluded, the packing of each column is transferred to a polyethylene container and the elements retained are determined by γ-spectrometry. Elimination of the elution stage considerably reduces the separation time. Separation of 40 elements into groups, usually consisting of a few components, using ten columns, takes 5–6 min. [698].

Improved or new ion-exchange separation schemes have recently been described for the determination of trace elements in biological materials by neutron-activation analysis [507a,510a,698a], systematic analysis of inorganic materials [698b], and determination of trace impurities in platinum [560], palladium [970] and rhodium [698c] by neutron-activation analysis, as well as for the determination of selected trace elements in rocks [698d].

Anion-exchange in aqueous and mixed solvents has been used for the determination of several major and trace constituents in manganese nodules [698e–698g].

Ion-exchange on hydrated manganese dioxide and diethyldithiocarbamate extraction have been used for simultaneous determination of As, Cr, Se, Sb, Cd, and Cu in environmental materials [698h]. Ion-exchangers have also been used for separating and analysing uranium fission products [699,700].

6.8.2.2 ISOLATION OF THE INDIVIDUAL COMPONENTS OF A MIXTURE

A group of elements belonging to a single group of the periodic table is usually the most difficult to separate. In group separation schemes the elements of one group of the periodic table (especially those belonging to A sub-groups) are usually isolated together. Therefore, discussions of possibilities of isolating particular elements from mixtures have been organized according to groups of the periodic table. It has not been possible to include all methods in the discussion, nor in the general summary of separation methods in Table 6.6: only the most common methods or those of growing importance have been included. It should also be mentioned that several methods which enable almost specific separation of certain elements from complex mixtures have been developed. Examples of this are copper [701], mercury [55], thorium [702], indium [73], uranium [704] and gold [560,570]. Separation schemes for almost specific isolation of traces of mercury from neutron irradiated biological materials employing exclusively ion-exchange [572a] or ion-exchange and extraction chromatography [704a] have been devised.

Group 1 Elements

Sub-group 1A. Alkali metals: Li, Na, K, Rb, Cs, Fr. Usually ions of alkali metals do not form complexes in aqueous solutions, and they are not normally retained on an anion exchanger. They may be separated by elution with acid solutions from sulphonate or phenolsulphonate cation-exchangers, separation being much better on phenolsulphonates, especially for heavy alkali metals [74,244,318,705–709]. The elution sequence is in accordance with increasing atomic number (see Figs. 6.43 and 6.51).

Separation of alkali metals on sulphonic acid ion-exchangers is better for more highly cross-linked resins [74,244]. Increase in temperature adversely affects the quality of separation [74,244]. Good separation can be obtained on cation-exchangers, in mixed, e.g. aqueous alcohol, solutions [705,

709,713]. The elution sequence is reversed when cation-exchangers with phosphonate functional groups are used in alkaline medium [715,716].

Very good separation is obtained on inorganic ion-exchangers, e.g. zirconium tungstate [95] or ammonium molybdophosphate [86]. In aqueous dioxan solution partial separation of alkali metals is possible on Dowex 1-X8 (NO_3^-) anion exchanger by elution with dilute nitric acid, the elution sequence being: Li, Na, Cs, Rb+K [712].

In aqueous alcohol solutions containing nitric acid, caesium is eluted from the anion exchanger before rubidium [713]. EDTA forms weak complexes with the lightest alkali metals, so caesium, sodium and lithium (in that order) can be separated on an anion-exchanger containing EDTA ions (i.e. in EDTA-form) by elution with aqueous EDTA solutions of pH 10·9 (Cs, Na) and 4·2 (Li) [714]. The elution sequence Na, Li, is also obtained on a sulphonic acid cation-exchanger in a $12M$ HBr solution [377]. An unusual sequence of selective elution (Li < K < Rb < < Cs < Na) and at the same time very good separation of alkali metals was obtained on a new inorganic ion-exchanger, antimonic acid [711].

Lithium was separated from alkali, alkaline-earth and other metals on a double column of Dowex 50W-X8 and antimonic acid [971]. Unusual elution sequences of alkali metals were also obtained on the conventional sulphonic acid cation-exchanger using eluents containing cyclic polyethers [719b].

Sub-group 1B: Cu, Ag, Au. The separation of metals of this sub-group is usually based on their differing abilities to form halide complexes. Copper(II) can be retained on a sulphonic acid exchanger from chloride solutions of pH 1·5 but gold(III) remains in solution [658]. Gold is very strongly retained by an anion-exchanger in chloride form from hydrochloric acid solutions, and can be separated in this way even from large amounts of copper, for which the partition coefficient is small under these conditions [562]. In concentrated HBr solutions bromoauric acid is adsorbed by the sulphonic acid cation-exchanger, whereas copper remains in the effluent. This method was used to detect traces of copper in spectrally pure gold [720]. Gold is easily removed from the cation-exchanger by washing with acetylacetone. Traces of silver can be separated from gold in a similar manner [377]. Silver has also been separated from gold on a hydrated zirconium oxide ion-exchanger by elution with hydrochloric acid. Gold appears first in the effluent [95]. Copper was separated from silver by precipitation chromatography on an anion-exchanger in chloride form. The silver retained in the column was subsequently eluted with an ammonia solution [721]. Copper was separated from silver in a similar manner on an anion-exchanger in arsenate form [722]. All three elements were separated on the weakly basic anionic-exchanger Amberlite IRA-68 (Br^-), by elution with $3 \cdot 2M$ $HBr + Br_2$ [589].

Group 2 Elements

Sub-group 2A. The alkaline-earth metals: Be, Mg, Ca, Sr, Ba, Ra. The alkaline-earth metals can be separated, like alkali metals, by simple elution with hydrochloric acid from a sulphonic acid cation-exchanger [245]. Calcium and strontium are best separated at low temperature on highly cross-linked ion-exchangers; strontium and barium at low temperature on a medium cross-linked ion-exchanger [245]. The results are better when organic acids or hydroxyacids, such as acetic, malic, tartaric, lactic, α-hydroxybutyric, malonic or citric acid and also acetylacetone are used

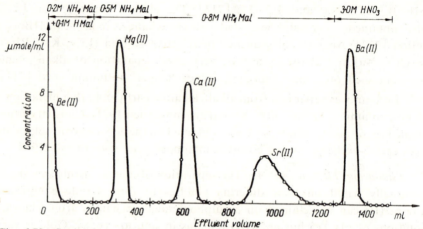

Fig. 6.72 Separation of alkaline-earth metals by elution from a cation-exchanger with ammonium malonate solutions (from [726], by permission of the copyright holders, Elsevier Publ. Co.). Column: $19 \cdot 1$ cm $\times 3 \cdot 14$ cm^2, cation-exchanger AG 50W-X8 (NH_4^+) (mesh 200–400), pH 7, flow rate $2 \cdot 0 \pm 0 \cdot 2$ cm/min

as eluents [723–727] (Fig. 6.72). Aqueous-alcohol solutions of hydrochloric acid [728], ammonium sulphate solutions [729] and buffered EDTA or DCTA solutions [674,730,731] were also used for elution from cation-exchangers. In all cases, the elution sequence was in order of increasing atomic numbers. The same sequence was obtained for separation of magnesium and calcium on Dowex 50-X4 cation-exchanger in concentrated $HClO_4$–HCl solutions [735]. In other cases, however, an unexpected elution sequence was obtained for concentrated solutions, eg. Be, Ba, Sr in $9M$ $HClO_3$ at 50°C [281] or Be + Mg, Ra, Ba, Sr + Ca in $12M$ HBr at 60°C [377]. A good separation of Ca, Sr, Ba and Ra (eluted in that sequence) was also obtained by using an inorganic ion-exchanger, i.e. zirconium molybdate and NH_4Cl–HCl solutions of different concentration as eluents [95]. Similarly, Mg, Ca and Sr were separated on the same ion-exchanger by using $(NH_4)_2SO_4$–CH_3OH and NH_4NO_3–HNO_3 solutions as eluents [733]. Alkaline-earth metals were also separated on anion-exchangers. Ba, Sr, Ca, Mg (in that order) were separated successfully on Dowex 1 ion-exchanger by using ammonium citrate solution as eluent [734]; elution with aqueous-alcohol solutions of nitric acid from Amberlite XN-1002 gave the following sequence: Mg, Ca, Sr, [735]. Different sequences of elution of the alkaline-earth elements from an anion-exchanger, can be obtained by varying the concentration of the 2,6-pyridinedicarboxylic acid solution (at pH 7·1–7·4) used as eluent [736]. On a column packed with a mixture of a cation- and an anion-exchanger, with tartaric acid–sodium chloride solution as eluent, the alkaline-earth elements were separated in the following order: Ca, Mg, Sr, Ba [737]. Beryllium forms an anionic complex with fluorides and it can be separated from cation on an anion-exchanger in fluoride form [634]. Beryllium has also been separated from most alkaline-earth elements in mixed solvent media [674,738]

Sub-group 2B. Zn, Cd, Hg. As for elements of sub-group 2A, the different abilities to form halide complexes are usually utilized here for separation. Cation- [590,741–744] and anion- [672,745–749] exchangers in aqueous and mixed solutions [499,549,745] were used.

Mercury can be separated selectively from 40 elements, including Zn and Cd, on a weakly alkaline DEAE cellulose anion-exchanger in ammonium thiocyanate solution [55]. Zn, Cd and Hg, were separated as dithizonates on a cation-exchanger by partition chromatography [397].

Separation of the elements of group 2B is also possible on an anion-exchanger in a buffered tartrate solution [749]. Good separation of the three elements was obtained on a weakly alkaline anion-exchanger in a hydrobromic acid medium [589]. Binary mixtures were also separated on cation- and anion-exchangers in aqueous–nonaqueous nitric acid solutions [409], on an anion-exchanger in an H_3PO_4 medium [582] and on inorganic ion-exchangers [109,111].

Group 3 Elements

Sub-group 3A. Sc, Y, La, the lanthanides, Ac, the actinides. Apart from elements of atomic numbers 90–94, the elements of subgroup 3A exist in solution almost always in the +III oxidation state, and their chemical properties are very similar. Because of their very limited ability to form complexes with simple inorganic ligands, Y, La, the lanthanides and Ac are not, for example, retained by an anion-exchanger in chloride form from aqueous solutions of hydrochloric acid. However, Sc and some of

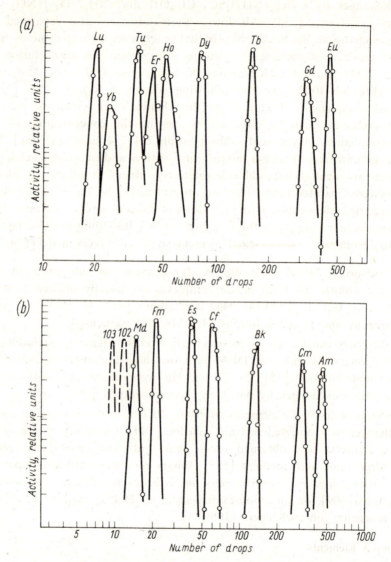

Fig. 6.73 Comparison of the elution of lanthanides (a) and actinides (b) from a 5 cm × 0·31 cm² column of Dowex 50-X12 with ammonium-hydroxybutyrate (from [764a], by permission of the copyright holders, Akad. Nauk SSSR)

the heavier transplutonium elements form weak complexes under these conditions, and this can be a basis for group separation [6,750]. A still better separation of lanthanides from actinides can be obtained in concentrated LiCl solution [3,307]. Scandium was separated from lanthanides on an anion-exchanger in a phosphoric-acid medium [582] and also on a cation-exchanger in concentrated HBr solutions [570].

Rare-earth elements were separated on anion-exchangers in concentrated nitrate solutions [547,752] or in nitrate solutions containing an organic solvent [419,642] in which case the lanthanides were eluted in the order of decreasing atomic numbers. Good separation of scandium from the rare-earth elements was obtained in aqueous-methanol solutions of nitric acid [754]. Anion exchange in a sulphate medium was used for separating rare-earth elements, scandium, thorium and uranium [755]. Rare-earth elements can be separated on anion-exchangers in aqueous-alcohol solutions of α-hydroxybutyric acid. In that case, lanthanides are eluted in the order of increasing atomic numbers [420]. An unusual order of elution: Lu, Yb, Tm, Sc, Er, Y, Ho, La, Dy, Ce, Tb, Pr, Nd, Gd, Pm, Eu, Sm (at room temperature) was obtained on an anion-exchanger in an EDTA solution [218,267,332,369,428,756] (see also Figs. 6.12, 6.21 and 6.41). A similar order of elution is obtained in DCTA solutions, the largest partition coefficients being, however, observed for Nd–Pm [333,371].

Since the affinities of rare-earth elements for cation-exchangers are very similar, a separation by elution with inorganic acids [351,352] is almost impossible. However, very good results can be obtained with complexing eluents such as buffered solutions of citric [1,355,358], lactic [358,532, 545,758,760,763], glycolic [757], α-hydroxyisobutyric [356,466,579,761,762] acids and others [362,432,764–767]. Lanthanides are eluted in the order of decreasing atomic numbers (see Fig. 6.45). Similar methods have been used for separating transplutonium elements: their ion-exchanger behaviour is analogous to that of the lanthanides [4,7,318,763,764a] (see Fig. 6.73). Such aminopolycarboxylic acids as EDTA [359,771], HEDTA (N-hydroxyethylethylenediaminetriacetic acid) [465], etc. [360–362], are useful as eluents for the separation of rare-earth elements on cation-exchangers. Separations involving complexing reagents give high separation factor values but unfavourable dynamics of column-process compared with those involving hydroxy acids [359]. However, this situation may be greatly improved by adding a neutral salt [465] or a hydroxy-acid [771] to the eluent. The relative position of yttrium, which usually accompanies heavy rare-earth elements, may change depending on the complexing agent used for elution, in accordance with the relative values of the stability constants of the yttrium and lanthanide complexes [772].

Elements of atomic numbers 90–94 have oxidation states different from +III and their chemical properties are different from those of the other elements of sub-group 3A. Thorium (IV) is retained very strongly by sulphonic acid cation-exchangers, so even concentrated acids do not remove it; this allows it to be separated from practically all elements [773, 774]. Thorium forms anionic complexes with nitrate, and this allows it to be separated, on an anion-exchanger, from rare-earth elements [410,775]. Uranium(VI) is retained on anion-exchangers from hydrochloric acid solutions, and thus can be separated from thorium and rare-earth elements [411,660]. Elements 90–94 have been separated from each other and from rare-earth elements by utilizing the differences in their abilities to form complexes with halides [377,587,768,776–779,784], nitrates [383,769, 780,781], sulphates [383,587,782], thiocyanates [785,786], etc. Aqueous-alcohol, aqueous-ketone, and other solutions have also been used [410,411, 704,973]. Ion-exchange chromatography has also been used for separating uranium [753] and plutonium [751] in different oxidation states.

Sub-group 3B. B, Al, Ga, In, Tl. In this group, boron is a typical non-metal, aluminium and gallium are amphoteric metals, indium and thallium are typical metals. Since boron usually occurs in anionic form in aqueous solution, its separation from the remaining elements of the group, e.g. on a sulphonic acid cation-exchanger column, presents no difficulties [793]. Methods for separating Al, Ga, In, and Tl are based on differences in their abilities to form complexes, especially halide ones [77,368,377,589,609,795,796]. Separation in aqueous-organic solutions gives good results [77,499,590,595,797,972]. The separation on Dowex 1 (Br$^-$) anion-exchanger is illustrated in Fig. 6.74. The type of ion-exchanger and

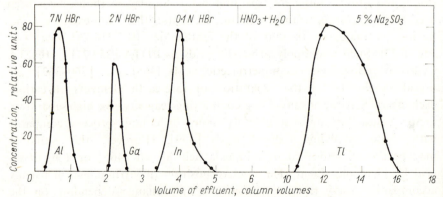

Fig. 6.74 Separation of aluminium, gallium, indium and thalium on a column of Dowex 1-X8 (Br$^-$) (200–400 mesh) anion-exchanger (from [368], by permission of the copyright holders, Acta Chem. Scand.)

the eluent composition have an influence on the order of elution of these elements (see Table 6.6). In hydrofluoric acid solutions, boron forms anionic complexes, and this might allow it to be separated from cations [633]. The amphoteric properties of aluminium and gallium were utilized to allow them to be separated from indium on an anion-exchanger in hydroxide form by retention from lactate solutions of pH 4. Gallium and aluminium were eluted with 10% NaOH as aluminate and gallate [798].

Group 4 Elements

Sub-group 4A. Ti, Zr, Hf, (Th). The +IV oxidation state is characteristic of the elements of this subgroup. In hydrochloric-acid solutions these elements (except for titanium) have very high affinity for sulphonic acid cation-exchangers [343,344,590,773]. This allows titanium to be separated from the other elements of the sub-group [390,799]. When there are no strong chelating agents in the solution, the very similar elements zirconium and hafnium undergo hydrolytic reactions and form multinuclear complexes [800], so the separation process will be very dependent on the history of the solution [591]. Titanium has a greater affinity for cation-exchangers in the form of its complex with hydrogen peroxide [801].

Although partial separation of hafnium and zirconium is possible in hydrochloric acid solution [802,806], most methods of separation of these elements depend on differences between the stabilities of the complexes with sulphuric [591,803,804], hydrofluoric [804], and organic hydroxy [807] acids. For separating zirconium and hafnium from titanium and a number of other elements, combined ion-exchange and extraction methods have been used [809]. A good separation of Zr and Hf was obtained on a cation-exchanger by using formic acid solution [810].

Sub-group 4B. C, Si, Ge, Sn, Pb. Carbon and silicon are typical non-metals, so they occur in solution as anions. There is little information available on their behaviour in ion-exchange chromatography. Silicic acid can be separated, after conversion into the α-form, from metal cations by passing an aqueous solution of pH 1 through a sulphonic acid cation-exchanger in the hydrogen form [812]. Germanium, tin, and lead have usually been separated by making use of the properties of their halide complexes, the ability of germanium to form a covalent, volatile tetrahalide, and the amphoteric properties of tin.

Germanium, as the volatile $GeCl_4$, can be removed from solution in a stream of nitrogen and subsequently trapped in a strongly alkaline anion-exchanger in the chloride form [813]. The volatility of germanium tetrachloride from concentrated hydrochloric-acid solutions may be the

Table 6.6 Some important methods of separating elements on ion-exchangers

Species	Ion-exchanger and its ionic form	Eluent	Ref.
1A			
Li-Na+many other elements	AG-50W X8 (H⁺)	$1M$ HNO₃–80% CH₃OH: Li	[710]
Li-Na	KU-1 (H⁺)	$0.25M$ HCl: Li, Na	[318]
Li-Na-K	Amberlite IR-120 ((CH₃)₄N⁺)	$0.1M$ uramil acetate (pH 7): Li, (pH 10): Na, $2M$ HCl: K	[717]
Li-Na-K-Rb-Cs	Bio-Rex 40 (H⁺)	$1.0M$ HCl–80% C₂H₅OH: Li, $0.2M$ HCl: Na, $0.7M$ HCl: K, Rb $4.0M$ HCl: Cs	[709]
Li-Na-K-Rb-Cs	Duolit C-3 (H⁺)	$2.4M$ HCl–80% CH₃OH: Li, Na, $1.5M$ HCl: K+Rb, $4M$ HCl: Cs	[705]
Na-K-Rb-Cs	MK-3 (H⁺)	$0.15M$ HCl: Na, K, Rb, Cs	[74]
Na-K-Rb-Cs	Dowex 50W-X16 (H⁺)	$0.64M$ HCl: Na, K, Rb, Cs	[244]
Rb-Cs-Fr	KU-1 (H⁺)	$5.5M$ HCl: Rb, Cs, Fr	[708]
Cs-Fr	Duolit C-3 (H⁺)	$4.5M$ HCl: Cs, Fr	[705]
Li-Na-K-Rb-Cs	zirconium tungstate (NH₄⁺)	$0.05M$ NH₄Cl: Li, $0.1M$ NH₄Cl: Na, $0.3M$ NH₄Cl: K, $0.75M$ NH₄Cl: Rb, $4.5M$ NH₄Cl: Cs	[95]
Na-K	molybdophosphate (NH₄⁺)	$0.02M$ NH₄NO₃: Na, $0.5M$ NH₄NO₃: K	[86]
K-Rb	molybdophosphate (NH₄⁺)	$0.2M$ NH₄NO₃: K, $3M$ NH₄NO₃: Rb	[86]
Rb-Cs	molybdophosphate (NH₄⁺)	$4M$ NH₄NO₃: Rb, sat. NH₄NO₃: Cs	[86]
Li-K-Cs-Rb-Na	antimonic acid (H⁺)	$1M$ HNO₃: Li, $2M$ HNO₃: K, $4M$ HNO₃: Cs, $6M$ HNO₃: Rb, $2M$ NH₄NO₃: Na	[711]
Li-Na-K-Rb-Cs	Dowex 1-X8 (NO₃⁻)	$0.4M$ KNO₃–78% dioxan (gradient): Li, Na+Cs+Rb+K (partial separation)	[708]
Cs-Na-Li	Dowex 1-X4 (EDTA)	$2.5 \times 10^{-3}M$ EDTA, pH 10·9: Cs, Na, $0.2M$ EDTA, pH 4·2: Li	[718]
Na-Li	Dowex 50-X4 (H⁺)	$12M$ HBr: Na, Li	[377]
Cs-many other elements	tungstophosphate (H⁺)	$0.005M$ HNO₃: contamination, $0.15M$ HNO₃: Cs,	[85]
alkali metals-	Dowex 50W-X8 ((CH₃)₄N⁺)	2% EDTA solution, pH 8 selective retention of alkali metals on column stripping $5M$ HCl	[718]

§ 6.8] Applications 445

earth metals		alkaline-earth metals	
Alkali metals-alkaline earth metals	zirconiumphosphate (NH_4^+)	$1M\ NH_4Cl$: alkali metals $1M\ HCl$: alkaline-earth metals	[95]
Li-Na-K-Rb-Cs	phosphoantimonic (V) acid (NH_4^+)	$0.1M\ NH_4NO_3$: Li+Na, K, Rb, $1M\ NH_4NO_3$: Cs	[719a]
K-Rb-Na-Cs	Aminex A-7 (H^+)	$0.01M$ (cis-syn-cis)dicyclohexyl-18-crown-6–$1M$ HCl–CH_3OH (80%): K, Rb+Na, $1M$ (cis-syn-cis)dicyclohexyl-18-crown-6–$1M$ HCl–CH_3OH (80%); Cs.	[719b]
K-Na-Rb-Cs	Aminex A-7 (H^+)	$0.01M$ (cis-syn-cis) dicyclohexyl-18-crown-6–$1M$ HCl–CH_3OH (80%): K, Na, Rb, $1M$ (cis-anti-cis)dicyclohexyl-18-crown-6–$1M$ HCl–CH_3OH (80%): Cs	[719b]
1B			
Au-Cu	Dowex 50W-X8 (H^+)	chloride solution, pH 1·5: Au, $3M$ HCl: Cu	[658]
Cu-Au	Amberlite IRA-400 (Cl^-)	$0.1M$ HCl: Cu, ion-exchanger incineration: Au	[562]
Cu-Au	Dowex 50W-X8 (H^+)	$6M\ HBr+Br_2$: Cu, $6M\ HBr+Br_2$, $t=75°C$, or acetylacetone: Au	[720]
Ag-Au	Dowex 50-X4 (H^+)	$9M\ HBr+Br_2$: Ag, $0.5M$ HBr: Au	[377]
Au-Ag	zirconium oxide (Cl^-)	$2M$ HCl: Au, Ag	[95]
Ag-Au	Dowex 1-X8 (Cl^-)	$9M$ HCl: Ag, 10% thiourea: Au	[679]
Cu-Ag	Wofatite L-150 (Cl^-)	$6M$-dilute HCl: Cu, 1–$10M\ NH_4OH$: Ag	[721]
Cu-Ag	Lewatit MN ($H_2AsO_4^-$)	$2M$ HCl: Cu, $(1{:}1)NH_3$(aq): Ag	[722]
Cu (II)-Ag (I)-Au (III)	Amberlite IRA-68 (Br^-)	$3·2M\ HBr$–$0.004M$ Br: Cu, Ag, Au	[589]
Cu-Ag	Dowex 1-X8 ($H_2PO_4^-$)	$1M\ H_3PO_4$: Cu, $6M$ HCl: Ag (traces)	[582a]
2A			
Be-Mg-Ca-Sr-Ba	AG 50-X8 (H^+, NH_4^+)	$0.2M$ ammonium malonate (NH_4Mal)–$0.1M\ NH_4Mal$: Be, $0.5M\ NH_4Mal$: Mg, $0.8M\ NH_4Mal$: Ca, Sr, $3M\ HNO_3$: Ba	[726]
Be-Mg-Ca-Sr-Ba-Ra	Dowex 50 (NH_4^+)	$0.55M$ ammonium lactate (NH_4Lac), pH 5, $t=78°C$: Be, $1.5M\ NH_4Lac$, pH 7, $t=78°C$: Mg, Ca, Sr, Ba, Ra	[723]

Table 6.6 (continued)

Species	Ion-exchanger and its ionic form	Eluent	Ref.
Be-Mg	KU-2 (H^+)	$1M$ HCl–50% acetone: Be, $3M$ HCl: Mg	[738]
Be-Mg, Ca, Sr, Ba + many other elements	Dowex 50W-X4 (H^+)	$0.25M$ NH_4SCN–20% dimethylsulphoxide–$0.01M$ HCl: Be	[647]
Ca-Mg-Sr-Ba	Diaion SK + Diaion SA (2:1)	$0.25M$ tartaric acid (pH 4·1)–$0.07M$ NaCl: Ca, Mg, Sr, Ba	[737]
Mg-Ca-Sr-Ba	KU-2 (NH_4^+)	ammonium acetate (NH_4Ac) 1.5–$4.0M$ (gradient): Mg, Ca, Sr, Ba	[725]
Ca-Sr-Ba	Dowex 50W-X16 (H^+)	$2M$ HCl, $t = 25°$: Ca, Sr, Ba	[245]
Ca-Sr-Ba	Amberlite IR-120 (NH_4^+)	$0.03M$ DCTA–$0.4M$ NH_4Ac, pH 5·1: Ca, $0.03M$ DCTA–$0.4M$ NH_4Ac, pH 7·2: Sr, $3M$ HCl: Ba	[731]
Ca-Sr-Ba	Dowex 50W-X8 (NH_4^+)	$0.1M$ EDTA, pH 5·3: Ca, $0.1M$ EDTA pH 6·0: Sr, $0.1M$ EDTA, pH 9·0: Ba	[730]
Ba-Ra	Dowex 50 (NH_4^+)	$0.3M$ ammonium citrate (NH_4Cit), pH 5·6: Ba, Ra	[739]
Be-Mg-Ra-Ba-Sr-Ca	Dowex 50-X4 (H^+)	$12M$ HBr: Be+Mg, Ra, Ba, $5M$ HBr: Sr+Ca	[15]
Ca-Sr-Ba-Ra	zirconium molybdate (NH_4^+)	$0.2M$ NH_4Cl–$0.005M$ HCl: Ca, $0.5M$ NH_4Cl–$0.005M$ HCl: Sr, $1M$ NH_4Cl–$0.005M$ HCl: Ba, sat. NH_4Cl–$0.01M$ HCl: Ra	
Mg-Ca-Sr	Amberlyst XN-1002 (NO_3^-)	$0.25M$ HNO_3–90% C_2H_5OH: Mg, $0.25M$ HNO_3–95% CH_3OH: Ca, H_2O: Sr	[735]
Ba-Sr-Ca-Mg	Dowex 1 (Cit)	$0.05M$ NH_4Cit, pH 7·5: Ba, Sr, Ca, $0.5M$ HCit: Mg	[734]
Sr-Ca	titanium tungstate (H^+)	$2M$ HNO_3: Sr, $0.5M$ NH_4NO_3–$0.05M$ HNO_3: Ca	[740]
Be-Mg	AG 50W-X8 (H^+)	$2M$ HNO_3–70% CH_3OH: Be, Mg	[740a]
2B			
Zn-Cd-Hg	Dowex 1 (Cl^-)	$0.01M$ HCl–25% CH_3OH: Zn, Cd, $0.01M$ HCl–$0.1M$ thiourea: Hg	[745]
Zn-Cd	Dowex 1-X8 (NO_3^-)	90% CH_3COOH–10% $1M$ HNO_3: Zn, $1M$ HNO_3: Cd	[748]
		$0.02M$ HCl: Zn, $1M$ NH_4OH:	[672]

§ 6.8] Applications 447

Elements	Resin	Conditions	Ref.
Hg-Zn-Cd + many other elements	DEAE (Cellex D) (SCN⁻)	0·02M NH₄Br: Cd	[747]
Hg-Cd-Hg		0·01M NH₄SCN-0·1M HCl: 40 elements, 0·5M HClO₄: Hg	[55]
Hg(II)-Cd(II)-Zn(II)	Amberlite IRA-68 (Br⁻)	0·1M HBr: Zn, 6·8M HBr: Cd, 0·5M HClO₄: Hg	[589]
	AG 50W-X8 (H⁺)	0·1M HBr-50% acetone: Hg, 0·2M HBr-50% acetone: Cd, 0·2M HBr-80% acetone: Zn	[589a]
Zn-Cd-Hg	cerium tungstate	H₂O: Zn+Cd, 0·4M NH₄Cl-0·05M HCl: Hg	[111]
Zn-Hg	Dowex 1-X8 (H₂PO₄⁻)	1M H₃PO₄: Zn, on the resin: Hg	[582]
Cd-Hg	titanium arsenate	0·1M NH₄NO₃: Cd, 1M HNO₃: Hg	[109]
Cd-Zn	DEAE (SCN⁻)	Methanol-1·1M NH₄SCN-0·0011M HCl (19 : 1): Cd, methanol-1·1M HCl: Zn	[746]
Zn-Hg	Amberlite IRA-400 (NO₃⁻)	60% tetrahydrofuran-0·6M HNO₃: Zn, 0·01M HCl-0·1M thiourea: Hg	[409]
Hg-Cd-Zn	Amberlite IR-120 (H⁺)	90% methanol-0·6M HNO₃: Hg, 60% tetrahydrofuran-0·6M HCl: Cd+Zn	[409]
Hg-Cd-Zn	AG 50W-X8 (H⁺)	0·1M HCl-80% C₂H₅OH: Hg, 0·5M HCl-20% C₂H₅OH: Cd, 0·5M HCl-80% C₂H₅OH: Zn	[499]
Zn-Cd-Hg	Dowex 2-X8 (tartrate)	8·5 × 10⁻²M ammonium tartrate, pH 2·5: Zn, Cd 1M HNO₃: Hg	[749]
Zn(II)-Cd(II)	AG 1-X8 (Br⁻, NO₃⁻)	0·1M HBr-0·5M HNO₃: Zn, 2M HNO₃: Cd	[749a]
Cd-Zn	AG 50W-X8 (H⁺)	0·5M HCl-40% C₂H₅OH: Cd, 0·5M HCl-80% C₂H₅OH: Zn	[590]
Hg-Cd	KU-2 (H⁺)	0·4M HNO₃-0·01M HCl: Hg, 0·5M HCl: Cd	[741]
Hg-Cd	Dowex 50W-X8 (H⁺)	0·1M HBr: Hg, 0·3M HBr: Cd	[742]
Zn-Cd	Dowex 50 (NH₄⁺)	0·25M NH₄ Cit, pH 4: Zn, 2M HCl: Cd	[743]
Cd-Zn	Dowex 50 (H⁺)	0·3M HI-0·075M H₂SO₄: Cd, 2M HCl: Zn	[744]
Hg-Zn-Cd	KU-2 X6 (H⁺) sat. 1M HBr	sample: Zn, Cd, Hg dithizonate in CCl₄ (C₂H₅)₂O sat. H₂O: Hg, (C₂H₅)₂O sat. HCNS: Zn, 1M HBr: Cd	[397]

Table 6.6 (*continued*)

Species	Ion-exchanger and its ionic form	Eluent	Ref.
3A			
Lu-Yb-Tn-Er-Ho-Dy-Tb-Gd-Eu-Sm-Pm-Nd-Pr-Ce-La	Dowex 50W-X12, (NH_4^+, H^+)	$0.5M$ α-hydroxyisobutyric acid, pH 2.7–4.8 (gradient): Lu, Yb, Tm, Er, Ho, Dy, Tb, Gd, Eu, Sm, Pm, Nd, Pr, Ce, La	[466]
Lu-Yb-Tm-Er-Ho-Y-Dy-Tb-Gd-Eu-Nd	KU-2 X6 (NH_4^+)	ammonium lactate, pH 5, 0.19–$0.5M$ (gradient): Lu, Yb, Tm, Er, Ho, Y, Dy, Tb, Gd, Eu, Nd	[758]
Y-Eu-Sm-Pm-Nd-Pr-Ce-La	Zerolit-225 (NH_4^+, H^+)	$1M$ ammonium lactate, $t = 87°C$, pH 3.25: Y, Eu, Sm, Pm, Nd, Pr, Ce, La	[763]
Lu-Yb-Tm-Ho-Tb-Gd-Eu-Sm-Pm-Nd-Pr-Ce	Dowex 50-X8 (H^+, K^+)	$10^{-2}M$ HEDTA–$1.5 \cdot 10^{-1}M$ KCl, $t = 80°C$, pH 2.5–4 (gradient): Lu, Yb, Tm, Ho, Tb, Gd+Eu, Sm, Pm, Nd, Pr, Ce	[465]
Y-Pm-Nd-Pr-Ce-La	Dowex 50-X12 (NH_4^+, H^+)	$0.4M$ lactic acid–$0.002M$ EDTA, pH 3.0–2.5 (gradient): Y, Pm, Nd, Pr, Ce, La	[771]
Y-Gd-Eu-Sm-Pm-Nd-Pr-Ce-La	Dowex 1 (NO_3^-)	65% CH_3OH–$0.01M$ HNO_3 gradient $LiNO_3$, $t = 50°C$: Y, Gd, Eu, Sm, Pm, Nd, Pr, Ce, La	[419]
La-Nd-Eu-Y-Yb	Dowex 1-X4 (α-hydroxyisobutyrate)	$0.02M$ α-hydroxyisobutyric acid in mixture CH_3OH–H_2O, 25% H_2O: La, 35% H_2O: Nd, 57.5% H_2O: Eu, 72.5% H_2O: Y, 100% H_2O: Yb	[420]
Y-Tb-La-Eu-Sm	Dowex 1-X4 (H_2Y^{2-})	$0.04M$ Na_2H_2Y (EDTA), $t = 75°C$: Y, Tb, La, Eu+Sm	[218]
Tm-Ho-La-Pr-Pm-Sm	Dowex 1-X4 (H_2Y^{2-})	$0.04M$ Na_2H_2Y, $t = 25°C$: Tm, Ho, La, Pr, Pm, Sm	[218]
Lu-Sc-Y	Amberlite IRA-400 (H_2Y^{2-})	$0.008M$ Na_2H_2Y: Lu, Sc, Y	[756]
Fm-Es-Cf-Bk-Cm-Am	Dowex 50-X12 (NH_4^+)	$0.4M$ ammonium lactate, pH 4.0–4.5, $t = 87°C$: Fm, Es, Cf, Bk, Cm, Am	[6]
Md-Fm-Es-Cf-Bk-Cm-Am	Dowex 50-X12 (NH_4^+)	$0.23M$ ammonium α-hydroxyisobutyrate, $t = 87°C$: Md, Fm, Es, Cf, Bk, Cm, Am	[764a]
Eu-Sm-Pm-Nd	AG 50-X12 (NH_4^+)	$0.25M$ α-hydroxy-α-methyl-isobutyric acid, pH 4.0: Eu, Sm, Pm, Nd	[764]
Lu-Tm-Er-Ho-Dy-Fb-Gd	Diaion SK+Diaion SA (1:2)	$0.5M$ lactic acid–$0.06M$ NaCl, pH 2.8: Lu, Tm, Er,	[766]

§ 6.8] Applications 449

System	Resin	Conditions	Ref.
La-Lu	Dowex 1-X8 ($H_2PO_4^-$)	$0.5M\ H_3PO_4$: La, $1M\ H_3PO_4$: Lu, $3M\ H_3PO_4$: Sc	[582]
Sc-La + rare earths	Dowex 50W-X2 (H^+)	$10M$ HBr: La + rare earths, $4M$ HBr: Sc	[570]
Lanthanides (III) + Sc(III)	Dowex 2-X8 (ascorbate)	$0.05M$ HCl: lanthanides, $2M$ HCl: Sc	[750a]
Lanthanides-actinides	Dowex 1-X8 (Cl^-)	$10M$ LiCl, $t = 87°C$: La-Lu, Pu(III)-Bk, Cf-Es	[307]
Lanthanides + Y-Sc-Th(IV)-U(VI)	Dowex 1-X8 (SO_4^{2-})	$0.1M\ (NH_4)_2SO_4$–$0.025M\ H_2SO_4$: lanthanides + YSc, $1M\ (NH_4)_2SO_4$–$0.025M\ H_2SO_4$: Th(IV), $1M\ HClO_4$: U(VI)	[755]
Fm-Es-Cf-Cm	Amberlite IRA-410 (EDTA)	$0.01M$ EDTA–$0.1M\ CH_3COONH_4$, pH 3·0, $t = 80°C$: Fm, Fs, Cf, Cm	[781]
Rare-earth elements-Th	AG 50W-X12 (H^+)	$4M$ HCl: rare-earth elements + Y, incineration of ionite: Th	[773]
Rare-earth elements-Th	Dowex 1-X8 (NO_3^-)	5–$8M\ HNO_3$ rare-earth elements + Y, $2\cdot 4M$ HCl: Th	[775]
Rare-earth elements-U-Th	Dowex 1-X10 (NO_3^-)	$8M\ HNO_3$: rare-earth elements, U(VI), $0.2M\ HNO_3$: Th(IV)	[383]
Rare-earth elements-U	Dowex 1-X8 (Cl^-)	CH_3COOH (90%)–HCl conc. (10%): rare-earth elements + Y, $0.5M$ HCl: U	[660]
Sc-lanthanides + Y	AG 50W-X8 (H^+)	$0.1M\ H_2C_2O_4$: Sc, $5M\ HNO_3$: lanthanides + Y	[787]
Rare-earth elements-Th	Dowex 1-X8 (Cl^-)	acetone (75%)–conc. HCl (25%): rare-earth elements + Y, Th	[788]
Th-Pa (V)-UO_2(II)	ASD-2 (Cl^-)	$11\cdot 7M$ HCl: Th, $9M$ HCl–$1M$ HF: Pa, $0.1M$ HCl: UO_2(II)	[776]
U(VI)-Th	DEAE (NO_3^-)	CH_3COOH–$7\cdot 6M\ HNO_3$ (19:1): U, $1.1M$ HCl: Th	[599]
U(VI)-many other elements	AG 50-X4 (H^+)	$0.5M$ HBr–86% acetone: many elements, $0.5M$ HCl–86% acetone: U, $5M\ HNO_3$ + Th + other elements	[704]
Rare-earth elements Th(IV)-Pa(V)-U(VI)	Dowex 1-X8 (Cl^-)	90% CH_3COOH + 10% ($12M$ HCl + $0.1M$ HF): rare-earth elements + Y, 60% CH_3COOH + 40% $7M$ HCl: Th, 60% CH_3COOH + 40% ($12M$ HCl + $1M$ HF): Pa, 30% CH_3COOH + 70% $0.1M$ HCl: U	[768]
U(VI)-Pa(V)-Th(IV)	Dowex 1-X8 (NO_3^-)	sample in 90% CH_3COOH + 10% ($14M\ HNO_3$ + $0.1M$ HF); 60% CH_3COOH + 40% $14M\ HNO_3$: U, 60% CH_3COOH + 40% ($14M\ HNO_3$ + $1M$ HF): Pa, 40% CH_3COOH + 60% $1M\ HNO_3$: Th	[769]

Table 6.6 (continued)

Species	Ion-exchanger and its ionic form	Eluent	Ref.
Th-UO_2(II)	AV-17 (CO_3^{2-})	sample pH 2·5; $1M$ $(NH_4)_2CO_3$: Th, $(NH_4)_2CO_3/NaCl$: UO_2(II)	[770]
Pa-Th	Dowex 50W-X8 (NH_4^+, H^+)	$1M$ NH_4SCN–$0·5M$ HCl: Pa, $2M$ NH_4SCN–$0·5M$ HCl: Th	[785]
La-Th-Pa-U	Dowex 1-X8 (SCN^-)	$1·5M$ NH_4SCN–$0·5M$ HCl: La, $8M$ HCl: Th, $8M$ HCl–$0·005M$ HF: Pa, $1M$ $HClO_4$: U	[786]
Th-rare-earths	Dowex 50 (H^+)	$0·1M$ TOPO in $CH_3OH + 5\%$ $12M$ HNO_3: Th, 4–$6M$ HCl: rare-earth elements	[789]
Sm-Th-U(VI)	Amberlite CG4B (Cl^-)	$11·4M$ HCl: Sm, $6M$ HCl: Th, $0·1M$ HCl: U	[600]
U(VI)-Np(IV)	Dowex 50-X4 (H^+)	$9M$ HBr: U(VI), $9M$ HCl: Np(IV)	[377]
Pu(III)-Np(IV)	Dowex 50-X4 (H^+)	$6M$ HBr, $t = 60°C$: Pu(III), $9M$ HBr–$0·2M$ HF: Np(IV)	[377]
Np(IV)-Pa(V)	Dowex 1-X8 (Cl^-)	$9M$ HCl, $5M$ HCl: Np, $2M$ HCl: Pa	[784]
Am(Pa, U)-Th-Pu-Np	Dowex 1 (NO_3^-)	$8M$ HNO_3: Am (Pa, U), $12M$ HCl: Th, $12M$ HCl–$0·1M$ NH_4I: Pu, $4M$ HCl: Np	[780]
Lanthanides + Sc + Y + Ac-Pu(IV)-Np(IV)-U(VI)	Dowex 1-X10 (Cl^-)	$t = 50°C$, $9M$ HCl: rare-earth elements + Ac, $9M$ HCl–$0·05M$ NH_4I: Pu(V→III), $4M$ HCl–$0·1M$ HF: Np(IV), $0·5M$ HCl–$1M$ HF: U(VI)	[778]
Rare-earths-Pu	Dowex 1-X2 (NO_3^-)	$7·2M$ HNO_3: rare-earths, $0·36M$ HCl–$0·01M$ HF: Pu	[790]
Ac-Th	Dowex 50 (H^+)	7% $H_2C_2O_4$: Th, 5% citric acid, pH 3: Ac	[791]
Ac-Th-Pa(V)	Dowex 1-X8 ($C_2O_4^{2-}$)	$0·5M$ $H_2C_2O_4$: Ac, $8M$ HCl: Th, $8M$ HCl–$0·1M$ HF: Pa(V)	[792]
Ac-La	Dowex 1-X8 (NO_3^-)	$4·4M$ $LiNO_3$: Ac, La	[752]
U(VI)-U(IV)	KU-2 X16 (H^+)	$0·5M$ H_2SO_4: U(VI), $2·5M$ H_2SO_4: U(IV)	[753]
Pu(colloid)-Pu(VI)-Re(III)-Pu(IV)	AG 50-X8 (H^+)	$5·25M$ HCl: Pu(colloid), Pu(VI), Pu(III), Pu(IV)	[751]

3B

Species	Ion-exchanger and its ionic form	Eluent	Ref.
Al-Ga-In-Tl(III)	Dowex 1-X8 (Br^-)	$7M$ HBr: Al, $2M$ HBr: Ga, $0·1M$ HBr: In, 5% Na_2SO_4: Tl	[368]

§ 6.8] Applications 451

System	Resin	Conditions	Ref.
Ga-In-Tl(III)	Amberlite IRA-68 (Br⁻)	0·75M HCl–90% C_2H_5OH: Ga, 3·0M HCl: Al 4M HBr: Ga, In, Tl	[589]
In(III)-Ga(III)-Al(III)	AG 50W-X8 (H⁺)	0·5M HBr–60% acetone: In, 0·5M–90% acetone: Ga, 3M HBr–30% acetone: Al	[589a]
Ga(III)-In(III)	Dowex 1-X4 ($H_2PO_4^-$)	1M H_3PO_4, t = 65°C: Ga, In	[246a]
Tl(III)-In-Ga-Al	AG 50W-X8 (H⁺)	0·1M HCl–50% acetone: Tl, 0·5M HCl–50% acetone: In, 2·0M HCl–70% acetone: Ga, 3·0M HCl: Al	[797]
Al-Ga-In	Dowex 1-X8 (Cl⁻)	2-methoxyethanol (90%)–6M HCl (10%): Al, 1M HCl: Ga, In	[796]
Ga-In-Al	Dowex 1-X8 (Cl⁻)	acetone (80%)–3M HCl (20%): Ga, acetone (90%)–6M HCl (10%): In, acetone (70%)–6M HCl (10%): Al	[796]
Al-Ga-In	Amberlite CG4B (Cl⁻)	6M HCl: Al, 4·5M HCl: Ga, 1M HCl: In	[600]
In-Ga-Tl(III)	Retardion 11A8 (H⁺, Cl⁻)	6M HCl: In, 1M HCl: Ga, acetylacetone: Tl	[77]
Al-Ga-In	Dowex 1-X8 (Br⁻)	CH_3OH (90%)–4·5M HBr (10%): Al, H_2O (90%)–4·5M HBr (10%): Ga+In	[595]
In-Ga-Al	AG 50W-X8 (H⁺)	0·2M HCl–90% C_2H_5OH: In, 0·75M HCl–90% C_2H_5OH: Ga, 3M HCl–50% C_2H_5OH: Al	[590]
Al-In-Ga	Dowex 50-X4 (H⁺)	9M HBr: Al, 9M HCl: In, 4M HBr: Ga	[377]
Al-In-Tl(III)	Dowex 50-X4 (H⁺)	9M HBr-Br_2: Al, 9M HCl-Cl_2: In, 1M HCl: Tl	[377]
Al-In-Ga	titanium tungstate (H⁺)	H_2O: Al+In, 1% HNO_3: Ga	[794]
4A			
Ti-Zr	KU-2 (H⁺)	1M HCl: Ti, 4M HCl: Zr	[799]
Hf-Zr	Dowex 50 (H⁺)	6M HCl: Hf, Zr	[802]
Zr-Hf	Zerolit 225 (H⁺)	1M H_2SO_4: Zr, 3M H_2SO_4: Hf	[803]
Zr-Hf	KU-2 X12 (H⁺)	0·65M H_2SO_4: Zr, Hf	[804]
Th-Hf-Zr	AG 1-X8 (SO_4^{2-}, HSO_4^-)	1·25M H_2SO_4+0·1% H_2O_2: Th, Hf, 1M H_2SO_4+0·1% H_2O_2: Zr	[591]
Zr-Ti-Th	Dowex 50 (H⁺)	1% ammonium citrate (pH 1·75): Zr, Ti, 0·05M ammonium citrate (pH 4·98): Th	[837]

Table 6.6 (continued)

Species	Ion-exchanger and its ionic form	Eluent	Ref.
Ti-Zr	AG 50W-X8 (H^+)	$1.25M$ HNO_3: Ti, $5.0M$ HCl: Zr	[811]
Zr-Th	Dowex 50W-X8 (H^+)	$0.05M$ HCOOH: Zr, $3M$ H_2SO_4: Th	[808]
Zr-Hf	Dowex 50W-X8 (H^+)	$1M$ HCOOH: Zr, $4M$ HNO_3: Hf	[810]
Zr-Hf	Amberlite IRA-400 (Cl^-)	sample: fluoride complexes of Zr and Hf, $0.1M$ HCl: Zr, Hf	[805]
Zr+Hf-Ti+other cations	Dowex 50W-X8 (H^+)	$0.1M$ TOPO in CH_3OH-$12M$ HNO_3(19 : 1): Zr+Hf	[809]
4B			
Pb(II)-Ge(IV)-Sn(IV)	Dowex 1-X10 (Cl^-)	$1M$ HCl–$1M$ HF: Pb, $6M$ HCl–$1M$ HF: Ge, $2M$ HCl–$3M$ HF: Sn	[576]
Ge(IV)-Pb(II)	AG 1-X8 (Br^-)	$0.1M$ HBr: Ge, $0.3M$ HNO_3–$0.25M$ HBr: Pb	[814]
Pb(II)-Sn(IV)	Amberlite IRA-400 (Cl^-)	$7M$ HCl: Pb, $6N$ NaOH: Sn	[816]
Ge(IV)-Sn(IV)-Pb(II)	Nalcite HCR X8 (H^+) + Amberlite IR-45 (OH^-)	aqueous solution, pH 2: Ge	[817]
Ge(IV)-Sn(IV)	Dowex 1-X8 (Br^-)	$3M$ HBr: Ge, $0.5M$ HBr: Sn	[368]
5A			
Pa(V)-Nb(V)-Ta(V)	Dowex 1 (Cl^-, F^-)	$9M$ HCl–$0.004M$ HF: Pa(V), $9M$ HCl–$0.18M$ HF: Nb(V), $1M$ HF–$4M$ NH_4Cl: Ta(V)	[815]
Nb-Ta	Dowex 1-X8 (Cl^-, F^-)	14% NH_4Cl–4% HF: Nb, 14% NH_4Cl–4% HF, pH 6: Ta	[819]
Nb-Ta	AV-17 X8 (SO_4^{2-}, F^-)	$2.5M$ H_2SO_4–$1M$ HF: Nb, $5M$ NH_4Cl–$1M$ HF: Ta	[820]
Nb-Ta	Dowex 1-X4 (Cl^-, $C_2O_4^{2-}$)	$1.5M$ HCl–$0.5M$ $H_2C_2O_4$–$0.09M$ H_2SO_4–$0.01M$ H_2O_2: Nb, $3.5M$ HCl–$0.5M$ $H_2C_2O_4$–$0.09M$ H_2SO_4–$0.01M$ H_2O_2: Ta	[464]
Nb-Ta	Dowex 1-X8 (Cl^-, $C_2O_4^{2-}$)	$0.5M$ $H_2C_2O_4$–$1M$ HCl t = 45°C: Nb, Ta	[823]
Ta-Nb	Dowex 1-X8 (Cl^-, $C_2O_4^{2-}$)	$0.01M$ $H_2C_2O_4$–$2M$ HCl: Ta, Nb	[823]
Ta-Nb	De Acidite FF (Cl^-)	$0.25M$ KCl–$0.01M$ KOH: Ta, $0.5M$ KCl–$0.01M$ KOH: Nb	[826]

§6.8] Applications

Ta(V)-Pa(V)/Ta(V)	KU-2 (H$^+$)	3% α-hydroxyisobutyric acid: Pa+Nb, 3M HCl–1M HF: Ta	[825]
Nb-Ta	Dowex 1-X8 (Cl$^-$, C$_2$O$_4^{2-}$)	1·5M HCl–0·5M H$_2$C$_2$O$_4$–0·007M H$_2$O$_2$: Nb, 3M HCl–0·5M H$_2$C$_2$O$_4$: Ta	[824]
Ta-Nb	AV-17 (Cl$^-$)	C$_2$H$_5$OH–HCl (4 : 6): Ta, 0·5M H$_2$SO$_4$–0·5M H$_2$O$_2$+40% C$_2$H$_5$OH: Nb	[827]
Nb-Ta	De Acidite FF (Cl$^-$, F$^-$)	0·2M HF–7M HCl: Nb, 1M NH$_4$F–4M NH$_4$Cl: Ta	[822]
5B			
NO$_3^-$-NO$_2^-$	KU-2 X6 (Na$^+$)	water–acetone mixture with acetone mole fraction $x = 0·7$: NO$_3^-$, NO$_2^-$	[830]
H$_2$PO$_2^-$–PO$_4^{3-}$–HPO$_3^{2-}$	Dowex 1-X8 (CH$_3$COO$^-$)	CH$_3$COONH$_4$ 0·1–0·5M (gradient): H$_2$PO$_2^-$, PO$_4^{3-}$, HPO$_3^{2-}$	[835]
PO$_4^{3-}$–P$_2$O$_7^{4-}$ -P$_3$O$_{10}^{5-}$–P$_4$O$_{13}^{6-}$ -P$_4$O$_{12}^{4-}$–P$_3$O$_9^{3-}$	AG 1-X8 (Cl$^-$)	0·25M KCl, pH 5: PO$_4^{3-}$, P$_2$O$_7^{4-}$, 0·4M KCl, pH 5: P$_3$O$_{10}^{5-}$, 0·5M KCl, pH 5: P$_4$O$_{13}^{6-}$+P$_4$O$_{12}^{4-}$, P$_3$O$_9^{3-}$	[832]
H$_2$PO$_2^-$–PO$_4^{3-}$–HPO$_3^{2-}$	Dowex 1-X8 (Cl$^-$)	KCl (gradient), pH 6·8, t = 18°C: H$_2$PO$_2^-$, PO$_4^{3-}$, HPO$_3^{2-}$	[839]
H$_2$PO$_2^-$–PO$_4^{3-}$–HPO$_3^{2-}$	Dowex 1-X8 (Cl$^-$)	KCl (gradient), pH 11·4, t = 2°C: H$_2$PO$_2^-$, HPO$_3^{2-}$, PO$_4^{3-}$	[839]
As(III)-Sb(III)-Bi(III)	Dowex 1-X10 (Cl$^-$)	3M HCl: As, 0·3M HCl–1M HF: Sb, 1M NH$_4$Cl–1M NH$_4$F: Bi(III)	[576]
As(III)-Sb(III)	Amberlite CG4B (Cl$^-$)	4·5M HCl: As, 1M NaOH: Sb	[600]
As(III)-Bi(III)	Amberlite CG4B (Cl$^-$)	4·5M HCl: As, 3M H$_2$SO$_4$: Bi	[600]
As(V)-Sb(V)	AV-17 (Cl$^-$)	11M HCl: As, 0·1M H$_2$C$_2$O$_4$–0·3M (NH$_4$)$_2$C$_2$O$_4$: Sb	[838]
As(III)-Sb(III)	Dowex 1-X8 (Cl$^-$)	3M HCl: As, 10% ascorbic acid–0·4M HCl: Sb	[836]
Sb(III)-Bi(III)	AG 1-X8 (Br$^-$)	0·1M HBr: Sb	[814]
As(V)-Bi(III)	KU-2 (H$^+$)+KB-4 (Na$^+$)	aqueous solution, (pH 1–1·7): As	[640]
As(V)-As(III)	Dowex 1-X10 (Cl$^-$)	9M HCl: As(V), 0·1M HCl: As(III)	[840]

Table 6.6 (continued).

Species	Ion-exchanger and its ionic form	Eluent	Ref.
Sb(III)-Sb(V)	Dowex 50-X4 (H^+)	$9M$ HBr: Sb(III); $9M$ HBr-$0.05M$ KI: Sb(V) → Sb(III)	[377]
As(III)-Sb(III)	Dowex 2 (OH^-)	sample: As and Sb sulphides dissolved in Na_2S, $1.2M$ KOH: As, $3.5M$ KOH: Sb	[841]
As(III)-Bi(III)-Sb(III)	Amberlyst A-26 (Cl^-)	$4M$ HCl: As, $1M$ HCl-$4.5M$ $HClO_4$: Bi, Sb	[480]
Sb(V)-As(V)	Amberlite IRA-400 (SO_4^{2-})	H_2O made alkaline with KOH to phenolphthalein: Sb, 8% KOH: As	[841]
6A			
$Cr_2O_7^{2-}$-CrO_4^{2-}	KU-2 X6 (Na^+)	water-acetone mixture with acetone mole fraction $x = 0.8$: $Cr_2O_7^{2-}$, $x = 0.6$: CrO_4^{2-}	[830]
Cr(III)-Mo(VI)-W(VI)	AG 1-X8 (SO_4^{2-})	$0.05M$ H_2SO_4-0.1% H_2O_2: Cr; $3M$ H_2SO_4-0.1% H_2O_2: Mo, $2M$ NH_4NO_3-$0.5M$ NH_4OH: W	[591]
Mo(VI)-W(VI)-Cr(III)	Dowex 50-X8 (H^+)	$0.25M$ H_2SO_4-1% H_2O_2: Mo+W, $4M$ HCl: Cr	[842]
U(VI)-W(VI)-Mo(VI)	Dowex 1-X10 (Cl^-)	$0.5M$ HCl-$1M$ HF: U, $7M$ HCl-$1M$ HF: W, $1M$ HCl: Mo	[843]
W(VI)-U(VI)	Dowex 1-X10 (Cl^-)	$7M$ HCl-$1M$ HF: W(VI), $0.1M$ HCl: U	[843]
W(VI)-Mo(VI)	AV-17 (Cl^-)	$4M$ HCl-$1M$ HF: W, $1M$ HCl: Mo	[673]
Cr(III)-Mo(VI)-Cr(VI)-W(VI)	AV-17 (Cl^-)	sample in $2.5M$ HF, H_2O: Cr(III), $1M$ HCl: Mo, $0.1M$ HCl+$5g/l$ Na_2SO_3: Cr(VI)→Cr(III), $1M$ NH_4Cl-5% NH_4OH: W	[844]
Mo(VI)-U(VI)	Dowex 50-X8 (H^+)	tetrahydrofuran (90%)+$6M$ HNO_3 (10%): Mo, tetrahydrofuran (80%)+$3M$ HCl (20%): U	[845]
W(VI)-Mo(VI)	Dowex 1-X8 ($C_2O_4^{2-}$, Cl^-)	sample solution in $1.5M$ HCl-$0.5M$ $H_2C_2O_4$-$0.007M$ H_2O_2, $4M$ HCl-$0.1M$ citric acid: W, $1.9M$ NH_4Cl-$0.4M$ ammonium citrate: Mo	[824]
W(VI)-Mo(VI)	Dowex 50-X4 (H^+)	$9M$ HBr, $t = 60°C$: W, Mo	[377]
Mo(VI)-W(VI)	Dowex 1-X8 (F^-)	$27.5M$ HF: Mo, W	[821]

§ 6.8] Applications 455

$SO_3^{2-}-S^{2-}-SO_4^{2-}-S_2O_3^{2-}$	Diaion SA 100 (NO_3^-)	$0.1M\ NH_4NO_3-NH_3(70\%\ H_2O-30\%$ acetone), pH 9.7; SO_3^{2-}, S^{2-}, $0.1M\ NaNO_3$; SO_4^{2-}, $1M\ NaNO_3$; $S_2O_3^{2-}$	[846]
$S_2O_6^{2-}-S_3O_6^{2-}-S_4O_6^{2-}-S_5O_6^{2-}$	Dowex 1-X2 (Cl^-)	$1M\ HCl: S_2O_6^{2-}$, $3M\ HCl: S_3O_6^{2-}$, $6M\ HCl: S_4O_6^{2-}$, $9M\ HCl: S_5O_6^{2-}$	[847]
$SeO_3^{2-}-SO_4^{2-}$	AV-17 (OH^-)	$0.01M\ HCl: SeO_3^{2-}$, $4M\ HCl: SO_4^{2-}$	[848]
S(VI)-Se(VI)-Te(VI)-Po	Dowex 1-X4 (Cl^-)	$12M\ HCl: SO_4^{2-}$, $6M\ HCl: Se$, $3M\ HCl: Te$, $1M\ HClO_4: Po(RaF)$	[849]
Se-Te	AV-17 (Cl^-)	$6.6M\ HCl: Se$, $0.2M\ H_2C_2O_4-0.5M\ (NH_2)_2CS: Te$	[852]
Se(IV)-Te(IV)	Anex S X8 (Cl^-)	$6M\ HCl: Se$, $1-2M\ HCl: Te$	[851]
$TeO_3^{2-}-SeO_3^{2-}-SO_3^{2-}$	Dowex 1-X8 (OH^-)	$0.5M\ NaOH-3M\ NH_3: TeO_3^{2-}$, SeO_3^{2-}, $2M\ NaOH: SO_3^{2-}$	[853]
Te(VI)-Se(VI)-Se(IV)	Amberlite IRA-400 (CH_3COO^-)	CH_3COOH, pH 2.8: Te	[854]
Se(IV)-Te(IV)	KU-2 (H^+)	$0.1M\ HCl: Se$	[855]
S(VI)-Se(IV)	Dowex 50W-X8 (H^+)	sample solution in HCl with excess Fe(III) brought to pH 1.8–2 with ammonia, $0.01M\ HCl: SO_4^{2-}$, $0.5M\ HCl-SeO_3^{2-}$, $6M\ HCl: Fe^{3+}$	[857]
Se(IV)-Te(IV)	Dowex 50W-X4 (H^+)	$12M\ HBr: Se, Te$	[377]
7A			
Mn(II)-Re(VII)-Tc(VII)	Amberlite IRA-400 (Cl^-)	$0.2M\ HCl: Mn$, $5\%\ NH_4SCN-0.2M\ HCl: Re$, $4M\ HNO_3: Tc$	[860]
Mn(II)-Re(VII)-Tc(VII)	Dowex 1-X8 (SCN^-)	$0.5M\ HCl-0.025M\ NH_4SCN: Mn$, $0.5M\ HCl-0.5M\ NH_4SCN: Re$, $4M\ HNO_3$ or $0.5M\ HClO_4: Tc$	[861][862]
Re(VIII)-Tc(VII)	Dowex 1-X8 (ClO_4^-)	$0.2M\ HClO_4: Re$, $0.2-2M\ HClO_4: Tc$	[864]
Re(VII)-Tc(IV)	AV-17 (Cl^-)	$10M\ HCl: Re$, $4M\ NH_4OH: Tc$	[865]
7B			
$ClO_2^--Cl^--ClO_3^--ClO_4^-$	Dowex 1 (HCO_3^-)	$0.5M\ KHCO_3: ClO_2^-$, Cl^-, $1M\ KNO_3: ClO_3^-$, $1M\ NaBF_4: ClO_4^-$	[866]

Table 6.6 (continued)

Species	Ion-exchanger and its ionic form	Eluent	Ref.
BrO_4^--BrO_3^--Br^-	Dowex 1 (HCO_3^-)	$0.5M$ $KHCO_3$: BrO_2^-, $1M$ $KHCO_3$: BrO_3^-, $1M$ KNO_3: Br^-	[867]
IO_4^--IO_3^--I^-	Amberlite IRA-400 (NO_3^-)	H_2O: IO_4^-, $2M$ NaOH-$2M$ KNO_3: IO_3^-, I^-	[869]
ClO_4^--ClO_3^--Cl^-	KU-2 X6 (Na^+)	water-acetone mixture with acetone mole fraction $x = 0.9$: ClO_4^-, $x = 0.7$: ClO_3^-, $x = 0.5$: Cl^-	[870]
F^--Cl^--Br^--I^-	Dowex 2 (NO_3^-)	$1M$ $NaNO_3$, pH 10.4: F^-, Cl^-, Br^-, I^-	[871]
Cl^--Br^--I^-	Dowex 1-X10 (NO_3^-)	$0.5M$ $NaNO_3$: Cl^-, $2.0M$ $NaNO_3$: Br^-, I^-	[872]
F^--Cl^--Br^--I^-	Wofatite SBW (NO_3^-)	$1M$ $NaNO_3$: F^-, Cl^-, Br^-, I^-	[873]
F^--Cl^--Br^-	Wofatite SBW ($C_2O_4^{2-}$)	$1M$ $Na_2C_2O_4$: F^-, Cl^-, Br^-	[873]
F^--Cl^--Br^-	Wofatite SBW (NO_2^-)	$1M$ $NaNO_2$: F^-, Cl^-, Br^-	[873]
F^--Cl^--Br^--I^-	Dowex 1-X4 (OH^-)	$0.045M$ KOH: F^-, $0.32M$ KOH: Cl^-, $0.7M$ KOH: Br^-, $1.63M$ KOH: I^-	[875]
I^--Br^--Cl^-	zirconium oxide (NO_3^-)	$0.15M$ KNO_3: I^-, $0.25M$ KNO_3: Br^-, $1M$ KNO_3: Cl^-	[96]
IO_3^--BrO_3^--ClO_3^-	Dowex 21K (NO_3^-)	$0.5M$ $NaNO_3$: IO_4^-, 0.5-$2.0M$ $NaNO_3$: BrO_3^-, ClO_3^-	[879]
ClO_3^--BrO_3^--IO_3^-	KU-2 X6 (Na^+)	water-acetone mixture with acetone mole fraction $x = 0.77$: ClO_3^-, $x = 0.6$: BrO_3^-, $x = 0.25$: IO_3^-	[870]
I^--Br^--Cl^--F^-	SBS (Na^+)	water-acetone mixture with acetone mole fraction $x = 0.8$: I^-, $x = 0.7$: Br^-, $x = 0.5$: Cl^-, $x = 0.3$: F^-	[876]
IO_3^--BrO_3^--ClO_3^-	Dowex 2 (OH^-)	$0.2M$ KOH: IO_3^-, BrO_3^-, $0.2M$ K_2SO_4: ClO_3^-	[878]
Cl^--Br^--I^-	MG-36 (CH_3COO^-)	$0.01M$ CH_3COONa: Cl^-, $0.025M$ CH_3COONa: Br^-, $0.1M$ NaOH: I^-	[877]

8

Species	Ion-exchanger and its ionic form	Eluent	Ref.
Ni(II)-Co(II)-Fe(III)	Dowex 1-X10 (Cl^-)	12 or $9M$ HCl: Ni, $4M$ HCl: Co, $0.5M$ HCl: Fe	[383], [890]
Ni-Co-Fe	Wofatite SBW (Cl^-)	$8M$ HCl-$0.1M$ H_2O_2 + 30% C_2H_5OH: Ni, $4M$ HCl-$0.1M$ H_2O_2 + 30% C_2H_5OH: Co, $0.5M$ HCl-$0.1M$ H_2O_2: Fe	[623]

§ 6.8] Applications 457

Mixture	Resin	Conditions	Ref.
Ni-Co-Fe(III)	Dowex 1-X8 (Cl⁻)	95% C_2H_5OH-0·3M HCl: Ni, 70% C_2H_5OH-0·3M HCl: Co, 40% C_2H_5OH-0·3M HCl: Fe	[888]
Fe(II)-Fe(III)	Amberlite IRA-400 (Cl⁻)	4M HCl: Fe(II), 0·5M HCl: Fe(III)	[882]
Fe(II)-Fe(III)	Dowex 2-X8	0·09M tartaric acid (pH 2): Fe(II), 1M HCl: Fe(III)	[883]
Fe(III)-Fe(II)	KU-2 (H⁺)	0·25M $H_2C_2O_4$: Fe(III), 2M HCl: Fe(II),	[881]
Fe(II)-Fe(III)	tin arsenate (H⁺)	1M NH_4NO_3: Fe(II), 3% HNO_3–1M NH_4NO_3: Fe(III)	[886]
Co(II)-Fe(III)	AG 50W-X8 (H⁺)	1·5M $HClO_4$: Co, Fe	[569]
Fe(III)-Ni(II)-Co(II)	Amberlite IRA-400 (nitroso-R-salt)	0·5M H_3PO_4: Fe, 0·3M HCl: Ni, 5M $HClO_4$: Co(II)	[894]
Ni(II)-Co(II)-Fe(III)	Dowex 1-X8 ($H_2PO_4^-$)	0·5M H_3PO_4: Ni, Co, 3M H_3PO_4: Fe	[582a]
Fe(III)-Co(II)-Fe(II)-Ni(II)	Amberlite 200 (H⁺)	0·64M HCl–75% acetone: Fe (III), 0·64M HCl–89% acetone: Co(II), 0·64M HCl–93% acetone: Fe(II), 1·07M HCl–57% acetone–36% dimethylformamide: Ni(II)	[54a]
Ni-Co-Fe(III)	Dowex 1-X8 (SCN⁻)	0·04M KSCN: Ni, 0·024M KSCN: Co, 12M HCl, H_2O: Fe	[891]
Co-Ni-Fe(III)	Dowex 2-X8	0·09M tartrate (pH 4): Co, Ni, 0·09M tartaric acid–0·1M HCl: Fe	[749]
Fe(III)-Co(II)- Ni(II)	Dowex 50-X8 (H⁺)	tetrahydrofuran (80%) + 3M HCl (20%): Fe, tetrahydrofuran (90%) + 6M HCl (10%): Co, 6M HCl; Ni	[884]
Fe(III)-Co(II)- Ni(II)	Dowex 2-X8 (Cl⁻)	sample: Fe(III), Co(III) and Ni(II)–EDTA complexes 0·1M KCl: Fe, Co, Ni	[892]
Ni-Co	Amberlite CG-120 (H⁺, Na⁺)	0·1M $H_2C_2O_4$–$Na_2C_2O_4$(pH 2·6): Ni, Co	[885]
Fe(III)-Co(II)-Ni(II)	Amberlite CG-120 (NH_4^+)	0·1M ammonium sulphosalicylate (pH 5, t = 80°C): Fe, 0·5M ammonium sulphosalicylate (pH 5, t = 80°C): Ni+Co	[496]
Fe(III)- Ni(II)-Co(II)	Dowex 1-X8 (Cl⁻)	6M HCl (10%)–acetone (90%): Fe, acetone (70%) 2M HCl (30%): Ni-Co	[889]

Table 6.6 (continued)

Species	Ion-exchanger and its ionic form	Eluent	Ref.
Pt-Pd-Rh-Ir	Dowex 50 (H^+)	sample evaporated from $HClO_4$, H_2O: Pt, 0·05–0·5M HCl: Pd, 2M HCl: Rh, 4–6M HCl: Ir	[905]
Rh(III)-Ir(IV)	Amberlite IRA-400 (Cl^-)	sample in 2% NaCl–0·1M HCl+5% Br_2aq, 2% NaCl–0·1M HCl+5% Br_2aq: Rh, 5M NH_4OH–1M NH_4Cl, 6M HCl: Ir	[900]
Rh(III)-Ir(IV)	Amberlite IRA-400 (Cl^-)	2M HCl: Rh, extraction with 6M HCl in Soxhlet apparatus: Ir	[898]
Rh(III)-Pd(II)-Pt(IV)	Amberlite IRA-400 (Cl^-)	sample in 2M HCl+NaCl: Rh, 9M HCl: Pd, 2·4M $HClO_4$: Pt	[895]
Ir(III)-Pd(II)	Dowex 1-X10 (Cl^-)	6M HCl: Ir, 6M HCl–1M $ZnCl_2$: Pd	[383]
Pt, Pd, Ir-Rh	Dowex 50W-X8 (H^+)	sample+EDTA+hydroquinone, precipitation of hydroxides brought to pH 2·8+Cl_2aq, passed through column 10% Cl_2aq: Pt+Pd+Ir, 3M HCl: Rh	[907]
Pt(IV)-Pd(II)	Amberlite IR-100 (NH_4^+)	sample: chloride solution+NH_4OH, 0·025M NH_4OH–0·025M NH_4Cl: Pt, 1M HCl: Pd	[908]
Ir(IV)-Rh(III)	Dowex 50W-X8 (H^+)	sample in 0·3M HCl+CS$(NH_2)_2$, 3M HCl: Ir, 6M HCl, $t = 74°C$: Rh	[906]
Pd(II)-Pt(IV)	Permutit ES (OH^-)	sample: chloride complexes, 1M NaOH: Pd, H_2O, 2·5M HNO_3: Pt	[909]
Ir(III)-Pt(IV)	Permutit ES (OH^-)	sample in dil. HCl+$Na_2C_2O_4$, 1M NaOH: Ir, 2·5M HNO_3: Pt	[909]
Rh(III)-Ir(IV)	Amberlite IRA-400 (Cl^-)	0·8M HCl–0·002M Ce$(SO_4)_2$: Rh, 16M HNO_3, $t = 74°C$: Ir(IV)	[907]
Rh(III)-Pt(IV)	Permutit ES (OH^-)	sample in dil. HCl, 1M NaOH: Rh, 2·5M HNO_3: Pt	[909]
Rh-Ir-Pd	Amberlite IR-4B (Cl^-)	sample in 0·1M NaOH, 1M CH_3COOH: Rh, 1M NaOH: Ir, dimethylglyoxime+10% NH_4OH: Pd	[910]
Rh, Ir, Pt-Pd	Amberlite IRC-50 (NH_4^+)	sample: neutral solution of chloride complexes+$NaNO_2$, H_2O: Rh+Ir+Pt, 1M HCl: Pd	[910]

Rh, Ir-Pd-Pt	Dowex 1-X4 (Br⁻)	conc. Na$_2$SO$_3$, 7M HCl: Pt(IV→II) 8·3M HBr+2% N$_2$H$_4$·HCl: Rh+Ir, 10M HBr+Br$_2$: Pd, 0·5M HBr+1% CS(NH$_2$)$_2$: Pt	[459]
Pt(II)-Pd(II)	sulphoguanidine exchanger	0·5M HCl: Pt, 3M HCl: Pd	[914]
Pd(II)-Pt(IV)	Amberlyst A-26 (Cl⁻)	0·2M HCl-4·9M HClO$_4$: Pd, Pt	[480]
Pt-Pd	Retardion 11A8 (H⁺, Cl⁻)	sample in 3M HCl+2% N$_2$H$_4$·HCl; 3M HCl: Pt, 1% CS(NH$_2$)$_2$ in 0·5M HCl: Pd	[77]
Pd(II)-Pt(IV)-Ir(IV)-Rh(III)	KU-2 (H⁺)	acetylacetone: Pd, 50% TBP in (C$_4$H$_9$)$_2$O Ir(IV)→Ir(III), 50% TBP in (C$_4$H$_9$)$_2$O sat. SO$_2$: Pt, CCl$_4$, CCl$_4$ sat. Cl$_2$: oxidized Ir(III)→Ir(IV), 50% TBP in CCl$_4$ sat. Cl$_2$: Ir, 4M HCl: Rh	[397]

cause of uncontrolled losses, and therefore the separations were conducted in HCl–HF aqueous mixtures in which case retention of germanium on the anion-exchanger takes place at lower acid concentrations [576]. Germanium can be separated from lead on an anion-exchanger from hydrobromic acid solutions [814]. In aqueous solutions of pH 2 germanium is not retained either by a sulphonic acid cation-exchanger or by a weakly alkaline anion-exchanger, so it can be separated from tin and lead [815]. An interesting principle was used for separating lead(II) and cadmium(II) ions. Pb(II) was found to diffuse into the deeper layers of the inorganic cation-exchanger, crystalline hafnium monohydrogenphosphate monohydrate [$Hf(HPO_4)_2 \cdot H_2O$], while Cd(II) remained near the surface. This property was used to improve the selectivity of the method by masking cadmium with a large ligand (1,10-phenanthroline), which prevents the metal from adsorbing with no effect on lead fixation [794].

Group 5 Elements

Sub-group 5A. V, Nb, Ta, (Pa). In minerals vanadium does not usually occur together with niobium and tantalum, and its chemical properties differ greatly from those of the other elements of this sub-group. In the literature, much attention has been paid to the separation of the very similar niobium and tantalum. These elements have strong tendencies to hydrolyse and can be kept in solution only in the presence of very strong complexing agents. Most of the separations have been done on strongly alkaline anion-exchangers by using fluoride [817–822] (Fig. 6.75), oxalate [464], (Fig. 6.75) [823,824] and other [825] complexes. They can also be separated as tantalates and niobates in alkaline solutions [826]. Tantalum and niobium have also been separated in an ethanol–HCl solution [827].

In acidic solutions vanadium is usually present as the VO_2^+ cation. Since it is singly charged, it can easily be separated from bivalent and multivalent metals on a cation-exchanger [828]. Because of the ease of reduction of V(V) to V(IV) such separations should be conducted in the presence of hydrogen peroxide.

Sub-group 5B. N, P, As, Sb, Bi. Inorganic nitrogen may exist as ammonium salts, nitrates or nitrites. Usually there is no need in the course of analysis to separate inorganic nitrogen compounds from the remaining elements of the subgroup. Nitrates and nitrites may be separated from each other by partition chromatography on a sulphonic-acid cation-exchanger in aqueous acetone solutions [830]. Inorganic phosphorus usually occurs as phosphates and as such it can be easily separated from metal

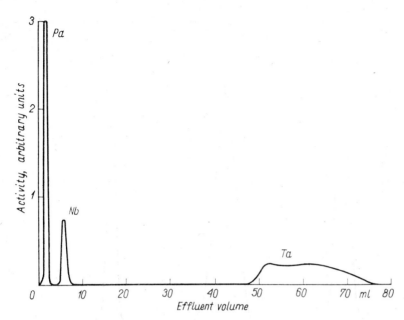

Fig. 6.75 Separation of protactinium, niobium and tantalum (from [817], by permission of the copyright holders, the American Chemical Society). Column: 12·5 cm × 0·021 cm², Dowex 1 (Cl^-, F^-) (200–230 mesh) eluent: 9M HCl + 0·05M HF, flow velocity 0·3 cm/min

cations on a strongly acidic cation-exchanger, even under static conditions [834]. Orthophosphates and polyphosphates of varying degrees of condensation were separated on a strongly basic anion-exchanger in a potassium chloride medium [832–834]. Hypophosphates, phosphites and orthophosphates were separated similarly [835,839].

Arsenic, antimony and bismuth have differing abilities to form complexes with halide ions, and this is the basis for separating them on anion-exchangers [576,600,835–839] (Fig. 6.76). In this way arsenic of different oxidation states can be separated [840]. Use has also been made of the absorption of antimony(V) on cation-exchangers in concentrated solutions of hydrogen halides [281,297,377], the formation of complexes with organic acids [836–838], the limited dissociation of arsenic acid [640], etc. The separation of arsenates and antimonates on an anion-exchanger in sulphate form [841] and of thio-salts of these elements on an anion-exchanger in a basic medium [841] have been described. Arsenic and antimony were also separated on tin dioxide [943].

Group 6 Elements

Sub-group 6A. Cr, Mo, W, U. Elements of this group usually occur in oxidation state +VI except for chromium, for which the +III state is

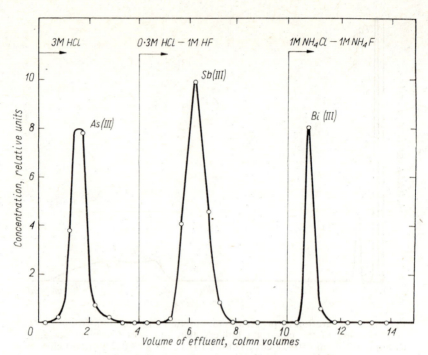

Fig. 6.76 Separation of arsenic, antimony and bismuth (from [576], by permission of the copyright holders, the American Chemical Society). Column: 4×0.25 cm^2, Dowex 1-X10 (Cl$^-$) (170–220 mesh)

more characteristic. Since molybdenum and tungsten usually occur as anions, in most methods for separation of these elements strongly basic anion-exchangers are used. Some difficulties may arise when weakly acidic solutions are used, since Mo(VI) and W(VI) have a tendency to hydrolyse. However, these difficulties can be avoided by adding appropriate complexing agents (e.g. F$^-$-ions).

Difficulties in ion-exchange separation of Cr(III) may be due to the ability of chromium to form inert complexes, and Cr(III) can be present in solution in the form of several co-existing complexes of differing charge [19,590].

Chromates can be separated from dichromates by partition chromatography on a sulphonic acid cation-exchanger in aqueous acetone solutions [830]. Chromium(III) was separated from Mo(VI) and W(VI) on a cation-exchanger in sulphuric acid medium containing hydrogen peroxide [842], Mo(VI) and W(VI), and also U(VI) can be separated on a strongly basic anion-exchanger from acidic solutions containing fluorides [673,677,678,821,822,843]. In similar conditions Mo and W can be separated from Cr(III) and Cr(VI) [844]. Separation on anion-exchangers was also done in a chloride–citrate medium [824] and also in sulphate solutions containing hydrogen peroxide [591].

W(VI) and Mo(VI) have been separated on a cation-exchanger in concentrated HBr solution [377].

Sub-group 6B. O, S, Se, Te, Po. The lighter elements of group 6B are typical non-metals, but with increasing atomic number the metallic character becomes more pronounced. Apart from oxygen, the elements of this sub-group typically appear in several different oxidation states.

Anion-exchange in aqueous-acetone solutions of nitrates was used for separating sulphates, sulphites, thiosulphates and sulphides [846]. Similarly, polythionates were separated by stepwise elution with HCl solutions [847]. Se(IV), Te(IV) and Po(IV) in concentrated hydrochloric-acid solutions have different affinities for the strongly basic anion-exchanger, and this may provide a basis for separating them from each other and from sulphur(VI) [838, 849–852]. A strongly basic anion-exchanger in hydroxide form was used for separating tellurites, selenites and sulphites [853]. Se(IV) and Se(VI) were separated from Te(VI) on an anion-exchanger in chloride form from a weakly acidic solution; here use was made of the fact that telluric acid is a relatively weak acid [854]. In weakly acidic solution tellurium is retained on cation-exchangers [855, 856], probably as the TeO^{2+} cation [855]; in this way selenium and tellurium have been separated from each other. In weakly acidic solutions containing a large excess of iron, selenium too can be retained on a cation-exchanger, evidently as the complex ion $FeHSeO_3^{2+}$, and it can be separated in this way from sulphates [877]. Selenium may subsequently be separated from iron by selective elution with dilute hydrochloric acid. Sulphates and selenites may be separated from a number of cations by passing the solution through a column of a strongly acidic cation-exchanger [856, 858].

In concentrated hydrobromic acid solutions quadrivalent selenium, tellurium and polonium are retained on a sulphonated cation-exchanger, the partition coefficients increasing with the atomic number of the elements [377].

Group 7 Elements

Sub-group 7A. Mn, Re, Tc. The most stable manganese oxidation state, especially in acidic solutions, is Mn(II), while for rhenium and technetium the oxidation state VII is most typical. The ReO_4^- and TcO_4^- ions have very high affinities for strongly basic anion-exchangers [276, 383, 859, 860] and they can be easily separated on an anion-exchanger of that type from Mn^{2+} [860]. Rhenium is easily eluted from an anion-exchanger with thiocyanate solutions [860–862]. The selective separation of rhenium from multicomponent mixtures has been based on that principle [862]. Quantitative elution of technetium from the column causes great difficulties and may be associated with the reduction of Tc(VII) on the ion-exchanger [861,863]. Rhenium was also separated from technetium by eluting with perchloric acid [863,864].

Re(VII) and Tc(IV) were also separated on an anion-exchanger in chloride form, by eluting Re with a concentrated HCl solution, then Tc(IV) with ammonia [865].

Sub-group 7B. The halogens: *F, Cl, Br, I, At*. The lighter elements of sub-group VIIB are typical non-metals, but the heaviest do have some metallic properties. The oxidation state $-I$ is most popular in this subgroup (in halides), but, with the exception of fluorine, the elements also occur as stable anions, in which the halogens have the oxidation states $+I$, $+III$, $+V$ and $+VII$. Chlorites, chlorates and chlorides can be separated on a strongly basic anion-exchanger in a weakly alkaline medium by elution with potassium bicarbonate, and subsequently with KNO_3 and $NaBF_4$ solutions [866].

Bromites, bromates and bromides were separated in a similar manner [867]. Iodides, iodates and periodates can be separated by elution with alkaline potassium nitrate solutions from anion-exchangers [868,869]. Perchlorates, chlorates and chlorides have also been separated by partition chromatography on a cation-exchanger in aqueous-acetone solutions [870].

Halides can readily be separated on strongly basic anion-exchangers by elution with nitrate, acetate or hydroxide solutions [871–875] (Fig. 6.77). In aqueous solutions, halides are eluted in the order of increasing atomic numbers. This order may be reversed in non-aqueous solutions or in

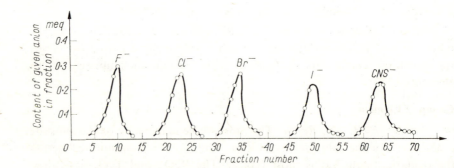

Fig. 6.77 Separation of halide ions by stepwise elution (from [875], by permission of the copyright holders, Akad. Nauk SSSR). Column: 50 cm × 0·79 cm², Dowex 1-X4(OH⁻) (100–200 mesh); flow velocity 1 cm/min; elution of F⁻ with 0·045N KOH, Cl⁻ with 0·32N KOH, Br⁻ with 0·70N KOH, I⁻ with 1·63N KOH, CNS⁻ with 1·93N KOH

mixed solutions, e.g. in dimethylformamide [874]. A reversed elution order was also obtained in elution of halides with potassium nitrate solution from the hydrated zirconium oxide inorganic exchanger [96]. When halogens are separated by partition chromatography on a cation-exchanger the order of elution is also I^- to F^- [876].

Halogens have been separated as anions in which the elements are present in oxidation state $+V$. In that case the first to be eluted from the anion-exchanger are iodates, then come bromates, and finally chlorates [878,879]. A reversed order of these anions has been obtained in partition chromatography on a cation-exchanger in aqueous acetone solutions [870].

Astatine in concentrated HCl and LiCl solutions is retained on a sulphonated cation-exchanger [880]. Thus, it can be separated from the other halogens.

Group 8 Elements

Iron triad: Fe, Co, Ni. Iron, cobalt and nickel are typical transition elements and can form complexes with many ligands. Complex forming is very important for most of the known methods of separating these metals on ion-exchangers. Iron(II) and iron(III), which are often present together can be separated both on cation-exchangers [881] and on ion-exchangers [882,883], by making use of their differing abilities to form chloride, tartrate or oxalate complexes.

For separating the iron-triad elements cation-exchangers have been much less used than anion-exchangers. Good separation on a sulphonic acid cation-exchanger was obtained when a mixture of hydrochloric acid and tetrahydrofuran was used as eluent [884]. Nickel was separated from cobalt in an oxalate medium [885]. Iron can be separated from nickel and cobalt by elution from a cation-exchanger with ammonium sulphosalicylate solutions [496].

For separating cobalt and iron, as well as Fe(II) and Fe(III) inorganic ion-exchangers were also used [886,887].

A number of separation methods of the iron-triad metals on anion-exchangers in hydrochloric acid [383,623,667,879,890] (cf. Fig. 6.59), thiocyanate [891], tartrate [749], etc., solutions have been described. These metals have also been separated as their EDTA complexes [892]. Iron has been separated from nickel and cobalt by precipitation chromatography on an anion-exchanger in molybdate form [893]. In many cases the use of non-aqueous and mixed solutions gives a better separation [623, 667,889]. Cobalt was separated from iron on a cation-exchanger in concentrated hydrobromic acid solutions [377]. Iron-triad metals were also separated on anion-exchangers substituted with nitroso-R-salt [894].

Ruthenium and osmium triads (platinum metals): *Ru, Rh, Pd, Os, Ir, Pt*. Separation of platinum metals is one of the most difficult problems in analytical chemistry. Most of the elements can occur in several oxidation states. For some of the platinum metals reduction to the metal has been observed on contact with the ion-exchanger [399,895]. Platinum metals have strong tendencies to form complexes, and often complicated equilibrium states occur in the solutions between aqueous complexes differing in composition and charge [896,897].

This applies especially to rhodium and iridium. Platinum metals can also form complexes with mixed ligands as well as multinuclear complexes. This means not only that separation of these elements is difficult, but also that the course of any separation depends on the history of the solution. Thus, data in the literature are often inconsistent and the separations described are seldom fully quantitative.

All platinum metals form strong chloride complexes, which are retained by anion-exchangers [383,985,897–902,974]. It is convenient, however, to separate osmium and ruthenium from the remaining elements by distillation as the tetraoxides [903,904]. Ion-exchange methods of separating platinum, palladium, rhodium and iridium were based mostly on the use of anion-exchangers in chloride solutions, where use was made of valency changes of the metals for reaching an adequate distribution coefficient (cf. Fig. 6.62). Good results in the separation both of trace and of milligram amounts of platinum metals have been obtained by using a strongly basic anion-exchanger in hydrobromic acid solutions (Fig. 6.36 and 6.78). Platinum, palladium, rhodium and iridium were also separated on a cation-exchanger from perchlorate solutions [905], but these results were later questioned. Rh(III) and Ir(IV) react with thiourea to form complexes of different signs which can be separated on a cation-exchanger [906]. Rhodium was separated from the remaining platinum metals on a cation-exchanger, after converting it into cation form [907]. The ability to form cationic complexes with ammonium was used to separate palladium from platinum, rhodium and iridium [908]. The platinum metals were also separated by ion-exchange by making use of the differing tendencies of their chloride complexes to hydrolyse [909,910]. Platinum and palladium were separated on an amphoteric ion-exchanger in HCl solution containing hydrazine [77]. Platinum metals were also separated by partition chromatography, with a sulphonic-acid cation-exchanger as carrier of the aqueous phase [396].

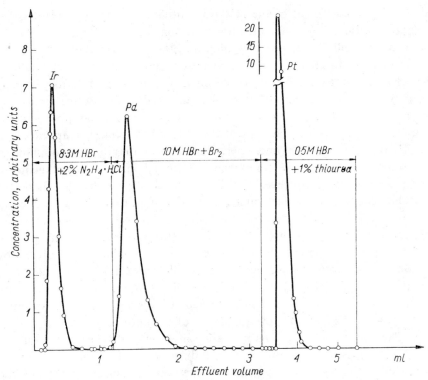

Fig. 6.78 Separation of iridium, palladium and platinum (from [459], by permission of the copyright holders, Akadémiai Kiadó). Column: 4 cm × 0·031 cm², Dowex 1-X4 (Br⁻) (2 μm ⩽ ⌀ ⩽ 53 μm grain size); temperature 75°C, flow velocity 0·6 cm/min

Platinum metals can be separated from base metals by retaining the latter on a sulphonic-acid cation-exchanger from hydrochloric acid or chloride solutions of suitable acidity [649,911,912]. Another method is selective retention of palladium(II) and platinum(IV) on a weakly basic cellulose anion-exchanger from thiocyanate solutions [630] or use of chelating ion-exchangers [913,914]. Os, Ir, Rh, Pt, Ru, Pd and also Au may be separated on a highly selective chelating ion-exchanger containing guanidine groups, available on the market as Srafion NMRR [69,72,644].

6.8.3 Methods Utilizing Reversal of Elution Order

As was already mentioned in Section 6.8.1.1, the effect of separation and time efficiency may depend on the sequence in which the separated elements appear in the effluent, and thus on the relative affinity of the ions for the ion-exchanger. When the quantities of elements separated differ greatly (e.g. separation of traces from a matrix), it is usually advantageous that the partition coefficient of the macroconstituent be small so that it is the first to appear in the effluent. However, sometimes, especially when the separation coefficient is not very high and the partition coefficient of the macroconstituent eluted first is greater than zero, it may be very difficult to achieve an adequate decontamination coefficient. In such a case, reversal of the elution sequence may greatly facilitate and accelerate the separation process [460]. It may happen that, in some chromatographic systems even when separating trace quantities, the component that appears first in the effluent gives an assymetric elution curve, so that its tail contaminates the fractions of successive components. Then, again, it is advantageous to reverse the elution sequence. The differences in the widths and shapes of the elution curves for a particular component of a mixture may be due not only to differences in the absolute weight proportions of the separated components in the given mixture, but also to the differences in the magnitude of the 'analytical signal' given by them when the effluent is analysed by a nonspecific instrumental analytical method. An example is the measurement of the radioactivity of the effluent when the sample is analysed by neutron-activation analysis, since the activation cross-sections may differ for different elements by several orders of magnitude.

The above considerations indicate that reversing the elution sequence may sometimes prove useful or even indispensable. The analyst or radiochemist should be well aware of the methods of achieving this. If the elements are separated, e.g. on a cation-exchanger, by elution with complexing agents under conditions where the differentiation of the partition coefficients depends mainly on the differences between the stability constants of the complexes, then use of an anion-exchanger in place of the cation-exchanger will reverse the elution sequence. This becomes clear, for instance, by examination of Eqs. (6.33) and (6.45).

For example, rare earths are eluted from a cation-exchanger by buffered citrate solutions in the order of decreasing atomic numbers [1,355, 358], but if an anion-exchanger is used in the same medium, the elution sequence is reversed [915]. Analogous changes in the elution sequence on passing from anion- to cation-exchanger have been observed for rare earth elements in α-hydroxybutyric acid solutions [420,466]. Similarly the elution sequence: Zr, Hf, obtained when eluting these elements from a cation-exchanger with sulphuric acid solution [803,804], is reversed when an anion-exchanger is used [531].

In accordance with the assumptions made in deriving Eqs. (6.33) and (6.45), the simple reversal of the elution sequence described above takes place at pH values where the cations and anionic complexes formed by the given cation with the eluate exist in solution in comparable quantities. At pH values where practically all the cations are bound as anionic complexes, the elution sequence depends on the relative affinities of the complex ions for anion-exchanger. And so, e.g. light lanthanides are eluted from an anion-exchanger substituted with EDTA anions, at pH 4·5–4·8, in the sequence: La, Ce, Pr, [218,332,369], i.e. in the reverse order to that observed when eluting lanthanides from a cation-exchanger with EDTA solutions [359]. The elution sequence for heavy rare-earth elements under the same conditions is: Lu, Tm, Ho, Tb, [218,332,369], and so it is the same as that when the elements are eluted from a cation-exchanger [359].

The elution sequence from a given type of ion-exchanger may depend on the type of eluent, and also on the factors affecting the differentiation of ion-exchange affinity of elements in the given system.

Rare-earth elements in a methanol solution of lithium nitrate are eluted from a strongly basic anion-exchanger in the sequence: Y, Eu, Nd, Pr, La [419], whereas when a methanolic α-hydroxybutyric acid solution is used, the elution sequence from an anion-exchanger of the same type is: La, Nd, Eu, Y, i.e. it is reversed [420].

The differing natures of the dependence of the partition coefficient on the concentration of hydrogen halide for anion-exchangers allows the elution sequence to be controlled in a number of cases. By changing the hydrochloric acid concentration, and possibly adding an organic solvent, we can obtain the following elution sequences of elements of sub-group 3B from a strongly basic anion-exchanger: Al, In, Ga [795], Ga, In, Al [796] and Al, Ga, In [796]. The elution sequence in the same ion-exchanger–solution system can be changed under favourable conditions by varying the temperature (cf. the elution curves for La and Tb in Fig. 6.50 and Figs. 6.11 and 6.12), and also, in the case of separating ions of differing charges, by varying the eluent concentration (cf. Fig. 6.9). An interesting reversal of selectivity can be obtained by passing from very dilute to very concentrated eluent solutions. The 'normal' elution sequence of alkaline-earth elements in dilute hydrogen halide solutions: Ca, Sr, Ba [245], is reversed in very concentrated solutions: Ba, Sr, Ca [377].

Reversal of the elution sequence can also be achieved by changing the type of functional group or, generally speaking, the type of ion-exchanger. For example the 'normal' elution sequence of alkali metals from a sulphonic-acid cation exchanger [74,244] or zirconium tungstate [95]: Na, Rb, Cs is reversed when antimonic acid is used as ion-exchanger [711]. Similarly, while halides are eluted with nitrate solutions from a strongly basic alkaline anion-exchanger in the sequence: Cl, Br, I [872], the same eluent gives a reversed sequence with zirconium oxide: I, Br, Cl [96].

Sometimes the elution sequence can be reversed with respect to the conventional ion-exchange method by using partition chromatography on ion-exchangers, as has been shown for separation of iodates, bromates and chlorates [870,871]. The selectivity can be reversed and thus the elution sequence changed by changing the oxidation state of one of the ions to be separated, as is possible with, e.g. platinum metals (cf. Fig. 6.35).

6.8.4 Qualitative Analysis

The uses of ion-exchangers in qualitative analysis have been widely discussed elsewhere [17,21], so only some aspects that are particularly relevant to trace analysis will be considered here.

Since many of the ion-exchangers now available are only slightly coloured or almost colourless small amounts or individual grains of exchanger can be used for spot tests: i.e. qualitative microchemical detection of given elements or ions in colour or precipitation reactions [916–922]. The ion-exchanger functions both as an agent for concentrating ions from solution and as the medium in which the colour reaction takes place. The detectability of elements in spot tests involving ion-exchangers is usually better than in the analogous conventional tests. Spot tests with use of ion-exchangers have also been used to detect some types of organic compounds, e.g. nitriles [923].

Another, chromatographic, application of ion-exchangers in qualitative trace analysis is the positive identification of an ion (element) on the basis of its retention volume in a given ion-exchanger–solution system. As was already mentioned, knowledge of the retention volume allows us to calculate the partition coefficient [Eq. (6.51)], and hence the selectivity coefficient (Eq. 6.15). If the effluent from the column is analysed by a non-selective instrumental method (e.g. measurement of radioactivity when the sample is subjected to neutron-activation analysis), small peaks are often observed on the elution curve; these peaks may be so small that unequivocal identification of the element (for instance by half-life measurement or γ-spectrometry in the case of radioactivity) is very difficult. In that case the partition coefficient or the selectivity coefficient determined from the peak position can serve as a basis for identification. The determination of the partition coefficient is not, however, a specific test, since several different ions may have similar selectivity coefficients. It has been shown, however, that the variation of the selectivity coefficient with temperature is characteristic for a given ion and may serve as basis for its identification [268]. This is carried out by determining the selectivity coefficient of the element from the retention volume of the given peak at several

§ 6.8] Applications 471

sufficiently remote temperatures, and comparing these data with the $k-1/T$ plots determined earlier for the elements whose presence in the sample is expected.

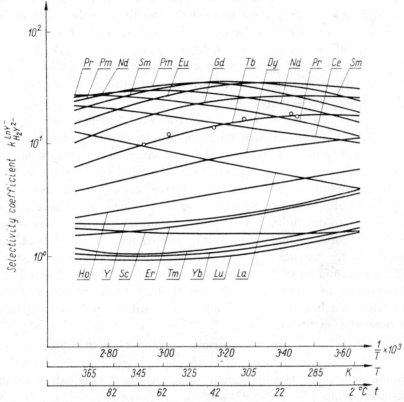

Fig. 6.79 Diagram illustrating qualitative identification of a sample (terbium) from chromatographic data obtained at different temperatures. System: Amberlite IRA-400 $(H_2Y^{2-})-Na_2H_2Y(aq)$. The solid lines represent the selectivity coefficient for the exchange reaction: $\frac{1}{2} R_2H_2Y + LnY^- = RLnY + \frac{1}{2} H_2Y^{2-}$, as a function of temperature. The empty circles represent experimental points for the unknown peak (from [268], by permission of the copyright holders, Elsevier Publ. Co.)

The above method has been used to identify ^{161}Tb in the products of irradiation of gadolinium oxide with neutrons in a nuclear reactor (Fig. 6.79) [268].

6.8.5 Quantitative Analysis

The trace quantities of elements separated by ion-exchange chromatography (ca. 100% in a given fraction) can be analysed by any method which is accurate and sensitive enough for the application.

The determination process is combined directly with the separation, when the effluent is analysed by an instrumental method continuously or after collecting the effluent fractions. In the latter case it usually suffices to analyse only some of the fractions collected, in order to draw the elution curve [452]. The quantity of element is defined by the area under the elution curve and can be calculated by integration, e.g. by using Eq. (6.54). The typical uses of ion-exchange chromatography, including its HPLC version and ion chromatography, as independent instrumental methods of analysis for quantitative determination of various ionic species in solution, have been discussed in Sections 6.7.2 and 6.7.3.

Quantities somewhat larger than traces of coloured ions can be determined by using narrow, long columns and the displacement method [924]. The quantity of element in the sample is defined in this case by the volume occupied in the column by the band of the given ion after the chromatogram is developed. On a narrow column (2-mm bore) filled with Amberlite IRA-400 saturated with 4-(2-pyridylazo)-resorcinol it was possible to determine 5–120 µg/ml of lead, with an error not exceeding 5%, in about 5 min, by measuring the length of the coloured layer appearing after passing the test solution [925]. A similar method was used for determining small amounts of mercury [926] and cadmium [927]. Individual grains of the ion-exchanger in Fe^{3+}-form or saturated with p-dimethylaminebenzylidenerhodanine were used as indicators in redox and precipitation titrations [928].

So-called ion-exchanger colorimetry, based on the direct measurement of the absorbance of the resin phase after sorption of coloured complex species, has recently been developed for the determination of some transition metals [928a,928b,975].

Among more specific applications of ion-exchangers in quantitative trace analysis, we can mention separation of anionic complexes of metals from their cationic forms after complexing the metal with a substoichiometric quantity of the reagent, in activation or isotope dilution [926] analysis.

It has been shown that individual ion-exchanger grains can be used as microstandards for extremely small (micro- and nanogram) quantities of various counter-ions [930,976,977]. Accurate determinations of such small quantities can be made on the basis of microscopic measurements of the diameter of the ion-exchanger bead. A similar principle was followed when preparing radioactive point standards [931].

References

[1] Marinsky, J. A., Glendenin, L. E. and Coryell, C. D., *J. Am. Chem. Soc.*, **69**, 2781 (1947).
[2] Katz, J. J. and Seaborg, G. T., *The Chemistry of the Actinide Elements*, Methuen, London, 1957.
[3] Seaborg, G. T., *Endeavour*, **18**, 15 (1959).
[4] Thompson, S. G., Cunningham, B. B. and Seaborg, G. T., *J. Am. Chem. Soc.*, **72**, 2798 (1950).
[5] Street, K., Jr., Thompson, S. G. and Seaborg, G. T., *J. Am. Chem. Soc.*, **72**, 4832 (1950).
[6] Thompson, S. G., Harvey, B. G., Choppin, G. R. and Seaborg, G. T., *J. Am. Chem. Soc.*, **76**, 6229 (1954).
[7] Ghiorso, A., Harvey, B. G., Choppin, G. R., Thompson, S. G. and Seaborg, G. T., *Phys. Rev.*, **98**, 1518 (1955).
[8] Nachod, F. C., (Ed.), *Ion Exchange, Theory and Application*, Academic Press, New York, 1949.
[9] Samuelson, O., *Ion Exchangers in Analytical Chemistry*, Wiley, New York, 1953.
[10] Nachod, F. C. and Schubert, J., (Eds.), *Ion Exchange Technology*, Academic Press, New York, 1956.
[11] Kitchener, J. A., *Ion Exchange Resins*, Methuen, London, 1957.
[12] Kunin, R., *Ion Exchange Resins*, Wiley, New York, 1958.
[13] Calmon, C. and Kressman, T. R. E., (Eds.), *Ion Exchangers in Organic and Biochemistry*, Interscience, New York, 1957.
[14] Salmon, J. E. and Hale, D. K., *Ion Exchange, A Laboratory Manual*, Butterworths, London, 1959.
[15] Helfferich, F., *Ion Exchange*, McGraw-Hill, London, 1962; University Microfilms International, Ann Arbor, MI, No. 2003413.
[16] Osborn, G. H., *Synthetic Ion Exchangers*, Champan and Hall, London, 1961.
[17] Samuelson, O., *Ion Exchange Separations in Analytical Chemistry*, Wiley, New York, 1963.
[18] Griessbach, R., *Austauschadsorption in Theorie und Praxis*, Akademie-Verlag, Berlin, 1957.
[19] Tatur, H. and Nowakowski, W., *Jonity, Teoria i zastosowanie w przemyśle* (*Ion Exchangers, Theory and Industrial Application*), PWT, Warsaw, 1955.
[20] Trémillon, B., *Jonity w procesach rozdzielczych* (*Ion Exchangers in Separation Processes*), PWN, Warsaw, 1970.
[21] Inczédy, J., *Analytische Anwendungen von Ionenaustauschern*, Akadémiai Kiadó, Budapest, 1964.
[22] Sabau, C., *Schimbul Ionic*, Editura Academiei Republicii Socialiste Romania, Bucharest, 1967.
[23] Amphlett, C. B., *Inorganic Ion Exchangers*, Elsevier, Amsterdam, 1964.
[24] Marinsky, J. A., (Ed.), *Ion Exchange*, Vol. 1., Dekker, New York, 1966.
[25] Shemyakin, F. M. and Stepin, V. V., *Ionoobmennyi khromatograficheskii analiz metallov* (*Ion-Exchange Chromatographic Analysis of Metals*), Metallurgiya, Moscow, 1965.
[26] Saldadze, K. M., Pashkov, A. B. and Titov, V. S., *Ionoobmennye vysokomolekulyarnye soedineniya* (*High-Molecular Ion Exchangers*), Gos. Nauch. Tekhn. Izd. Khim. Lit., Moscow, 1960.
[27] Bogatyrev, V. L., *Ionity v smeshannom sloe* (*Ion Exchangers in a Mixed Bed*), Izd. Khimiya, Leningrad, 1968.

[28] Rieman, W., III and Walton, H. F., *Ion Exchange in Analytical Chemistry*, Pergamon, Oxford, 1970.
[29] Hering, R., *Chelatbildene Ionenaustauscher*, Akademie-Verlag, Berlin, 1967.
[30] Chmutov, K. P., (Ed.), *Ionnyi obmen i ego primenenie* (*Ion Exchange and Its Application*), Izd. Akad. Nauk SSSR, Moscow, 1959.
[31] Samsonov, G. V., Prostyanskaya, E. B. and El'kin, G. E., *Ionnyi obmen. Sorbtsiya organicheskikh veshchestv* (*Ion Exchange. Sorption of Organic Compounds*), Izd. Nauka, Leningrad, 1969.
[32] Rabek, T. I., *Teoretyczne podstawy syntezy polielektrolitów i wymieniaczy jonowych* (*Theoretical Foundations of Synthesis of Polyelectrolytes and Ion Exchangers*), PWN, Warsaw, 1960.
[33] Morozov, A. A., *Khromatografiya v neorganicheskom analize* (*Chromatography in Inorganic Analysis*), Izd. Vysshaya Shkola, Moscow, 1972.
[34] Kokotov, Yu. A. and Pacechkin, V. A., *Ravnovesie i kinetika ionnogo obmena* (*Equilibrium and Kinetics in Ion Exchange Processes*), Izd. Khimiya, Leningrad, 1970.
[35] Egorov, E. V. and Makarova, S. B., *Ionnyi obmen v radiokhimii* (*Ion Exchange in Radiochemistry*), Atomizdat, Moscow, 1971.
[36] Marinsky, J. A., (Ed.), *Ion Exchange, Vol 2.*, Dekker, New York, 1969.
[37] Marinsky, J. A. and Marcus, Y., (Eds.), *Ion Exchange and Solvent Extraction*, Vol. 4, Dekker, New York, 1973.
[38] Marinsky, J. A. and Marcus, Y., (Eds.), *Ion Exchange and Solvent Extraction*, Vol. 5, Dekker, New York, 1973.
[39] Dorfner, K., *Ionenaustauscher*, Walter de Gruyter Co., Berlin, 1970.
[40] Korkisch, J., *Modern Methods for the Separation of Rarer Metal Ions*, Pergamon, Oxford, 1969.
[41] Marcus, Y. and Kertes, A. S., *Ion Exchange and Solvent Extraction of Metal Complexes*, Wiley-Interscience, London, 1969.
[42] Patterson, R., *An Introduction to Ion Exchange*, Heyden/Sadtler, London, 1970.
[43] Soldatov, V. S., *Prostye ionoobmennye ravnovesiya* (*Simple Ion-Exchange Equilibria*), Izd. Nauka i Tekhnika, Minsk, 1972.
[44] Kraus, K. A., *Ion Exchange. Trace Analysis*, Yoe, J. H. and Koch, H. J., Jr., (Eds.), Wiley, New York, 1957.
[45] Mizuike, A., in *Trace Analysis: Physical Methods*, Morrison, G. H., (Ed.), Interscience, New York, 1965.
[45a] Dybczyński, R., *Chromatografia jonitowa* (*Ion-exchange Chromatography*) Chapter 14.3 in *Poradnik fizykochemiczny* (*Physical-chemical Manual*), WNT, Warsaw, 1974, pp. 460-496.
[45b] Dybczyński, R., *Nukleonika*, **21**, 547 (1976).
[46] Świętosławski, W., *Przem. Chem.*, 1950, (29) VI, 41.
[46a] Steinnes, E., *Int. J. Appl. Rad. Isotop.*, **26**, 595 (1975).
[47] Bio-Rad Laboratories, Catalogue Y, August 1973.
[48] Lur'e, A. A., *Sorbenty i khromatograficheskie nositeli* (*Sorbents and Chromatographic Carriers*), Izd. Khimiya, Moscow, 1972.
[49] Gregor, H. P. and Frederick, M., *Ann. New York Acad. Sci.*, **57**, 87 (1953).
[50] Marhol, M., *At. Energy Rev.*, **4**, 63 (1966).
[51] Hodgman, C.D. (Ed.), *CRC Handbook of Chemistry and Physics*, The Chemical Rubber Publishing Co., Cleveland, 1961, p. 3479.
[52] *New Isoporous Resins*, Ref. IE. 22, Zerolit Ltd., 1964.

[53] Kun, K. H. and Kunin, R., *J. Polymer Sci.*, **B2**, 587, 839 (1964).
[54] *Amber-hi-lites*, No. 90, Rohm and Haas Co., November 1965.
[54a] Kawazu, K., *J. Chromatogr.*, **137**, 381 (1977).
[55] Kuroda, R., Kiriyama, T. and Ishida, K., *Anal. Chim. Acta*, **40**, 305 (1968).
[56] Remport-Horváth, Zs. and Ördögh, M., *Mikrochim. Acta*, **1972**, 491.
[57] *Dowex, Ion Exchange*, Dow Chemical Co., 1964.
[58] Millar, J. R., *Chem. Ind. (London)*, **1957**, 606.
[59] Hale, D. K., *Analyst (London)*, **83**, 3 (1958).
[60] Schmuckler, G., *Talanta*, **12**, 281 (1965).
[61] Bauman, A. J., Weetall, H. H. and Weliky, W., *Anal. Chem.*, **39**, 932 (1967).
[62] Klyachko, V. A., *Teoriya i praktika primeneniya ionoobmennykh materialov (Theory and Practice of Use of Ion Exchange Materials)*, Izd. Akad. Nauk SSSR, Moscow, 1955, p. 48.
[63] Sinyakovskii, V. G., *Selektivnye ionity (Selective Ion Exchangers)*, Izd. Tekhnika, Kiev, 1967.
[64] Hirsch, R. F., Gancher, E. and Russo, F. R., *Talanta*, **17**, 483 (1970).
[65] Luttrell, G. H., Jr., More, C. and Kenner, C. T., *Anal. Chem.*, **43**, 1370 (1971).
[66] Blasius, E. and Brosio, B., in *Chelates in Analytical Chemistry*, Flaschka, H. A. and Barnard, A. J., Jr. (Eds.), Arnold, London, 1967.
[67] Myasoedova, R. V., Eliseeva, O. P. and Savvin, S. B., *Zh. Analit. Khim.*, **26**, 2172 (1972).
[68] Loewenschuss, H. and Schmuckler, G., *Talanta*, **11**, 1399 (1964).
[68a] Jewett, G. L., Himes, R. P. and Anders, O. U., *J. Radioanal. Chem.*, **37**, 813 (1977).
[68b] Kramer, H. J. and Neidhart, B., *Radiochem. Radioanal. Lett.*, **22**, 209 (1975).
[68c] Vernon, F. and Nyo, K. M., *J. Inorg. Nucl. Chem.*, **40**, 887 (1978).
[68d] Orf, G. M. and Fritz, J. S., *Anal. Chem.*, **50**, 1328 (1978).
[68e] Burba, P. and Lieser, K. H., *Z. Anal. Chem.*, **291**, 205 (1978).
[68f] Phillips, R. J. and Fritz, J. S., *Anal. Chem.*, **50**, 1504 (1978).
[68g] Moyers, E. M. and Fritz, J. S., *Anal. Chem.*, **49**, 418 (1977).
[69] Koster, G. and Schmuckler, G., *Anal. Chim. Acta*, **38**, 179 (1967).
[70] Law, S. L., *Science*, **174**, 285 (1971).
[71] Law, S. L., *Int. Lab.*, **1973**, Sept./Oct., 53.
[72] Nadkarni, R. A. and Morrison, G. H., *Anal. Chem.*, **46**, 232 (1974).
[73] Dybczyński, R., *Rocz. Chem.*, **41**, 1689 (1967); Inst. of Nuclear Research, Report 'P', No. 810 (VIII) C, 1967.
[74] Dybczyński, R., *Nukleonika*, **12**, 927 (1967); Inst. of Nuclear Research, Report 'P' No. 809 (VIII) C, 1967.
[75] Saldadze, K. M. and Kopylova, V. D., *Zh. Analit. Khim.*, **27**, 956 (1972).
[75a] Fritz, J. S., *Pure Appl. Chem.*, **49**, 1547 (1977).
[75b] Blasius, E., Adrian, W., Klautke, G., Lorscheider, R., Maurer, P., Nguyen, V. B., Nguyen Tien, T., Scholten, G. and Stockemer, J., *Z. Anal. Chem.*, **284**, 337 (1977).
[75c] Grossman, P. and Simon, W., *Anal. Lett.*, **10**, 949 (1977).
[75d] Chmielowiec, J. and Simon, W., *Chromatographia*, **11**, 99 (1978).
[76] Hatch, M. J., Dillon, J. A. and Smith, H. B., *Ind. Eng. Chem.*, **49**, 1812 (1957).
[77] Dybczyński, R. and Sterlińska, E., *J. Chromatogr.*, **102**, 263 (1974).
[77a] Murr, H. and Hering, R., *Z. Anorg. Allgem. Chem.*, **427**, 180 (1976).
[78] Gregor, H. P., Hoeschele, G. K., Potenza, J., Tsuk, A. G., Feinland, R., Shida, M. and Teyssié, Ph., *J. Am. Chem. Soc.*, **87**, 5525 (1965).

[79] Tsuk, A. G. and Gregor, H. P., *J. Am. Chem. Soc.*, **87**, 5538 (1965).
[80] Barrer, R. M., *Proc. Chem. Soc.*, **1958**, 99.
[81] Barrer, R. M., *Endeavour*, **23**, 12 (1964).
[82] Barycka, I., *Wiad. Chem.*, **21**, 549 (1967).
[83] Sherry, H. S., in *Ion Exchange*, Vol. 2, Marinsky, J. A. (Ed.), Dekker, New York, 1969.
[84] Pekárek, V. and Veselý, V., *Talanta*, **19**, 1245 (1972).
[85] Caron, H. L. and Sugihara, T. T., *Anal. Chem.*, **34**, 1082 (1962).
[86] Smit, J. van R., Robb, W. and Jacobs, J. J., *J. Inorg. Nucl. Chem.*, **12**, 104 (1959).
[87] Krtil, J. and Křivý, I., *J. Inorg. Nucl. Chem.*, **25**, 1191 (1963).
[88] Van R. Smit, J., Robb, W. and Jacobs, J. J., *Nucleonics*, **17**, 116 (1959).
[89] Broadbank, R. W. C., Hands, J. D. and Harding, R. D., *Analyst (London)*, **88**, 43 (1963).
[90] Morgan, A. and Arkell, G. M., *Health Phys.*, **9**, 857 (1963).
[91] Broadbank, R. W. C., Dhabanandana, S. and Harding, R. D., *J. Inorg. Nucl. Chem.*, **23**, 311 (1961).
[92] Lavrukhina, A. K., Kourzhim, V. and Rodin, S. S., *Acta Chim. Acad. Sci. Hung.*, **33**, 309 (1962).
[93] Smit, J. van R., Jacobs, J. J. and Robb, W., *J. Inorg. Nucl. Chem.*, **12**, 95 (1959).
[94] Veselý, V. and Pekárek, V., *Talanta*, **19**, 219 (1972).
[95] Kraus, K. A., Philips, H. O., Carlson, T. A. and Johnson, J. S., *Proc. 2nd Intern. Conf. Peaceful Uses At. Energy, Geneva*, **28**, 3 (1958).
[95a] De, A. K. and Das, S. K., *Sep. Sci.*, **13**, 465 (1978).
[96] Tustanowski, S., *J. Chromatogr.*, **31**, 268 (1967).
[97] Ahrland, S., Grenthe, I. and Noren, B., *Acta Chem. Scand.*, **14**, 1059 (1960).
[98] Vydra, F., *Anal. Chim. Acta*, **38**, 201 (1967).
[98a] Abe, M., Nasir, B. A. and Yoshida, T., *J. Chromatogr.*, **153**, 295 (1978).
[98b] Jaffrezic-Renault, N., *J. Inorg. Nucl. Chem.*, **40**, 539 (1978).
[99] Amphlett, C. B., *Proc. 2nd Intern. Conf. Peaceful Uses At. Energy, Geneva*, **28**, 17 (1958).
[100] Gal, I. J. and Gal, O. S., *Proc. 2nd Intern. Conf. Peaceful Uses At. Energy, Geneva*, **28**, 24 (1958).
[101] Inoue, Y., *J. Inorg. Nucl. Chem.*, **26**, 2241 (1964).
[102] Clearfield, A., Smith, G. H. and Hammond, B., *J. Inorg. Nucl. Chem.*, **30**, 277 (1968).
[103] Larsen, E. M. and Cilley, W. A., *J. Inorg. Nucl. Chem.*, **30**, 287 (1968).
[104] Qureshi, M. and Rowat, J. P., *J. Inorg. Nucl. Chem.*, **30**, 305 (1968).
[104a] Qureshi, M. and Thakur, J. S., *Sep. Sci.*, **11**, 467 (1976).
[104b] Qureshi, M., Varshney, K. G. and Fatima, N., *Sep. Sci.*, **11**, 509 (1976).
[104c] Qureshi, M., Varshney, K. G. and Fatima, N., *Sep. Sci.*, **12**, 321 (1977).
[104d] Qureshi, M., Varshney, K. G., Gupta, S. P. and Gupta, M. P., *Sep. Sci.*, **12**, 649 (1977).
[104e] Qureshi, M., Varshney, K. G. and Fatima, N., *Sep. Sci.*, **13**, 321 (1978).
[104f] Qureshi, M. and Sharma, S. D., *Sep. Sci.*, **13**, 723 (1978).
[105] Alberti, G., Constantino, U., di Gregorio, F., Galli, P. and Toracca, E., *J. Inorg. Nucl. Chem.* **30**, 295 (1968).
[105a] Alberti, G., Bernasconi, M. G., Constantino, U. and Gill, J. S., *J. Chromatogr.*, **132**, 323 (1977).

[105b] Alberti, G., Bernasconi, M. G., Cascola, M. and Constantino, U., *J. Chromatogr.*, **160**, 109 (1978).
[106] Qureshi, M. and Nabi, S. A., *J. Inorg. Nucl. Chem.*, **32**, 2059 (1970).
[107] Qureshi, M. and Rathore, H. S., *J. Chem. Soc.*, (*A*), **1969**, 2515.
[108] Qureshi, M. and Kumar, V., *J. Chem. Soc.*, (*A*), **1970**, 1488.
[109] Qureshi, M., Zehra, N., Nabi, S. A. and Kumar, V., *Talanta*, **20**, 609 (1973).
[110] Gill, J. S. and Tandon, S. N., *Talanta*, **19**, 1355 (1972).
[111] Tandon, S. N. and Gill, J. S., *Talanta*, **20**, 585 (1973).
[112] Qureshi, M. and Nabi, S. A., *Talanta*, **19**, 1033 (1972).
[113] Clearfield, A., Nancollas, G. H. and Blessing, R. H., in *New Inorganic Ion Exchangers in Ion Exchange and Solvent Extraction, Vol.* 5, Marinsky, J. A. and Marcus, Y., (Eds.), Dekker, New York, 1973.
[114] Qureshi, M., Qureshi, S. Z., Gupta, J. P. and Rathore, H. S., *Sep. Sci.*, **7**, 615 (1972).
[115] Huys, D. and Baetsle, L. H., *J. Inorg. Nucl. Chem.*, **27**, 2459 (1965).
[116] Krtil, J., *J. Inorg. Nucl. Chem.*, **27**, 233 (1965).
[117] Prout, W. E., Russell, E. R. and Groh, H. J., *J. Inorg. Nucl. Chem.*, **27**, 473 (1965).
[118] Kawamura, S., Kuraku, H. and Kurotaki, K., *Anal. Chim. Acta*, **49**, 317 (1970).
[119] Phillips, H. O. and Kraus, K. A., *J. Chromatogr.*, **17**, 549 (1965).
[120] L'vovich, B. I. and Volikhin, V. V., *Zh. Neorgan. Khim.*, **15**, 520 (1970).
[121] Baetsle, L. and Pelsmaekers, J., *J. Inorg. Nucl. Chem.*, **21**, 124 (1961).
[122] Girardi, F., Pietra, R. and Sabbioni, E., *J. Radioanal. Chem.*, **5**, 141 (1970).
[123] Girardi, F. and Sabbioni, E., *J. Radioanal. Chem.*, **1**, 169 (1968).
[124] Caletká, R., Konečny, C. and Simková, M., *J. Radioanal. Chem.*, **10**, 5 (1972).
[125] De Soete, D., Gijbels, R. and Hoste, J., *Neutron Activation Analysis*, Wiley-Interscience, London, 1972.
[125a] De, A. K. and Sen, A. K., *Sep. Sci.*, **13**, 517 (1978).
[126] Doremus, R. H., in *Ion Exchange, Vol.* 2, Marinsky, J. A., (Ed.), Dekker, New York, 1969.
[127] Vydra, F. and Marková, V., *J. Inorg. Nucl. Chem.*, **26**, 1319 (1964).
[128] Kokotov, Yu. A. and Popova, R. F., *Radiokhimicheskie metody opredeleniya mikroelementov* (*Radiochemical Methods of Determining Microelements*), Izd. Nauka, Moscow, 1965, p. 76.
[129] Coleman, C. F., Blake, C. A., Jr. and Brown, K. B., *Talanta*, **9**, 297 (1962).
[130] Green, H., *Talanta*, **11**, 1561 (1964).
[131] Cerrai, E., *Chromatogr. Rev.*, **6**, 129 (1964).
[132] Cerrai, E. and Ghersini, G., in *Advances in Chromatography, Vol.* 9, Giddings, J. C. and Keller, R. A., (Eds.), Dekker, New York, 1970.
[133] Brinkman, U. A. Th. and de Vries, G., *J. Chem. Educ.*, **49**, 244 (1972).
[134] Green, H., *Talanta*, **20**, 139 (1973).
[135] Lederer, M., *Anal. Chim. Acta*, **12**, 142 (1955).
[136] Witkowski, H., *Rocz. Chem.*, **30**, 549 (1956).
[137] Hale, D. K., *Chem. Ind.* (*London*), **1955**, 1147.
[138] Testa, C., *J. Chromatogr.*, **5**, 236 (1961).
[139] Alberti, G. and Grassini, G., *J. Chromatogr.*, **4**, 83 (1960).
[140] Peixoto Cabral, J. M., *J. Chromatogr.*, **4**, 86 (1960).
[141] Alberti, G. and Grassini, G., *J. Chromatogr.*, **4**, 423 (1960).
[142] Cerrai, E. and Testa, C., *J. Chromatogr.*, **5**, 442 (1961).
[143] Lewandowski, A., *Anal. Chim. Acta*, **23**, 317 (1960).

[144] Arnold, R. and Ritchie, J. F., *J. Chromatogr.*, **10**, 205 (1963).
[145] Alberti, G., Dobici, F. and Grassini, G., *J. Chromatogr.*, **8**, 103 (1962).
[146] Alberti, G., *J. Chromatogr.*, **31**, 177 (1967).
[147] Lederer, M., *J. Chromatogr.*, **29**, 306 (1967).
[148] Sherma, J., *Talanta*, **9**, 775 (1962).
[149] Lederer, M., *J. Chromatogr.*, **2**, 209 (1959).
[150] Prášilová, J. and Šebesta, F., *J. Chromatogr.*, **4**, 555 (1960).
[151] Lederer, M. and Ossicini, L., *J. Chromatogr.*, **13**, 188 (1964).
[152] Cerrai, E. and Testa, C., *J. Chromatogr.*, **8**, 232 (1962).
[153] Cerrai, F. and Ghersini, G., *J. Chromatogr.*, **13**, 211 (1964).
[154] Fuller, J., *Chromatogr. Rev.*, **14**, 45 (1971).
[155] Alberti, G., *Chromatogr. Rev.*, **8**, 246 (1966).
[156] Cerrai, E., *Report* CISE-103, October 1966.
[157] Cerrai, E. and Ghersini, G., *J. Chromatogr.*, **15**, 236 (1964).
[158] Cerrai, E. and Ghersini, G., *J. Chromatogr.*, **22**, 425 (1966).
[159] Cerrai, E. and Ghersini, G., *J. Chromatogr.*, **18**, 124 (1965).
[160] Cerrai, E. and Ghersini, G., *Energ. Nucl. (Milan)*, **11**, 441 (1964).
[161] Cerrai, E. and Ghersini, G., *J. Chromatogr.*, **16**, 258 (1964).
[162] Cerrai, E. and Ghersini, G., *Analyst (London)*, **94**, 599 (1969).
[163] Lederer, M. and Marini-Bettolo, R., *J. Chromatogr.*, **43**, 149 (1969).
[164] Husain, W. S., *Analusis*, **1**, 314 (1972).
[165] Yamabe, T., Falk, E., and Takai, N., *Seisan Kenkyu*, **24**, 358 (1972); *Chem. Abstr.*, **78**, 37475v (1973).
[166] Lepri, L., Desideri, P. G. and Mascherini, R., *J. Chromatogr.*, **70**, 212 (1972).
[167] Dubuquoy, J., Gusmini, S. and Poupard, D., *J. Chromatogr.*, **70**, 216 (1972).
[168] Muchová, A. and Jokl, V., *Chem. Zvesti*, **26**, 303 (1972).
[169] Lederer, M. and Majani, C., *Chromatogr. Rev.*, **12**, 239 (1970).
[170] Kuroda, R. and Kojima, N., *Bull. Chem. Soc. Japan*, **45**, 3211 (1972).
[170a] Kuroda, R., Saita, T., Oguma, K. and Takemoto, M., *J. Chromatogr.*, **139**, 355 (1977).
[171] Kroschwitz, H., Pungor, E. and Ferenczi, S., *Talanta* **19**, 695 (1972).
[172] Husain, W., and Kazmi, S. K., *Experientia*, **28**, 988 (1972).
[173] Brinkman, U. A. Th., de Vries, G. and Kuroda, R., *J. Chromatogr.*, **85**, 187 (1973).
[174] Kuroda, R. and Kondo, T., *J. Chromatogr.*, **80**, 241 (1973).
[175] Kirk, R. E. and Othmer, D. F., *Encyclopedia of Chemical Technology*, Vol. 8, Interscience Encyclopedia Inc., New York, 1952, p. 13.
[176] *Operation and Control of Ion-Exchange Processes for Treatment of Radioactive Wastes*, Technical Reports Series No. 78, IAEA, Vienna 1967, p. 118.
[177] Błaszkowska, Z. and Dybczyński, R., *Przem. Chem.*, **38**, 168 (1959).
[177a] Armitage, G. M., Lyle, S. J. and Nair, V. C., *Talanta*, **23**, 58 (1976).
[178] Fischer, S. and Kunin, R., in *Analytical Chemistry of Polymers, Part 1*, Kline, G. M., (Ed.), Interscience, New York, 1959.
[179] Hamilton, P. B., *Anal. Chem.*, **30**, 914 (1958).
[180] Gregor, H. F., Belle, J. and Marcus, R. A., *J. Am. Chem. Soc.*, **76**, 1984 (1954).
[181] Strobel, H. A. and Gable, R. W., *J. Am. Chem. Soc.*, **76**, 5911 (1954).
[182] Kopylova, V. D., Asambadze, T. D. and Saldadze, K. M., *Zavodsk. Lab.*, **35**, 1180 (1969).
[182a] Parrish, J. R., *Anal. Chem.*, **47**, 1999 (1975).
[182b] Bunzl, K. and Sansoni, B., *Anal. Chem.*, **48**, 2279 (1976).

References

[183] Helfferich, F., *Ion Exchange*, McGraw-Hill, London, 1962, Section 4-4.
[184] Tatur, H. and Nowakowski, W., *Jonity. Teoria i zastosowanie w przemyśle (Ion-Exchange Resins. Theory and Industrial Applications)*, PWT, Warsaw, 1955, p. 102.
[185] Kunin, R., *Ion-Exchange Resins*, Wiley, New York, 1958, p. 358.
[186] Kunin, R., *Ion-Exchange Resins*, Wiley, New York, 1958, p. 359, 367.
[187] Saldadze, K. M., Pashkov, A. B. and Titov, V. S., *Ionoobmennye vysokomolekulyarnye soedineniya (High-Molecular Ion-Exchange Compounds)*, Gos. Nauch. Tekhn. Izd. Khim. Lit., Moscow, 1960, pp. 100, 119, 127, 134.
[188] Prokhorov, F. G., *Teoriya i praktika primeneniya ionoobmennykh materialov (Theory and Practice of Use of Ion-Exchange Compounds)*, Izd. Akad. Nauk SSSR, Moscow, 1955, p. 57.
[189] Saldadze, K. M., Pashkov, A. B. and Titov, V. S., *Ionoobmennye vysokomolekulyarnye soedineniya (High-Molecular Ion-Exchange Compounds)*, Gos. Nauch. Tekhn. Izd. Khim. Lit., Moscow, 1960, pp. 110, 120, 134.
[190] Vasil'ev, A. A., *Issledovaniya v oblasti ionoobmennoi khromatografii (Studies in Ion-Exchange Chromatography)*, Izd. Akad. Nauk SSSR, Moscow, 1957, p. 89.
[191] Kraus, K. A. and Raridon, R. J., *J. Phys. Chem.*, **63**, 1901 (1959).
[192] Dybczyński, R., *Przem. Chem.*, **38**, 216 (1959).
[193] Polyanskii, N. G., *Issledovaniya v oblasti ionoobmennoi, raspredelitel'noi i osadochnoi khromatografii (Studies in Ion-Exchange, Partition and Deposition Chromatography)*, Izd. Akad. Nauk SSSR, Moscow, 1959, p. 95.
[194] Wilson, A. L., *J. Appl. Chem.*, **11**, 151 (1961).
[195] Jakubovic, A. O., *J. Chem. Soc.*, **1960**, 4820.
[196] Nachod, F. C. and Schubert, J. (Eds.), *Ion-Exchange Technology*, Academic Press, New York, 1956, p. 442.
[197] Wiley, R. H. and De Venuto, G., *J. Appl. Polymer. Sci.*, **9**, 2001 (1965).
[198] Nikashina, V. A., Slovokhotova, N. A. and Senyavin, M. M., *Trudy II vsesoyuznogo soveshchaniya po radiatsionnoi khimii (Proceedings of the 2nd All-Soviet Conference on Radiation Chemistry)*, Izd. Akad. Nauk SSSR, Moscow, 1962, p. 596.
[199] Narębska, A., Basiński, A. and Litowska, M., *Nukleonika*, **15**, 177 (1969).
[200] Narębska, A. and Mazurkiewicz, B., *Nukleonika*, **17**, 169 (1972).
[201] Litowska, M. and Basiński, A., *Nukleonika*, **18**, 195 (1973).
[202] Peryshkina, N. G., Sudarikova, N. I. and Soldatov, V. S., *Vestsi Akad. Navuk Belarusk. SSR, Ser. Khim. Navuk*, **1965** (1), 20.
[203] Bauman, W. C., Skidmore, J. R. and Osmun, R. H., *Ind. Eng. Chem.*, **40**, 1350 (1948).
[204] Dybczyński, R., *Wpływ temperatury na zdolność wymienną Wofatytu P (Effect of Temperature on Wofatite P Exchange Capacity)*, IChO Report Nr. BC/IV-2-14 (unpublished).
[205] Wheaton, R. M. and Bauman, W. C., *Ind. Eng. Chem.*, **43**, 1088 (1951).
[206] Frisch, N. W. and Kunin, R., *Ind. Eng. Chem.*, **49**, 1365 (1957).
[207] Kunin, R., *Ion Exchange Resins*, Wiley, New York, 1958, p. 362.
[208] Amphlett, C. B., *Treatment and Disposal of Radioactive Wastes*, Pergamon, Oxford, 1961.
[209] Lindsay, E. K. and D'Amico, J. S., *Ind. Eng. Chem.*, **43**, 1085 (1951).
[210] Alberti, G. and Conte, A., *J. Chromatogr.*, **5**, 244 (1961).

[211] Qureshi, M., Rathore, H. S. and Kumar, R. J., *J. Chromatogr.*, **54**, 269 (1971).
[212] Kolosova, G. M. and Senyavin, M. M., *Zh. Fiz. Khim.*, **38**, 2819, (1964).
[213] Pepper, K. W., *J. Appl. Chem.*, **1**, 124 (1951).
[214] Pepper, K. W., Reichenberg, D. and Hale, D. K., *J. Chem. Soc.*, **1952**, 3129.
[215] Howe, P. G. and Kitchener, J. A., *J. Chem. Soc.*, **1955**, 2143.
[216] Kressman, T. R. E. and Millar, J. R., *Chem. Ind.*, (*London*), **1961**, 1833.
[217] Pepper, K. W., Paisely, H. M. and Young, M. A., *J. Chem. Soc.*, **1953**, 4097.
[218] Dybczyński, R., *J. Chromatogr.*, **50**, 487 (1970).
[219] Boyd, G. E. and Soldano, B. A., *J. Am. Chem. Soc.*, **75**, 6091 (1953).
[219a] Naveh, J. and Marcus, Y., *J. Chromatogr.*, **148**, 495 (1978).
[220] Soldano, B. A. and Boyd, G. E., *J. Am. Chem. Soc.*, **75**, 6099 (1953).
[221] Griessbach, R., *Austauschadsorption in Theorie und Praxis*, Akademie-Verlag, Berlin, 1957, p. 378.
[222] Freeman, D. H., in *Ion Exchange*, Vol. 1, Marinsky, J. A. (Ed.), Dekker, New York, 1966.
[223] Gregor, H. P., Held, K. M. and Berlin, J., *Anal. Chem.*, **23**, 620 (1951).
[224] Pollio, F. X., *Anal. Chem.*, **35**, 2164 (1963).
[225] Sharma, H. D. and Subramanian, N., *Anal. Chem.*, **42**, 1287 (1970).
[226] Grieser, M. D., Wilks, A. D. and Pietrzyk, D. J., *Anal. Chem.*, **44**, 671 (1972).
[227] Blasius, E. and Schmitt, R., *Z. Anal. Chem.*, **241**, 4 (1968).
[228] Pelzbauer, Z. and Foršt, V., *Collect. Czech. Chem. Commun.*, **31**, 2338 (1966).
[229] Parrish, J. R., *Anal. Chem.*, **45**, 1659 (1973).
[230] Blasius, E., Lohde, H. and Hausler, H., *Z. Anal. Chem.*, **264**, 290 (1973).
[231] Reichenberg, D., Pepper, K. W. and McCauley, D. J., *J. Chem. Soc.*, **1951**, 493.
[232] Reichenberg, D. and McCauley, D. J., *J. Chem. Soc.*, **1955**, 2741.
[233] Bonner, O. D. and Smith, L. L., *J. Phys. Chem.*, **61**, 326 (1957).
[234] Gregor, H. P., Hamilton, M. J., Oza, R. J. and Bernstein, F., *J. Phys. Chem.*, **60**, 263 (1956).
[235] Soldano, B. A. and Chesnut, D., *J. Am. Chem. Soc.*, **77**, 1334 (1955).
[236] Boyd, G. E., Lindenbaum, S. and Myers, G. E., *J. Phys. Chem.*, **65**, 577 (1961).
[237] Myers, G. E. and Boyd, G. E., *J. Phys. Chem.*, **60**, 521 (1956).
[238] Soldano, B., Larson, Q. V. and Myers, G. E., *J. Am. Chem. Soc.*, **77**, 1339 (1955).
[239] Helfferich, F., *Ion Exchange*, McGraw-Hill, New York, 1962, pp. 255, 285.
[240] Mann, C. K. and Swanson, C. L., *Anal. Chem.*, **33**, 459 (1961).
[241] Sigodina, A. B., Nikolaev, N. I. and Tunitskii, N. N., *Usp. Khim.*, **33**, 439 (1964).
[242] Herber, R. H., Tonguc, K. and Irvine, J. W., Jr., *J. Am. Chem. Soc.*, **77**, 5840 (1955).
[243] Senyavin, M. M., *Izotopy i izlucheniya v khimii* (*Isotopes and Radiation in Chemistry*), Izd. Akad. Nauk SSSR, Moscow, 1958, p. 186.
[244] Dybczyński, R., *J. Chromatogr.*, **71**, 507 (1972).
[245] Dybczyński, R., *Proc. 3rd Analytical Conference*, Vol. 1, Akadémiai Kiadó, Budapest, 1970.
[246] Wódkiewicz, L. and Dybczyński, R., *J. Chromatogr.*, **68**, 131 (1972).
[246a] Dybczyński, R., Polkowska-Motrenko, H. and Shabana, R. M., *J. Chromatogr.*, **134**, 285 (1977).
[247] Fisher, S. and Kunin, R., *J. Phys. Chem.*, **60**, 1030 (1956).
[248] Suryaraman, M. G. and Walton, H. F., *Science*, **131**, 829 (1960).
[249] Bortel, E., *Przem. Chem.*, **44**, 255 (1965).

[250] Bortel, E., *Przem. Chem.*, **46**, 723 (1967).
[251] Bortel, E., *Przem. Chem.*, **47**, 557 (1968).
[252] Wiegner, G. and Jenny, H., *Kolloid Z.*, **42**, 268 (1927).
[253] Vageler, P. and Wolterdorf, J., *U. Pflanzenernähr. Düng. Bodenkd.*, **A15**, 329 (1930).
[254] Rothmund, V. and Kornfeld, G., *Z. Anorg. Chem.*, **103**, 129 (1918); **108**, 215 (1919).
[255] Lowen, W. K., Stoenner, R. W., Argersinger, W. J., Jr., Davidson, A. W. and Hume, D. N., *J. Am. Chem. Soc.*, **73**, 2666 (1951).
[256] Argersinger, W. J., Jr. and Davidson, A. W., *J. Phys. Chem.*, **56**, 92 (1952).
[257] Bonner, O. D., Argersinger, W. J. and Davidson, A. W., *J. Am. Chem. Soc.*, **74**, 1044 (1952).
[258] Högfeldt, E., Ekedahl, E. and Sillén, L. G., *Acta Chem. Scand.*, **4**, 828 (1950).
[259] Bonner, O. D., *J. Phys. Chem.*, **58**, 318 (1954).
[260] Davidson, A. W. and Argersinger, W. J., Jr., *Ann. N. Y. Acad. Sci.*, **57**, 105 (1953).
[261] Bonner, O. D., and Livingston, F. L., *J. Phys. Chem.*, **60**, 530 (1956).
[262] Bauman, W. C. and Eichorn, J., *J. Am. Chem. Soc.*, **69**, 2830 (1947).
[263] Kraus, K. A. and Nelson, F., in *The Structure of Electrolytic Solutions*, Hamer, W. J. (Ed.), Wiley, New York, 1959.
[264] Harned, H. S. and Owen, B. B., *The Physical Chemistry of Electrolytic Solutions*, Reinhold, New York, 1958.
[265] Robinson, R. A. and Stokes, R. H., *Electrolyte Solutions*, Butterworths, London, 1959.
[266] Gregor, H. P., Belle, J. and Marcus, R. A., *J. Am. Chem. Soc.*, **77**, 2713 (1955).
[267] Dybczyński, R., *Wpływ temperatury i usieciowania jonitu na reakcje wymiany jonowej i przebieg rozdzielań chromatograficznych* (*Effect of Temperature and Cross-Linking of the Ion-Exchange Resin on the Ion-Exchange Reactions and Chromatographic Separation*), Inst. Nucl. Res. (Warsaw) Report No. 1115 (VIII) C.
[268] Dybczyński, R., *Anal. Chim. Acta*, **29**, 369 (1963).
[269] Cerrai, E., Ghersini, G. Lederer, M. and Mazzei, M., *J. Chromatogr.*, **44**, 161 (1969).
[270] Gregor, H. P., *J. Am. Chem. Soc.*, **70**, 1293 (1948).
[271] Gregor, H. P., *J. Am. Chem. Soc.*, **73**, 642 (1951).
[272] Pepper, K. W. and Reichenberg, D., *Z. Elektrochem.*, **57**, 183 (1953).
[273] Reinchenberg, D., in *Ion Exchange*, Vol. 1, Marinsky, J. A. (Ed.), Dekker, New York, 1966.
[274] Eisenman, G., *Biophys. J. Suppl.*, **2**, 259 (1962).
[275] Eisenman, G., *Membrane Transport and Metabolism*, Academic Press, New York, 1961, p. 163.
[276] Diamond, R. M. and Whitney, D. C., in *Ion Exchange*, Vol. 1, Marinsky, J. A. (Ed.), Dekker, New York, 1966.
[277] Chu, B., Whitney, D. C. and Diamond, R. M., *J. Inorg. Nucl. Chem.*, **24**, 1405 (1962).
[278] Whitney, D. C. and Diamond, R. M., *Inorg. Chem.*, **2**, 1284 (1963).
[279] Whitney, D. C. and Diamond, R. M., *J. Phys. Chem.*, **68**, 1884 (1964).
[280] Whitney, D. C. and Diamond, R. M., *J. Inorg. Nucl. Chem.*, **27**, 219 (1965).
[281] Nelson, F., Murase, T. and Kraus, K. A., *J. Chromatogr.*, **13**, 503 (1964).

[282] Lindenbaum, S. and Boyd, G. E., *J. Phys. Chem.*, **69**, 2374 (1965).
[283] Kennedy, J. and Davies, R. V., *Chem. Ind. (London)*, **1956**, 378.
[284] Persoz, J. and Rosset, R., *Bull. Soc. Chim. Fr.*, **1964**, 2197.
[285] Kennedy, J., Davies, R. V., Small, H. and Robinson, B. K., *J. Appl. Chem.*, **9**, 32, (1959).
[286] Riley, J. P. and Taylor, D., *Anal. Chim. Acta*, **40**, 479 (1968).
[287] Biechler, D. G., *Anal. Chem.*, **37**, 1054 (1965).
[288] Van der Reyden, A. J. and van Lingen, R. L. M., *Z. Anal. Chem.*, **187**, 241 (1962).
[289] Rosset, R., *Bull. Soc. Chim. Fr.*, **1966**, 59.
[290] Nickless, G. and Marshall, G. R., *Chromatogr. Rev.*, **6**, 154 (1964).
[291] Fritz, J. S. and King, J. N., *J. Chromatogr.*, **153**, 507 (1978).
[292] Saldadze, K. M., Demonterik, Z. G. and Klimova, Z. B., *Issledovaniya v oblasti ionoobmennoi khromatografii (Studies in Ion-Exchange Chromatography)*, Izd. Akad. Nauk SSSR, Moscow, 1957, p. 48.
[293] Walton, H. F. and Martinez, J. M., *J. Phys. Chem.*, **63**, 1318 (1959).
[294] Kressman, T. R. E. and Kitchener, J. A., *J. Chem. Soc.*, **1949**, 1208.
[295] Feitelson, J., in *Ion Exchange, Vol. 2*, Marinsky, J. A. (Ed.), Dekker, New York, 1969.
[296] Nemethy, G. and Scheraga, H. A., *J. Chem. Phys.*, **36**, 3382, 3401 (1962).
[297] Kraus, K. A., Michelson, D. C. and Nelson, F., *J. Am. Chem. Soc.*, **81**, 3204 (1959).
[298] Wang Fu-Chun, Norseev, Yu. V., Khalkin, V. A. and Chao, Tao-Nan, *Radiokhimiya*, **5**, 661 (1963).
[299] Titze, H. and Samuelson, O., *Acta Chem. Scand.*, **16**, 678 (1962).
[300] Dybczyński, R. and Maleszewska, H., *Analyst (London)*, **94**, 527 (1969).
[300a] Pfrepper, G., *J. Chromatogr.*, **110**, 133 (1975).
[300b] Pfrepper, G., *J. Chromatogr.*, **116**, 407 (1976).
[300c] Pfrepper, G., *J. Chromatogr.*, **120**, 399 (1976).
[301] Kitchener, J. A., in *Modern Aspects of Electrochemistry, No. 2*, Bockris, J. O'M. (Ed.), Butterworths, London, 1959.
[302] Soldano, B. A. and Larson, Q. V., *J. Am. Chem. Soc.*, **77**, 1331 (1953).
[303] Lindenbaum, S., Jumper, C. F. and Boyd, G. E., *J. Phys. Chem.*, **63**, 1924 (1959).
[304] Walton, H. F., Jordan, D. E., Samedy, S. R. and McKay, W. W., *J. Phys. Chem.*, **65**, 1447 (1961).
[305] Kitchener, J. A., in *Ion Exchange and its Applications*, Society of Chemical Industry, London, 1955, p. 24.
[306] Marcus, Y. and Maydan, D., *J. Phys. Chem.*, **67**, 983 (1963).
[307] Hulet, E. K., Gutmacher, R. G. and Coops, M. S., *J. Inorg. Nucl. Chem.*, **17**, 350 (1961).
[308] Millar, J. R., Smith, D. G., Marr, W. E. and Kressman, T. R. E., *J. Chem. Soc.*, **1964**, 2740.
[309] Talašek, V., *Collect. Czech. Chem. Commun.*, **33**, 35 (1968).
[310] Hartler, N., *Acta Chem. Scand.*, **11**, 1162 (1957).
[311] Richardson, R. W., *J. Chem. Soc.*, **1951**, 910.
[312] Partridge, S. M., in *Ion Exchange and its Applications*, Society of Chemical Industry, London, 1955, p. 131.
[313] Kressman, T. R. E., *J. Phys. Chem.*, **56**, 118 (1952).
[314] Kunin, R. and Myers, R. J., *Disc. Faraday Soc.*, **7**, 114 (1949).
[315] Hale, D. K., Packham, D. I. and Pepper, K. W., *J. Chem. Soc.*, **1953**, 844.

[316] Blanchard, J. and Nairn, J., *J. Phys. Chem.*, **72**, 1204 (1968).
[317] Bregman, J. I., *Ann. N. Y. Acad. Sci.*, **57**, 125 (1953).
[318] Preobrazhenskii, B. K., in *Radiokhimia i khimiya yadernykh protsessov* (*Radiochemistry and Chemistry of Nuclear Reactions*), Murin, A. N., Nefedov, V. D. and Shvedov, V. P., (Eds.), Goskhimizdat, Leningrad, 1960.
[319] Alexa, J., *Collect. Czech. Chem. Commun.*, **33**, 1933 (1968).
[320] Strelow, F. W. E. and Gricius, A. J., *Anal. Chem.*, **44**, 1898 (1972).
[321] Ross, R., Beulich, H. and Huller, J., *Report. No. ZfK*-197, Zentralinstitut für Kernforschung, Rossendorf, Dec. 1969.
[322] Bonner, O. D. and Pruett, R. R., *J. Phys. Chem.*, **63**, 1417 (1959).
[323] Bonner, O. D. and Pruett, R. R., *J. Phys. Chem.*, **63**, 1420 (1959).
[324] Cruickshank, E. H. and Meares, P., *Trans. Faraday Soc.*, **53**, 1289, 1299 (1957).
[325] Kraus, K. A., Raridon, R. J. and Holcomb, D. L., *J. Chromatogr.*, **3**, 178 (1960).
[326] Kressman, T. R. E. and Kitchener, J. A., *J. Chem. Soc.*, **1949**, 1190.
[327] Duncan, J. F. and Lister, B. A. J., *J. Chem. Soc.*, **1949**, 3285.
[328] Bonner, O. D. and Smith, L. L., *J. Phys. Chem.*, **61**, 1614 (1957).
[329] Wóycicki, W., *Rocz. Chem.*, **34**, 1413, 1901 (1959).
[330] Boyd, G. E., Vaslow, F. and Lindenbaum, S., *J. Phys. Chem.*, **68**, 590 (1964).
[331] Dybczyński, R., *Rocz. Chem.*, **37**, 1411 (1963).
[332] Dybczyński, R., *J. Chromatogr.*, **14**, 79 (1964).
[333] Dybczyński, R. and Wódkiewicz, L., *J. Inorg. Nucl. Chem.*, **31**, 1495 (1969).
[334] Aleksandrova, L. S. and Elovich, S. Yu., *Kolloid. Zh.*, **20**, 687 (1955).
[335] Matorina, N. N. and Popov, A. N., *Zh. Fiz. Khim.*, **32**, 2772 (1958).
[336] Soldatov, V. S. and Starobinets, G. L., *Issledovanie svoistv ionoobmennykh materialov* (*Investigation of the Properties of Ion Exchange Materials*), Izd. Nauka, Moscow, 1964, p. 36.
[337] Shataeva, L. K. and Samsonov, G. B., *Termodinamika ionnogo obmena* (*Thermodynamics of Ion Exchange*), Izd. Nauka i Tekhnika, Minsk, 1968, p. 193.
[338] Soldatov, V. S., Novitskaya, L. V., Bespal'ko, M. S. and Kazan, Z. I., *Sintez i svoistva ionoobmennykh materialov* (*Synthesis of Ion Exchange Resins and Their Properties*), Izd. Nauka, Moscow, 1968, p. 216.
[339] Davidov, A. T. and Lisovina, G. M., *Khromatografiya, ee teoriya i primenenie* (*Chromatography, Theory and Application*), Izd. Akad. Nauk SSSR, Moscow, 1960, p. 122.
[340] Baetsle, L., *J. Inorg. Nucl. Chem.*, **25**, 271 (1963).
[341] Dybczyński, R., *J. Chromatogr.*, **31**, 155 (1967).
[342] Kraus, K. A. and Raridon, R. J., *J. Am. Chem. Soc.*, **82**, 3271 (1960).
[343] Strelow, F. W. E., *Anal. Chem.*, **32**, 1185 (1960).
[344] Strelow, F. W. E., Rethemeyer, R. and Bothma, C. J. C., *Anal. Chem.*, **37**, 106 (1965).
[345] Welcher, F. J., *Analityczne zastosowanie kwasu wersenowego*, WNT, Warsaw, 1963; *The Analytical Uses of Ethylenediamine Tetra-acetic Acid*, Van Nostrand, Princeton, N. J., 1958.
[346] Přibil, R., *Kompleksony v khimicheskom analize* (*Complexing Agents in Chemical Analysis*), Izd. Inostrannoi Literatury, Moscow, 1960.
[347] Wheelwright, E. J., Spedding, F. H. and Schwarzenbach, G., *J. Am. Chem. Soc.*, **75**, 4196 (1953).
[348] Schwarzenbach, G., Gut, R. and Anderegg, G., *Helv. Chim. Acta*, **37**, 937 (1954).

[349] Fuger, J., *J. Inorg. Nucl. Chem.*, **5**, 332 (1958).
[350] Fuger, J., *J. Inorg. Nucl. Chem.*, **18**, 263 (1961).
[351] Kolosova, G. M. and Radionova, N. M., *Zh. Fiz. Khim.*, **42**, 2873 (1968).
[352] Surls, J. P., Jr. and Choppin, G. R., *J. Am. Chem. Soc.*, **79**, 855 (1957).
[353] Baybarz, R. D., *J. Inorg. Nucl. Chem.*, **28**, 1055 (1966).
[354] Johnson, W. C., Quill, L. L. and Daniels, F., *Chem. Eng. News*, **25**, 2494 (1947).
[355] Ketelle, B. H. and Boyd, G. E., *J. Am. Chem. Soc.*, **69**, 2800 (1947).
[356] Choppin, G. R. and Silva, R. J., *J. Inorg. Nucl. Chem.*, **3**, 153 (1956).
[357] Vickery, R. C., *Analytical Chemistry of the Rare Earths*, Pergamon, New York, 1961.
[358] Mayer, S. W. and Freiling, E. C., *J. Am. Chem. Soc.*, **75**, 5647 (1953).
[359] Cornish, F. W., Phillips, G. and Thomas, A., *Can. J. Chem.*, **34**, 1471 (1956).
[360] Fitch, F. T. and Russell, D. G., *Can. J. Chem.*, **29**, 363 (1951).
[361] Wolf, L., *Chem. Tech. (Berlin)*, **10**, 590 (1958).
[362] Vickery, R. C., *J. Chem. Soc.*, **1952**, 4357.
[363] Bjerrum, J., Schwarzenbach, G. and Sillén, L. G., *Stability Constants*, The Chemical Society, London, 1957.
[364] Fritz, J. S. and Umbreit, G. R., *Anal. Chim. Acta*, **19**, 509 (1958).
[365] Povondra, P., Přibil, R. and Šulcek, Z., *Talanta*, **5**, 86 (1960).
[366] Ibbett, R. D., *Analyst (London)*, **92**, 417 (1967).
[367] Chwastowska, J. and Szymczak, S., *Chem. Anal. (Warsaw)*, **14**, 1161 (1969).
[368] Andersen, T. and Knutsen, A. B., *Acta Chem. Scand.*, **16**, 849 (1962).
[369] Minczewski, J. and Dybczyński, R., *J. Chromatogr.*, **7**, 98 (1962).
[370] Nelson, F., Day, R. A., Jr. and Kraus, K. A., *J. Inorg. Nucl. Chem.*, **15**, 140 (1960).
[371] Wódkiewicz, L. and Dybczyński, R., *J. Chromatogr.*, **32**, 394 (1968).
[372] Högfeldt, E., *Science*, **128**, 1435 (1958).
[373] Freeman, D. H., Patel, V. C. and Smith, M. E., *J. Polymer Sci.*, **A3**, 2893 (1955).
[374] Goldring, L. S., in *Ion Exchange*, Marinsky, J. A. (Ed.), Dekker, New York, 1966.
[375] Pokrovskaya, A. and Soldatov, V. S., *Termodinamika ionnogo obmena* (*Thermodynamics of Ion Exchange Processes*), Nauka i Tekhnika, Minsk, 1968, p. 84.
[376] Boyd, G. E. and Lindenbaum, S., *J. Phys. Chem.*, **69**, 2378 (1965).
[377] Nelson, F. and Michelson, D. C., *J. Chromatogr.*, **25**, 414 (1966).
[378] Jensen, C. H. and Diamond, R. M., *J. Phys. Chem.*, **69**, 3440 (1965).
[379] Kraus, K. A. and Nelson, F., *Proc. 1st Intern. Conf. Peaceful Uses At. Energy, Geneva*, **7**, 113, 131 (1956).
[380] Faris, J. P., *Anal. Chem.*, **32**, 520 (1960).
[381] Nikitin, M. K., *Dokl. Akad. Nauk SSSR*, **148**, No. 3, 595 (1963).
[382] Faris, J. P. and Buchanan, R. F., *Anal. Chem.*, **36**, 1157 (1964).
[383] Kraus, K. A. and Nelson, F., *Symposium on Ion Exchange and Chromatography in Analytical Chemistry*, ASTM Special Technical Publication No. 195, Philadelphia, 1958, p. 27.
[384] Lindenbaum, S. and Boyd, G. E., *J. Phys. Chem.*, **66**, 1383 (1962).
[385] Pfrepper, G. and Chi, L. T., *J. Chromatogr.*, **44**, 594 (1969).
[386] Irving, H. and Woods, G. T., *J. Chem. Soc.*, **1963**, 939.
[387] Boyd, G. E., Lindenbaum, S. and Larsen, Q. V., *Inorg. Chem.*, **3**, 1437 (1964).
[388] Kennedy, J., Marriot, J. and Wheeler, V. J., *J. Inorg. Nucl. Chem.*, **22**, 269 (1961).

[389] Maleszewska, H. and Dybczyński, R., *Nukleonika*, **12**, 1181 (1967).
[390] Levin, H., Diamond, W. J. and Brown, B. J., *Ind. Eng. Chem.*, **51**, 313 (1959).
[391] Ayres, J. A., *Ind. Eng. Chem.*, **43**, 1526 (1951).
[392] Smits, R., Van den Winkel, P., Massart, D. L., Juillard, J., and Morel, J. P., *Anal. Chem.*, **45**, 339 (1973).
[393] Davies, C. W. and Owen, B. D. R., *J. Chem. Soc.*, **1956**, 1676.
[394] Ohtaki, H., Kakihana, H. and Yamasaki, K., *Z. Phys. Chem. Frankfurt*, **21**, 224 (1959).
[395] Ohtaki, H., *Z. Physik. Chem. Frankfurt*, **27**, 209 (1961).
[396] Small, H., *J. Inorg. Nucl. Chem.*, **19**, 160 (1961).
[397] Moskvin, L. N., Preobrazhenskii, B. K. and Rzhanitsyna, L. N., *Radiokhimiya*, **5**, 299 (1963).
[398] Moskvin, L. N. and Preobrazhenskii, B. K., *Radiokhimiya*, **6**, 237 (1964).
[399] Chuveleva, E. A., Nazarov, P. P. and Chmutov, K. V., *Zh. Fiz. Khim.*, **36**, 1022 (1962).
[400] Grigorescu-Sabau, C., *J. Inorg. Nucl. Chem.*, **24**, 195 (1962).
[401] Grigorescu-Sabau, C. and Spiridon, S., *Acad. Rep. Populare Romine, Studii Cercetari Chim.*, **10**, 235 (1962).
[402] Peterson, S. F., Tera, F. and Morrison, G. H., *J. Radioanal. Chem.*, **2**, 113 (1969).
[403] Fessler, R. G. and Strobel, H. A., *J. Phys. Chem.*, **67**, 2562 (1963).
[404] Panchenkov, G. M., Gorshkov, V. I. and Kilanova, M. V., *Zh. Fiz. Khim.*, **32**, 361, 616, (1958).
[405] Gable, R. W. and Strobel, H. A., *J. Phys. Chem.*, **60**, 513 (1956).
[406] Janauer, G. E., Van Wart, H. E. and Carrano, J. T., *Anal. Chem.*, **42**, 215 (1970).
[407] Vojtěch, O., Brožek, V. and Neumann, L., *Collect. Czech. Chem. Commun.*, **27**, 2535 (1962).
[408] Gorshkov, V. I., Korolev, Yu. Z. and Shabanov, A. A., *Zh. Fiz. Khim.*, **40**, 1878 (1966).
[409] Gupta, C. B. and Tandon, S. N., *Sep. Sci.*, **7**, 513 (1972).
[410] Korkisch, J. and Janauer, G. E., *Talanta*, **9**, 957 (1962).
[411] Korkisch, J. and Hazan, I., *Talanta*, **11**, 1157 (1964).
[412] Korkisch, J. and Ahluwalia, S. S., *Talanta*, **14**, 155 (1967).
[413] Korkisch, J. and Khakl, H., *Talanta*, **15**, 339 (1968).
[414] Popov, A. N., Kononov, Yu. S. and Gorbachev, V. M., *Izv. Sib. Otd. Akad. Nauk SSSR*, **11**, No. 3, 141 (1965).
[415] Korkisch, J., Antal, P. and Hecht, F., *Z. Anal. Chem.*, **172**, 401 (1960).
[416] Tera, F. and Korkisch, J., *Anal. Chim. Acta*, **25**, 222 (1961).
[417] Korkisch, J. and Tera, F., *Anal. Chem.*, **33**, 1265 (1961).
[418] Korkisch, J. and Antal, P., *Z. Anal. Chem.*, **171**, 22 (1959).
[419] Molnár, F., Horváth, A. and Khalkin, V. A., *J. Chromatogr.*, **26**, 215, 225 (1967).
[420] Faris, J. P., *J. Chromatogr.*, **26**, 232 (1967).
[421] Korkisch, J. *Nature*, **210**, 626 (1966).
[422] Korkisch, J. and Ahluwalia, S. S., *Anal. Chem.*, **38**, 497 (1966).
[423] Marple, L. W., *J. Inorg. Nucl. Chem.*, **27**, 1693 (1965).
[424] Penciner, J., Eliezer, I. and Marcus, Y., *J. Phys. Chem.*, **69**, 2955 (1965).
[424a] Poitrenaud, C., *J. Chromatogr.*, **133**, 15 (1977).
[424b] Poitrenaud, C., *J. Chromatogr.*, **124**, 197 (1976).
[425] Bhatnagar, R. P. and Garde, S. L., *J. Indian Chem. Soc.*, **43**, 259 (1966).

[426] Bodamer, G. W. and Kunin, R., *Ind. Eng. Chem.*, **45**, 2577 (1953).
[427] Wilson, S. and Lapidus, L., *Ind. Eng. Chem.*, **48**, 992 (1956).
[428] Dybczyński, R., unpublished work.
[429] Marsh, J. K., *J. Chem. Soc.*, **1955**, 451.
[430] Dybczyński, R. and Wódkiewicz, L., unpublished work.
[431] Danon, J., *J. Inorg. Nucl. Chem.*, **7**, 422 (1958).
[432] Yamaguchi, H., Okuchi, A., Onuma, N. and Kuroda, R., *J. Chromatogr.*, **16**, 396 (1964).
[433] Yamatera, H., *J. Inorg. Nucl. Chem.*, **7**, 299 (1958).
[434] Boyd, G. E., Adamson, A. W. and Myers, L. S., Jr., *J. Am. Chem. Soc.*, **69**, 2836 (1947).
[435] Helfferich, F., in *Ion Exchange*, Vol. 1, Marinsky, J. A. (Ed.), Dekker, New York, 1966.
[436] Turse, R. and Rieman, W., III, *J. Phys. Chem.*, **65**, 1821 (1961).
[437] Soldano, B. A., *Ann. N. Y. Acad. Sci.*, **57**, 116 (1953).
[438] Reichenberg, D., *J. Am. Chem. Soc.*, **75**, 589 (1953).
[439] Helfferich, F., *J. Phys. Chem.*, **69**, 1178 (1965).
[440] Dal Nogare, S. and Juvet, R. S., Jr., *Gas-Liquid Chromatography*, Interscience, New York, 1962.
[441] Opieńska-Blauth, J., Waksmundzki, A. and Kański, M. (Eds.), *Chromatografia (Chromatography)*, PWN, Warsaw, 1957.
[442] Spedding, F. H. and Powell, J. E., in *Ion Exchange Technology*, Nachod, F. C. and Schubert, J., (Eds.), Academic Press, New York, 1956.
[443] Cornet, C., Coursier, J. and Huré, J., *Anal. Chim. Acta*, **19**, 259 (1958).
[444] Manalo, C. D., Turse, R., and Rieman W., III, *Anal. Chim. Acta*, **21**, 383 (1959).
[445] Glueckauf, E., *Trans. Faraday Soc.*, **51**, 34 (1955).
[446] Dybczyński, R., unpublished work.
[447] Giddings, J. C., *Dynamics of Chromatography*, Part 1, Dekker, New York, 1965.
[448] Tompkins, E. R., *J. Chem. Educ.*, **26**, 92 (1949).
[449] Martin, A. J. P. and Synge, R. L., *Biochem. J.*, **35**, 1385 (1941).
[450] Mayer, S. W. and Tompkins, E. R., *J. Am. Chem. Soc.*, **69**, 2866 (1947).
[451] Beukenkamp, J., Rieman, W., III, and Lindenbaum, S., *Anal. Chem.*, **26**, 505 (1954).
[452] Minczewski, J. and Dybczyński, R., *Chem. Anal. (Warsaw)*, **6**, 725 (1961).
[453] Czechowski, T., Fisz, M., Iwiński, T., Lange, O., Sadowski, W. and Zasępa, R., *Tablice Statystyczne (Statistical Tables)*, PWN, Warsaw, 1957.
[454] Glueckauf, E., in *Ion Exchange and Its Applications*, Society of Chemical Industry, London, 1958, p. 34.
[455] Hamilton, P. B., Bogue, D. C. and Anderson, R. A., *Anal. Chem.*, **32**, 1782 (1960).
[456] Van Deemter, J. J., Zuiderweg, F. J. and Klinkenberg, A., *Chem. Eng. Sci.*, **5**, 271 (1956).
[457] Cornish, F. W., *Analyst (London)*, **83**, 634 (1958).
[458] Ambrose, D., James, A. T., Keulemans, A. I. M., Kováts, E., Röck, H., Rouit, C. and Stross, F. H., *Pure Appl. Chem.*, **1**, 169 (1960).
[458a] De Corte, F., Van Acker, P. and Hoste, J., *Anal. Chim. Acta*, **75**, 246 (1975).
[459] Dybczyński, R. and Maleszewska, H., *J. Radioanal. Chem.*, **21**, 229 (1974).
[460] Dybczyński, R., *J. Radioanal. Chem.*, **31**, 115 (1976).
[461] Timmins, R. S., Mir, L. and Ryan, J. M., *Chem. Eng.*, **76**, 170 (1969).

[462] Betts, R. H., Dahlinger, O. F. and Munro, D. M., in *Radioisotopes in Scientific Research*, Vol. 2, Exterman, R. C. (Ed.), Pergamon, London, 1958, p. 326.
[463] Maleszewska, H. and Dybczyński, R., *Radiochem. Radioanal. Lett.*, **6**, 7 (1971).
[464] Fołdzińska, A., *Chem. Anal. (Warsaw)*, **16**, 821 (1971).
[465] Merciny, E. and Duyckaerts, G., *J. Chromatogr.*, **22**, 164 (1966).
[465a] Rahm, J., *J. Chromatogr.*, **115**, 455 (1975).
[466] Aubouin, G. and Laverlochère, J., *J. Radioanal. Chem.*, **1**, 123 (1968).
[467] Dybczyński, R. and Wódkiewicz, L., *Pr. Nauk. Inst. Technol. Org. Tworzyw Sztucznych Politech. Wrocław.*, **13**, 119 (1973).
[468] Scott, C. D., in *Modern Practice of Liquid Chromatography*, Kirkland, J. J. (Ed.), Wiley-Interscience, New York, 1971.
[469] Horvath, C., in *Ion Exchange and Solvent Extraction*, Vol. 5, Marinsky, J. A. and Marcus, Y., (Eds.), Dekker, New York, 1973.
[470] Majors, R. E., *Int. Lab.*, **1972**, (July/August), 25.
[470a] Majors, R. E., *Int. Lab.*, **1975** (November/December), 11.
[470b] Majors, R. E., *J. Chromatogr. Sci.*, **15**, 334 (1977).
[471] McNair, H. M. and Chandler, C. D., *J. Chromatogr. Sci.*, **12**, 425 (1974).
[472] Kirkland, J. J. (Ed.), *Modern Practice of Liquid Chromatography*, Wiley-Interscience, New York, 1971.
[472a] Rajcsanyi, P. M. and Rajcsanyi, E., *High-Speed Liquid Chromatography*, Dekker, New York, 1975.
[472b] Snyder, L. R. and Kirkland, J. J., *Introduction to Modern Liquid Chromatography*, Wiley, New York, 1974.
[472c] Knox, J. H., Done, J. N., Fell, A. T., Gilbert, M. T., Pryde, A. and Wall, R. A., *High-Performance Liquid Chromatography*, Edinburgh University Press, Edinburgh, 1978.
[472d] Huber, J. F. K. (Ed.), *Instrumentation for High-Performance Liquid Chromatography*, Elsevier, Amsterdam, 1978.
[473] Campbell, D. O. and Ketelle, B. H., *Inorg. Nucl. Chem. Lett.*, **5**, 533 (1969).
[474] Campbell, D. O. and Buxton, S. R., *Ind. Eng. Chem., Process Des. Dev.*, **9**, 89 (1970).
[475] Campbell, D. O., *Ind. Eng. Chem. Process Des. Dev.*, **9**, 95 (1970).
[476] Sisson, D. H., Mode, A. A. and Campbell, D. O., *J. Chromatogr.*, **66**, 129 (1972).
[477] Story, J. N. and Fritz, J. S., *Talanta*, **21**, 892 (1974).
[477a] Quaim, S. M., *Radiochem. Radioanal. Lett.*, **25**, 335 (1976).
[477b] Campbell, D. O., *Sep. Purific. Methods*, **5**, 97 (1976).
[478] Huber, J. F. K. and Van Urk-Schoen, A. M., *Anal. Chim. Acta*, **58**, 395 (1972).
[479] Seymour, M. D., Sickafoose, J. P. and Fritz, J. S., *Anal. Chem.*, **43**, 1734 (1971).
[480] Seymour, M. D. and Fritz, J. S., *Anal. Chem.*, **45**, 1394 (1973).
[481] Seymour, M. D. and Fritz, J. S., *Anal. Chem.*, **45**, 1632 (1973).
[482] Scott, C. D. and Lee, N. E., *J. Chromatogr.*, **42**, 263 (1969).
[483] Horvath, C. G., Preiss, B. A. and Lipsky, S. R., *Anal. Chem.*, **39**, 1422 (1967).
[484] Kirkland, J. J., *J. Chromatogr. Sci.*, **7**, 361 (1969).
[485] Kirkland, J. J., *J. Chromatogr. Sci.*, **8**, 72 (1970).
[486] Kirkland, J. J., *Analyst (London)*, **99**, 859 (1974).
[487] Skafi, M. and Lieser, K. H., *Z. Anal. Chem.*, **249**, 182 (1970).
[488] Skafi, M. and Lieser, K. H., *Z. Anal. Chem.*, **250**, 306 (1970).

[489]   Hansen, L. C. and Gilbert, T. W., *J. Chromatogr. Sci.*, **12**, 458 (1974).
[490]   Hansen, L. C. and Gilbert, T. W., *J. Chromatogr. Sci.*, **12**, 464 (1974).
[491]   Morozovich, *J. Chromatogr. Sci.*, **12**, 453 (1974).
[492]   Brust, O. E., Sebestian, I. and Halász, I., *J. Chromatogr.*, **83**, 15 (1973).
[493]   Knox, J. H. and Vasvari, G., *J. Chromatogr. Sci.*, **12**, 449 (1974).
[494]   Pryde, A., *J. Chromatogr. Sci.*, **12**, 486 (1974).
[494a] Leyden, D. E. and Luttrell, G. H., *Anal. Chim. Acta*, **84**, 97 (1976).
[494b] Fritz, J. S. and King, J. N., *Anal. Chem.*, **48**, 570 (1976).
[495]   Brücher, E. and Szarvas, P., *Proc. 3rd Analytical Conference, Vol. 1*, Akadémiai Kiadó, Budapest, 1970, p. 27.
[496]   Inczédy, J., Gábor-Klatsmányi, R. and Erdey, L., *Proc. Conf. Application of Physico-Chemical Methods in Chemical Analysis, Vol. II*, Budapest 1966, p. 21.
[497]   Steward, D. C., *J. Inorg. Nucl. Chem.*, **4**, 131 (1957).
[497a] Liteanu, C. and Gocan, S., *Gradient Liquid Chromatography*, Horwood, Chichester, 1974, p. 107.
[498]   Preobrazhenskii, B. K. and Lilova, O. M., *Radiokhimiya*, **8**, 73 (1966).
[499]   Strelow, F. W. E., in *Ion Exchange and Solvent Extraction, Vol. 5*, Marinsky, J. A. and Marcus, Y., (Eds.), Dekker, New York, 1973.
[500]   Giddings, J. C., *Anal. Chem.*, **32**, 1707 (1960).
[501]   Christophe, A. B., *Chromatographia*, **4**, 455 (1971).
[502]   Kaiser, R., *Gas Phase Chromatography, Vol. 1*, Butterworths, London, 1963, p. 39.
[503]   Carle, G. C., *Anal. Chem.*, **44**, 1905 (1972).
[503a] Mikeš, O., *Laboratory Handbook of Chromatographic and Allied Methods*, Horwood, Chichester, 1979.
[504]   Vassiliou, B. and Kunin, R., *Anal. Chem.*, **35**, 1328 (1963).
[505]   Bagg, J. and Gregor, H. P., *J. Am. Chem. Soc.*, **86**, 3626 (1964).
[506]   Mathers, W. G. and Hoelke, C. W., *Anal. Chem.*, **35**, 2064 (1963).
[507]   Tjioe, P. S., de Goeij, J. J. M. and Houtman, J. P. W., *J. Radioanal. Chem.*, **16**, 153 (1973).
[507a] Tjioe, P. S., De Goeij, J. J. M. and Houtman, J. P. W., *J. Radioanal. Chem.*, **37**, 511 (1977).
[508]   Girardi, F., Merlini, M., Pauly, J. and Pietra, R., *Radiochemical Methods of Analysis, Vol. II*, IAEA, Vienna, 1965, p. 3.
[508a] Girardi, F. and Pietra, R , *At. Energy Rev.*, **14**, 521 (1976).
[509]   Girardi, F., Guzzi, G., Pauly, J. and Pietra, R., *Proc. 2nd Int. Conf. Modern Trends in Activation Analysis*, Texas A and M Univ., 1965, p. 341.
[510]   Samsahl, K., *Nukleonika*, **8**, 252 (1966).
[510a] Iyengar, G. V., *Berichte der Kernforschungsanlage Jülich* No. 1308, June 1976.
[511]   Kirkland, J. J., *Anal. Chem.*, **43**, No. 12, 36A (1971).
[512]   Stahl, K. W., Schlimme, E. and Bojanowski, D., *J. Chromatogr.* **83**, 395 (1973).
[513]   Scott, C. D., Chilcote, D. D., Katz, S. and Pitt, W. W. Jr., *J. Chromatogr. Sci.*, **11**, 96 (1973).
[514]   Scott, C. D. and Lee, N. E., *J. Chromatogr.*, **83**, 383 (1973).
[515]   Shimomura, K. and Walton, H. F., *Anal. Chem.*, **37**, 1012 (1965).
[516]   Spackman, D. H., Stein, W. H. and Moore, S., *Anal. Chem.*, **30**, 1190 (1958).
[517]   Piez, K. A. and Morris, L., *Anal. Biochem.*, **1**, 187 (1960).
[518]   Kawazu, K. and Fritz, J. S., *J. Chromatogr.*, **77**, 397 (1973).

[519] Boyd, G. E., Myers, L. S., Jr. and Adamson, A. W., *J. Am. Chem. Soc.*, **69**, 2849 (1947).
[520] Schutte, L., *J. Chromatogr.*, **72**, 303 (1972).
[521] Wickbold, R., *Z. Anal. Chem.*, **132**, 401 (1951).
[522] Drake, B., *Ark. Kemi*, **8**, No. 18, 159 (1955).
[523] Rebertus, R. L., Cappell, R. J. and Bond, G. W., *Anal. Chem.*, **30**, 1825 (1958).
[524] Lewis, J. A. and Overton, K. C., *Analyst (London)*, **79**, 293 (1954).
[525] Chmielowiec, J. and Kemula, W., *J. Chromatogr.*, **102**, 197 (1974).
[526] Johnson, D. C. and Larochelle, J., *Talanta*, **20**, 959 (1973).
[527] Takata, Y. and Muto, G., *Anal. Chem.*, **45**, 1864 (1973).
[528] Taylor, L. R. and Johnson, D. C., *Anal. Chem.*, **46**, 262 (1974).
[529] Azaki, S., Suzuku, S. and Yamada, M., *Talanta*, **19**, 577 (1972).
[530] Gilbert, T. W. and Dobbs, R. A., *Anal. Chem.*, **45**, 1390 (1973).
[530a] Dorsey, J. G., Denton, M. S. and Gilbert, T. W., *Anal. Chem.*, **50**, 1330 (1978).
[530b] Small, H., Stevens, T. S. and Bauman, W. C., *Anal. Chem.*, **47**, 1801 (1975).
[530c] Small, H. and Solc, J., *Proc. Intern. Conf. held at Churchill College, Univ. of Cambridge, 25–30 July*, 1976, Soc. Chem. Ind., London, 1976, p. 32–1.
[530d] Stevens, T. S., Turkelson, V. T. and Albe, W. R., *Anal. Chem.*, **49**, 1176 (1977).
[530e] Otterson, D. A., *Abstracts 29th Pittsburgh Conference, Feb. 27–March 3, 1978*, Cleveland, Ohio, USA, No. 360.
[530f] Bogen, D. C., Nagourney, S. and Welford, G. A., *Abstracts 29th Pittsburgh Conference, Feb. 27–March 3, 1978*, Cleveland, Ohio, USA, No. 361.
[530g] Fulmer, M. A., Nadalin, R. J. and Penkrot, J., *Abstracts 29th Pittsburgh Conference, Feb. 27–March 3, 1978*, Cleveland, Ohio, USA, No. 362.
[530h] Carlson, G. L. and Fulmer, M. A., *Abstracts 29th Pittsburgh Conference, Feb. 27–March 3, 1978*, Cleveland, Ohio, USA, No. 363.
[530i] Koop, D. J., Silvester, M. D., Abercrombie, F. N. and Thomson, I., *Abstracts 29th Pittsburgh Conference, Feb. 27–March 3, 1978*, Cleveland, Ohio, USA, No. 364.
[530j] Chang, R. C. and Woodruff, L. A., *Abstracts 29th Pittsburgh Conference, Feb. 27–March 3, 1978*, Cleveland, Ohio, USA, No. 365.
[530k] Johnson, S. J., *Abstracts 29th Pittsburgh Conference, Feb. 27–March 3, 1978*, Cleveland, Ohio, USA, No. 366.
[530l] Rich, W. E., *Anal. Instrum.*, **15**, 113 (1977).
[530m] Bouyoucos, A. S., *Anal. Chem.*, **49**, 401 (1977).
[530n] Colaruotolo, J. F. and Eddy, R. S., *Anal. Chem.*, **49**, 884 (1977).
[531] Mosen, A. W., Schmitt, R. A. and Vasilevskis, J., *Anal. Chim. Acta*, **25**, 10 (1961).
[532] Freiling, E. C. and Bunney, L. R., *J. Am. Chem. Soc.*, **76**, 1021 (1954).
[533] Drake, B., *Ark. Kemi*, **8**, 1 (1955).
[534] Snyder, L. R., *Chromatogr. Rev.*, **7**, 1 (1965).
[535] Piez, K. A., *Anal. Chem.*, **28**, 1451 (1956).
[536] Freiling, E. C., *J. Am. Chem. Soc.*, **77**, 2967 (1955).
[537] Schwab, H., Rieman, W., III and Vaughan, P. A., *Anal. Chem.*, **29**, 1357 (1957).
[538] Koguchi, K., Waki, H. and Ohashi, S., *J. Chromatogr.*, **25**, 398 (1955).
[539] Ohashi, S. and Koguchi, K., *J. Chromatogr.*, **27**, 214 (1967).
[540] Massart, D. L. and Bossaert, W., *J. Chromatogr.*, **32**, 195 (1968).
[541] Hagdahl, L., Williams, R. J. P. and Tiselius, A., *Ark. Kemi*, **4**, 193 (1952).

[542] Giddings, J. C., *Anal. Chem.*, **39**, 1027 (1967).
[543] Horvath, C. G. and Lipsky, S. R., *Anal. Chem.*, **39**, 1893 (1967).
[544] Grande, J. A. and Beukenkamp, J., *Anal. Chem.*, **28**, 1497 (1956).
[545] Nervik, W. E., *J. Phys. Chem.*, **59**, 690 (1955).
[546] Maslova, G. B., Nazarov, P. P. and Chmutov, K. V., *Ionoobmennye sorbenty v promyshlennsoti* (*Ion Exchange Adsorbents in Industry*), Izd. Akad. Nauk SSSR, Moscow, 1963, p. 103.
[547] Marcus, Y. and Nelson, B., *J. Phys. Chem.*, **63**, 77 (1959).
[548] Zeligman, M. M., *Anal. Chem.*, **37**, 524 (1965).
[549] Pollard, F. H., Nickless, G. and Spincer, D., *J. Chromatogr.*, **11**, 215 (1963).
[550] Pollard, F. H., Nickless, G. and Spincer, D., *J. Chromatogr.*, **11**, 542 (1963).
[550a] Liteanu, C., Gocan, S., Hodisan, T., Nascu, H. and Marutoiu, E., *Rev. Roum. Chim.*, **17**, 497 (1972).
[550b] Liteanu, C. and Gocan, S., *Pure Appl. Chem.*, **31**, 455 (1972).
[551] Minczewski, J., *Pure Appl. Chem.*, **10**, 567 (1965).
[552] Marczenko, Z., *Chem. Anal. (Warsaw)*, **11**, 347 (1966).
[553] Tölg, G., *Talanta*, **19**, 1489 (1972).
[554] Brooks, R. R., *Analyst (London)*, **85**, 745 (1960).
[555] Matthews, A. D. and Riley, J. P., *Anal. Chim. Acta*, **48**, 25 (1961).
[556] Sterliński, S. and Dybczyński, R., *Nukleonika*, **11**, 533 (1966).
[557] Kulus, E., Molnár, F. and Szabó, E., *J. Radioanal. Chem.*, **7**, 347 (1971).
[558] Hirose, A. and Ishii, D., *Nippon Kagaku Kaishi*, **1972**, 2364; *Chem. Abstr.*, **78**, 66522p (1973).
[559] Dybczyński, R. and Sterliński, S., *Chem. Anal. (Warsaw)*, **17**, 1275 (1972).
[560] Maleszewska, H. and Dybczyński, R., *J. Radioanal. Chem.*, **31**, 177 (1976).
[561] Lur'e, Yu. Yu., *Zavodsk. Lab.*, **14**, 176 (1948).
[562] Mizuike, A., Ida, Y., Yamada, K. and Hirano, S., *Anal. Chim. Acta*, **32**, 428 (1965).
[563] Blount, C. W., Leyden, D. E., Thomas, T. L. and Guill, S. M., *Anal. Chem.*, **45**, 1045 (1973).
[564] Hołyńska, B., *Radiochem. Radioanal. Lett.*, **17**, 313 (1974).
[565] Eisner, U. and Mark, H. B., Jr., *Talanta*, **16**, 27 (1969).
[566] Hołyńska, B., Leszko, M. and Nahlik, E., *J. Radioanal. Chem.*, **13**, 401 (1971).
[567] James, H., *Analyst (London)*, **98**, 274 (1973).
[568] Bunney, L. R., Ballou, N. E., Pascual, J. and Foti, S., *Anal. Chem.*, **31**, 324 (1959).
[569] Strelow, F. W. E. and Sondorp, H., *Talanta*, **19**, 1113 (1972)
[570] Fołdzińska, A and Dybczyński, R., *J. Radioanal. Chem.*, **21**, 507 (1974).
[571] Maeck, W. J., Kussy, M. E. and Rein, J. E., *Anal. Chem.*, **35**, 2086 (1963).
[572] Hirsch, R. F. and Portock, J. D., *Anal. Chim. Acta*, **49**, 473 (1970).
[572a] Fołdzińska, A. and Dybczyński, R., *J. Radioanal. Chem.*, **31**, 89 (1976).
[573] Van den Winkel, P., De Corte, F., Specke, A. and Hoste, J., *Anal. Chim. Acta*, **42**, 340 (1968).
[574] Van den Winkel, P., De Corte, F. and Hoste, J., *Anal. Chim. Acta*, **56**, 241 (1971).
[575] Van den Winkel, P., De Corte, F. and Hoste, J., *J. Radioanal. Chem.*, **10**, 139 (1972).
[576] Nelson, F., Rush, R. M. and Kraus, K. A., *J. Am. Chem. Soc.*, **82**, 339 (1960).
[577] De Corte, F., Van Acker, P. and Hoste, J., *Anal. Chim. Acta*, **64**, 177 (1973).

[578] Strelow, F. W. E., Weinert, C. H. S. W. and Eloff, C., *Anal. Chem.*, **44**, 2352 (1972).
[579] Danielsson, L., *Acta Chem. Scand.*, **19**, 1859 (1965).
[580] Danielsson, L., *Acta Chem. Scand.*, **19**, 670 (1965).
[581] Freiling, E. C., Pascual, J. and Delucchi, A. A., *Anal. Chem.*, **31**, 330 (1959).
[582] Polkowska-Motrenko, H., and Dybczyński, R., *J. Chromatogr.*, **88**, 387 (1974).
[582a] Polkowska-Motrenko, H. and Dybczyński, R., *Chem. Anal. (Warsaw)*, **22**, 1021 (1977).
[583] Bhatnagar, R. P., Trivedi, R. G. and Bala, Y., *Talanta*, **17**, 249 (1970).
[584] Nikitin, M. K., Ostyk-Narbut, E. and Tomilev, S. B., *Vestnik Leningr. Gos. Univ., Ser. Fiz. i Khim.*, **1**, No. 4, 149 (1966).
[585] De Corte, F., van den Winkel, P., Specke, A. and Hoste, J., *Anal. Chim. Acta*, **42**, 67 (1968).
[586] Usubakunov, M. Yu., Bleshinskii, S. V. and Kakeeva, M. K., *Izv. Akad. Nauk Kirg. SSR*, **1971** (2), 47; *Chem. Abstr.*, **75**, 70951h (1971).
[586a] Oguma, K., Maruyama, T. and Kuroda, R., *Anal. Chim. Acta*, **74**, 339 (1975).
[586b] Akatsu, E. and Watanabe, H., *Anal. Chim. Acta*, **93**, 317 (1977).
[586c] Strelow, F. W. E., *Anal. Chem.*, **50**, 1359 (1978).
[587] Kawabuchi, K., Ito, T. and Kuroda, R., *J. Chromatogr.*, **39**, 61 (1969).
[588] Kuroda, R., Oguma, K., Kono, N. and Takahashi, Y., *Anal. Chim. Acta*, **62**, 343 (1972).
[589] Wódkiewicz, L. and Dybczyński, R., *J. Chromatogr.*, **102**, 277 (1974).
[589a] Strelow, F. W. E., Hanekom, M. D., Victor, A. H. and Eloff, C., *Anal. Chim. Acta*, **76**, 377 (1975).
[590] Strelow, F. W. E., van Zyl, C. R. and Bothma, C. J. C., *Anal. Chim. Acta*, **45**, 81 (1969).
[591] Strelow, F. W. E. and Bothma, C. J. C., *Anal. Chem.*, **39**, 595 (1967).
[592] Strelow, F. W. E., Victor, A. H., van Zyl, C. R. and Eloff, C., *Anal. Chem.*, **43**, 870 (1971).
[593] Korkisch, J., and Khater, M. M., *Talanta*, **19**, 1654 (1972).
[594] Korkisch, J. and Huber, A., *Talanta*, **15**, 119 (1968).
[595] Korkisch, J. and Hazan, I., *Anal. Chem.*, **37**, 707 (1965).
[596] Klakl, E. and Korkisch, J., *Talanta*, **16**, 1177 (1969).
[597] Kuroda, R. and Yoshikuni, N., *Talanta*, **18**, 1123 (1971).
[598] Kuroda, R., Ono, T. and Ishida, K., *Bunseki Kagaku*, **20**, 1142 (1971); *Chem. Abstr.*, **76**, 121121t (1972).
[599] Kuroda, R., Kondo, T. and Oguma, K., *Talanta*, **20**, 533 (1973).
[600] Kuroda, R., Ishida, K. and Kiriyama, T., *Anal. Chem.*, **40**, 1502 (1968).
[601] Wódkiewicz, L. and Dybczyński, R., *Chem. Anal. (Warsaw)*, 1981 (in press).
[602] Gimesi, O., Bányai, É., Csajka, M. and Szabadházy, A., *Talanta*, **17**, 1183 (1970).
[603] Renault, N. and Deschamps, N., *Radiochem. Radioanal. Lett.*, **13**, 207 (1973).
[604] Qureshi, M., Gupta, J. P. and Sharma, V., *Talanta*, **21**, 102 (1974).
[605] Potter, E. C. and Moresby, J. F., in *Ion Exchange and Its Applications*, Society of Chemical Industry, London, 1955, p. 454.
[606] Aleskovskiĭ, V. B., Libina, R. I. and Miller, A. D., *Tr. Leningr. Tekhnol. Inst. im. Lensoveta*, **48**, 5 (1948).
[607] McNutt, N. S. and Maier, R. H., *Anal. Chem.*, **34**, 276 (1962).
[608] Linstedt, K. D. and Kruger, P., *Anal. Chem.*, **42**, 113 (1970).

[608a] Brits, R. J. N. and Smit, M. C. B., *Anal. Chem.*, **49**, 67 (1977).
[609] Korkisch, J. and Dimitriadis, D., *Talanta*, **20**, 1287 (1973).
[610] Korkisch, J. and Dimitriadis, D., *Talanta*, **20**, 1295 (1973).
[611] Korkisch, J. and Dimitriadis, D., *Talanta*, **20**, 1303 (1973).
[612] Korkisch, J. and Sorio, A., *Talanta*, **22**, 273 (1975).
[612a] Korkisch, J. and Sorio, A., *Anal. Chim. Acta*, **76**, 393 (1975).
[613] Korkisch, J., Gödl, L. and Gross, H., *Talanta*, **22**, 281 (1973).
[614] Korkish, J., Gödl, L. and Gross, H., *Talanta*, **22**, 289 (1973).
[615] Pankov, J. F. and Janauer, G. E., *Anal. Chim. Acta*, **69**, 97 (1974).
[615a] Hirose, A., Kobori, K. and Ishii, D., *Anal. Chim. Acta*, **97**, 303 (1978).
[615b] Lenvik, K., Steinnes, E. and Pappas, A. C., *Anal. Chim. Acta*, **97**, 295 (1978).
[615c] Gladney, E. S. and Owens, J. W., *Anal. Chem.*, **48**, 2220 (1976).
[616] Senegačnik, M. and Paljk, Š., *Z. Anal. Chem.*, **232**, 409 (1967).
[617] Ohno, S. and Tsutsui, T., *Analyst (London)*, **95**, 396 (1970).
[617a] MacKenzie, A. B., Baxter, M. S., McKinley, I. G., Swan, D. S. and Jack, W., *J. Radioanal. Chem.*, **48**, 29 (1979).
[618] Paez, D. M. and Vasquez, J. A., *Rev. Asoc. Bioquim. Argent.*, **33**, 152 (1968); *Chem. Abstr.*, **72**, 15539t (1970).
[619] Riley, J. P. and Taylor, D., *Anal. Chim. Acta*, **41**, 175 (1968).
[619a] Abdullah, M. I., El-Rayis, O. A. and Riley, J. P., *Anal. Chim. Acta*, **84**, 363 (1976).
[620] Florence, T. M. and Batley, G. E., *Talanta*, **22**, 201 (1975).
[620a] Lamathe, J., *Anal. Chim. Acta*, **104**, 307 (1979).
[620b] Lee, C., Kim, N. B., Lee, I. C. and Chung, K. S., *Talanta*, **24**, 241 (1977).
[621] Kühn, G. and Hering, R., *Z. Chem.*, **5**, 316 (1965).
[622] Bosholm, J., *J. Chromatogr.*, **21**, 286 (1966).
[623] Grossmann, O., Döge, H. G. and Grosse-Ruyken, H., *Z. Anal. Chem.*, **219**, 48 (1966).
[624] Davankov, A. B. and Laifer, V. M., *Zavodsk. Lab.*, **22**, 294 (1957).
[625] Ezerskaya, N. A. and Markova, N. V., *Zh. Prikl. Khim.*, **30**, 1071 (1957).
[626] Beyermann, K., *Z. Anal. Chem.*, **200**, 183 (1964).
[627] Nikanorov, G. S. and Kist, A. A., *Zh. Analit. Khim.*, **27**, 2438 (1972).
[628] Podorvanova, N. F., *Zh. Analit. Khim.*, **26**, 818 (1971).
[629] Wald, M., Soyka, W. and Kaysser, B., *Talanta*, **20**, 405 (1973).
[630] Ishida, K., Kiriyama, T. and Kuroda, R., *Anal. Chim. Acta*, **41**, 537 (1968).
[631] McCrackan, J. D., Vecchione, M. C. and Longo, S. L., *At. Absorption Newsl.*, **8**, 102 (1969); *Chem. Abstr.*, **72**: 8858t (1970).
[632] Tikhomirov, B. K. and Petrukhin, N. V., *Radiokhimiya*, **10**, 111 (1968).
[633] Semov, M. P., *Zh. Analit. Khim.*, **23**, 245 (1968).
[634] Éristavi, V. D., Éristavi, D. I. and Brouchek, F. I., *Zh. Analit. Khim.*, **23**, 782 (1968).
[635] Govinduraju, K., *Anal. Chem.*, **40**, 24 (1968).
[636] Smales, A. A. and Salmon, L., *Analyst (London)*, **80**, 37 (1955).
[637] Andersen, N. R. and Hume, D. N., *Anal. Chim. Acta*, **40**, 207 (1968).
[638] Bosholm, J., *Anal. Chim. Acta*, **34**, 71 (1966).
[639] Hesselbach, H., *Z. Anal. Chem.*, **248**, 289 (1969).
[640] Otmakhova, Z. I., Chashina, O. V. and Slezko, N. I., *Zavodsk. Lab.*, **35**, 685 (1969).
[641] Gal, I. J. and Ruvarac, A., *J. Chromatogr.*, **13**, 549 (1964).
[642] Brunfelt, A. O. and Steinnes, E., *Analyst (London)*, **94**, 979 (1969).

[643] Strelow, F. W. E. and Jackson, P. F. S., *Anal. Chem.*, **46**, 1481 (1974).
[644] Das, H. A., Janssen, R. and Zonderhuis, J., *Radiochem. Radioanal. Lett.*, **8**, 257 (1971).
[645] Dingman, J. F., Jr., Gloss, K. M., Milano, E. A. and Siggia, S., *Anal. Chem.*, **46**, 774 (1974).
[646] Kraft, G., Dosch, H. and Gabbert, K., *Z. Anal. Chem.*, **267**, 106 (1973).
[647] Janauer, G. E. and Madrid, E. O., *Mikrochim. Acta*, **1974**, 769.
[648] Myasoedova, G. V. and Bol'shakova, L. I., *Zh. Analit. Khim.*, **23**, 504 (1968).
[649] Chwastowska, J., Dybczyński, R. and Kucharzewski, B., *Chem. Anal. (Warsaw)*, **13**, 721 (1968).
[650] Prout, W. E., Russell, E. R. and Groh, H. J., *J. Inorg. Nucl. Chem.*, **27**, 473 (1965).
[651] Boni, A. L., *Anal. Chem.*, **38**, 89 (1966).
[652] Derecki, J., Geisler, J., Jankovic, D. and Jaworowski, Z., *Nukleonika*, **8**, 69 (1963).
[653] Sugihara, T. T., James, H. I., Froianello, E. J. and Bowen, V. T., *Anal. Chem.*, **31**, 44 (1959).
[654] Porter, C. R. and Kahn, B., *Anal. Chem.*, **36**, 676 (1964).
[655] Rane, A. T. and Bhatbi, K. S., *Anal. Chem.*, **38**, 1598 (1966).
[656] Blake, W. E., Oldham, G. and Sumpter, D., *Nature (London)*, **203**, 862 (1964).
[657] Senegaćnik, M., Paljk, Š. and Kristan, J., *Z. Anal. Chem.*, **249**, 39 (1970).
[658] Pitts, A. E. and Beamish, F. E., *Anal Chem.*, **41**, 1107 (1969).
[659] Krawczyk, I., *Nukleonika*, **5**, 649 (1960).
[660] Wódkiewicz, L. and Dybczyński, R., *Chem. Anal. (Warsaw)*, **19**, 175 (1974).
[661] Steinnes, E., *Radiochim. Acta*, **17**, 119 (1972).
[662] Evans, H. B., Bloomquist, C. A. A. and Hughes, J. P., *Anal. Chem.*, **34**, 1692 (1962).
[663] Sen Gupta, J. G., *Anal. Chem.*, **39**, 18 (1967).
[664] Sen Gupta, J. G., *Anal. Chim. Acta*, **42**, 481 (1968).
[665] Minczewski, J. and Dybczyński, R., *Chem. Anal. (Warsaw)*, **10**, 1113 (1965).
[666] Dybczyński, R., Sterliński, S. and Golian, C., *J. Radioanal. Chem.*, **16**, 105 (1973).
[667] Vink, J. J., *Analyst (London)*, **95**, 399 (1970).
[668] Tera, F., Ruch, R. R. and Morrison, G. H., *Anal. Chem.*, **37**, 358 (1965).
[669] Jervis, R. E. and Wong, K. Y., *Nuclear Activation Techniques in the Life Sciences*, IAEA, Vienna, 1967, p. 137.
[670] Kalinin, A. I., Kuznetsov, R. A., Moiseev, V. V. and Tsepurnek, V. É., *Radiokhimicheskie metody opredeleniya mikroelementov (Radiochemical Methods of Determining Microelements)*, Izd. Nauka, Moscow, 1965, p. 161.
[671] Zasukhin, E. N., Kalinin, A. I., Kuznetsov, R. A. and Moiseev, V. V., *Radiokhimicheskie metody opredeleniya mikroelementov (Radiochemical Methods of Determining Microelements)*, Izd. Nauka, Moscow, 1965, p. 168.
[672] Kalinin, A. I., Kuznetsov, R. A. and Moiseev, V. V., *Radiokhimicheskie metody opredeleniya mikroelementov (Radiochemical Methods of Determining Microelements)*, Izd. Nauka, Moscow, 1965, p. 171.
[673] Kalinin, A. I., Kuznetsov, R. A. and Moiseev, V. V., *Radiokhimicheskie metody opredeleniya mikroelementov (Radiochemical Methods of Determining Microelements)*, Izd. Nauka, Moscow, 1965, p. 176.
[674] Kalinin, A. I., Kuznetsov, R. A., Moiseev, V. V. and Sokolova, M. N., *Radiokhimicheskie metody opredeleniya mikroelementov* (Radiochemical Methods of Determining Microelements), Izd. Nauka, Moscow, 1965, p. 180.

[675] Yoshimura, J. and Waki, H., *Bull. Chem. Soc. Japan*, **35**, 416 (1962).
[676] Blaedel, W. J., Olsen, E. D. and Buchanan, R. F., *Anal. Chem.*, **32**, 1866 (1960).
[677] Wilkins, D. H., *Talanta*, **2**, 355 (1959).
[678] Aubouin, G. and Laverlochère, J., *C.E.A. Report* DR/AR/G/63-18.
[679] Van den Winkel, P., Specke, A. and Hoste, J. J., *Nuclear Activation Techniques in the Life Sciences*, IAEA, Vienna, 1967, p. 159.
[680] Van Erkelens, P. C., *Anal. Chim. Acta*, **25**, 42 (1961).
[681] Plantin, L. O., *Nuclear Activation Techniques in the Life Sciences*, Proc. Symposium, Bled 1972, IAEA, Vienna, 1972, p. 73.
[682] Morrison, G. H. and Potter, N. M., *Anal. Chem.*, **44**, 839 (1972).
[683] Velandia, J. A. and Perkons, A. K., *J. Radioanal. Chem.*, **20**, 473 (1974).
[684] Ahrens, L. H., Edge, R. A. and Brooks, R. R., *Anal. Chim. Acta*, **28**, 551 (1963).
[685] Boni, A. L., *Anal. Chem.*, **32**, 599 (1960).
[686] Fritz, J. S. and Latwesen, G. L., *Talanta*, **17**, 81 (1970).
[687] Morrison, G. H., Gerard, J. T., Travesi, A., Currie, R. L., Peterson, S. F. and Potter, N. M., *Anal. Chem.*, **41**, 1633 (1969).
[688] Morrison, G. H., Gerard, J .T., Kashuba, A. T., Gangadharan, E. V., Rothenberg, A. M., Potter, N. M. and Miller, G. B., *Science*, **167**, 505 (1970).
[689] Turekian, K. K. and Kharkar, D. P., *Science*, **167**, 507 (1970).
[689a] Morrison, G. H., Nadkarni, R. A., Potter, N. M., Rothenberg, A. M. and Wong, S. F., *Radiochem. Radioanal. Lett.*, **11**, 251 (1972).
[690] Schmidt, D. H. and Fritz, J. S., *Talanta*, **15**, 515 (1968).
[691] Ryabinin, A. I., Artyuzhik, P. I. and Bogatyrev, V. L., *Radiokhimiya*, **9**, 335 (1967).
[692] Lavrukhina, A. K. and Mil'nikova, Z. K., *Zh. Analit. Khim.*, **24**, 1714 (1969).
[693] Vasil'ev, I. Ya., Razumova, G. N. and Shuba, I. D., *Radiokhimiya*, **11**, 573 (1969).
[694] Orlandini, R. A. and Korkisch, J., *Anal. Chim. Acta*, **43**, 459 (1968).
[695] Malvano, R., Grosso, P. and Zanardi, M., *Anal. Chim. Acta*, **41**, 251 (1968).
[696] Samsahl, K., Brune, D. and Wester, P. O., *Int. J. Appl. Radiat. Isot.*, **16**, 273 (1965).
[697] Samsahl, K., Wester, P. O. and Landström, O., *Anal. Chem.*, **40**, 181 (1968).
[698] Samsahl, K., *Aktiebolaget Atomenergi Report* AE-389, 1970.
[698a] Lievens, P., Cornelis, R. and Hoste, J., *Anal. Chim. Acta*, **80**, 97 (1975).
[698b] Kalinin, A. I., *Zh. Analit. Khim.*, **32**, 21 (1977).
[698c] Sterliński, S., Maleszewska, H. and Dybczyński, R., *J. Radioanal. Chem.*, **31**, 61 (1976).
[698d] Strelow, F. W. E., Victor, A. H. and Weinert, C. H. S. W., *Geostandards Newsl.*, **2**, 49 (1978).
[698e] Korkisch, J., Hübner, H., Steffan, I., Arrhenius, G., Fisk, M. and Frazer, J., *Anal. Chim. Acta*, **83**, 83 (1976).
[698f] Korkisch, J., Stefan, I., Arrhenius, G., Fisk, M. and Frazer, J., *Anal. Chim. Acta*, **90**, 151 (1977).
[698g] Korkisch, J., Stefan, I. and Arrhenius, G., *Anal. Chim. Acta*, **94**, 237 (1977).
[698h] Gallorini, M., Greenberg, R. R. and Gills, T. E., *Anal. Chem.*, **50**, 1479 (1978).
[699] Natsume, H., Umezawa, H., Suzuki, T., Ichikawa, F., Sato, T., Baba, S. and Amano, H., *J. Radioanal. Chem.*, **7**, 189 (1971).
[700] Vdovenko, V. M., Krivokhatskii, A. S. and Skovorodkin, N. V., *Radiokhimiya*, **13**, 416 (1971).

References

[701] Strelow, F. W. E., Victor, A. H. and Weinert, C. H. S. W., *Anal. Chim. Acta*, **69**, 105 (1974).
[702] Strelow, F. W. E. and Boshoff, M. D., *Anal. Chim. Acta*, **62**, 351 (1972).
[703] Strelow, F. W. E., Weinert, C. H. S. W. and Van der Walt, T. N., *Talanta*, **21**, 1183 (1974).
[704] Strelow, F. W. E. and Weinert, C. H. S. W., *Talanta*, **20**, 1127 (1973).
[704a] Polkowska-Motrenko, H. and Dybczyński, R., *Radiochem. Radioanal. Lett.*, **26**, 217 (1976).
[705] Nelson, F., Michelson, D. C., Phillips, H. O. and Kraus, K. A., *J. Chromatogr.*, **20**, 107 (1965).
[706] Brooksbank, W. A. and Leddicotte, G. W., *J. Phys. Chem.*, **57**, 819 (1953).
[707] Lilova, O. M. and Preobrazhenskii, B. K., *Radiokhimiya*, **2**, 728 (1960).
[708] Lavrukhina, A. K., Poznyakov, A. A., Rodin, S. S. and Moskaleva, A. P., *Dokl. Akad. Nauk SSSR*, **130**, 88 (1960).
[709] Strelow, F. W. E., Lindenberg, C. J. and Toerien, F. von S., *Anal. Chim. Acta*, **43**, 465 (1968).
[710] Strelow, F. W. E., Weinert, C. H. S. W. and Van der Walt, T. N., *Anal. Chim. Acta*, **71**, 123 (1974).
[711] Abe, M. and Ito, T., *Bull. Chem. Soc. Japan*, **40**, 1013 (1967).
[712] Ruch, R. R., Tera, F. and Morrison, G., *Anal. Chem.*, **36**, 2311 (1964).
[713] I Liang Sun and Lan-Chiang Lu, *Hua Hsueh Pao*, **30**, 117 (1964); *Chem. Abstr.*, **61**: 6433a (1964).
[714] Nelson, F., *J. Am. Chem. Soc.*, **77**, 813 (1955).
[715] Li Tien, H., *J. Phys. Chem.*, **68**, 1021 (1964).
[716] Bregman, J. I. and Murata, Y., *J. Am. Chem. Soc.*, **74**, 1867 (1952).
[717] Buser, W., *Helv. Chim. Acta*, **34**, 1635 (1951).
[718] Olsen, F. O. and Sobel, H. R., *Talanta*, **12**, 81 (1965).
[719] Samuelson, O. and Sjöstrom, E., *Anal. Chem.*, **26**, 1908 (1954).
[719a] Abe, M., *J. Chromatogr.*, **134**, 507 (1977).
[719b] Delphin, W. H. and Horwitz, E. P., *Anal. Chem.*, **50**, 843 (1978).
[720] Dybczyński, R. and Maleszewska, H., unpublished work.
[721] Kemula, W., Brajter, K., Cieślik, S. and Lipińska-Kostrowicka, H., *Chem. Anal. (Warsaw)*, **5**, 225 (1960).
[722] Kiełczewski, W., *Chem. Anal. (Warsaw)*, **8**, 691 (1963).
[723] Milton, G. M. and Grummitt, W. E., *Can. J. Chem.*, **35**, 541 (1957).
[724] Lerner, M. and Rieman, W., III, *Anal. Chem.*, **26**, 610 (1954).
[725] Lilova, O. M. and Preobrazhenskii, B. K., *Radiokhimiya*, **2**, 731 (1960).
[726] Strelow, F. W. E., van Zyl, C. R. and Nolte, C. R., *Anal. Chim. Acta*, **40**, 145 (1968).
[727] Strelow, F. W. E. and Weinert, C. H. S. W., *Talanta*, **17**, 1 (1970).
[728] Strelow, F. W. E. and van Zyl, C. R., *Anal. Chim. Acta*, **41**, 529 (1968).
[729] Khristova, R. and Kruschevska, A., *Anal. Chim. Acta*, **36**, 392 (1966).
[730] Bouquiaux, J. J. and Gillard, J. H. C., *Anal. Chim. Acta*, **30**, 273 (1964).
[731] Šulcek, Z., Povondra, P. and Štangl, R., *Talanta*, **9**, 647 (1962).
[732] Nelson, F., Holloway, J. H. and Kraus, K. A., *J. Chromatogr.*, **11**, 258 (1963).
[733] Campbell, M. H., *Anal. Chem.*, **37**, 252 (1965).
[734] Nelson, F. and Kraus, K. A., *J. Am. Chem. Soc.*, **77**, 801 (1955).
[735] Fritz, J. S., Waki, H. and Garralda, B. B., *Anal. Chem.*, **36**, 900 (1964).
[736] Bennett, W. E. and Skovlin, D. O., *Anal. Chem.*, **38**, 518 (1966).

[737] Yamabe, T. and Hayashi, T., *J. Chromatogr.*, **102**, 273 (1974).
[738] Belyavskaya, T. A., and Nemirovskaya, I. A., *Zh. Analit. Khim.*, **28**, 2337 (1973).
[739] Power, W. H., Kirby, H. W., Cluggage, W. C., Nelson, G. D. and Payne, J. H., *Anal. Chem.*, **31**, 1077 (1959).
[740] Qureshi, M. and Gupta, J. P., *J. Chem. Soc.*, *A*, **1969**, 1755.
[740a] Strelow, F. W. E. and Weinert, C. H. S. W., *Anal. Chim. Acta*, **83**, 179 (1976).
[741] Preobrazhenskii, B. K. and Saikov, Yu. P., *Radiokhimiya*, **2**, 68 (1960).
[742] Fritz, J. S. and Garralda, B. B., *Anal. Chem.*, **34**, 102 (1962).
[743] Gierst, L. and Dubru, L., *Bull. Soc. Chim. Belg.*, **63**, 379 (1954).
[744] Kallman, S., Oberthin, H. and Liu, R., *Anal. Chem.*, **32**, 58 (1960).
[745] Berg, E. W. and Truemper, J. T., *Anal. Chem.*, **30**, 1827 (1958).
[746] Kuroda, R., Kondo, T. and Oguma, K., *Talanta*, **19**, 1043 (1972).
[747] Strelow, F. W. E., Louw, W. J. and Weinert, C. H. S. W., *Anal. Chem.*, **40**, 2021 (1968).
[748] Korkisch, J. and Feik, F., *Anal. Chim. Acta*, **32**, 110 (1965).
[749] Morie, G. P. and Sweet, T. R., *J. Chromatogr.*, **16**, 201 (1964).
[749a] Strelow, F. W. E., *Anal. Chim. Acta*, **100**, 577 (1978).
[750] Diamond, R. M., Street, K., Jr. and Seaborg, G. T., *J. Am. Chem. Soc.*, **76**, 1461 (1954).
[750a] Chakravotry, M. and Khopkar, S. M., *Chromatographia*, **10**, 100 (1977).
[751] Specht, S. and Höhlein, G., *Radiochim. Acta*, **12**, 38 (1969).
[752] Danon, J., *J. Inorg. Nucl. Chem.*, **7**, 422 (1958).
[753] Kavalova, Z. K., Shibaeva, N. P. and Pyzhova, Z. I., *Zh. Analit. Khim.*, **23**, 655 (1968).
[754] Faris, F. P. and Warton, J. W., *Anal. Chem.*, **34**, 1077 (1967).
[755] Hamaguchi, H., Ohuchi, A., Shimizu, T., Onuma, N. and Kuroda, R., *Anal. Chem.*, **36**, 2304 (1964).
[756] Minczewski, J. and Dybczyński, R., *J. Chromatogr.*, **7**, 568 (1962).
[757] Stewart, D. C., *Proc. 1st Intern. Conf. Peaceful Uses Atomic Energy, Geneva*, **7**, 321 (1956).
[758] Preobrazhenskii, B. K., Kalyamin, A. V. and Lilova, O. M., *Zh. Analit. Khim.*, **2**, 1164 (1957).
[759] Foti, S. C. and Wish, L., *J. Chromatogr.*, **29**, 203 (1967).
[760] Almásy, A., *Acta Chim. Acad. Sci. Hung.*, **17**, 55 (1958).
[761] Wolfsberg, K., *Anal. Chem.*, **34**, 518 (1962).
[762] Deelstra, H. and Verbeek, F., *J. Chromatogr.*, **17**, 558 (1965).
[763] Cuninghame, J. G., Sizeland, M. L., Willis, H. H., Eakins, J. and Mercer, E. R., *J. Inorg. Nucl. Chem.*, **1**, 163 (1955).
[764] Karol, P. J., *J. Chromatogr.*, **79**, 287 (1973).
[764a] Seaborg, G. T., *Atomnaya Energiya*, **6**, 21 (1959).
[765] Gatti, R. C., Phillips, L., Sikkeland, T., Muga, M. L. and Thompson, S. G., *J. Inorg. Nucl. Chem.*, **11**, 251 (1959).
[766] Yamabe, T., *J. Chromatogr.*, **83**, 59 (1973).
[767] Glass, R. A., *J. Am. Chem. Soc.*, **77**, 807 (1955).
[768] Kim, J. I. and Born, H. J., *Radiochim. Acta*, **14**, 35 (1970).
[769] Kim, J. I. and Born, H. J., *Radiochim. Acta*, **14**, 65 (1970).
[770] Eristavi, V. D., Kutsyava, N. A. and Kekeliya, R. A., *Zh. Analit. Khim.*, **29**, 158 (1974).

References

[771] Brücher, E. and Szarvas, P., *Acta Chim. Acad. Sci. Hung.*, **52**, 31 (1967).
[772] Powell, J. E. and Spedding, F. H., *Trans. Am. Inst. Min. Metall. Pet. Eng.*, **215**, 457 (1959).
[773] Strelow, F. W. E., *Anal. Chem.*, **31**, 1201 (1959).
[774] Strelow, F. W. E., *Anal. Chim. Acta*, **34**, 387 (1966).
[775] Danon, J., *J. Inorg. Nucl. Chem.*, **5**, 237 (1958).
[776] Gąsior, M., Mikulski, J. and Stroński, I., *Nukleonika*, **6**, 757 (1961).
[777] Kraus, K. A., Moore, G. E. and Nelson, F., *J. Am. Chem. Soc.*, **78**, 2692 (1956).
[778] Nelson, F., Michelson, D. C. and Holloway, J. H., *J. Chromatogr.*, **14**, 258 (1964).
[779] Bubernak, J., Lew, M. S. and Matlack, G. M., *Anal. Chim. Acta*, **48**, 233 (1969).
[780] Roberts, E. P. and Brauer, F. P., *U.S. AEC Report* HW-60552, 1959.
[781] Baybarz, R. D., *J. Inorg. Nucl. Chem.*, **28**, 1723 (1966).
[782] Nagle, R. A. and Murphy, T. K. S., *Analyst (London)*, **84**, 37 (1959).
[783] Kuroda, R. and Hikawa, I., *J. Chromatogr.*, **25**, 408 (1966).
[784] Cesarano, C. and Pungetti, G., *C.N.E.N. Report* RT/CHI (63) 31.
[785] Kuroda, R. and Ishida, K., *J. Chromatogr.*, **18**, 438 (1965).
[786] Hamaguchi, H., Ishida, K. and Kuroda, R., *Anal. Chim. Acta*, **33**, 91 (1965).
[787] Orlandini, K. A., *Inorg. Nucl. Chem. Lett.*, **5**, 325 (1969).
[788] Cummings, T. F. and Korkisch, J., *Anal. Chim. Acta*, **40**, 520 (1968).
[789] Korkisch, J. and Orlandini, K. A., *Anal. Chem.*, **40**, 1952 (1968).
[790] Kressin, I. V. and Waterburg, G. R., *Anal. Chem.*, **34**, 1598 (1962).
[791] Radhakrishna, P., *J. Chim. Phys.*, **51**, 354 (1954).
[792] Nakanishi, T. and Sakanone, M., *Radiochim. Acta*, **11**, 119 (1969).
[793] Wenzel, A. W. and Pietri, C. E., *Anal. Chem.*, **36**, 2083 (1964).
[794] Balsene, L. R. and Simona, M. G., *Anal. Chim. Acta*, **104**, 319 (1979).
[795] Kraus, K. A., Nelson, F. and Smith, G. W., *J. Phys. Chem.*, **58**, 11 (1954).
[796] Korkisch, J. and Hazan, I., *Anal. Chem.*, **36**, 2308 (1964).
[797] Strelow, F. W. E. and Victor, A. H., *Talanta*, **19**, 1019 (1972).
[798] Pyatnitskii, I. V. and Kolomets, L. L., *Zh. Analit. Khim.*, **25**, 479 (1970).
[799] Belyavskaya, T. A., Alimarin, I. P. and Kolosova, I. F., *Zh. Analit. Khim.*, **13**, 668 (1958).
[800] Larsen, E. M. and Pei Wang, *J. Am. Chem. Soc.*, **76**, 6223 (1954).
[801] Alimarin, I. P. and Medvedeva, A. M., *Khromatografiya. Ee teoriya i primenenie (Chromatography. Theory and Application)*, Izd. Akad. Nauk SSSR, Moscow, 1960, p. 379.
[802] Street, K., Jr. and Seaborg, G. T., *J. Am. Chem. Soc.*, **70**, 4268 (1948).
[803] Lister, B. A. J. and Hutcheon, J. M., *Research (London)*, **5**, 291 (1952).
[804] Kolosova, G. M., Chen Yun-Pan and Senyavin, M. M., *Zh. Analit. Khim.*, **15**, 364 (1960).
[805] Huffman, E. H. and Lilly, R. C., *J. Am. Chem. Soc.*, **73**, 2902 (1951).
[806] Huffman, E. H., Iddings, G. M. and Lilly, R. C., *J. Am. Chem. Soc.*, **73**, 4474 (1951).
[807] Brown, W. E. and Rieman, W., III, *J. Am. Chem. Soc.*, **74**, 1278 (1952).
[808] Qureshi, M., Husain, W. and Israili, A. H., *Talanta*, **15**, 789 (1968).
[809] Korkisch, J. and Orlandini, K. A., *Talanta*, **16**, 45 (1969).
[810] Qureshi, M. and Husain, K., *Anal. Chem.*, **43**, 447 (1971).
[811] Strelow, F. W. E., Liebenberg, C. J. and Toerien, F. von S., *Anal. Chim. Acta*, **47**, 251 (1969).

[812] Nemodruk, A. A., Palei, P. N. and Bezrogova, E. V., *Zh. Analit. Khim.*, **25**, 319 (1970).
[813] Nelson, F. and Kraus, K. A., *J. Chromatogr.*, **3**, 279 (1960).
[814] Strelow, F. W. E. and Toerien, F. von S., *Anal. Chem.*, **38**, 545 (1966).
[815] Cabbell, T. R., Orr, A. A. and Hayes, J. R., *Anal. Chem.*, **32**, 1602 (1960).
[816] Ariel, M. and Kirowa, E., *Talanta*, **8**, 214 (1961).
[817] Kraus, K. and Moore, G., *J. Am. Chem. Soc.*, **73**, 2900 (1951).
[818] Wilkins, D. H., *Talanta*, **2**, 355 (1959).
[819] Kallmann, S., Oberthin, H. and Liu, R., *Anal. Chem.*, **34**, 609 (1962).
[820] Pakholkov, V. S. and Maksimov, I. E., *Zh. Analit. Khim.*, **39**, 1179 (1966).
[821] Ferraro, T. A., *Talanta*, **16**, 669 (1969).
[822] Dixon, E. J. and Headridge, J. B., *Analyst (London)*, **89**, 185 (1964).
[823] Specke, A. and Hoste, J., *Talanta*, **2**, 332 (1959).
[824] Bandi, W. R., Buyok, E. G., Lewis, L. L. and Melnick, L. M., *Anal. Chem.*, **33**, 1275 (1961).
[825] Stroński, I. and Zieliński, A., *Nukleonika*, **9**, 802 (1964).
[826] Smith, F. W., *J. Chromatogr.*, **22**, 580 (1966).
[827] Nidrutskaya, L. N. and Nabivanets, B. I., *Zh. Analit. Khim.*, **23**, 1643 (1968).
[828] De, A. K. and Majumdar, S. K., *Z. Anal. Chem.*, **191**, 40 (1962).
[829] Fritz, J. S. and Abbink, J. E., *Anal. Chem.*, **34**, 1980 (1962).
[830] Starobinets, G. L. and Mechkovski, S. A., *Zh. Analit. Khim.*, **18**, 298 (1963).
[831] Schafer, H. N. S., *Anal. Chem.*, **35**, 53 (1963).
[832] Peters, T. A., Jr. and Rieman, W., III, *Anal. Chim. Acta*, **14**, 131 (1956).
[833] Ohashi, S. and Takada, S., *Bull. Chem. Soc. Japan*, **34**, 1516 (1961).
[834] Pollard, F. H., Nickless, G. and Murray, J. D., *J. Chromatogr.*, **22**, 139 (1966).
[835] Koguchi, K., Waki, H. and Ohashi, S., *J. Chromatogr.*, **25**, 398 (1966).
[836] Kiesl, W., *Z. Anal. Chem.*, **227**, 13 (1967).
[837] De, A. K. and Chakrabarty, T., *Indian J. Chem.*, **7**, 180 (1969).
[838] Preobrazhenskii, B. K. and Moskvin, L. N., *Radiokhimiya*, **3**, 309 (1961).
[839] Pollard, F. H., Rogers, D. E., Rothwell, M. T. and Nickless, G., *J. Chromatogr.*, **9**, 227 (1962).
[840] Nelson, F. and Kraus, K. A., *J. Am. Chem. Soc.*, **77**, 4508 (1955).
[841] Klement, R. and Kühn, A., *Z. Anal. Chem.*, **152**, 146 (1956).
[842] Fritz, J. S. and Dahmer, L. H., *Anal. Chem.*, **37**, 1272 (1956).
[843] Kraus, K. A., Nelson, F. and Moore, G. E., *J. Am. Chem. Soc.*, **77**, 3972 (1955).
[844] Pakholkov, V. S. and Panikarovskii, V. E., *Izv. Vyssh. Uchebn. Zaved., Khim. Khim. Tekhnol.*, **10**, 168 (1967).
[845] Feik, F. and Korkisch, J., *Mikrochim. Acta*, **1967**, 900.
[846] Iguchi, A., *Bull. Chem. Soc. Japan*, **31**, 600 (1958).
[847] Iguchi, A., *Bull. Chem. Soc. Japan*, **31**, 597 (1958).
[848] Ioffe, V. P. and Rubinshtein, R. N., *Sb. Nauch. Tr., Gos. Nauchno-Issled. Inst. Tsvetn. Met.*, **1971**, (34), 75; *Chem. Abstr.*, **78**, 52092m (1973).
[849] Sasaki, Y., *Bull. Chem. Soc. Japan*, **28**, 89 (1955).
[850] Kleeman, E. and Hermann, G., *J. Chromatogr.*, **3**, 275 (1960).
[851] Šimek, M., *Chem. Listy*, **60**, 817 (1966).
[852] Vasil'ev, I. Ya., Razumova, G. N. and Shuba, I. D., *Radiokhimiya*, **11**, 573 (1969).
[853] Iguchi, A., *Bull. Chem. Soc. Japan*, **31**, 748 (1958).
[854] Veale, C. R., *J. Inorg. Nucl. Chem.*, **10**, 333 (1959).

[855] Gaibakyan, D. S. and Darbinyan, M. V., *Izv. Akad. Nauk Arm. SSR*, **16**, 211 (1963); *Chem. Abstr.*, **59**, 14863h (1963).
[856] Strel'nikova, N. P. and Lystsova, G. G., *Zavodsk. Lab.*, **26**, 142 (1960).
[857] Yamamoto, M. and Sakai, H., *Anal. Chim. Acta*, **32**, 370 (1965).
[858] Fritz, J. S. and Yamamura, S. S., *Anal. Chem.*, **27**, 1461 (1955).
[859] Zelikman, A. N. and Meisner, G., *Zh. Prikl. Khim.*, **39**, 992 (1966).
[860] Pirš, M. and Magee, J., *Talanta*, **8**, 395 (1961).
[861] Hamaguchi, H., Kawabuchi, K. and Kuroda, R., *Anal. Chem.*, **36**, 1654 (1964).
[862] Kawabuchi, K., Hamaguchi, H. and Kuroda, R., *J. Chromatogr.*, **17**, 567 (1965).
[863] Sen Sarma, R. N., Andres, E. and Miller, J. M., *J. Phys. Chem.*, **63**, 559 (1959).
[864] Dams, R. and Hoste, J., *Anal. Chim. Acta*, **41**, 197 (1968).
[865] Pozdnyakov, A. A. and Ryabchikov, D. I., *Radiokhimicheskie metody opredeleniya mikroelementov* (*Radiochemical Methods of Determining Microelements*), Izd. Nauka, Moscow, 1965, p. 130.
[866] Boyd, G. E. and Larsen, Q. V., *J. Am. Chem. Soc.*, **90**, 5092 (1968).
[867] Boyd, G. E. and Larsen, Q. V., *J. Am. Chem. Soc.*, **90**, 254 (1968).
[868] Good, M. L., Purdy, M. B. and Hoering, Th. C., *J. Inorg. Nucl. Chem.*, **6**, 73 (1958).
[869] Mikulski, J. and Stroński, I., *Rocz. Chem.*, **34**, 721 (1960).
[870] Starobinets, G. L. and Mechkovskii, S. A., *Zh. Analit. Khim.*, **18**, 255 (1963).
[871] Atterberry, R. W. and Boyd, G. E., *J. Am. Chem. Soc.*, **72**, 4805 (1950).
[872] De Geiso, R. C., Rieman, W., III and Lindenbaum, S., *Anal. Chem.*, **26**, 1840 (1954).
[873] Holzapfel, H. and Gürtler, O., *J. Prakt. Chem.*, **35**, 113 (1967).
[874] Starobinets, G. L. and Bulat-skaya, G. N., *Vestsi Akad. Navuk BSSR, Ser. Khim. Navuk*, **1969** (3), 106.
[875] Zalevskaya, T. L. and Starobinets, G. L., *Zh. Analit. Khim.*, **24**, 721 (1969).
[876] Starobinets, G. L. and Mechkovskii, S. A., *Zh. Analit. Khim.*, **16**, 319 (1961).
[877] Nabivanets, B. I., *Ukr. Khim. Zh.*, **22**, 816 (1956).
[878] Kikindai, M., *Compt. Rend.*, **240**, 1100 (1955).
[879] Skloss, J. L., Hudson, J. A. and Cummiskey, C. J., *Anal. Chem.*, **37**, 1240 (1965).
[880] Wang, Fu-Chun, Kang, Meng-Hua and Khalkin, V. A., *Radiokhimiya*, **4**, 94 (1962).
[881] Sagortschew, B. (Zagorchev, B.), *Acta Chim. Acad. Sci. Hung.*, **26**, 289 (1961).
[882] Pollard, F. H., McOmie, J. F. W., Nickless, G. and Hansen, P., *J. Chromatogr.*, **4**, 108 (1960).
[883] Morie, G. P. and Sweet, T. R., *Anal. Chem.*, **36**, 140 (1964).
[884] Korkisch, J. and Ahluwalia, S. S., *Anal. Chim. Acta*, **34**, 308 (1966).
[885] Inczédy, J., Gábor-Klatsmányi, P. and Erdey, L., *Acta Chim. Acad. Sci. Hung.*, **61**, 261 (1969).
[886] Qureshi, M., Kumar, R. and Rathore, H. S., *J. Chem. Soc. (A)*, **1970**, 272.
[887] Qureshi, M. and Husain, W., *J. Chem. Soc. (A)*, **1970**, 1204.
[888] Fritz, J. S. and Pietrzyk, D. J., *Talanta*, **8**, 143 (1961).
[889] Lavrukhina, A. K. and Sazhina, N. K., *Zh. Analit. Khim.*, **24**, 870 (1969).
[890] Wilkins, D. H., *Anal. Chim. Acta*, **20**, 271 (1959).
[891] Turner, J. B., Philp, R. H. and Day, R. A., Jr., *Anal. Chim. Acta*, **26**, 94 (1961).
[892] Vanderdeelen, J., *Anal. Chim. Acta*, **49**, 360 (1970).
[893] Gera, J., *Chem. Anal. (Warsaw)*, **14**, 581 (1969).
[894] Brajter, K., *Chem. Anal. (Warsaw)*, **18**, 125 (1973).

[895] Berman, S. S. and McBryde, W. A. E., *Can. J. Chem.*, **36**, 835 (1958).
[896] Blasius, E. and Preetz, W., *Chromatogr. Rev.*, **6**, 191 (1964).
[897] MacNevin, W. M. and McKay, E. S., *Anal. Chem.*, **29**, 1220 (1957).
[898] Berman, S. S. and McBryde, W. A. E., *Can. J. Chem.*, **36**, 845 (1958).
[899] Coufalík, F. and Svach, M., *Z. Anal. Chem.*, **173**, 113 (1960).
[900] Cluett, M. L., Berman, S. S. and McBryde, W. A. E., *Analyst (London)*, **80**, 204 (1955).
[901] Busch, D. D., Prospero, J. M. and Naumann, R. A., *Anal. Chem.*, **31**, 884 (1959).
[902] Taylor, H. and Beamish, F. E., *Talanta*, **15**, 497 (1968).
[903] Gijbels, R. and Hoste, J., *Anal. Chim. Acta*, **41**, 419 (1968).
[904] Chung, K. S. and Beamish, F. E., *Talanta*, **15**, 823 (1968).
[905] Stevenson, P. C., Franke, A. A., Borg, R. and Nervik, W., *J. Am. Chem. Soc.*, **75**, 4876 (1953).
[906] Berg, E. W. and Senn, W. D., Jr., *Anal. Chem.*, **27**, 1255 (1955).
[907] Kanert, G. A. and Chow, A., *Anal. Chim. Acta*, **78**, 375 (1975).
[908] MacNevin, W. M. and Crummet, W. B., *Anal. Chem.*, **25**, 1628 (1953).
[909] Blasius, E. and Wachtel, U., *Z. Anal. Chem.*, **142**, 341 (1954).
[910] Blasius, E. and Rexin, D., *Z. Anal. Chem.*, **179**, 105 (1961).
[911] Marks, A. G. and Beamish, F. E., *Anal. Chem.*, **30**, 1462 (1958).
[912] Pshenitsyn, N. K., Gladyshevskaya, K. A. and Ryakhova, L. M., *Zh. Analit. Khim.*, **2**, 1057 (1967).
[913] Kühn, G. and Hoyer, E., *Z. Chem. (Leipzig)*, **7**, 113 (1967).
[914] Gulko, A., Feigenbaum, H. and Schmuckler, G., *Anal. Chim. Acta*, **59**, 397 (1972).
[915] Huffman, E. H. and Oswalt, R. L., *J. Am. Chem. Soc.*, **72**, 3323 (1950).
[916] Fujimoto, M., *Chemist-Analyst*, **49**, 4 (1960).
[917] Fujimoto, M., *Bull. Chem. Soc. Japan*, **33**, 864 (1960).
[918] Miller, W. E., *Anal. Chem.*, **29**, 1891 (1957).
[919] Kakihana, H., *Mikrochim. Acta*, **1956**, 682.
[920] Fujimoto, M. and Nakatsukasa, Y., *Mikrochim. Acta*, **1968**, 551.
[921] Fujimoto, M., *Bull. Chem. Soc. Japan*, **29**, 646 (1956).
[922] Fujimoto, M., Nakayama, H., Ito, M. and Suga, T., *Mikrochim. Acta*, **1974**, 151.
[923] Qureshi, M., Qureshi, S. Z. and Zehra, N., *Anal. Chim. Acta*, **47**, 169 (1969).
[924] Quinche, J. P. and Quinche-Sax, S., *J. Chromatogr.*, **32**, 162 (1968).
[925] Ogawa, T. and Tsukahara, S., *Denki Kagaku*, **39**, 888 (1971); *Chem. Abstr.*, **76**, 148451n (1972).
[926] Ogawa, T. and Miura, Y., *Denki Kagaku*, **39**, 960 (1972); *Chem. Abstr.*, **76**, 148494d (1972).
[927] Ogawa, T. and Tani, A., *Denki Kagaku*, **40**, 729 (1972); *Chem. Abstr.*, **78**, 92194c (1973).
[928] Qureshi, M., Qureshi, S. Z. and Zehra, N., *Talanta*, **19**, 377 (1972).
[928a] Yoshimura, K. and Ohashi, S., *Talanta*, **25**, 103 (1978).
[928b] Yoshimura, K., Waki, H. and Ohashi, S., *Talanta*, **23**, 449 (1976); **25**, 579 (1978).
[929] Růžička, J. and Starý, J., *Substoichiometry in Radiochemical Analysis*, Pergamon, Oxford, 1968.
[930] Freeman, D. H., Currie, L. A., Kuehner, E. C., Dixon, H. D. and Paulson, R. A. *Anal. Chem.*, **42**, 203 (1970).
[931] Hahn, P. B. and Shleien, B., *Anal. Chem.*, **42**, 1608 (1970).

[932] Kuroda, R., Seki, T., Misu, Y., Oguma, K. and Saito, T., *Talanta*, **26**, 211 (1979).
[933] Novikov, Yu. P., Mikheeva, M. N., Myasoedov, B. F., Akhmanova, M. V. and Komarevskiĭ, V. M., *Radiokhimiya*, **22**, 336 (1980).
[934] Grote, M. and Kettrup, A., *Z. Anal. Chem.*, **295**, 366 (1979).
[935] Sugii, A. and Ogawa, N., *Talanta*, **26**, 970 (1979).
[936] Tscholakowa, J., Burba, P., Gleitsmann, B. and Lieser, K. H., *Z. Anal. Chem.*, **300**, 121 (1980).
[937] Blasius, E., Janzen, K. P., Luxenburger, H., Hguyen, V. B., Klotz, H. and Stockemer, J., *J. Chromatogr.* **167**, 307 (1978).
[938] Suzuki, N., Saitoh, K. and Hamada S., *Radiochem. Radioanal. Lett.*, **32**, 121 (1978).
[939] Hafez, M. B., Nazmy, A. F., Salem, F. and Eldesoki, M., *J. Radioanal. Chem.*, **47**, 115 (1978).
[940] Jain, A. K., Agrawal, S. and Singh, R. P., *J. Radioanal. Chem.*, **54**, 171 (1979).
[941] Singh, N. J. and Tandon, S. N., *J. Radioanal. Chem.*, **49**, 195 (1979).
[942] Srivasatava, S. K., Jain, A. K., Kumar, S., Singh, R. P. and Agrawal, S., *J. Radioanal. Chem.*, **53**, 49 (1979).
[943] Rengan, K., Haushalter, J. P. and Jones, J. D., *J. Radioanal. Chem.*, **54**, 347 (1979).
[944] Bhattacharyya, D. K. and Basu, S., *J. Radioanal. Chem.*, **52**, 267 (1979).
[945] Heinonen, O. J., *Radiochem. Radioanal. Lett.*, **43**, 293 (1980).
[946] Bhattacharyya, D. K. and Basu, S., *J. Radioanal. Chem.*, **47**, 105 (1978).
[947] Akilimali, K., Lumu, B. and Mwamba, W., *Anal. Chem.*, **51**, 166 (1979).
[948] Sipos-Galiba, I. and Lieser, K. H., *Radiochem. Radioanal. Lett.*, **42**, 329 (1980).
[949] Jain, A. K., Agrawal, S. and Singh, R. P., *J. Radioanal. Chem.*, **60**, 111 (1980).
[950] Fedorov, V. A., Polyanskiĭ, N. G. and Reikshfel'd, V. O., *Zh. Prikl. Khim.*, **52**, 1042 (1979).
[951] Van Acker, P., *Anal. Chim. Acta*, **113**, 149 (1980).
[952] Polkowska-Motrenko, H. and Dybczyński, R., *J. Radioanal. Chem.*, **59**, 31 (1980).
[953] Rabel, F. M., in *Advances in Chromatography*, Vol. 17, Giddings, J. C., Grushka, E., Cazes, J. and Brown, P. R. (Eds.), Dekker, New York, 1979.
[954] Ishii, D., Hirose, A. and Iwasaki, Y., *J. Radioanal. Chem.*, **46**, 41 (1978).
[955] Elchuk, S. and Cassidy, R. M., *Anal. Chem.*, **51**, 1434 (1979).
[956] Lipski, A. J. and Vairo, C. J., *Canadian Research* **13**, 45 (1980).
[957] Wetzel, R. A., Anderson, C. L., Schleicher, H. and Crook, G. D., *Anal. Chem.*, **51**, 1532 (1979).
[958] Sawicki, E., Mulik, J. D. and Wittgenstein, E., *Ion Chromatographic Analysis of Environmental Pollutants*, Ann Arbor Science, Ann Arbor, 1978.
[959] Anlauf, K., Barrie, L. A., Wiebe, H. A. and Fellin, P., *Canadian Research*, **13**, 49 (1980).
[960] Dulski, T. R., *Anal. Chem.*, **51**, 1439 (1979).
[961] Kuroda, R. and Seki, T., *Z. Anal. Chem.*, **300**, 107 (1980).
[962] Singh, D. and Tandon, S. N., *Talanta*, **26**, 163 (1979).
[963] Nelson, F. and Kraus, K. A., *J. Chromatogr.*, **178**, 163 (1979).
[964] Sava, S., Zikovsky, L. and Poisvert, J., *J. Radioanal. Chem.*, **57**, 23 (1980).
[965] Litman, R., Mallet, J. and Notini, B. R., *Radiochem. Radioanal. Lett.*, **45**, 347 (1980).
[966] Victor, A. H. and Strelow, F. W. E., *Geostandards Newsl.* **4**, 217 (1980).

[967] Croudace, I. W., *J. Radioanal. Chem.*, **59**, 323 (1980).
[968] Yamashita, M., and Suzuki, N., *J. Radioanal. Chem.*, **60,** 73 (1980).
[969] Nakaoka, A., Yokoyama, H., Fukushima, M. and Takagi, S., *J. Radioanal. Chem.*, **56**, 13 (1980).
[970] Sterliński, S., Maleszewska, H., Szopa, Z. and Dybczyński, R., *J. Radioanal. Chem.*, **59**, 141 (1980).
[971] Abe, M., Ichsan, E. A. A. and Hayashi, K., *Anal. Chem.*, **52**, 524 (1980).
[972] Strelow, F. W. E., *Anal. Chim. Acta*, **113**, 323 (1980).
[973] Guseva, L. I. and Tikhomirova, G. S., *J. Radioanal. Chem.*, **52**, 369 (1979).
[974] Brajter, K., Słonawska, K. and Vorbrodt, Z., *Chem. Anal. (Warsaw)* **24**, 763 (1979).
[975] Toshimitsu, Y., Yoshimura, K. and Ohashi, S., *Talanta*, **26**, 273 (1979).
[976] Kayasth, S. R., Iyer, R. K. and Sankar Das, M., *J. Radioanal. Chem.*, **59**, 373 (1980).
[977] Kayasth, S. R. and Desai, H. B., *Radiochem. Radioanal. Lett.*, **44**, 403 (1980).

Chapter 7

EXTRACTION CHROMATOGRAPHY

7.1 GENERAL

Extraction chromatography, also known as reversed-phase partition chromatography, is a modern separation method increasingly used in radiochemistry and analytical chemistry. In the method, the stationary phase is an extractant which coats or is bonded to a porous hydrophobic support, and the mobile phase is a suitable solution of an acid, base or salt. Although the method is fairly recent [1] many reviews [2–13], monographs [14–16] and bibliographies [17–19] are available.

Among the materials used as stationary-phase supports are silica gel, diatomaceous earth (Celite, Chromosorb W, Hyflo Super Cel) silanized with dimethyldichlorosilane (DMCS), various organic polymers such as polytrifluorochloroethylene (Kel-F, Hostaflon C2, Ftoroplast-3, Daiflon M-300, Plascon CTFE-2300, Voltalef), polytetrafluoroethylene (Teflon, Algoflon, Tee Six, Ftoroplast-4, Chromosorb T, Fluoropak-80, Haloport, etc.), polyethylene, vinyl chloride–vinyl acetate copolymer (Corvic), polyisopropylene, grafted styrene copolymers (S-D.V.B., Chromosorb 101 and 102, Bio Beads SM-1 and SM-2, Amberlyst XAD-2), and cellulose powder. Cellular plastics and rubbers [20,168], and glass fiber filter paper [169] have recently been added to the list.

The range of the extractants used for impregnating the support is virtually unlimited. Those most frequently used for extraction chromatography are tributyl phosphate (TBP), methyl isobutyl ketone (MIBK), tri-n-octylphosphine oxide (TOPO), tri-n-octylamine (TNOA), di(2-ethylhexyl)orthophosphoric acid (HDEHP), 2-ethylhexylphenyl phosphonic acid (HEHΦP), Aliquat-336 (methyltri-n-alkylammonium chloride), and dinonylnaphthalenesulphonic acid (DNNS). Some other extractants also used [21,22] include chelating compounds such as α-benzoinoxime [23], dithizone [24,25,170] etc., as well as binary mixtures, used to obtain synergic effects [26]. If tetrachlorohydroquinone is deposited on a hydrophobic support, the electron exchanger obtained can be used for some redox reactions [27,28]. The redox reaction may be combined with an extraction by preparing a column packed with both an electron exchanger and a suitable extractant [29].

Extraction chromatography combines high extraction selectivity (which occasionally can be enhanced by addition of complexing agents to the aqueous phase) with the fundamental advantage of chromatography, the multiple partition of the substances being separated, between the stationary and mobile phases. Thus excellent separations of similar elements can be accomplished in shorter times and with much less complicated apparatus than in static extraction.

The quantity of extractant in the column is also generally much lower than that needed for an analogous static separation, and the column can usually be reused. Extractants that give emulsions under static extraction conditions may be suitable for extraction chromatography.

However, extraction chromatography has a number of limitations. The stability of the column used is very variable. Although some extractant–support systems are known to stay unchanged in the column for scores of runs [30–32], cases where only a single run is possible have been reported [33–35]. The fact that the extractant is gradually bleeding from the bed at a definite rate into the aqueous phase is one of the reasons for the dependence of the distribution coefficients for the elements, in a given system, on the column history.

For such supports as silica gel or diatomaceous earth made hydrophobic with dimethyldichlorosilane, the hydrophobic film has been observed to decompose with time, and adsorption effects become important in the chromatographic process [36]. Even Teflon, a supposedly inert support, has recently been shown by gas-chromatographic experiments [37] to be capable of sorbing some organic compounds. Therefore, sorption of compounds of inorganic ions e.g. chelates, cannot be disregarded. Extraction chromatography often fails if slowly reacting extractants are used for the stationary phase (e.g. thenoyltrifluoroacetone, TTA). Highly polar organic compounds are not sufficiently retained by hydrophobic supports and should not be used as stationary phases. Another drawback is the relatively low extractive capacity of the columns (when a certain extractant to support ratio is exceeded the column efficiency sharply declines). Finally, the preparation of adequate columns is still difficult, and columns capable of high performance and repeatability cannot always be made routinely.

In spite of these restrictions, extraction chromatography is a valuable technique of ever-increasing importance. This method and ion-exchange chromatography complement one another and are frequently used jointly in complex separation schemes [23,38–42]. A separation scheme for more than ten elements based exclusively on extraction chromatography has been reported [43].

7.2 THEORY

7.2.1 The Distribution-Reaction Equilibrium

The theory of the process of distribution of a substance between two immiscible phases is covered in Chapter 5. The treatment that follows will be concerned only with the extraction of inorganic ions with liquid ion-exchanger extractants.

A solution of an amine that is insoluble in water, e.g. TNOA, in an organic solvent, equilibrated with a solution of an acid, extracts metal complexes according to the following equation

$$(M_a A_m^{p-})_{aq} + (pR_3 NH^+ A^-)_o \rightleftharpoons (pA^-)_{aq} + [(R_3 NH)_p M_a A_m]_o \quad (7.1)$$

The exchange reaction for a liquid anion-exchanger deposited on a solid support has an analogous form. Accordingly, the ion-exchange extraction of uranyl chloride complexes from hydrochloric acid solutions is as follows

$$(UO_2 Cl_4^{2-})_{aq} + (2R_3 NH^+ Cl^-)_o \rightleftharpoons 2Cl^-)_{aq} + [(R_3 NH)_2 UO_2 Cl_4]_o \quad (7.2)$$

The overall picture of variations of the distribution coefficients of individual ions with hydrochloric acid concentration is similar to the analogous relations for strongly basic anion-exchange resins [2].

The extraction of metal cations by water-insoluble phosphoric acids (e.g. HDEHP) is also of an ion-exchange character. HDEHP occurs in organic solution as a dimer, and the ion-exchange extraction reaction can be written as

$$(M^{p+})_{aq} + (pH_2 A_2)_o \rightleftharpoons [M(HA_2)_p]_o + (pH^+)_{aq} \quad (7.3)$$

The concentration equilibrium constant of reaction (7.3) is defined by the equation

$$K = \frac{[M(HA_2)_p]_o [(H^+)^p]_{aq}}{[M^{p+}]_{aq}[(H_2A_2)^p]_o} = D \frac{[(H^+)^p]_{aq}}{[(H_2A_2)^p]_o} \quad (7.4)$$

where D is the distribution coefficient (ratio of the concentration in the organic phase to the concentration in the aqueous phase) for a given ion.

The distribution coefficient can be found from the elution curves as described in Chapter 6, by using the formula

$$D = [U_{max} - (U_0 + V)]/V_0 \qquad (7.5)$$

where U_{max} is the retention volume of the ion considered, U_0 is the dead volume of the column, V is the free volume of the bed and V_0 is the stationary-phase volume (volume of extractant adsorbed on support).

The distribution coefficients, and the separation factor

$$\alpha_1^2 = D_2/D_1 \qquad (7.6)$$

determined by extraction chromatography are usually very close to the corresponding values found by static extraction [44–49]. A theoretical treatment of this has been reported [50].

Owing to this coincidence the values of the distribution coefficients listed for use in and obtained by static extraction are applicable to planning extraction-chromatography separations. In some recent studies the distribution coefficients were determined by a static method with an extractant deposited on a hydrophobic support; that is, in a system identical to extraction chromatography. Examples of distribution coefficients determined thus are given in Figs. 7.1–7.3 [51].

Liquid ion-exchangers have also been widely used for separation of metal ions by paper chromatography. The results are commonly reported in terms of values of R_f, defined as the ratio of the distance of travel of the centre of the spot to that of the eluent front [52].

If the R_f values in a given system are known, it is possible to estimate the potential of the system for a particular extraction-chromatography separation, since the following relation holds [53]:

$$D = A_1/A_s(R_f^{-1} - 1) = \text{const}(R_f^{-1} - 1) \qquad (7.7)$$

where A_1 and A_s are the cross-sectional areas of the mobile and stationary phases respectively. The R_f values for most of the elements of the periodic table for paper impregnated with HDEHP and TOPO are shown in Figs. 7.4–7.7 [5,54,55].

By taking the logarithm of Eq. (7.4) the following relation is obtained:

$$\log D = p \log [H_2A_2]_o - p \log [H^+]_{aq} + \text{const} \qquad (7.8)$$

It follows from this formula that a plot of $\log D$ or $\log(R_f^{-1} - 1)$ vs. pH should be a straight line of slope equal to the charge on the ion extracted (p).

Theory

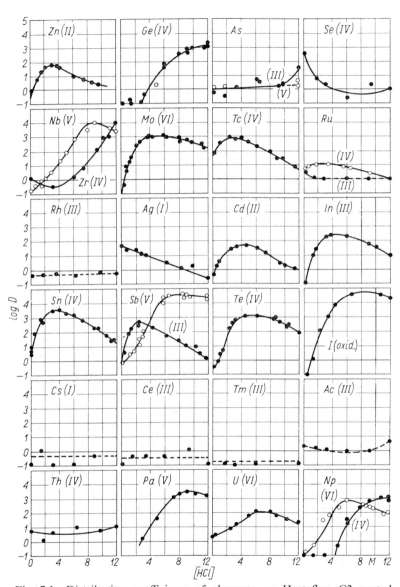

Fig. 7.1 Distribution coefficients of elements on Hostaflon C2 coated with TBP-HCl(aq) (from [51], by permission of the copyright holders Akadémiai Kiadó)

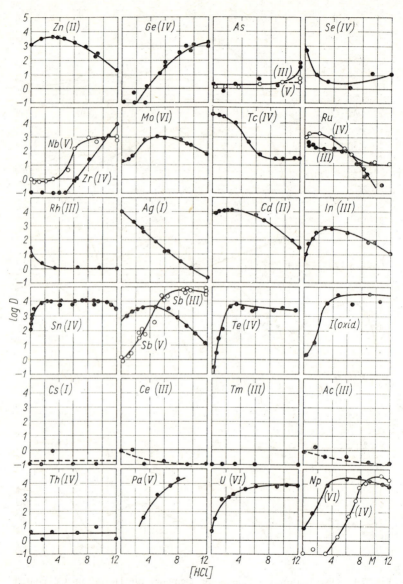

Fig. 7.2 Distribution coefficients for the elements on TNOA-impregnated Hostaflon C2–HCl(aq) (from [51], by permission of the copyright holders, Akadémiai Kiadó)

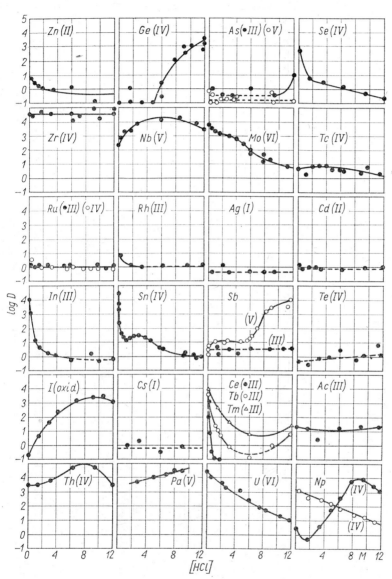

Fig. 7.3 Distribution coefficients of elements on HDEHP-impregnated Hostaflon C2–HCl(aq) (from [51], by permission of the copyright holders, Akadémiai Kiadó)

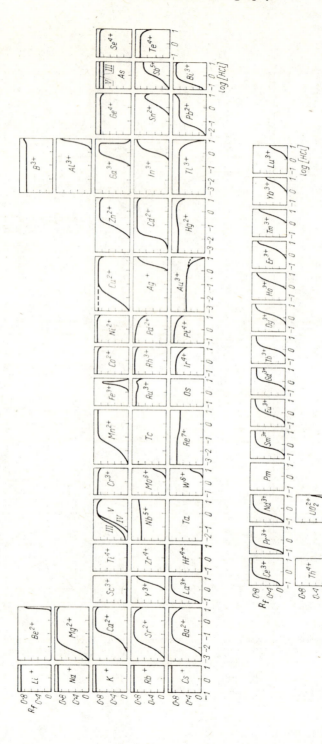

Fig. 7.4 Chromatography on paper impregnated with 0.1M HDEHP in cyclohexane; the R_f values for the elements vs. hydrochloric acid (eluent) concentration (from [5] and [54], by permission of the copyright holders, C.I.S.E. and Elsevier Publ. Co.)

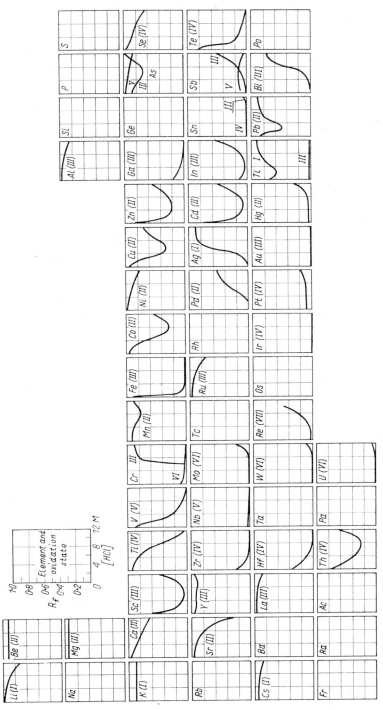

Fig. 7.5 Chromatography on paper impregnated with 0·025M TOPO in cyclohexane; the R_f values for the elements vs. hydrochloric acid (eluent) concentration (from [5] and [54], by permission of the copyright holders, C.I.S.E. and Elsevier Publ. Co.)

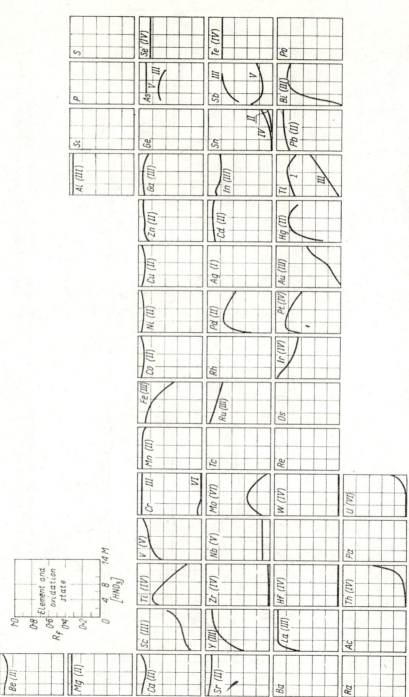

Fig. 7.6 Chromatography on paper impregnated with 0·025M TOPO in cyclohexane; the R_f values of elements vs. nitric acid (eluent) concentration (from [5] and [55], by permission of the copyright holders, C.I.S.E. and Elsevier Publ. Co.)

§ 7.2] **Theory** 513

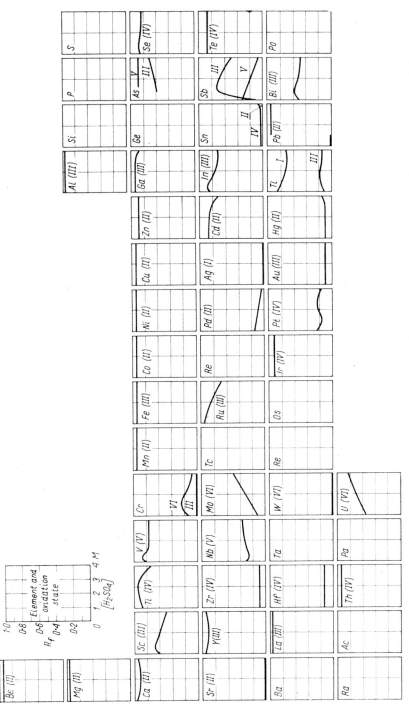

Fig. 7.7 Chromatography on paper impregnated with 0·025M TOPO in cyclohexane; the R_f values of elements vs. sulphuric acid (eluent) concentration (from [5] and [55], by permission of the copyright holders, C.I.S.E. and Elsevier Publ. Co.)

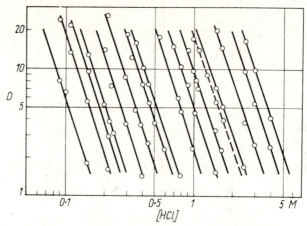

Fig. 7.8 Distribution coefficients of rare earths, determined by extraction chromatography on HDEHP-impregnated silica gel–HCl(aq) *vs.* HCl concentration (from [44], by permission of the copyright holders, Johann Ambrosius Barth Verlag)

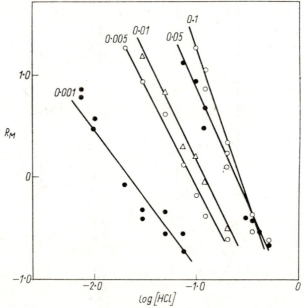

Fig. 7.9 The R_M values for La(III) for paper impregnated with cyclohexane HDEHP solutions of various concentrations *vs.* hydrochloric acid concentration (eluate) (from [56], by permission of the copyright holders, Elsevier Publ. Co.)

This relation is generally obeyed roughly (Fig. 7.8) [44], although quite often the slopes found experimentally are significantly lower than expected from the formal charge on the ions. A possible explanation for this is that complexes of type $MCl_n^{(p-n)+}$ are formed in the aqueous solution, particularly at high concentration of acid, or that analogous hydrolysis products are formed [54].

It has also been shown that the slope depends on the concentration of the liquid ion-exchanger on the support, and that it attains the value predicted by theory only if that concentration is not too low [56] (Fig. 7.9).

From Eq. (7.8) it follows that the plot of log D vs. log $[H_2A_2]_o$ at constant acidity of the solution is also a line of slope p, and this has been observed experimentally [54,57].

As in the ion-exchange process, the heats of the extraction reactions may differ even for chemically related elements like lanthanides or actinides [58–60]. Thus, not only the distribution coefficients but also the separation factors vary (rise or fall) with temperature.

7.2.2 The Column Separation Process

The general discussion of the factors that affect column efficiency and a quantitative description of that efficiency, provided in Chapter 6 for ion-exchange chromatography, holds also for extraction chromatography. However, the relative extents of the individual mechanisms of chromatographic-band broadening (cf. Eq. 6.68) and the specific form of some of the terms of that equation (particularly those responsible for the mass-transfer process between the two phases, cf. [61]) may differ substantially.

Studies of the effects of various factors on column performance and the quality of the separations in extraction chromatography are rather scarce [60,62–69]. Some of the data available are not completely reliable, because the measurement of the height of a theoretical plate as a function of a definite parameter was not always made at a constant value of the distribution coefficient. Nevertheless, some generalizations appear to be feasible, although they hold good only for certain extraction systems and for a relatively narrow range of experimental variables such as temperature, flow-rate, etc.

Better (lower) values of the theoretical plate height are generally achieved for silanized silica gel or diatomaceous earth than for organic-polymer or cellulose supports [32]. The column performance is affected by the way silica gel is silanized, and by how much dimethylchlorosilane was taken for the treatment. Impaired mechanical strength of the packing, and tailing, have been observed when large excess of dimethylchlorosilane is employed. [69]. The height of a theoretical plate was found to be independent of the pore diameter of the silica gel used, above a certain value (~35 Å) [69]. Other authors suggested, however, that the better theoretical-plate height observed for diatomaceous earths compared with silica gels arises primarily from the difference in the mean pore diameters [70].

The theoretical plate height (H) depends to a large degree on the quantity of extractant used in relation to a given weight of support. Usually there is an optimum value of that ratio, above or below which H increases. For HDEHP supported on silica gel the optimum value of the ratio is 0·6–0·8 cm^3/g of silica gel [63–69].

If a solution of an extractant in a low-volatility solvent is taken instead of the pure extractant, the minimum value of the plate height is generally observed for a certain definite concentration of that solution [63], although exceptions to this rule are not uncommon [32]. Likewise the manner of loading the stationary phase on to the support may have an effect on the plate height. For HDEHP and silica gel, a better plate height was obtained when the extractant was deposited from a chloroform solution, followed by the removal of the solvent by evaporation first in air then under reduced pressure, than when neat HDEHP was added dropwise into silanized silica gel with stirring until a dry powder was produced [69].

The plate height as a rule increases with the grain size of the support [60,63,65–67]; it also generally increases with the flow-rate [60,62–64, 66,67], although at low rates occasionally a minimum on the curve of H vs. u occurs [62]; this could be indicative of increasing importance of longitudinal diffusion. On the basis of investigations made to date it is established that higher temperatures decrease the plate height [60,62–64, 66,68,71]. In one study, however, there was a minimum in the plot of H vs. temperature [68]. Occasionally, if a certain temperature is exceeded the extractant is rapidly eluted from the column, resulting in impairment of column performance [66]. The method of packing the column and the compactness of packing have an important influence on H. There is generally an optimum degree of packing for which the plate height is smallest [30,63]. If the bed is packed too tightly, tailing may occur [30]. The theoretical plate height may also depend on the composition of the mobile phase. For instance, in the separation of rare earths on columns packed with HDEHP on diatomaceous earth, substantial differences were observed in plate heights for heavy-earth elements, depending on the mineral acid used for elution [62].

As in ion-exchange chromatography, overloading of the column with ions results in increased plate height [63].

7.3 TECHNIQUES IN EXTRACTION CHROMATOGRAPHY

In general, the apparatus for extraction chromatography, operation of the column, and eluate analysis, etc. are analogous to those described in Chapter 6. Glass columns used in extraction chromatography are sometimes silanized with dimethyldichlorosilane [38,72]. Various techniques of silanization of silica gel or diatomaceous earth have been reported. Dimethyldichlorosilane vapours may be used at atmospheric [32,71] or reduced [60] pressure, or else the support may be shaken with an ethereal solution of dimethyldichlorosilane [73] or boiled in a carbon tetrachloride solution of it [107]. The silanized material is then dried at 100–200°C [32,60]. Porous silica microspherical supports such as Zorbax-SIL were even dried and silanized with trimethylchlorosilane in a vacuum system [74]. Cellulose powder, before being impregnated with extractant, is dried at 40°C [75]. The organic polymers used as supports generally require no special treatment. A recommended practice, however, is to wash them in acetone then ether, followed by drying in air before use [78].

Many procedures have been reported for coating the support material with the extractant. Under static conditions this can be done by shaking the support with a mixture of the extractant and an aqueous phase of definite composition [1], by mixing the support with a solution of the extractant in a volatile solvent that is subsequently removed by evaporation [32,71], by addition of pure extractant to the support [69], by shaking the support with a solution of the extractant in a heavy solvent (here the stationary phase is the solution) [63,76], etc.

An alternative procedure is to pack the column with the support, impregnate it with extractant equilibrated with aqueous phase, then remove the excess of extractant by washing it off with extractant-saturated eluent [1,38,77].

Any air bubbles can be removed from the column by prolonged washing with an eluent solution under pressure [32]. Although dry column packing is common practice, particularly in earlier work, there is much evidence that better results can be achieved by packing the column with an aqueous suspension of the support already impregnated with extractant [67,70,74].

The linear flow-rates used in extraction chromatography seldom exceed 1 cm/min, but some separations at rates higher by an order of magnitude or more have been attempted [65,67,79].

In very recent years, reversed-phase HPLC systems have become very popular for organic separations, particularly as a result of the introduction of packing materials with the extractant bonded to the support material by Si—O—Si—C— type linkages [80–82]. It has been reported that 80% of the analytical columns used in HPLC are packed with microparticulate octadecyl-, octyl- or phenyl-silica [81a]. As yet, little use has been made of high performance reversed-phase systems for inorganic separations, but some work has been reported [74], and it seems likely that this field will soon expand.

Recently an apparatus for high speed liquid chromatography has been constructed in which column, tubing and valves are all constructed either of glass or of plastic so that concentrated acids and other corrosive liquids can be used as eluents. The metal ions in the effluent are automatically detected by a spectrophotometric detector coupled to a recorder[12a]. Similar apparatus has been used by other workers for the separation of numerous radionuclides including lanthanides and actinides [74,82a]. In this case effluent was collected in drops on stainless steel or tantalum discs, the activity of which was then automatically counted with an end window proportional counter.

7.4 APPLICATIONS

Extraction chromatography has been used more often in radiochemistry than in analytical chemistry. Much of the work is concerned with the separation of elements of similar chemical properties present in trace quantities, for which this method is particularly good (an example is shown in Fig. 7.10). The method has also been used, however, for macro–micro separations, such as isolation of carrier-free gold isotope from an irradiated platinum target [84] or for an analogous isolation of carrier-free manganese from iron [85], arsenic from germanium [47], protactinium from thorium [86], neptunium from uranium [87,88], and radium from thorium [89], separation of uranium from fission products [87,90–92], separation of plutonium from uranium and fission products [92] (and separation of fission products from each other [91,93–96]), separation of traces of some rare earths from other rare earths present in macro quantities [32,34,97], etc. Extraction chromatography has also been used to prepare ultrapure gadolinium and dysprosium oxides [98]. The high selectivity of the separation obtainable by extraction chromatography was found helpful for trace analysis in the development of high-selectivity isolation procedures for copper [75], silver [99,100], tin [101], bismuth [102], molybdenum [23,103], iron [104], mercury [105] and gold [106] from composite mixtures.

Other typical analytical applications include the isolation of trace impurities of certain rare earths from pure samples of other rare earths, for their determination by methods such as activation analysis [107,108], spectrography [107], and spectrophotometry [109], the analogous determination of tungsten in steels by activation analysis [110], spectrographic determination of nanogram quantities of beryllium and rare earths in uranium [111], and the isolation of fission products from spent nuclear fuels to facilitate their subsequent determination by radiometry [112], activation analysis [113], or mass-spectrometry [114]. Extraction chromatography was employed together with activation analysis for the determination of trace impurities in niobium [115], and in a highly activating matrix such as germanium [171], gallium [172], gallium arsenide [116], or antimony [117, 172], for which high decontamination factors were achieved (10^6–10^{10}). Trace gold in platinum was determined by selective isolation of the matrix before activation in the nuclear reactor [118]. Trace amounts of indium in semiconductor alloys could be determined by atomic-absorption spectrometry after the main constituents of the alloy (Te, Sn, and Pb) were removed by extraction chromatography [118a]. Rare earths and scandium employed as additions in various aluminium alloys were separated by extraction chromatography from the macro component as well as other elements (Ca, Mg, Mn, Cu, Ni, Cr, Zn, Cd) and determined spectrophotometrically in the eluate [118b].

A variant of extraction chromatography, in which the elements retained on the column were displaced selectively by means of other elements was used in the determination of traces of certain metals by activation analysis [119–122], and by spectrophotometry [123]. As mentioned previously, extraction chromatography has been used in conjunction with ion-exchange chromatography for the separation of multicomponent mixtures into groups and individual components [23,38–42,173].

The method appears to be very useful for separating and concentrating elements in trace amounts, and radionuclides. Typical examples are rapid selective isolation of ^{131}I from large volumes of water using I_2-toluene immobilized on open-cell polyurethane foam [123a], analogous separation of ^{131}I from milk using I_2 in toluene diluted Alamine 336 on the same support [123b], rapid isolation of copper ions from large volumes of water or sodium chloride solutions using the liquid ion exchanger 'Kelex 100', supported on the macroporous resin Amberlite XAD-7 [123c] etc. High speed extraction chromatography enabled the separation of ^{225}Ac and its short-lived daughters (^{221}Fr, ^{213}Bi, ^{209}Tl, and ^{209}Pb) in approximately 1 min, as well as purification and recovery of a sample of ^{257}Fm from other transplutonium elements and yttrium [82a].

A listing of the major separation procedures for elements of various groups of the periodic table is given in Table 7.1, and some typical examples are illustrated in Figs. 7.10–7.14.

Table 7.1 Extraction chromatography systems for the separation of the elements within individual groups of the periodic table

Elements isolated	Support	Stationary phase	Mobile phase (Eluent)	Ref.
1A				
Na, K, Rb, Cs	Hyflo Super Cel	monododecylphosphoric acid + chlorobenzene	extractant-sat. $0.04M$ LiNO$_3$–$0.01M$ HCl: Na, K, Rb, Cs	[76]
Li, Na, K, Rb, Cs	Hyflo Super Cel	monododecylphosphoric acid + chlorobenzene	extractant, sat. with H$_2$O: Li, Na, K, Rb, $0.5M$ HNO$_3$: Cs	[76]
Li, Na, K, Rb, Cs	Hyflo Super Cel	acyclic polyether (1,13 bis(8-quinolinyl)1,4,7,10,13-pentaoxytridecane)	$0.01M$ NaSCN, pH 12, $t = 40°C$: (Li+Na), K, Rb, Cs	[153]
Cs, K, Rb, Na, Li	Hyflo Super Cel	cyclic polyether (dibenzo-18-crown-6)	$0.01M$ NaSCN, pH 7, $t = 40°C$: Cs, K, Rb, Na, Li	[153]
Li, other alkali metals	polytetrafluoroethylene	$0.1M$ dibenzoylmethane–$0.1M$ TOPO in dodecane	$3.2M$ NH$_4$OH: Cs+Rb, K+Na, $0.6M$ HCl: Li	[124]
Li, Na, K, Rb, Cs	Kel F	nitrobenzene + (I$_2$+NH$_4$I) (3:1)	nitrobenzene sat. aq. solution (1.4g I$_2$ + 1.5 g NH$_4$I)/1: Li+Na, K, Rb, retained on column: Cs	[125]
Rb, Cs	Kel F	nitrobenzene + (I$_2$+NH$_4$I) (3:1)	$1M$ HCl: Rb, $6M$ HCl: Cs	[125]
Cs, fission products	Kel F	$1M$ Ph$_4$B in amyl acetate	$0.0025M$ EDTA, pH 6·8: fission products, $1M$ HCl-acetone: Cs	[33]
Cs, fission products	Kel F	dipicrylamine in nitrobenzene	$0.0025M$ EDTA, pH 7·0: fission products, $0.6M$ HNO$_3$: Cs	[126]
1B				
Ag, Au	Ftoroplast-4	TBP	sorption from $0.5M$ HCl, $5M$ HCl: Ag, retained on column: Au	[127]
Ag, Cu + numerous elements,	Polyfluorocarbon	Tri-iso-octylthiophosphate	$8M$ HNO$_3$, $t = 60°$: Cu + many other elements, $1M$ HNO$_3$: Ag	[99]

Applications

Elements	Support	Extractant	Conditions	Ref.
Ba, Sr, Ca, Mg	Kel F	1·5M TTA in MIBK	Acetate buffer, pH 6·5: Ba, pH 5·5: Sr+Ca, 0·1M HCl: Mg	[26]
Sr, Ca	Kel F	HDEHP	0·30M HNO$_3$: Sr, 1M HNO$_3$: Ca	[128]
Ba, Sr, Ca	SiO$_2$	HDEHP	0·5M NaNO$_3$, pH 3·0: Ba, pH 2·0: Sr, pH 0·8: Ca	[129]
Ba, Sr, Ca	Hyflo Super Cel	cyclic polyether (dibenzo-18-crown-6)	0·01M NaClO$_4$, pH 7·05, $t = 25°C$: Ba, Sr, Ca	[153a]
2B				
Zn, Cd	Teflon	TBP	0·3M HBr: Zn, Cd	[7]
Zn, Cd	silica gel	TNOA	0·25M HCl+0·5M HNO$_3$: Zn, 0·5M HNO$_3$: Cd	[66]
Zn, Cd	Celite 545	dithizone in CCl$_4$	0·1M H$_2$C$_2$O$_4$+Na$_2$CO$_3$, pH 4·1: Zn, 0·1M H$_2$C$_2$O$_4$: Cd	[25]
Zn, Hg	styrene–divinylbenzene	lead dithizonate	Bi^{3+} solution in acetate buffer: Zn, retained on column: Hg	[119]
3A				
Ce, Pm, Sm, Eu, Gd, Tb, Sm, Gd, Tb, Dy, Ho, Er, Y, Tm, Yb, Lu	Hyflo Super Cel	TBP	15·1M HNO$_3$; Ce, Pm, Sm, Eu, Cd, Tb	[30]
	Hyflo Super Cel	TBP	11·5M HNO$_3$: Sm, Gd, Tb, Dy, Ho, Er, Y, Tm, Yb, Lu	[83]
La, Ce, Pr, Nd, Pm, Sm, Eu	Kel F	HDEHP	$t = 85°C$, HCl 0·265–0·565M (gradient): La, Ce, Pr, Nd, Pm, Sm, Eu	[131]
La, Pr, Sm, Eu, Gd, Tb, Dy, Ho, Er, Tm, Yb, Lu	Kel F	HDEHP	$t = 85°C$, HCl 0·310–4·74M (gradient): La, Pr, Sm, Eu, Gd, Tb, Dy, Ho, Er, Tm, Yb, Lu	[131]
La, Ce, Nd, Gd, Tb, Tm	cellulose	HDEHP	0·25M HCl: La, Ce, 0·8M HCl: Nd, Gd, Tb, 6M HCl: Tm	[130]
La, Ce, Pr, Nd, Pm, Sm, Eu, Gd, Tb, Dy, Ho, Y, Er, Tm, Yb, Lu	Celite	HDEHP	$t = 70°C$, HCl 0·194–3·8M (gradient elution): La, Ce, Pr, Nd, Pm, Sm, Eu, Gd, Tb, Dy, Y+Er, Tm, Yb, Lu	[71]
Eu, Tb, Tm	SiO$_2$	TBHP+TBPP	14·5M HNO$_3$: Eu, Tb, Tm	[45]

Table 7.1 (continued)

Elements isolated	Stationary phase	Support	Mobile phase (Eluent)	Ref.
Rare earths, Sc	TBP	Silica gel KSK	$9M$ HCl: rare earths, $0.1M$ HCl: Sc	[118b]
Sm, Pm	Aliquat-336	Kel F	$3M$ LiNO$_3$: Sm, Pm	[132]
Er, Ho	Aliquat-336	Kel F	$5.8M$LiNO$_3$: Er, Ho	[132]
Eu, Sm, Nd, Pr, Ce, La	methyldioctylamine	Teflon	$1.6M$ Al(NO$_3$)$_3$: Eu, Sm, $1.2M$ Al(NO$_3$)$_3$: Nd, Pr, Ce, $1.0M$ Al(NO$_3$)$_3$: La	[174]
La, Ce, Pr, Nd	HEHФP	Hyflo Super Cel	$0.42M$ HCl: La, Ce, Pr, Nd	[133]
Pm, Eu, Gd	HEHФP	Hyflo Super Cel	$0.95M$ HCl: Pm, Eu, Gd	[133]
Tb, Dy, Ho, Er	HEHФP	Hyflo Super Cel	$7.4M$ HNO$_3$: Tb, Dy, Ho, Er	[133]
Tm, Yb, Lu	HEHФP	Hyflo Super Cel	$7.4M$ HNO$_3$: Tm, Yb, Lu	[133]
La, Ce (IV)	HDEHP	Fluoroplast-4	$0.5M$ HNO$_3$ + $0.4M$ (NH$_4$)$_2$S$_2$O$_8$ + $0.1M$ AgNO$_3$: La, $6M$ HCl: Ce (IV→III)	[134]
Eu(II), rare earths	DOPA	Ftoroplast-4 + SiO$_2$	$0.2M$ HCl + $3M$ NH$_4$Cl: Eu (II)	[97]
Ac, La (III),	TBP	Ftoroplast-4	$12M$ HCl: rare earths (III)	[135]
actinides (III), lanthanides (III),	HDEHP in heptane	Tee Six	$10M$ NH$_4$NO$_3$ + $0.1M$ HNO$_3$: Ac, $0.1M$ HNO$_3$: La	[160]
			$0.25M$ diethylenetriaminepenta-acetic acid + $1M$ CH$_2$ClCOOH buffered to pH 3: actinides, $6M$ HCl: lanthanides	
Pu, Np	TOPO + 2·5-*tert*-phenylhydroquinone	Microthene	$6M$ HCl: Pu (III), $1M$ HF: Np (IV)	[29]
Bk (III), Cf (III)	HDEHP	Celite	$0.37M$ HNO$_3$: $t = 75°$: Bk, Cf	[65]
Bk (III), Es (III), Fm (III)	HDEHP	Celite	$0.4M$ HNO$_3$: $t = 60°$: Bk, Es, $1M$ HNO$_3$, $t = 60°$: Fm	[65]
Ac, Pa, Th, U	TBP	Teflon-4	$6M$ HNO$_3$: Ac + Pa, $6M$HCl: Th, $0.1M$ HNO$_3$: U	[175]
Es (III), Fm (III), Md (III)	HDEHP	Celite	$t = 60°$: $1M$ HNO$_3$: Es + Fm, $2M$ HNO$_3$: Md	[65]
Np (V), Np (IV) Np (VI)	HDEHP	Ftoroplast-4	$0.5M$ HCl: Np (V), $0.3M$ H$_2$C$_2$O$_4$: Np (IV), $7M$ HCl, $t = 20$-$85°$C: Np (VI)	[161]

§ 7.4] Applications 523

Elements	Support	Extractant	Conditions	Ref.
fission products	—	—	$1M$ HNO_3: transplutonides + the majority of rare earths	[162]
Am, Cm, Cf, Fm, Md	SiO_2	TBP	$13 \cdot 1M$ HNO_3: Am, Cm, Cf, Fm, Md	[136]
Es, Fm	Celite	HDEHP	$0 \cdot 41M$ HNO_3: Es, Fm	[60]
Cm, Cf	Teflon-6	HDEHP	$0 \cdot 2M$ HNO_3: Cm, $4M$ HNO_3: Cf	[137]
Ce, Bk	Teflon-6	HDEHP	$0 \cdot 15M$ HNO_3: Ce, $4M$ HNO_3: Bk	[137]
rare earths, Am	Celite	Aliquat-336	$1M$ $NH_4SCN + 0 \cdot 05M$ H_2SO_4: rare earths, $0 \cdot 1M$ $NH_4SCN + 0 \cdot 05M$ H_2SO_4: Am	[73]
La, Ce, Pm, Eu, Tm, Am	Celite	Aliquat-336	$9M$ $NH_4SCN + 0 \cdot 01N$ H_2SO_4: La, $7M$ $NH_4SCN + 0 \cdot 01N$ H_2SO_4: Ce, $5M$ $NH_4SCN + 0 \cdot 01N$ H_2SO_4: Pm, $1M$ $NH_4SCN + 0 \cdot 01N$ H_2SO_4: Tm, $0 \cdot 1M$ $NH_4SCN + 0 \cdot 05M$ H_2SO_4: Am	[73]
Am, Pu	Teflon	TNOA	$4M$ HCl: Am, $0 \cdot 1M$ HCl: Pu	[72]
Th, Pa	Hyflo Super Cel	TBP	$3M$ HCl: Th, 5% $H_2C_2O_4$: Pa	[86]
Th, Pa, U	Kel F	Aliquat-336	$10M$ HCl: Th, $2N$ HCl: Pa $0 \cdot 1N$ HCl: U	[138]
Th, U	Teflon	TOPO	$1N$ H_2SO_4: Th, $9 \cdot 1M$ $HClO_4$: U	[139]
Am, Cm, Bk, Cf, Es, Fm, Y	Zorbax-SIL	HDEHP-dodecane	$0 \cdot 51M$ HNO_3, $t = 50°C$: Am+Cm, Bk, Cf+Es, Fm, $5M$ HNO_3: Y	[74]
Th, U	Hyflo Super Cel	TBP	$8M$ HCl: Th, $0 \cdot 5M$ HCl: U	[140]
Th(IV), U(VI)	cellulose	TNOA	$10M$ HCl: Th, $0 \cdot 05M$ HNO_3: U	[141]
U(VI), Th(IV)	cellulose	TNOA	$6M$ HNO_3: U, $8M$ HCl: Th	[141]
Eu, Cf, Fm, Tb, Md	Hyflo Super Cel	HDEHP	$t = 64°C$, $1 \cdot 02M$ HCl: Eu+Cf, Fm, Tb, Md	[142]
Np(VI), Th(IV), U(VI)	diatomaceous earth	TBP	$0 \cdot 2M$ $N_2H_4 + 4M$ HNO_3: Np (VI→V), $4M$ HNO_3: Th, U	[88]
Y, Th, Sc	cellulose	TOPO	$8M$ HCl: Y, $1M$ HCl: Th, $4M$ H_3PO_4: Sc	[143]
Pu(III), Pu(colloid.)	diatomaceous earth	TBP	$0 \cdot 42M$ HNO_3: Pu (III) + Pu (colloid.)	[144]
Pu(VI), Pu(IV)			Pu (VI), Pu (IV)	
Np(V), Np(IV), Np(VI)	Hyflo Super Cel	TBP	$0 \cdot 5M$ HNO_3: Np (V), Np (IV), Np (VI)	[145]

Table 7.1 (continued)

Elements isolated	Support	Stationary phase	Mobile phase (Eluent)	Ref.
3B				
Tl (III), Ga (III)	cellulose	HDEHP	$0.3M$ HCl: Tl, $0.7M$ HCl: Ga,	[146]
In (III), Al (III)			$3M$ HCl: In, $8M$ HCl: Al	
Ga (III), In (III), Tl (III)	Amberlyst XAD-2	di-isopropyl ether	$5M$ HBr: Ga, $1M$ HBr: In, $3M$ HNO_3: di-isopropyl ether: Tl	[163]
Ga, In	Ftoroplast-4	TBP	$0.8M$ HBr: Ga, H_2O: In	[147]
In, Ga	Ftoroplast-4	TBP	$3M$ HCl: In, $1M$ HCl: Ga	[147]
In, Tl	Ftoroplast-4	isoamyl acetate	sorption from $7M$ HBr, $6M$ HCl: In, $0.1M$ HCl: Tl	[148]
4A				
Hf, Zr	cellulose	TNOA	$8M$ HCl + 5% HNO_3: Hf, Zr	[140]
Zr, Th	cellulose	TNOA	$10M$ NH_4NO_3: Zr, $8M$ HCl: Th	[141]
Th, Zr	cellulose	TNOA	$10M$ HCl: Th, $6M$ HCl: Zr	[141]
Zr, Hf	Teflon-6	KMIB	$4M$ $NH_4SCN + 0.2M$ $(NH_4)_2SO_4$: Zr, $1M$ H_2SO_4: Hf	[149]
Hf, Zr	Celite	TBP	$7.1N$ HCl: Hf, Zr	[150]
Th, Zr, Hf	XAD-2	TOPO-cyclohexane	$12M$ HCl: Th, $1M$ HCl: Zr + Hf	[38]
4B				
Sn (IV), Pb (II) + other cations	Teflon-6	KMIB	$8M$ HCl: Pb + (Bi + Cd + Cu + Hg + Zn), retained on column: Sn	[101]
Sn (II), Sn (IV)	SiO_2	TBP	$11.7M$ HCl: Sn (II), $1M$ HCl: Sn (IV)	[151]
5A				
V(IV), V(V)	Ftoroplast-4	benzohydroxamic acid in TBP	$5M$ H_3PO_4: V (IV), H_3PO_4 conc.: V (V)	[152]
Pa(V), Nb(V), Ta(V)	Teflon	N-benzoyl-N-phenyl-hydroxylamine in	$6M$ HCl + $0.05M$ HF: Pa, $2M$ HCl + $0.4M$ HF: Nb, $2M$ HF: Ta	[35]

§7.4] Applications

Element(s)	Support	Reagent	Conditions/Results	Ref.
Nb, Ta	Teflon-6	MIBK	3M HCl+1M HF: Nb, MIBK: Ta	[78]
5B				
As(V), As(III)	Hyflo Super Cel	TBP	7M HCl: As(V), As(III)	[47]
As(III), Sb(III)	cellulose	TOPO	1M HCl: As, 5M HNO₃: Sb	[143]
As, Bi	Ftoroplast-4	di-isoamylphosphoric acid	0·5M HNO₃ As+many other elements 13·5M HNO₃: Bi	[102]
6A				
W, Mo	Teflon-6	MIBK	6M HCl–3M HF–1M H₂SO₄: W, 3M HCl–10M HF–1M H₂SO₄: Mo	[78]
W(VI), Mo(VI)	Amberlyst XAD-2	10-hydroxyeicosan-9-one-oxime in toluene	0·005M H₂SO₄: W, NH₃ aq., pH 12: Mo	[103]
W, Mo	Teflon-6	MIBK	7M HCl–2M HF: W, 6M HCl–6M HF: Mo	[78]
W, Mo	Kel F	MIBK	1+1 HCl–1M H₂SO₄–0·1M HF: W 1+1 HCl: Mo	[154]
Cr(III), W(VI)	Kel F	50% solution of TNOA in xylene	10M HCl: Cr, 7M HCl: W	[110] [138]
Cr(VI), W(VI), Mo(VI)	Kel F	Aliquat-336	8M HCl: Cr, 2M HCl: W, Mo	[143]
Cr(III), Cr(VI)	cellulose	TOPO	2M HCl: Cr(III), 10M HCl: Cr(VI)	[23]
Cr(III), Mo(VI)	Algoflon F	α-benzoinoxime	2M HCl: Cr+many other elements, retained on column: Mo	
6B				
Se(VI), Te(VI)	Teflon	TBP	sorption from 13M HCl+Cl₂, 5M HCl+Cl₂: Se, 1M HCl: Te	[155]
7A				
Re, Tc	Ftoroplast-4	TBP	sorption from 0·5M HCl, 11M HCl: Re, retained on column: Tc	[27]
Re(VII), Tc(VII)	Chromosorb W	N-benzoyl-N-phenyl-hydroxylamine in CHCl₃	5M HClO₄: Re, 0·03M HClO₄: Tc	[164]

526 Extraction Chromatography [Ch. 7

Table 7.1 (continued)

Elements isolated	Support	Stationary phase	Mobile phase (Eluent)	Ref.
8				
Ni (II), Co (II), Fe (III)	cellulose	TNOA	8M HCl: Ni, 3M HCl: Co, 0·2M HNO$_3$: Fe	[141]
Co (II), Fe (III)	cellulose	HDEHP	1M HCl: Co, 6M HCl: Fe	[156]
Co (II), Fe (III)	Kel F	Aliquat-336	10M HCl: Co, 1M HCl: Fe	[138]
Ni (II), Co (II), Fe (III)	cellulose	TOPO	7M HCl: Ni, 3M HCl: Co, 0·5M H$_2$SO$_4$: Fe	[143]
Co (II), Fe (III)	Ftoroplast-4	TBP	11M HCl: Co, 0·1M HCl: Fe	[148]
Co, Ni, Fe (III)	Haloport-4	Octanone-2	8M HCl: Co+Ni+many other elements, CH$_3$OH, (C$_2$H$_5$)$_2$O: Fe	[104]
Fe (II), Fe (III)	cellulose	TOPO	2M HCl: Fe (II), 0·5M H$_2$SO$_4$: Fe (III)	[143]
Fe (III), many metals	hydrophobic silica gel	TBP	3–6M HCl	[156a]
Fe (CN)$_6^{3-}$, Fe (CN)$_6^{4-}$	Kel F	TBP	3M HCl: Fe (CN)$_6^{3-}$, H$_2$O: Fe (CN)$_6^{4-}$	[157]
Co (II), Fe (III)	Hyflo Super Cel	TBP	9M HCl: Co, 0·5M HCl: Fe	[158]
Co, Ni, Fe (III)	Hyflo Super Cel	TBHP+TBPP	9M HCl: Ni+Co, 0·5M HCl: Fe	[159]
Ni (II), Co (II), Fe (III)	silica gel	TNOA	7M HCl: Ni, 1·7M HCl: Co, 0·25M HCl+0·2M HNO$_3$: Fe	[66]
Pd (II), Pt (IV)	Daiflon	TBP	4M HCl: Pd, Pt	[165]
Rh, Pd, Pt	Porasil C	TBP	4M H$_2$SO$_4$: Rh, 5M HCl: Pd, 0·1M HCl: Pt+Ir	[166]
Pt, Pd, Ir, Rh	cellulose	TBP	sorption from 5M HCl, TBP–toluene (1 : 1): Pt, Pd, oxidation (Cl$_2$ in TBP–toluene), TBP–toluene (1 : 1): Ir, H$_2$O: Rh	[167]
Pd (II), Pt (IV)	cellulose	TOPO	8M HCl: Pd, 2M HNO$_3$: Pt	[143]
Pd, Ru, Os Ir, Rh, Pt	Ftoroplast	isoamyl acetate	sample in 6M HCl– 0·5M SnCl$_2$ or 6M HCl– 0·05M SnCl$_2$: Pd+Ru+Os+Ir; 3·5M HCl– 0·05M SnCl$_2$: Rh, 1M HCl:Pt	[176]

§ 7.4]   **Applications**   527

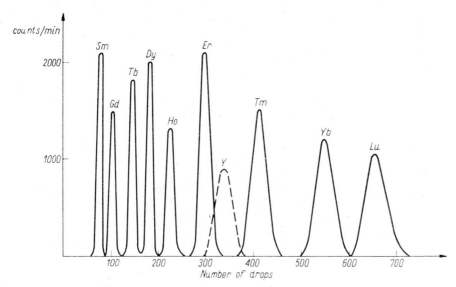

Fig. 7.10 Separation of heavy rare earths. Column 11 cm × 3 mm, Hyflo Super Cell (TBP) ($\leqslant 0.08$ mm), eluent: $11.5M$ HNO$_3$; flow-rate 0.4 cm/min (from [83], by permission of the copyright holders, Elsevier Publ. Co.)

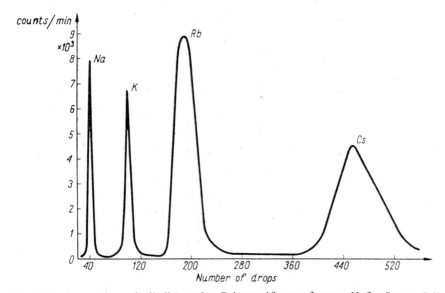

Fig. 7.11 Separation of alkali metals. Column 10 cm × 3 mm, Hyflo Super Cell impregnated with a solution of dodecylphosphoric acid in chlorobenzene, eluent: $0.04M$ LiNO$_3$–$0.01M$ HNO$_3$, flow-rate 0.3–0.5 cm/min (from [76], by permission of the copyright holders, Elsevier Publ. Co.)

528 Extraction Chromatography [Ch. 7

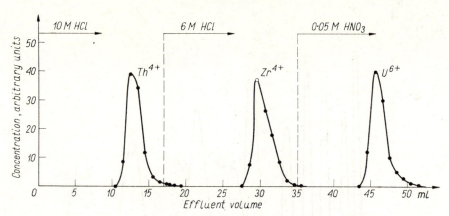

Fig. 7.12 Separation of thorium, zirconium and uranium; column 150 mm × 5·6 mm, cellulose (TNOA), flow-rate 0·25 cm/min (from [141] by permission of the copyright holders, Elsevier Publ. Co.)

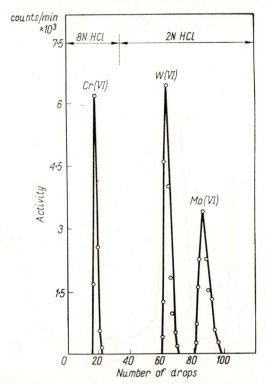

Fig. 7.13 Separation of chromium, tungsten and molybdenum. Column 6 cm × 5 mm, Kel-F (Aliquat-366) (250–325 mesh) flow-rate 0·25 cm/min (from [138], by permission of the copyright holders, Akadémiai Kiadó)

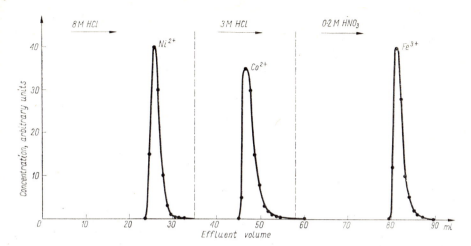

Fig. 7.14 Separation of nickel, cobalt and iron, Column 250 mm × 5·6 mm cellulose (TNOA), flow-rate 0·2 cm/min (from [141], by permission of the copyright holders, Elsevier Publ. Co.)

References

[1] Siekierski, S. and Kotlińska, B., *At. Energ.* **7**, 160 (1959).
[2] Cerrai, E., *Chromatogr. Rev.*, **6**, 129 (1964).
[3] Green, H., *Talanta*, **11**, 1561 (1964).
[4] Katykhin, G. S., *Zh. Analit. Khim.*, **20**, 615 (1965).
[5] Cerrai, E., *Report* CISE 103, October 1966.
[6] Stroński, I., *Österr. Chem. Ztg.*, **68**, 5 (1967).
[7] Alimarin, I. P. and Bolshova, T. A., *Pure Appl. Chem.*, **31**, 493 (1972).
[8] Katykhin, G. S., *Zh. Analit. Khim.*, **27**, 849 (1972).
[9] Brinkman, U. A. Th. and de Vries, G., *J. Chem. Educ.*, **49**, 244 (1972).
[10] Green, H., *Talanta*, **20**, 139 (1973).
[11] Werner, G., *J. Chromatogr.*, **102**, 69 (1974).
[12] Ghersini, G., *J. Chromatogr.*, **102**, 299 (1974).
[12a] Fritz, J. S., *Pure Appl. Chem.*, **49**, 1547 (1977).
[13] Stroński, I., *Inst. Nucl. Physics (Cracow) Report* No. 803/C (1972).
[14] Korkisch, J., *Modern Methods for the Separation of Rarer Metal Ions*, Pergamon, Oxford, 1969.
[15] Cerrai, E. and Ghersini, C., in *Advances in Chromatography*, Vol. 9, Giddings, J. C. and Keller, R. A., (Eds.), Dekker, New York, 1970.
[16] Braun, T. and Ghersini, G., (Eds.), *Extraction Chromatography*, Akadémiai Kiadó, Budapest, 1975.
[17] Hedrick, C. E. and Fritz, J. S., *Bibliography of Reversed-Phase Partition Chromatography*, Institute for Atomic Research and Department of Chemistry, Iowa State Univ., Report IS-950, Chemistry (UC-4); TID-4500, June 1964.
[18] Eschrich, H. and Drent, W., *Bibliography on Applications of Reversed-Phase Partition Chromatography to Inorganic Chemistry and Analysis*, Eurochemic Report ETR 211, Mol, Belgium, November 1967.

[19] Eschrich, H. and Drent, W. in *Extraction Chromatography*, Braun, T. and Ghersini, G., (Eds.), Akadémiai Kiadó, Budapest, 1975.
[20] Braun, T. and Farag, A. B., in *Extraction Chromatography*, Braun, T. and Ghersini, G. (Eds.), Akadémiai Kiadó, Budapest, 1975.
[21] Leene, H. R., de Vries, G. and Brinkman, U. A. Th., *J. Chromatogr.*, **57**, 173 (1971).
[22] Brinkman, U. A. Th., de Vries, G. and Leene, H. R., *J. Chromatogr.*, **69**, 181 (1972).
[23] Malvano, G., Grosso, P. and Zanardi, M., *Anal. Chim. Acta*, **41**, 251 (1968).
[24] Šebesta, F., *J. Radioanal. Chem.*, **6**, 41 (1970).
[25] Šebesta, F., *J. Radioanal. Chem.*, **7**, 41 (1971).
[26] Akaza, I., *Bull. Chem. Soc. Japan*, **39**, 980 (1966).
[27] Braun, T., Farag, A. B. and Klimes-Szmik, A., *Anal. Chim. Acta*, **64**, 71 (1973).
[28] Cerrai, E. and Testa, C., *Anal. Chim. Acta*, **28**, 205 (1963).
[29] Testa, C. and Delle Site, A., *J. Chromatogr.*, **102**, 293 (1974).
[30] Siekierski, S. and Fidelis, I., *J. Chromatogr.*, **4**, 60 (1960).
[31] Hamlin, A. G., Roberts, B. J., Loughlin, W. and Walker, S. G., *Anal. Chem.*, **33**, 1547 (1961).
[32] Sochacka, R. J. and Siekierski, S., *J. Chromatogr.*, **16**, 376 (1964).
[33] Cesarano, C., Pugnetti, G. and Testa, C., *J. Chromatogr.*, **19**, 589 (1965).
[34] Kotlińska-Filipek, B. and Siekierski, S., *Nukleonika*, **8**, 607 (1963).
[35] Šebesta, F. and Pošta, S., *Radiochem. Radioanal. Lett.* **14**, 183 (1973).
[36] Mikulski, J., Gavrilov, K. A. and Knoblokh, V., *Preprint Ob'edin. Inst. Yadernykh Issledov.* (Dubna), **1964**, 1775.
[37] Conder J. R., *Anal. Chem.*, **43**, 367 (1971).
[38] Fritz, J. S. and Latwesen, G. L., *Talanta*, **17**, 81 (1970).
[39] Samsahl, K., Wester, P. O., and Landstrom, O., *Anal. Chem.*, **40**, 181 (1968).
[40] Samsahl, K., *Aktiebolaget Atomenergi Report*, AE-247, 1966.
[41] Fołdzińska, A. and Dybczyński, R., *J. Radioanal. Chem.*, **21**, 507 (1974).
[41a] Polkowska-Motrenko, H., Dybczyński, R., *Radiochem. Radioanal. Lett.*, **26**, 217 (1976).
[42] Samsahl, K., *Aktiebolaget Atomenergi Report*, AE-389, 1970.
[43] Akaza, I., Tajima, T. and Kiba, T., *Bull. Chem. Soc. Japan*, **46**, 1199 (1973).
[44] Bosholm, J. and Grosse-Ruyken, H., *J. Prakt. Chem.*, **26**, 83 (1964).
[45] Mikulski, J., Gavrilov, K. A. and Knoblokh, V., *Nukleonika*, **9**, 785 (1964).
[46] Mikulski, J., Gavrilov, K. A. and Knoblokh, V., *Nukleonika*, **10**, 81 (1965).
[47] Fidelis, I., Gwóźdź, R. and Siekierski, S., *Nukleonika*, **8**, 319 (1963).
[48] Akaza, I. in *Extraction Chromatography*, Braun, T. and Ghersini, G., (Eds.), Akadémiai Kiadó, Budapest, 1975.
[49] Pierce, T. B., Peck, P. F. and Hobbs, R. S., *J. Chromatogr.*, **12**, 81 (1963).
[50] Siekierski, S. in *Extraction Chromatography*, Braun, T. and Ghersini, G., (Eds.), Akadémiai Kiadó, Budapest, 1975.
[51] Dening, R., Trautman, N. and Herrmann, G., *J. Radioanal. Chem.*, **5**, 223 (1970).
[52] Opieńska, J., Kański, M. and Waksmundzki, A., (Eds.), *Chromatografia*, PWN, Warsaw, 1957.
[53] Consden, R., Gordon, A. H. and Martin, A. J. P., *Biochem. J.*, **38**, 224 (1944).
[54] Cerrai, E. and Ghersini, G., *J. Chromatogr.*, **24**, 383 (1966).
[55] Cerrai, E. and Testa, C., *J. Chromatogr.*, **7**, 112 (1962).

[56] Cerrai, E., Ghersini, G., Lederer, M. and Mazzei, M., *J. Chromatogr.*, **44**, 161 (1969).
[57] Cerrai, E. and Ghersini, G., *J. Chromatogr.*, **18**, 124 (1965).
[58] Fidelis, I., *Nukleonika*, **12**, 477 (1967).
[59] Fidelis, I., *Inst. Nucl. Res. (Warsaw), Rept.* No. 1394/V/C (1972).
[60] Horwitz, E. P., Bloomquist, C. A. A. and Henderson, D. J., *J. Inorg. Nucl. Chem.*, **31**, 1149 (1969).
[61] Giddings, J. C., *Dynamics of Chromatography, Part 1*, Dekker, New York, 1965.
[62] Siekierski, S. and Sochacka, R. J., *J. Chromatogr.*, **16**, 385 (1964).
[63] Grosse-Ruyken, H. and Bosholm, J., *J. Prakt. Chem.*, **25**, 79 (1964).
[64] Pierce, T. B. and Hobbs, R. S., *J. Chromatogr.*, **12**, 74 (1963).
[65] Horwitz, E. P. and Bloomquist, C. A. A., *J. Inorg. Nucl. Chem.*, **35**, 271 (1973).
[66] Neef, B. and Grosse-Ruyken, H., *J. Chromatogr.*, **79**, 275 (1973).
[67] Horwitz, E. P. and Bloomquist, C. A. A., *J. Inorg. Nucl. Chem.*, **34**, 3851 (1972).
[68] Becker, R. and Hecht, F., *Mikrochim. Acta*, **1973**, 625.
[69] Herrmann, E., *J. Chromatogr.*, **38**, 498 (1968).
[70] Pszonicka, M. and Siekierski, S., *Chem. Anal. (Warsaw)*, **17**, 1321 (1972).
[71] Winchester, J. W., *J. Chromatogr.*, **10**, 502 (1963).
[72] Mikulski, J., *Nukleonika*, **11**, 57 (1966).
[73] Barbano, P. G. and Rigali, L., *J. Chromatogr.*, **29**, 309 (1967).
[74] Horwitz, E. P., Bloomquist, C. A. A. and Delphin, W. H., *J. Chromatogr. Sci.*, **15**, 41 (1977).
[75] Cerrai, E. and Ghersini, G., *Analyst (London)*, **94**, 599 (1969).
[76] Smulek, W. and Siekierski, S., *J. Chromatogr.*, **19**, 580 (1965).
[77] Moskvin, L. N. and Preobrazhenskii, B. K., *Radiokhimicheskie metody opredeleniya mikroelementov (Radiochemical Methods of Determining Microelements)*, Izd. Nauka, Moscow, 1965, p. 85.
[78] Fritz, J. S. and Dahmers, L. H., *Anal. Chem.*, **40**, 20 (1968).
[79] Riccato, M. T. and Herrmann, G., *Radiochim. Acta*, **14**, 107 (1970).
[80] Aslin, D., *Int. Lab.*, **1977** (July/August), 59.
[81] Scott, R. P. W., *Analyst (London)*, **103**, 37 (1978).
[81a] Horváth, C. and Melander, W., *Int. Lab.*, **1977** (Nov./Dec.), 11.
[82] Snyder, L. R. and Kirkland, J. J., *Introduction to Modern Liquid Chromatography*, Wiley-Interscience, New York, 1974.
[82a] Horwitz, E. P., Delphin, W. H., Bloomquist, C. A. A. and Vandergrift, G. F., *J. Chromatogr.*, **125**, 203 (1976).
[83] Fidelis, I. and Siekierski, S., *J. Chromatogr.*, **5**, 161 (1961).
[84] Fidelis, I., Gwóźdź, R. and Siekierski, S., *Nukleonika*, **8**, 327 (1963).
[85] Smulek, W., *Nukleonika*, **11**, 635 (1966).
[86] Fidelis, I., Gwóźdź, R. and Siekierski, S., *Nukleonika*, **8**, 245 (1963).
[87] Narbutt, J. and Smulek, W., *Nukleonika*, **18**, 375 (1973).
[88] Lis, S., Józefowicz, E. T., and Siekierski, S., *J. Inorg. Nucl. Chem.*, **28**, 199 (1966).
[89] Šebesta, F. and Starý, J., *J. Radioanal. Chem.*, **21**, 151 (1974).
[90] Tomažić, B. and Siekierski, S., *J. Chromatogr.*, **21**, 98 (1966).
[91] Denig, R., Trautmann, N. and Herrmann, G., *J. Radioanal. Chem.*, **6**, 331 (1970).
[92] Hübener, S. and Hermann, A., *J. Radioanal. Chem.*, **21**, 157 (1974).

[93] Denig, R., Trautmann, N. and Herrmann, G., *J. Radioanal. Chem.*, **6**, 57 (1970).
[94] Fidelis, I. and Smułek, W., *Inst. Nucl. Res. (Warsaw), Rept.* 'P', No 1357/V/C (1971).
[95] Smułek, W. and Borkowski, M., *Nukleonika*, **19**, 995 (1974).
[96] Bonnevie-Svendsen, M. and Joon, K., in *Extraction Chromatography*, Braun, T. and Ghersini, G., (Eds.), Akadémiai Kiadó, Budapest, 1975.
[97] Moskvin, L. N. and Tomikov, S. B., *Radiokhimicheskie metody opredeleniya mikroelementov (Radiochemical Methods of Determining Microelements)*, Izd. Nauka, Moscow 1965, p. 93.
[98] Grosse-Ruyken, H., Bosholm, J. and Reinhardt, G. G., *Isotopenpraxis*, **1**, 124 (1965).
[99] Nelson, F., *J. Chromatogr.*, **20**, 378 (1965).
[100] Braun, T. and Farag, A. B., *Anal. Chim. Acta*, **69**, 85 (1974).
[101] Fritz, J. S. and Latwesen, G., *Talanta*, **14**, 529 (1967).
[102] Kalinin, S. K., Katykhin, G. S., Nikitin, M. K. and Yakovleva, G. A., *Zh. Analit. Khim.*, **23**, 1481 (1968).
[103] Fritz, J. S. and Beuerman, D. R., *Anal. Chem.*, **44**, 692 (1972).
[104] Fritz, J. S. and Hedrick, C. E., *Anal. Chem.*, **34**, 1411 (1962).
[105] Braun, T. and Farag, A. B., *Anal. Chim. Acta*, **71**, 133 (1974).
[106] Warshawsky, A., *Talanta*, **21**, 962 (1974).
[107] Zemskova, M. G., Lebedev, N. A., Melamed, Sh. G., Saunkin, O. F., Sukhov, G. V., Khalkin, V. A., Kherrmann, E. and Shmanenkova, G. I., *Zavodsk. Lab.*, **33**, 667 (1967).
[108] Grosse-Ruyken, H. and Bosholm, J., *Kernenergie*, **8**, 224 (1965).
[109] Grosse-Ruyken, H. and Bosholm, J., *J. Prakt. Chem.*, **30**, 77 (1965).
[110] Espanol, C. E. and Mazafuschi, A. M., *J. Chromatogr.*, **29**, 31 (1967).
[111] Krefeld, R., Rossi, G. and Hainski, Z., *Mikrochim. Acta*, **1965**, 133.
[112] Smułek, W. and Zelenay, T., *Analytical Chemistry of Nuclear Fuels*, IAEA, Vienna, 1972, p. 119.
[113] Zelenay, T., *Radiochem. Radioanal. Lett.*, **2**, 33 (1969).
[114] Marsh, S. F., *Anal. Chem.*, **39**, 641 (1967).
[115] Shmanenkova, G. I., Firsova, V. I., Shchelkova, V. P. and Shchulepnikov, M. N., *Zh. Analit. Khim.*, **28**, 323 (1973).
[116] Shchulepnikov, M. N., Shmanenkova, G. I., Yakovlev, Yu. V. and Dogadkin, N. N., *Zh. Analit. Khim.*, **26**, 1167 (1971).
[117] Shchulepnikov, M. N., Shmanenkova, G. I., Shchelkova, V. P. and Yakovlev, Yu. V., *Zh. Analit. Khim.*, **28**, 608 (1973).
[118] Špeváčkova, V. and Křivánek, M., *J. Radioanal. Chem.*, **21**, 485 (1974).
[118a] Alimarin, I. P., Skobelkina, E. V., Bol'shova, T. A. and Zorov, N. B., *Zh. Analit. Khim.*, **33**, 1318 (1978).
[118b] Shmanenkova, G. I., Toldova, L. M. and Pleshakova, G. P., *Zh. Analit. Khim.*, **33**, 1129 (1978).
[119] Špevačkova, V. and Křivánek, M., *Proc. 3rd Analytical Chemical Conference*, Budapest 1970, Vol. 1, p. 121.
[120] Alimarin, I. P., Yakovlev, Yu. V. and Stepanets, O. V., *J. Radioanal. Chem.*, **16**, 227 (1973).
[121] Yakovlev, Yu. V. and Stepanets, O. V., *Zh. Analit. Khim.*, **25**, 578 (1970).
[122] Erdtmann, G. and Aboulwafa, O., *Z. Anal. Chem.*, **270**, 1 (1974).

[123] Yakovlev, Yu. V., Stepanets, O. V. and Sukhanovskaya, A. I., *Zh. Analit. Khim.*, **28**, 173 (1973).
[123a] Palágyi, Š. and Bilá, E., *Radiochem. Radioanal. Lett.*, **32**, 87 (1978).
[123b] Palágyi, S. and Markusova, R., *Radiochem. Radioanal. Lett.*, **32**, 103 (1978).
[123c] Parrish, J. R., *Anal. Chem.* **49**, 1189 (1977).
[124] Lee, D. A., *J. Chromatogr.*, **26**, 342 (1967).
[125] Akaza, I., *Bull. Chem. Soc. Japan*, **39**, 585 (1966).
[126] Testa, C. and Cesarano, C., *J. Chromatogr.*, **19**, 594 (1965).
[127] Kalyamin, A. V., *Radiokhimiya*, **5**, 749 (1963).
[128] Lieser, K. H. and Bernhard, H., *Z. Anal. Chem.*, **219**, 401 (1966).
[129] Jaskólska, H., *Chem. Anal. (Warsaw)*, **14**, 285 (1969).
[130] Cerrai, E., Testa, C. and Triluzi, C., *Energ. Nucl. (Milan)*, **9**, 377 (1962).
[131] Cerrai, E. and Testa, C., *J. Inorg. Nucl. Chem.*, **25**, 1045 (1963).
[131a] Story, J. N. and Fritz, J. S., *Talanta*, **21**, 894 (1974).
[132] Stroński, I., *Chromatographia*, **2**, 285 (1969).
[133] Fidelis, I. and Siekierski, S., *J. Chromatogr.*, **17**, 542 (1965).
[134] Moskvin, L. N. and Novikov, V. T., *Radiokhimicheskie metody opredeleniya mikroelementov (Radiochemical Methods of Determining Microelements)*, Izd. Nauka, Moscow, 1965, p. 95.
[135] Zub, D. M., Shestakov, B. J. and Shestakova, I. A., *Radiokhimiya*, **10**, 738 (1968).
[136] Taube, M., Gwóźdź, E., Gawriłow, K. A., Mały, J., Brandstetr, I. and Van Tun-Sen, *Nukleonika*, **7**, 479 (1962).
[137] Moore, F. L. and Jurriaanse, A., *Anal. Chem.*, **39**, 733 (1967).
[138] Stroński, I., *Radiochem. Radioanal. Lett.*, **1**, 191 (1969).
[139] Stroński, I., Bittner, M. and Kruk, J., *Nukleonika*, **11**, 47 (1966).
[140] Stroński, I., *Kernenergie*, **8**, 175 (1965).
[141] Cerrai, E. and Testa, C., *J. Chromatogr.*, **6**, 443 (1961).
[142] Gavrilov, K. A., Gvuzdz, E., Starý, J. and Seng, W. T., *Talanta*, **13**, 471 (1964).
[143] Cerrai, E. and Testa, E., *Energ. Nucl. (Milan)*, **8**, 510 (1961).
[144] Gwóźdź, R. and Siekierski, S., *Nukleonika*, **5**, 671 (1960).
[145] Eschrich, H., *Z. Anal. Chem.*, **226**, 100 (1967).
[146] Cerrai, E. and Ghersini, E., *J. Chromatogr.*, **16**, 258 (1964).
[147] Bol'shova, T. A., Alimarin, I. P. and Litvincheva, A. S., *Vestn. Mosk. Gos. Univ.*, *Ser. II*, **21** (6), 59 (1966).
[148] Preobrazhenskii, B. K. and Katykhin, G. S., *Radiokhimiya*, **4**, 536 (1962).
[149] Fritz, J. S. and Frazee, R. T., *Anal. Chem.*, **37**, 1358 (1965).
[150] Ueno, K. and Hoshi, M., *Bull. Chem. Soc. Japan*, **39**, 2183 (1966).
[151] Mikulski, J. and Stroński, I., *Nukleonika*, **6**, 775 (1961).
[152] Cherkovnitskaya, I. A. and Luchinin, V. A., *Vestn. Mosk. Gos. Univ.*, *Ser. Fiz. i Khim.*, **1969**, (10), 162.
[153] Smułek, W. and Łada, W., *Radiochem. Radioanal. Lett.*, **30**, 199 (1977).
[153a] Łada, W. A., and Smułek, W., *Radiochem. Radioanal. Lett.*, **34**, 41 (1978).
[154] Fritz, J. S. and Hedrick, C. E., *Anal. Chem.*, **36**, 1324, (1964).
[155] Moskvin, L. N., *Radiokhimiya*, **6**, 110 (1964).
[156] Lyle, S. J. and Nair, V. C., *Talanta*, **16**, 813 (1969).
[157] Akaza, I., Kiba, T. and Taba, M., *Bull. Chem. Soc. Japan*, **42**, 1291 (1969).
[158] Mikulski, J. and Stroński, I., *Nukleonika*, **6**, 295 (1961).
[159] Mikulski, J. and Stroński, I., *Nature*, **207**, 749 (1965).

[160] Filer, T. D., *Anal. Chem.*, **46**, 608 (1974).
[161] Ushatskii, V. N., Preobrazhenskaya, L. D., and Omelchenko, M. G., *Radiokhimiya*, **14**, 892 (1972).
[162] Levakov, B. I. and Timofeev, G. A., *Zh. Analit. Khim.*, **29**, 1023 (1974).
[163] Fritz, J. S., Frazee, R. T. and Latwesen, G. L., *Talanta*, **17**, 857 (1970).
[164] Šebesta, F., Posta, S. and Randa, Z., *Radiochem. Radioanal. Lett.*, **11**, 359 (1972).
[165] Akaza, I., Kiba, Toshiyasu and Kiba, Tomoe, *Bull. Chem. Soc. Japan*, **43**, 2063 (1970).
[166] Pohlandt, C. and Steele, T. W., *Talanta*, **19**, 839 (1972).
[167] Pohlandt, C. and Steele, T. W., *Talanta*, **21**, 919 (1974).
[168] Moody, G. J., Thomas, J.D.R., *Analyst*, **104**, 1 (1979).
[169] Akaza, I. and Inamura, J., *J. Radioanal. Chem.*, **54**, 27 (1979).
[170] Howard, A. G. and Arbab-Zavar, M. H., *Talanta*, **26**, 895 (1979).
[171] Veriovkin, G. V., Gilbert, E. N., Mikhailov, V. A, and Yakhina, V.A., *J. Radioanal. Chem*, **59**, 361 (1980).
[172] Gil'bert, E. N., Veriovkin, G. V., Yakhina, V. A. and Gureev, E. S., *Zh. Analit. Khim*, **35**, 656 (1980).
[173] Bigot, S. and Treuil, M., *J. Radioanal. Chem.* **59**, 341 (1980).
[174] Khuzhaev, S. and Gureev, E. S., *Radiokhimiya*, **21** 276 (1979).
[175] Sinitsyna, G. S., Shestakova. I. A., Shestakov, B. I., Plyushcheva, N. A., Malyshev, N. A., Belyatskii, A. F. and Tsirlin, V. A., *Radiokhimiya*, **21**, 172 (1979).
176] Kalinin, S. K. and Yakovleva, G. A,. *Zh. Analit. Khim.*, **33**, 1995 (1978).

INDEX

NOTE: references to individual elements in the tables have not been included in the index.

Acetylacetone (AA) 183
　table 185
Acids for trace analysis 31
Actinides **439,** 135, 145, 340, 518
Actinium **439,** 519
Adsorption colloid flotation 52
Aging 37
Air-borne metals 16, 17
Air-borne particulates 17
Alkali metals **436,** 135, 328, 346, 347, 350, 352, 353, 354, 381, 383, 385, 390, 392, 394, 396, 469, 525
Alkaline-earth metals **437,** 146, 327, 335, 347, 353, 381, 469
Aluminium **442,** 43, 47, 49, 116, 425, 426, 430, 469
Amalgam formation 49, 51
Amberlite, titration curve for 310
Americium **439,** 53, 137, 145, 335
Amines, high-molecular-weight 227
　extraction with 164
　　effect of amine salt concentration on 165
　　influence of solvent on 164
　tables 226, 228
Anion-exchangers, organic 302
　table 290
Anionic complexes, extraction of 147
Antimony **460,** 49, 69, 70, 71, 75, 116, 117, 329, 340, 348, 424, 428, 518
Anodic stripping voltammetry 56
Anthracite, wet ashing of 88
Arsenic **460,** 46, 49, 69, 70, 71, 75, 424, 518
Ashing 68, 76
　acceleration of 84
　　oxygen-bomb 84
　　Schöniger-flask 84

Ashing (*continued*)
　additives for
　　H_2SO_4, HNO_3 84
　　NH_4NO_3, $NaNO_3$, KNO_3, $MgNO_3$, $CaNO_3$ 85
　　Na_3PO_4, MgO, Na_2CO_3 85
　contamination by boron in 84
　dry 79, 83
　　additive-modified 83
　　vessels for 83
　　losses on 80, 84, 85
　Middleton–Stuckey procedure 80
　procedure for individual trace elements, table 90
　recoveries after, table 82
　wet 79, 80, 85
　　Carius 86
　　Kjeldahl 86
　　of oxine, organic reagents, plastics etc. 88
　　with hydrogen peroxide 88, 89
　　with nitric acid 86
　　with nitric + perchloric acid 87
　　with perchloric acid 86
　　with sulphuric + nitric acid 86
　　with sulphuric + nitric + perchloric acid 88
　　with sulphuric + perchloric acid 88
Ashing methods, requirements for 78
Astatine **464**
Atmospheres, urban 16, 17

Barium **437,** 335, 343, 349, 425, 427
Bases for trace analysis 31
Basicity of solvents 152
Bed-distribution coefficient 368

Bed-distribution coefficients of elements
  for HBr/Dowex 50-X4 419
  for HCl/Dowex 1-X10 414
  for HCl/Dowex 50-X4 417
  for HClO$_4$/Dowex 50-X4 418
  for HF/AV-17 X14 416
  for HNO$_3$/Dowex 1-X10 415
Benzoylphenylhydroxamic acid 121
N-Benzoyl-N-phenylhydroxylamine, BPHA 203
  dependence of distribution constants for on acidity 119, 120
  table 204
Berkelium **439**
Beryllium **437**, 47, 71, 343, 426, 519
BET specific surface 316
Bethge's apparatus for distillation 81, 87
Biological materials 11
Bismuth **460**, 43, 46, 49, 71, 116, 425, 426, 518, 519
Bleeding 504
Boron **442**, 56, 70, 71, 76, 79, 426, 427
Break-through capacity 356
Break-through curve 356
  effect of ion-exchange reaction equation on 361
  theory 360
Bromides, extraction of 238
Bromine **464**, 76

Cadmium **439**, 46, 49, 51, 69, 71, 128, 131, 145, 335, 340, 347, 393, 424, 425, 426, 428, 472, 519
Caesium **436**, 52, 143, 144, 305, 354, 424, 427, 428
Calcium **437**, 47, 83, 145, 146, 335, 343, 425, 430, 519
Californium **439**, 335
Carbon **443**, 76
Carbon as collector of impurities in liquids 69
Carrier distillation 71
Cation-exchangers, table 287
Cerium **439**, 105, 155, 162
Chelates 104
  see also complexes
Chlorides, extraction of, table 235
Chlorine **464**, 49, 76

Chromatography
  extraction (reversed phase) 503
    supports for 503
    theory 505
  ion-exchange 283
Chromium **461**, 69, 70, 76, 146, 341, 424, 519, 528
Cleaning of laboratory ware 27
Clean rooms 18
Cobalt **465**, 56, 51, 105, 131, 133, 139, 145, 146, 155, 156, 164, 328, 335, 343, 393, 408, 424, 425, 429, 528
Cocoa beans, for study of ashing 81
Co-crystallization 52
Co-extraction 145, 153
  causes of 146
Collector 37, 39, 40, 42
Collectors
  carbon, activated or electrode, as 56
  hydroxide-type, table 48
  inorganic 45
    plus 8-hydroxyquinoline, thionalide and tannic acid 51
  miscellaneous, table 50
  mixed 51
  organic 52
    table 54
  requirements for 45
  sulphide-type, table 47
Colorimetry, ion-exchanger 472
Column dynamics, effect on break-through 363
Columns, ion-exchange 401, 402
Commercial ion-exchangers, table 288
Comminution 21
Complex equilibria 334
Complexes
  anisotropic 105
  co-ordinatively saturated 104, 106
  co-ordinatively unsaturated 104, 106
  effect of on ion-exchange selectivity 332
  hydrated 104
  inner 104
  ligand-bridged heteronuclear 106
  mixed-donor 105
  mixed-ligand 105
  outer-sphere 105
  stability constants of 111
Concentration, ranges of 2

Index 537

Construction materials for sampling vessels 28
Containers 22
Contamination 16, 397
  avoidance of 18
Co-ordinatively solvated salts, extraction of 154
Copper **437**, 47, 49, 51, 58, 113, 125, 126, 131, 135, 145, 328, 362, 363, 364, 365, 393, 408, 423, 424, 425, 428, 429, 518, 519
Co-precipitation 37, 39, 43
  colloidal-chemical 53
  mechanism of 39
  mixed-crystal formation 39
  occlusion 41
Co-precipitation of trace elements 45, 47, 48, 50, 51, 52, 54
Counter-ions 284, 285
Cross-linking, degree of 287, 329
Crushing 21
Crystal growth, mechanism of 38
Cupferron table 200, 201
Curium **439**, 335

Dead volume 360
Degree of cross-linking, 287
  effect on ion-exchange selectivity 329
Degree of separation 102, 107
Development techniques 357
Dibenzoylmethane, DBM 184
  table 192
Di-n-butylphosphoric acid, HDBP 215
  table 214
Di (2-ethylhexyl) phosphoric acid, HDEHP or D2EHP 215
  table 216
ß-Diketones
  as extractants 183
  pK_{HA} values for 118
Displacement development 357, 359
Distillation
  direct, for separation of traces 66
  of hydrochloric acid, preconcentration on, table 68
  of sample matrices, table 72
  of traces 71
    table 77

Distribution coefficient 99, 106, 107, 108, 124, 320, 368, 506
  effect of chelating reagent on 118, 341
  effect of hydrolysis on 110
  effect of metal ion on 111
    concentration 111
  effect of pH on 112
    influence on extraction curves 113
  effects of reactions in the aqueous phase 115
  effect of solvent 121
  effect of temperature 128
Distribution coefficients for elements
  on Hostaflon C2/HDEHP–HCl 509
  on Hostaflon C2/TBP–HCl 507
  on Hostaflon C2/TNOA–HCl 508
Distribution coefficients of rare earths on HDEHP/silica gel–HCl 514
Distribution constant 97, 98, 99
  $vs.$ solubility parameter 124
Distribution function, normal 373
Distribution law 97, 98
Dithiocarbamates 207
Dithizone, H_2Dz 203
  table 208
  use in group separations 176
Dysprosium **439**, 424

EDTA, purification of 32
EDTA complexes of metals, retention on Dowex 50-X4 338
Effluent analysis 405
Einsteinium **439**
Electrochemical stripping analysis 56
Electrodeposition 56
Electroselectivity 323
Electrostatic interactions, effect on ion-exchange selectivity 325
Elution curves 358
  shape 365
Elution development 357, 358
  technique 364, 407
    gradient elution 409
    simple elution 407
Elution order reversal 468
Equipment 20
  for ion-exchange 401
Enthalpy, ion-exchange 330

Entropy effects in ion-exchange 328
Erbium **439**
Europium **439**, 139, 145, 424
Extraction, liquid-liquid 97
   applications for separation 176
   as chemical reaction 106
   equations for 106
   equilibrium constant for 106
   interactions between elements in 153
   suppression of 153
Extraction chromatography 503
   applications 518
   separation factor 506
   techniques 517
   theory 505
Extraction coefficient, conditional 107
Extraction constants, effect of temperature on, table 129
Extraction curves, typical 109
Extraction efficiency 100
Extraction kinetics 129
Extraction rate formula 130
Extraction systems
   chelate 104
   classification 103
   inorganic 102
   ion-association 147

Fermium **439**, 519
Fluorides, extraction of 240
Fluorine **464**, 76, 105, 425
Fraction collectors 405
Francium **436**, 519
Freundlich adsorption isotherm 42
Frontal analysis 357

Gadolinium **439**
Gallium **442**, 49, 71, 116, 153, 329, 335, 340, 344, 347, 348, 383, 469
Germanium **443**, 69, 70, 71, 75, 340, 347, 518
Germanium tetrachloride, impurities in 69
Glass vessels 24
Glove boxes 19
Gold **437**, 47, 49, 50, 329, 347, 348, 426, 428, 518, 519
Gradient elution 409
   equipment 410

Grinding 21
Group isolation of elements, table 174
Group separations 430, 431, 432
Getzeit method 1, 75

Hafnium **443**, 53, 116
Halides, extraction with 227
   tables
      bromide 238
      chloride 235
      fluoride 240
      iodide 239
      thiocyanate 241
Halogens **464**, 76, 469
Holmium **439**
HPLC 388, 389, 404, 407, 517
Hydrobromic acid, use of in pre-concentration 67
Hydrochloric acid
   impurities in 69
   use of in pre-concentration 67, 68
Hydrofluorination of oxides, table 70
Hydrogen-bonding in water 148
Hydrogen fluoride gas, volatilization of metal oxides by 70
Hydroxamic acids 200
8-Hydroxyquinoline (and derivatives) 124, 126, 128, 193, 199
   dependence of distribution constants on acidity 119, 120
   in group separation 174
   table 196

Indium **442**, 49, 71, 151, 152, 340, 348, 383, 469, 519
Inhomogeneity of ion-exchangers 344
Inorganic materials 11
Iodides, extraction of 239
Iodine **464**, 76, 519
Ion-association extraction systems 147
   dependence of distribution constants on metal-ion concentration 149
   dependence of properties on dielectric constant of solvent 148
   effect of acid concentration on distribution coefficients 150
   theory of 148, 151

Index

Ion-chromatography 406
Ion-exchange 283
  bed-distribution cofficient 368
  bed-distribution coefficients of the elements
    for HBr/Dowex 50-X4 419
    for HCl/Dowex 1-X10 414
    for HCl/Dowex 50-X4 417
    for HClO$_4$/Dowex 50-X4 418
    for HF/AV-17 X14 416
    for HNO$_3$/Dowex 1-X10 415
  break-through capacity 356
  break-through curves 356
    effect of ion-exchange reaction equation on 361
  chromatography 283
  column 401, 402
  column process 355
  column size 378
  complex equilibria and 334
  distribution coefficient 320, 368
    in presence of EDTA etc 341
  equilibrium 316
    fundamental definitions 316
      activity in ion-exchanger phase 317
      equilibrium constant 317
      reaction 317
      selectivity coefficient, rational 317
      selectivity coefficient, corrected 318
      standard state 317
  equilibrium theories 367
  molal selectivity coefficient 319
  electroselectivity 323
  equipment 401
  group separations 430, 431, 432
  in concentrated electrolyte solutions 345
  in non-aqueous and mixed solvents 349
  in very dilute solutions 348
  isotherms 365
  kinetics 355
  plate theories 367
  qualitative analysis 470
  quantitative analysis 472
  rate theories 370
  resolution 371
    influences on 372
  selectivity 320
    effect of complexes on 332

Ion-exchange (*continued*)
    effect of degree of cross-linkage on 329
    effect of electrostatic interactions on 325
    effect of enthalpy on 330
    effect of entropy on 328
    effect of functional groups on 327
    effect of solvation of ions on 324
    effect of temperature on 330
    effect of water structure on 326
    prediction of 323
    sources of 323
  separation factor 321, 322, 343
    variables affecting 374
  separation function 398
  separation of groups of elements
    actinides 439
    alkali metals 436
    alkaline-earth metals 437
    coinage metals 437
    group 1A 436
    group 1B 437
    group 2A 437
    group 2B 439
    group 3A 439
    group 3B 442
    group 4A 443
    group 4B 443
    group 5A 460
    group 5B 460
    group 6A 461
    group 6B 463
    group 7A 463
    group 7B 464
    group 8 465
    halogens 464
    iron triad 465
    platinum metals 466
    rare-earth metals 439
  sieve action 329
  techniques 399
    development 357
  theoretical plate height, factors affecting
    column size 378
    cross-linking 379
    distribution coefficients 389
    flow-rate 388
    particle size 384

Ion-exchange *(continued)*
  sample size 396
  temperature 396
  weight-distribution coefficients for elements
    on KU-2X6/HF 420
    on zirconium phosphate at varying pH 421

Ion-exchange paper 307
Ion-exchanger colorimetry 472
Ion-exchangers 283
  amphoteric 304
    table 294
  anionic, organic 302
    table 290
  cationic, table 287
  cellulose 296
  chelating, 303
    table 290
  commercial, table 288
  for HPLC 298
  heteropoly acid salt 305
  heteroporous 303
  hydrous oxide 306
  inorganic 284, 304
    table 298
  insoluble salt 306
  isoporous 303
  liquid 166, 307
  macroporous 303
  natural 284
  oleophilic 304
  organic 284
  properties of
    bed density 315
    capacity and polyfunctionality 308
      determination of 309
    characteristics 307
    cross-linkage and swelling 314
    decomposition 313
    dryness 308
    homogeneity 315, 344
    mechanical and chemical stability 311
    particle size 308
    pK, determination of 311
    pore-size distribution 316
    resin density 315
    solvation 314
    swelling 314

Ion-exchangers *(continued)*
    thermal stability 312
    volumes swollen and dry 314
  pretreatment of 399
  purification of 401
  zeolite 304
Iridium **466,** 49, 376
Iron **465,** 56, 183, 106, 116, 145, 150, 155, 164, 328, 329, 335, 347, 348, 393, 408, 423, 425, 426, 428, 429, 518, 528
Isolation of individual compounds 436
Isotherms, ion-exchange 365

Khlopin's law 40
Kinetics, ion-exchange 355
Kuznetsov 52

Lanthanides (rare earths) **439,** 135, 144, 145, 146, 164, 330, 332, 341, 343, 346, 347, 351, 353, 354, 377, 380, 382, 384, 387, 390, 392, 393, 394, 395, 427, 429, 469, 518, 519, 527
Lanthanum **439,** 46, 335, 349, 424
Lead **443,** 43, 46, 47, 49, 71, 76, 86, 328, 340, 424, 425, 428, 429, 472, 517
Lewatit, titration curves of 313
Lithium **436**
Logarithmic distribution law 39
Lutetium **439**

Macroporous resins 288
Macroreticular resins 287
Magnesium **437,** 47, 83, 335, 425, 430, 519
Major components, extractive separation of, table 168
Major constituents, separation by distillation 72
Manganese **463,** 46, 75, 83, 145, 343, 393, 408, 424, 425, 429, 518, 519
Marsh test 75
Masking 5
Materials for sampling vessels 28
Matrices, distillation of, table 72
Maxima in plots of log D vs. log C 153
Mendelevium **439,** 283

Index

Mercury **439**, 43, 47, 49, 52, 69, 71, 75, 86, 87, 328, 335, 340, 347, 351, 424, 428, 472, 518

Mesh sizes, conversion to mm 287

Metals and alloys, sampling 12

Methyldioctylamine 227
  table 226

Methylhydroxyquinoline, extraction with, table 199

Minima in extraction curves 116, 117

Molybdenum **461**, 43, 46, 5, 56, 76, 424, 425, 429, 518, 528

Mono-2-ethylhexylphosphoric acid, H_2MEHP 213
  table 213

Neodymium **439**

Neptunium **439**, 122, 518

Nickel **465**, 46, 71, 105, 131, 133, 134, 145, 146, 155, 328, 393, 408, 425, 428, 429, 519, 528

Niobium **460**, 116, 117, 132, 146, 518

Nitric acid, impurities in 69

Nitrogen **460**, 75

Nitrosoarylhydroxylamines 200

Nobelium **439**

Normal distribution function 373

Nucleation 37

Organic materials 11

Organophosphorus acids 210

Organophosphorus compounds as extractants 156, 210

Osmium **466**, 69, 75

Oxygen **463**

Oxygen-bomb for ashing 84

Palladium **466**, 49, 105, 155, 347, 351, 426, 429

Paneth, Fajans and Hahn's adsorption rule 41

Paper, ion-exchange 307

Peak shape 365

Perchloric acid, hazards of 87

Phase rule 97

1-Phenyl-3-methyl-4-benzoyl-5-pyrazolone, PMBP 193
  table 194

Phosphine oxides 219

Phosphorus **460**, 71, 378

Plastic vessels 26

Plate theories 367

Platinum **466**, 347, 376, 426, 518, 519

Platinum-group metals 375, 429

Platinum vessels 25, 27
  in dry ashing 83

Plutonium **439**, 53, 111, 112, 135, 257, 518

Polonium **463**, 348

Polyethylene vessels 26

Pore size, apparent 287

Potassium **436**, 52

Praseodymium **439**

Pre-concentration 3, 360, 361, 412
  choice of conditions for 412
  on distillation of hydrochloric acid, table 68

Precipitates
  colloidal 38
    hydrophilic 38
    hydrophobic 39
  crystalline 37

Precipitation 37
  from homogeneous solution 38, 43
  separation by 42
  theory 37

Preservation procedures 34

Pretreatment of ion-exchangers 399

Promethium **439**, 137, 283

Protactinium **439**, 518

Pseudohalides, extraction with 227

Purification of ion-exchangers 401

Qualitative analysis by ion-exchange 470

Quantitative analysis by ion-exchange 472

Quartz vessels 25, 27

Radium **437**, 343, 518

Rare-earth metals **439**, 135, 144, 145, 146, 164, 330, 332, 341, 343, 346, 347, 351, 353, 354, 377, 380, 382, 384, 387, 390, 392, 393, 394, 395, 427, 429, 469, 518, 519, 527

Index

Rate constants 131
Rate theories, ion-exchange 370
Reagents, need for purity 344
Reagents, ultrapure, trace element concentration in 32
Resolution in ion-exchange 371
  influences on 372
Reversion techniques 207
R_f-values for paper-chromatography of elements
  HDEHP/cyclohexane–HCl 510
  TOPO/cyclohexane–HCl 511
  TOPO/cyclohexane–HNO_3 512
  TOPO/cyclohexane–H_2SO_4 513
Rhenium **463**, 69, 75
Rhodium **466**, 347, 423
Rubidium **436**, 52, 427, 430
Ruthenium **466**, 46, 49, 69, 75, 146

Salting out 160, 162, 163
  by nitrates 161
Samarium **439**, 145, 383, 424
Sampling 11, 12
  geological materials 15
  liquids 15
  metals 12
  organic materials 15
  other materials 16
  powders 15
Scandium **439**, 71, 124, 125, 519
Schöniger method 84
Segregation 12
Selectivity, ion-exchange 320
  effect of complexes on 332
  effect of degree of cross-linkage on 329
  effect of electrostatic interactions on 325
  effect of enthalpy on 330
  effect of entropy on 328
  effect of functional groups on 327
  effect of solvation of ions on 324
  effect of temperature on 330
  effect of water structure on 326
  prediction of 323
  sources of 323
Selenium **463**, 49, 69, 70
Separation factor 102, 321, 322, 343
  variables affecting 374

Separation function 398
Separation into groups by ion-exchange 430, 431, 432
Separation methods, general principles 3
Sieve action 329
Sieve sizes, US standard, conversion to mm 287
Silanization techniques 517
Silicon **443**, 49, 70, 76, 427
Silver **437**, 43, 56, 47, 49, 71, 141, 328, 424, 425, 426, 428, 430, 518
Sodium **436**, 69, 335, 363
Sodium diethyldithiocarbamate, NaDD-TC 209
  extraction of metals with, table 211
  for group separation 174, 175
  reactivity of elements with, table 280
Solutions for trace analysis 33
Solvation of ions, effect on selectivity 324
Solvent
  classification of 122
  effect on extraction 121
  purity of 127
Solvent basicity 152
Solvents, oxygen-containing organic, classes I and II 154
Sorption of traces onto vessels 23
Sorption of traces, selective 423
Spectrochemical analysis 19
Stability constants 111
  of ligand–metal complex in ion-exchange 341
Storage of samples 23
Strontium **437**, 335, 343, 424, 425, 427
Styrene–divinylbenzene co-polymer 286
Sulphur **463**, 71, 75
Sulphuric acid, impurities in 69
Supersaturation 38
Swelling of resins 287
Synergic coefficient 136
Synergic effect, mechanism of 145
Synergism 135
Synergist 135

Tantalum **460**, 116, 117, 132, 425
Technetium **463**
Technique of ion-exchange chromatography 399

Index

Teflon vessels 26, 27
Tellurium **463**, 49, 71, 519
Temperature effects in solvation extraction processes 161
Terbium **439**, 383, 471
Thallium **442**, 46, 71, 113, 114, 120, 121, 122, 328, 329, 340, 347, 426, 519
Thenoyltrifluoroacetone, TTA 184
    table 188
Theoretical plate height 367
    dependence on (in ion-exchange)
        column size 378
        cross-linking 379
        distribution coefficients 389
        flow rate 388
        particle size 384
        sample size 396
        temperature 396
    in extraction chromatography 515
Thiocyanates, extraction of 243
Thiohydroxamic acids 202
Thorium **443**, 141, 145, 155, 162, 163, 328, 335, 351, 424, 426, 518, 528
Thulium **439**
Tin **443**, 51, 69, 70, 71, 340, 347, 426, 518, 519
Titanium **443**, 46, 70, 348, 425
Trace analysis
    general outline 10
    working techniques 9
Trace elements, ashing procedures for 90
Trace impurities, separation by distillation, table 77
Traces
    distillation of 71
    selective sorption of 423
Trialkyl phosphates 219
Tri-n-butyl phosphate 219
    table 220
Trichlorosilane, impurities in 69
Tri-n-octylphosphine oxide, TOPO 219
    table 224
Tungsten **461**, 46, 424, 425, 524

Ultrapure reagents, trace-element concentration in 32
Ultrasonics 56

Uranium **461,** 52, 53, 75, 113, 122, 125, 135, 138, 141, 145, 155, 156, 157, 158, 161, 162, 166, 328, 347, 351, 424, 429, 518, 519, 528

Vanadium **460**, 45, 46, 56, 76, 146, 328, 341, 347, 424, 430
Vessels 22
Void bed volume 360
Volatilization 66
    of elements as chlorides or bromides, table 67
    losses in 70
        source of 34

Water for trace analysis 30
    trace-element concentration in, distilled 30
Weight-distribution coefficients for elements
    on KU-2X6/HF 420
    on zirconium phosphate at varying pH 421
Wet ashing 79, 80, 85
    Carius 86
    Kjeldahl 86
    of oxine, organic reagents, plastics, etc 88
    with hydrogen peroxide 88, 89
    with nitric acid 86
    with nitric + perchloric acid 87
    with perchloric acid 86
    with sulphuric + nitric acid 86
    with sulphuric + nitric + perchloric acid 88
    with sulphuric + perchloric acid 88
Wofatit, titration of 310

Ytterbium **439**
Yttrium **439**, 386

Zeolites 302
Zinc **439**, 46, 49, 51, 69, 71, 83, 113, 114, 125, 131, 132, 134, 137, 141, 144, 145, 340, 347, 393, 408, 424, 425, 519
Zirconium **443**, 43, 46, 53, 116, 117, 155, 528
Zone refining 12